Physical Chemistry in Action

Physical Chemistry in Action presents volumes which outline essential physicochemical principles and techniques needed for areas of interdisciplinary research. The scope and coverage includes all areas of research permeated by physical chemistry: organic and inorganic chemistry; biophysics, biochemistry and the life sciences; the pharmaceutical sciences; crystallography; materials sciences; and many more. This series is aimed at students, researchers and academics who require a fundamental knowledge of physical chemistry for working in their particular research field. The series publishes edited volumes, authored monographs and textbooks, and encourages contributions from field experts working in all of the various disciplines.

More information about this series at https://link.springer.com/bookseries/10915

Mohamed M. Chehimi · Jean Pinson ·
Fatima Mousli
Editors

Aryl Diazonium Salts and Related Compounds

Surface Chemistry and Applications

Springer

Editors
Mohamed M. Chehimi
Université Paris Cité, CNRS, ITODYS
Paris, France

Jean Pinson
Université Paris Cité, CNRS, ITODYS
Paris, France

Fatima Mousli
Faculté des Sciences, LPCM
Mouloud Mammeri University
of Tizi-Ouzou
Tizi-Ouzou, Algeria

ISSN 2197-4349 ISSN 2197-4357 (electronic)
Physical Chemistry in Action
ISBN 978-3-031-04397-0 ISBN 978-3-031-04398-7 (eBook)
https://doi.org/10.1007/978-3-031-04398-7

This Springer imprint is published by the registered company Springer Nature Switzerland AG
The registered company address is: Gewerbestrasse 11, 6330 Cham, Switzerland

Preface

Nearly, 30 years after the introduction of the grafting reactions of diazonium salts and 10 years after the first book *Aryl Diazonium Salts. New Coupling Agents in Polymer and Surface Science*, time had come to bring together the numerous results obtained in the field. The surface modification by reaction of diazonium salts was first observed on glassy carbon and soon after on diamond; this implied the—not so surprising—formation of C-C bonds but the observation of the same reaction on metals was more puzzling. At the beginning, it was necessary to establish the reality of the grafting reaction and the main characteristics of the organic film. The stability of the film and the formation of a covalent bond between metals and aryl groups, a distinctive feature of this reaction, had to be demonstrated. The structure and the formation mechanism of the organic film had to be established. At the same time, the first possible applications appeared, for example, in organic electronics while industrialization concerned specialty pigments and fillers.

During the last decade, the fundamental characteristics of the reaction were refined; for example, advances have appeared in the control and the molecular observation of the film. On the one hand, the experimental findings were supported by theoretical simulations that provided the nature and the energy of the bond between the substrate and the film. On the other hand, the scope of the reaction was considerably enlarged, involving new activation processes (e.g., by photochemistry, plasmon excitation, or direct formation of capped nanoparticles). New materials (graphene, carbon nanotubes, MXenes, natural clay, nanocatalysts, and polymers) were involved in the reaction, and new potential applications have appeared. The process is now of common use in different research areas: analytical chemistry, composite materials, organic electronics, catalysis, forensics, and nanomedicine to name but a few. Finally, some industrial applications have appeared, and several start-ups are investigating other possible developments.

As the grafting reaction is closely related to the other organic reactions of diazonium salts, we have added two chapters that examine the chemistry of diazonium salts including their solution chemistry; they should permit to include the surface modification in a larger landscape and hopefully open new perspectives. We also included a chapter concerning the surface modification with iodonium salts; they are closely related to diazonium salts. The book is split into two major sections: the first chapters (1–9) describe the advances in the fundamentals of the process (monolayer formation, nucleation, theoretical simulation, and new grafting process), while the remaining chapters (10–23) cover a vast panorama of possible applications from analytical to material and energy sciences.

We would like to take this opportunity to warmly thank all contributors from the five continents for their time and willingness to share their knowledge and latest developments in the field. With 23 chapters for this second book compared to 14 chapters in 2012, clearly the surface chemistry of diazonium salts and the related compounds is investigated worldwide by remarkably increasing number of laboratories.

We, editors, do hope that this book will be useful to the readers interested in this topic; they will find both an overview of the process and its development in current and emerging fields.

Paris, France Mohamed M. Chehimi
Paris, France Jean Pinson
Tizi-Ouzou, Algeria Fatima Mousli

Contents

Contributors

Daniel Bélanger Département de Chimie, Université du Québec À Montréal, Montréal, QC, Canada

Karim Benzarti Univ Gustave Eiffel, Marne la Vallée, France

Avni Berisha Chemistry Department of Natural Sciences Faculty, University of Prishtina, Prishtina, Kosovo;
Materials Science-Nanochemistry Research Group, NanoAlb—Unit of Albanian Nanoscience and Nanotechnology, Tirana, Albania

Arvind K. Bhakta St. Joseph's College, Bangalore, Karnataka, India;
Université Paris Cité, CNRS, ITODYS, Paris, France

Pascale Blond Laboratoire de Chimie Organique, Service de Chimie et PhysicoChimie Organiques, Université libre de Bruxelles (ULB), Brussels, Belgium

Alexander A. Bondarev Altai State University, Barnaul, Russian Federation

Carlos Bravo-Díaz Physical Chemistry Department, Universidad de Vigo, Vigo, Spain

Tony Breton Univ Angers, CNRS, MOLTECH-Anjou, SFR MATRIX, Angers, France

Gilles Bruylants Engineering of Molecular NanoSystems, Ecole Polytechnique de Bruxelles, Université libre de Bruxelles (ULB), Brussels, Belgium

Marta Cerruti Department of Mining and Materials Engineering, McGill University, Montreal, Canada

Gaëlle Charrier DRF-IRAMIS-UMR NIMBE, Université Paris-Saclay, CEA Saclay, Gif-sur-Yvette, France

Mohamed M. Chehimi Université Paris Cité, CNRS, ITODYS, Paris, France

Steven De Feyter Department of Chemistry, Division of Molecular Imaging and Photonics, KU Leuven, Leuven, Belgium

Guy Deniau DRF-IRAMIS-UMR NIMBE, Université Paris-Saclay, CEA Saclay, Gif-sur-Yvette, France

Denis Doizi DRF-IRAMIS-UMR NIMBE, Université Paris-Saclay, CEA Saclay, Gif-sur-Yvette, France

Aurélien Doublet DRF-IRAMIS-UMR NIMBE, Université Paris-Saclay, CEA Saclay, Gif-sur-Yvette, France

Roman Elashnikov Department of Solid State Engineering, University of Chemistry and Technology, Prague, Czech Republic

Nordin Felidj Université Paris Cité, CNRS, ITODYS, Paris, France

Victor D. Filimonov National Research Tomsk Polytechnic University, Tomsk, Russian Federation

Charlène Gadroy Normandie Université, UNIROUEN, INSA Rouen, CNRS, COBRA, UMR CNRS 6014, Normandie University, Evreux, France

Christelle Gautier Univ Angers, CNRS, MOLTECH-Anjou, SFR MATRIX, Angers, France

Théo Geronimi Université Paris Cité, CNRS, ITODYS, Paris, France

Elisa González-Romero Analytical and Food Chemistry Department, Universidad de Vigo, Vigo, Spain

Nebewia Griffete Sorbonne Université, Paris, France

Olga Guselnikova Research School of Chemistry and Applied Biomedical Science, Tomsk Polytechnic University, Tomsk, Russian Federation;
JST-ERATO Yamauchi Materials Space-Tectonics Project and International Center for Materials Nanoarchitectonics (WPI-MANA), National Institute for Materials Science (NIMS), Tsukuba, Ibaraki, Japan

Luke C. Henderson Deakin University, Geelong, VIC, Australia

Ivan Jabin Laboratoire de Chimie Organique, Service de Chimie et PhysicoChimie Organiques, Université libre de Bruxelles (ULB), Brussels, Belgium

Cheng Jiang Department of Chemistry, University of Oxford, Oxford, UK

Khouloud Jlassi Qatar University, Doha, Qatar

Ahmed M. Khalil National Research Centre, Giza, Egypt

Elena A. Krasnokutskaya National Research Tomsk Polytechnic University, Tomsk, Russian Federation

Jean Christophe Lacroix Université Paris Cité, CNRS, ITODYS, Paris cedex 13, France

Corinne Lagrost Université de Rennes, CNRS, ISCR-UMR 6226, Rennes, France

Franck Le Derf Normandie Université, UNIROUEN, INSA Rouen, CNRS, COBRA, UMR CNRS 6014, Normandie University, Evreux, France

Da Li Université Paris Cité, CNRS, LCBPT, Paris, France

Guozhen Liu School of Medicine, The Chinese University of Hong Kong, Shenzhen, P.R. China

Yun Luo Université Paris Cité, CNRS, LCBPT, Paris, France

Oleksiy Lyutakov Department of Solid State Engineering, University of Chemistry and Technology, Prague, Czech Republic;
Research School of Chemistry and Applied Biomedical Science, Tomsk Polytechnic University, Tomsk, Russian Federation

Kunal S. Mali Department of Chemistry, Division of Molecular Imaging and Photonics, KU Leuven, Leuven, Belgium

Claire Mangeney Université Paris Cité, CNRS, LCBPT, Paris, France

Alice Mattiuzzi X4C, Montigny-le-Tilleul, Belgium

Elena Miliutina Department of Solid State Engineering, University of Chemistry and Technology, Prague, Czech Republic;
Research School of Chemistry and Applied Biomedical Science, Tomsk Polytechnic University, Tomsk, Russian Federation

Ahmed A. Mohamed Department of Chemistry, University of Sharjah, Sharjah, UAE

Fatima Mousli Université Mouloud Mammeri, Tizi-Ouzou, Algeria;
Université Paris Cité, CNRS, ITODYS, Paris, France

Philippe Nizard Université Paris Cité, CNRS, LCBPT, Paris, France

Delphine Onidas Université Paris Cité, CNRS, LCBPT, Paris, France

Jean Pinson Université Paris Cité, CNRS, ITODYS, Paris, France

Fetah I. Podvorica Academy of Sciences and Arts of Kosova, Prishtina, Kosovo;
Department of Chemistry, Faculty of Mathematical-Natural Sciences, University of Prishtina "Hasan Prishtina", Prishtina, Kosovo;
NanoAlb-Unit of Albanian Nanoscience and Nanotechnology, Tirana, Albania

Pavel S. Postnikov Research School of Chemistry and Applied Biomedical Science, Tomsk Polytechnic University, Tomsk, Russian Federation;
Department of Solid State Engineering, University of Chemistry and Technology, Prague, Czech Republic

Olivia Reinaud Université Paris Cité, CNRS, LCBPT, Paris, France

Maurice Retout Engineering of Molecular NanoSystems, Ecole Polytechnique de Bruxelles, Université libre de Bruxelles (ULB), Brussels, Belgium

Miriam C. Rodríguez González Department of Chemistry, Division of Molecular Imaging and Photonics, KU Leuven, Leuven, Belgium

Hatem Ben Romdhane Université Tunis El-Manar, Manar II, Tunisia

Ahmed Saad Department of Mining and Materials Engineering, McGill University, Montreal, Canada

Brahim Samir Normandie Université, UNIROUEN, INSA Rouen, CNRS, COBRA, UMR CNRS 6014, Normandie University, Evreux, France

Mahamadou Seydou Université Paris Cité, CNRS, ITODYS, Paris, France

Youssef Snoussi Université Paris Cité, CNRS, ITODYS, Paris, France

Natalia S. Soldatova Research School of Chemistry and Applied Biomedical Science, Tomsk Polytechnic University, Tomsk, Russian Federation

Melissa K. Stanfield Deakin University, Geelong, VIC, Australia

Beata Strzemiecka Poznan University of Technology, Poznań, Poland

Vaclav Svorcik Department of Solid State Engineering, University of Chemistry and Technology, Prague, Czech Republic

Ludovic Troian-Gautier Laboratoire de Chimie Organique, Service de Chimie et PhysicoChimie Organiques, Université libre de Bruxelles (ULB), Brussels, Belgium

Christine Vautrin-Ul Laboratoire Interfaces, Confinement, Materials and Nanostructures (ICMN)—UMR 7374, University of Orleans/CNRS, Orléans, France

Julien Vieillard Normandie Université, UNIROUEN, INSA Rouen, CNRS, COBRA, UMR CNRS 6014, Normandie University, Evreux, France

Robert Wojcieszak Université de Lille & CNRS, Lille, France

Yusuf Yagci Department of Chemistry, Istanbul Technical University, Maslak, Istanbul, Turkey

Cédric Zobrist DRF-IRAMIS-UMR NIMBE, Université Paris-Saclay, CEA Saclay, Gif-sur-Yvette, France

Principle, General Features and Scope of the Reaction, Recent Advances, Future Prospects

Jean Pinson and Fetah I. Podvorica

Abstract This first chapter gives a general overview of the surface modification by diazonium salts: the different methods that permit to attach aryl groups to surfaces, the characterization of the bond between the surface and the organic film, the mechanism of the grafting reaction, the structure of the organic film, the different groups that can functionalize the substrates, the different emerging substrates. In each case, we emphasize the most recent results and we refer the reader to the different chapters for additional information. In a final section, we describe an original method that permits, starting from a specially designed diazonium salt, grafting of a wide variety of different molecules such as, for example, organic halides, acetonitrile, methylamine. This method is based on atom abstraction and could apply to large range of molecules.

1 Introduction

Materials are characterized by their structural properties, but also by their surface properties; for example, iron is widely used in building and car industry for its very interesting mechanical properties hardness, toughness, yield and tensile strength, ductility, but its surface reacts with ambient atmosphere and rusts leading to the corrosion of the structure and loss of structural properties. To prevent this phenomenon, the surface must be protected from contact with atmosphere and very sophisticated processes based on cataphoresis are established that are so efficient that a car body does not rust during its lifetime. Surface coating is ubiquitous in our daily

J. Pinson (✉)
Université Paris Cité, CNRS, ITODYS, F-75013 Paris, France
e-mail: jean.pinson@u-paris.fr

F. I. Podvorica
Academy of Sciences and Arts of Kosova, Rr. "Agim Ramadani" nr 305, 10000 Prishtina, Kosovo

Department of Chemistry, Faculty of Mathematical-Natural Sciences, University of Prishtina "Hasan Prishtina", 10000 Prishtina, Kosovo

NanoAlb-Unit of Albanian Nanoscience and Nanotechnology, 1000 Tirana, Albania

M. M. Chehimi et al. (eds.), *Aryl Diazonium Salts and Related Compounds*, Physical Chemistry in Action, https://doi.org/10.1007/978-3-031-04398-7_1

life from our cars, as explained above, to chromium-coated watches, gold-coated jewels, and polymer-coated houseware. In addition to preventing oxidation of iron, many other properties of a surface can be modified: its adhesion to another material for composite materials, but conversely its non-adhesion to another material (lubrication); its hydrophily/hydrophobicity, its modification with a drug that can be released...

For these purposes, many different methods were established: silanes, phosphonates, carboxylates, catechols, alkynes and alkenes, amines, hydroxamic acids react on oxide surfaces [1]; long-chain thiols react on gold surfaces to give organized monolayers [2]; many different chemicals (carboxylic acids, benzophenone derivatives, carbenes...) can functionalize polymer surfaces [3, 4]. These surface reactions can be triggered thermally, photochemically, by plasma, ... In addition, to these methods, grafting of diazonium salts [5] (and related compounds) has emerged as a very interesting method that provides robust, strongly bonded coatings, the properties of which can be tuned nearly at will [6–12]. Since the initial electrografting reaction, new methods, and many applications were developed some of which are industrial, and new compounds were electrografted by reduction: iodonium (ArI^+Ar') (see Chap. 4) [13], sulfonium (Ar_3S^+) [14], phosphonium, ammoniums, [15] halides (RX, X=I, Br), [16] arylazosulfones ($Ar–N=N–S(=O)_2CH_3$) [17], or by oxidation: amines (RNH2) [18], carboxylic acids [19]. It is the purpose of this book to review the different facets of this reaction. The first part deals with fundamentals (Chaps. 2–6): synthesis and stability of diazonium salts (Chap. 2), their solution chemistry (Chap. 3), the electrografting of iodoniums (Chap. 4), the mechanistics aspects and control of films growth (Chap. 5) and the computational chemistry related to the grafting of diazonium salts (Chap. 6). The modification of materials is reported in Chaps. 7–11. The surface modification of sp^2 carbons such as nanotubes (Chap. 7) and graphene (Chap. 8), capped gold nanoparticles (Chap. 9), synthetic and biobased polymeric materials (Chap. 10) and plasmonic nanomaterials (Chap. 11). The numerous applications [20] are dealt with in Chaps. (12–23) dealing with: organic microelectronics (Chap. 12), the use of calixarenediazonium (Chap. 13), the surface modification of implants (Chap. 14), hybrid materials and coatings for catalysis (Chap. 15), the growth of polymers from surfaces (Chap. 16), the surface modification of gold gratings (Chap. 17), the fabrication of sensors (Chap. 18) and biosensors (Chap. 19), the modified reinforcing fillers (Chap. 20), the formation of composites from carbon fibers (Chap. 21), the problems of energy storage and conversion (Chap. 22) and industrial applications (Chap. 23).

2 The Principles of the Reaction

Electrografting of diazonium salts [21, 22] (easily prepared from aromatic amines, see Chap. 2 for the synthesis and stability) is summarized by the reaction of Scheme 1.

Electrochemical initiation is the most demonstrative case; a conducting surface immerged in an aqueous or aprotic solution of a diazonium salt is connected to

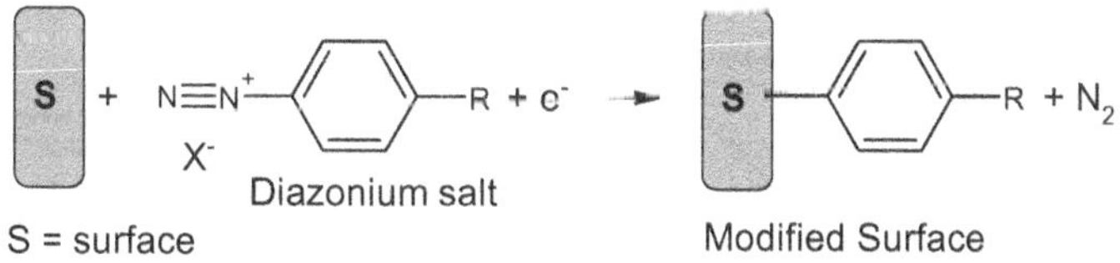

Scheme 1 Electrografting of diazonium salts

a potential source as a cathode, a potential is applied that permits the reduction and cleavage of the diazonium salt to an aryl radical and dinitrogen (a homolytic dediazonation). This radical reacts with the surface to give a functionalized surface.

Further investigations indicate that many other methods permit triggering this reaction: by using a reducing surface or a reducing agent, by changing the pH (see Chap. 3 for solution chemistry), by thermal, photochemical, plasmon (see Chap. 11), and ultrasonic activation.

Electrochemical initiation requires that the surface be conductive (metals, carbon in its different forms (see Chaps. 7–8) or semiconductive (Si, Ge), but other methods permit the modification of non-conductive oxides and polymers (see Chap. 10).

Scheme 1 is oversimplified as the very reactive intermediate aryl radical reacts not only with the surface but also with the first grafted phenyl group to give oligomers bonded to the surface, which are generally referred as multilayers. The formation of these multilayers can be favorable for some applications (composite materials, see Chap. 21) but should be avoided for some others such as biosensors (the multilayers slow down the electron transfer to the detecting group) and different methods have been described that limit the film to a monolayer (see Chap. 5).

A distinct characteristic of this reaction at the origin of its success is the strong covalent bonding between the surface and the organic film that provide very robust assemblies. In this context, one can consider the question of using alkyldiazonium instead of aryldiazonium salts to tether strongly alkyl layers to materials surfaces. This is practically almost impossible due to very high instability of alkyldiazonium salts unless the nitrite reagent and the diazonium itself are synthetized and simultaneously electrochemically grafted on a copper electrode [23].

In this chapter we review these different points and refer the reader to other chapters when necessary, we will also describe other reactions similar to the grafting of diazonium salts and finally a novel grafting reaction also deriving from diazonium salts.

3 Electrografting

Diazonium salts are reduced at potentials close to 0 V/SCE (both in aprotic media and in acidic solutions, their voltammogram (Fig. 1) shows an irreversible cathodic wave (due to the loss of dinitrogen) that corresponds to the transfer of one electron along

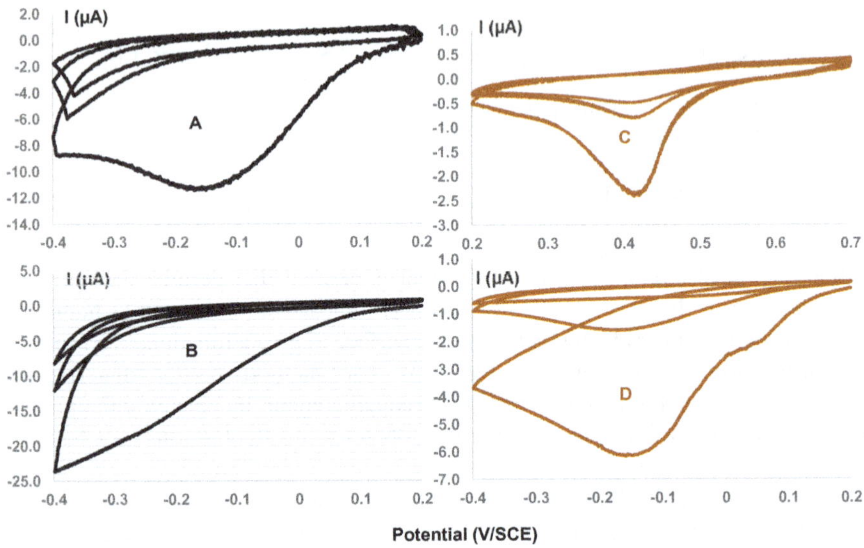

Fig. 1 Cyclic voltammogram of 4-nitro (A,C) and 4-methylbenzenediazonium (B,D) salts in acetonitrile (ACN) on a glassy carbon (GC)(A,B) and a gold (C,D) electrode

with the formation of an aryl radical and dinitrogen. Upon repetitive scanning, the cathodic peak intensity decreases coming close to the background after a few cycles. This is a characteristic feature of electrografting reactions; as the electrochemical scan proceeds, the surface is covered by a more and more compact, thicker, and thicker film that prevents the electron transfer and thus reduces the current.

Electrografting can be performed by multiple scans voltammetry, [24, 25] but also by chronoamperometry. On some occasions, a prewave is observed on the voltammogram, on gold it has been assigned to different crystallographic facets [26] and to a catalytic reduction on glassy carbon (GC) [27].

The intermediate aryl radical is a key species in the grafting reaction. This radical formed by one electron reduction of the diazonium salt is reduced, at more negative potentials ($Ar\cdot + 1e^- \leftrightarrow Ar^-$, $E°_{Ar\bullet\ /Ar^-}$) to the anion (leading to an overall two electrons reaction). By measuring the number of transferred electrons from $n = 1$ to $n = 2$, as a function of the reduction potential of different precursors leading to an aryl radical (as diazonium salts, iodonium, aryl iodides) a sigmoidal curve is obtained. It is possible to obtain $E°_{Ar}\ {}_{-/Ar^-} = -0.87$ V/SCE at the inflexion point [28]. This value is nearly identical ($E°_{Ar\bullet\ /Ar^-} = -0.91$ V/SCE) to that calculated by DFT methods giving confidence in this experimental measurement [29]. As a consequence, at negative potentials, the radical should be reduced to the anion and no grafting should occur from iodobenzene Ep ≈ -2.23 V/SCE (direct two electrons transfer ($ArI + 2e^- \rightarrow Ar^- + I^-$); nevertheless a slow grafting reaction is observed, this may be due to the negative shift of the reduction potential of $Ar\cdot$ to Ar^- as the electrode is modified by a thicker and thicker, more and more resistant organic film [30].

4 Non-electrochemical Grafting

Electrons responsible for the reduction are also obtained by photochemical methods and by plasmon excitation. Diazonium salts are very easily reducible, as shown by electrochemistry and many reducing agent or surfaces lead to the aryl radical with ensuing grafting. Even mechanical energy is used to break the aryl nitrogen bond as in ball milling and ultrasonication. Finally, there is at least one clearly identified case where energy transfer and not electron transfer is responsible for the grafting (azosulfones) [17].

Photochemical Dediazonation. Diazoniums are involved in photosynthetic redox reactions that provide aryl radicals [31, 32]; the same principle permits surface reactions. For example, visible irradiation of a gold or PVC surface dipped into a solution of diversely substituted aryldiazonium salts in the presence of a photosensitizer ($Ru(bipy)_3{}^{2+}$ or eosin Y) leads to a modification of the surface by a nanometer-thick polyphenylene film [33]. The photografting is also accomplished without sensitizer under visible or UV irradiation on gold, iron, and copper [34].

This principle is applied to diazonium and iodonium salts $Ar–I^+–Ar'$ in order to obtain a pattern on the surface of cellulose and gold [35–37], but also to inkjet printing, in this case, the reaction is performed in the presence of a vinylic that polymerizes (GraftFast Process™, see Chap. 23) [38]. A phenazinediazonium is grafted on graphene to give a structured monolayer [39]. Visible light irradiation of diazonium derivatives azosulfones $R–C_6H_4–N = N–S(=O)_2CH_3$ leads to bifunctionalized surfaces through the grafting of both the aryl and sulfone groups [17].

Photochemical excitation of surface plasmons also permits the grafting of diazonium salts as detailed in Chap. 11.

Reducing agents and surfaces. Due to the easy reduction of diazonium salts, many reducing agents are used for surface grafting: ascorbic acid [40], hypophosphorous acid [41], iron powder [42], sodium borohydride [43], iodide [44]. Reducing surfaces also initiate the grafting reaction: copper, nickel, iron, and zinc substrates but also GC in aprotic and aqueous acid solution [45–48], carbon nanotubes, graphene (see Chaps. 7 and 8), Al [49] as well as Cr [50] even if covered by their oxide, as well as carbon-coated $LiFePO_4$ powder [51].

Neutral or basic aqueous solutions. As explained in Chap. 3 diazonium salts are stable in acetonitrile and aqueous solutions but decompose rapidly in neutral and basic aqueous solutions. This permits to produce the radical that reacts with the surface. This method is used to prepare SERS (Surface Enhanced Raman Spectroscopy) encoded gold nanoparticles [52] and nanodiamonds with surface sulfonic acid groups in water [53] and water-soluble magnetic Fe_3O_4 nanocrystals in basic medium [54].

Ball Milling. Ball milling of graphite in the presence of an aryl diazonium salts permits the simultaneous exfoliation and surface modification of the resulting graphene nanoplatelets; [55] however this method is somewhat dangerous as diazonium salts may be explosives (see Chap. 2 concerning the stability of diazonium salts).

Microwaves. 4-chlorophenylbenzenediazonium is grafted on carbon nanotubes by microwave activation [56].

5 Bonding of the Organic Film

The interest in grafting diazonium salts relies on the robust attachment of the aryl groups to the surface. Figure 1 presents the most commonly admitted structure of grafted aryl groups a Surface-C(aryl) strong covalent bond between the surface and the organic film, its existence is demonstrated in different ways.

The ambient and thermal stability of the diazonium-modified surface supports the formation of a strong bond. For example, black phosphorus presents a poor ambient stability that is improved upon aryl functionalization using iodonium and diazonium salts, the first one giving better stabilization [57]. Aging experiments in ambient air of the H-terminated Si(111) surface show that the nitro and bromophenyl layers (obtained by diazonium chemistry) improve the stability against oxidation and defect formation [58]. The thermal stability also points to the formation of a strong bond. On stainless steel grafted with nitro, methoxy, and fluorophenyl groups, the band intensities of the PMIRRAS spectra decrease as the temperature increases (250, 300, and 350 °C) and completely disappear at 350–400 °C [59]. On HOPG, the bond with the aryl moieties breaks close to 200 °C [60]. During the grafting of SWCNT (Single-Wall Carbon Nanotube) by 4-iodobenzenediazonium the covalent SWCNT-aryl bond breaks at 286 °C as demonstrated by TGA/MS and TGA/GC/MS (TGA: thermogravimetry, GC: gas chromatography, MS: mass spectroscopy [61]).

The bond between a gold nanoparticles and the 4-nitrophenyl groups is observed by SERS through a band at 412 cm^{-1} assigned to the Au-C(aryl) bond through DFT calculations [62]. This is confirmed during the grafting of other diazonium salts (410 cm^{-1}) [63] and alkyliodides (387 cm^{-1}) [64]. 4-Nitrobenzenediazonium is synthetized with a ^{13}C label on the ipso position ($^{13}C{-}N^{+}{\equiv}N$) and used to derivatize gold nanoparticles; these nanoparticles are examined by ^{13}C cross-polarization/magic angle spinning solid-state nmr. A ^{13}C nmr shift at 165 ppm is identified as the aromatic carbon linked to the gold surface. This result is confirmed by ToF–SIMS through the observation of the fragment $Au_2{}^{13}CC_5H_4NO_2$ (Fig. 2) [65].

On HOPG, evidence of a C(HOPG)-C(aryl) bond is obtained from the transformation of the sp^2 carbon to sp^3 carbons (characterized by the observation of a Raman D band) upon reaction of the aryl radical (see Chap. 8) [66–68].

On maghemite, γ-Fe_2O_3 nanoparticles modified by reaction with $^{+}N_2$-C_6H_4-$(CH_2)_2$-NH_2, the formation of an O-aryl bond is demonstrated by ^{57}Fe Mössbauer spectroscopy that permits to exclude a Fe–C bond [69]. In addition, theoretical simulations with both modified gold and HOPG have confirmed the existence of a bond

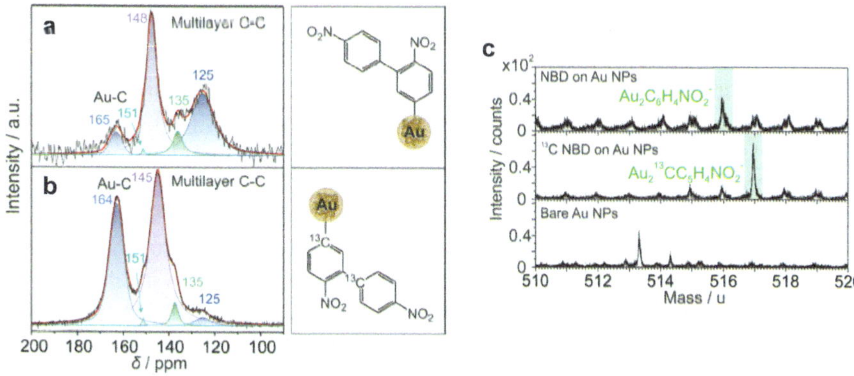

Fig. 2 4-nitrobenzenediazonium modified Au nanoparticles, and ^{13}C labeled analogs (**a**) **a**, **b** CP/MASssNMR spectra, and **c** negative high-resolution ToF–SIMS spectra showing the fragments of $Au_2C_6H_4NO_2$ and $Au_2{}^{13}CC_5H_4NO_2$. Reused with permission from Reference [65]. Copyright (2020) American Chemical Society

between the surface and the aryl group (see Chap. 6). Therefore there is little doubt about the formation of covalent bonds on different substrates: C(surface)-C(aryl) on carbons, Au-C(aryl) bond on gold and precious metals, and O(surface)-C(aryl) bond on oxidizable metals.

However, a different type of bonding has been evidenced in some cases on the surface of gold: Au–N=N–C(aryl) structures. Such structures have been observed by SERS on gold nanoparticles [70] and by XPS [71], but are not detected in other studies indicating that the formation of this weak bond [64] is highly dependent on the conditions. This structure could originate from the direct attack of a diazonium salt ($Ar\text{-}N^+{\equiv}N$) but also from the aryldiazenyl radical ($Ar–N = N\cdot$).

6 The Grafting Mechanism

It was suggested that diazonium salts could be adsorbed on a surface, prior to their grafting. The energy of planar adsorption of diazonium salts on graphene oxide is calculated in the range of −22 up to −45 kcal mol^{-1} [72, 73] (see Chap. 6). In this case, adsorption is due to π-π interactions between the aromatic molecules and the aromatic rings of the carbon substrate. The resulting corresponding aromatics also adsorb on the surface (ArH, obtained by H abstraction from the solvent) during the grafting of diazonium salts $ArN_2{}^+$, on carbon surfaces. For example, the voltammograms [74] and XPS spectra [75] of adsorbed Azure A and 9,10 anthraquinone, on carbon nanotubes and Black Pearls® carbon, are different from the corresponding electrografted aryl groups. Similarly, iodobenzene adsorbed and iodophenyl groups grafted on SWCNT could be distinguished by TGA [61]. But no experimental evidence is reported of adsorption prior to grafting (see however Chap. 6); conversely, on an iron substrate, the absence of adsorption of a diazonium salt could be established [76].

The reductive grafting of diazonium (ArN_2^+) [5–12], iodonium (ArI^+Ar') [13], sulfonium (Ar_3S^+) [14], phosphonium, ammoniums, [15] halides (RX, X=I, Br), [16] azosulfones ($Ar\text{-}N{=}N\text{–}S({=}O)_2CH_3$) [17] and oxidative grafting of amines (RNH_2) [18] and carboxylates ($RCOO^-$) [77] is assigned to the reaction of an intermediate radical. The aryl radical obtained by electroreduction on gold of 4-nitrobenzenediazonium is characterized, by esr, through its adduct with 2-methyl-2-nitrosopropane dimer, [78]. Upon spontaneous reaction of 4-bromobenzenediazonium on single-wall carbon nanotubes (SWCNT) in aqueous solution an esr signal is assigned to the radical resulting from the coupling of the aryl radical with a SWCNT. When the same experiment is performed in absence of SWCNT, a very small signal is assigned to the 4-bromophenyl radical [79]. A continuous-wave X-band ESR spectrum of pristine (6,5)-SWCNTs is recorded at 10 K, no signal is observed while a signal is present after modification with 4-nitrobenzenediazonium cation. These results demonstrate the generation of unpaired electrons in the functionalized SWCNTs [80].

Another possible intermediate is the aryldiazenyl radical Ar–N–=N·, this radical was evidenced from the photolysis of azosulfides at −110 °C, [81] it could lead to Surface –N=N-Ar modification as observed in some instances [70, 71], although its esr was never recorded during the grafting of diazonium salts.

A mechanism is proposed in Scheme 2 that accounts for the grafting of diazonium salts, the formation of multilayers, and the presence of azo bonds inside the film. It can be easily transposed to iodonium salts as the reduction of ArI^+Ar' also leads to Ar·.

The mechanism starts with the monoelectronic reduction of two moieties of diazonium salts that give two aryl radicals **(R1),** Scheme 2. The first one binds to the substrate (**R2**), while the second one attacks the already grafted phenyl group **2** to give a cyclohexadienyl radical **3**, (**R3**). At this point, two ways are opened: Path A leads to a pure polyphenylene layer **5** (a multilayer). Reaction **R4** is an electron exchange leading to the reoxidation of the cyclohexadienyl radical and reduction of a benzenediazonium (this reaction permits the formation of multilayers in absence of a reducing agent). The formation of azo bonds (path B) in the polymeric chains begins with an attack of a diazonium on the cyclohexadienyl radical **3**. Reaction **R6** gives rise to a radical cation **6,** reactions **R7** + **R8** leading from **6** to **8**, amount to the addition of one electron and the loss of two hydrogen atoms. Reaction **R8** involves the reoxidation of a cyclohexadiene; the driving force for this reaction being the restoration of the aromaticity and of the conjugation of the two substituents (Scheme 2) [82].

A different initiation takes place under visible light in the presence a photosensitizer (Scheme 3) [33].

Under visible light $Ru(bipy)_3^{2+}$ gives an excited state, it acts as a reducing agent that reduces the diazonium salt to a radical that reacts with the surface (gold or polyvinyl chloride, PVC) to give a modified film and $Ru(bipy)_3^{3+}$; further attack of another aryl radical leads to the formation of small oligomers, meanwhile the starting

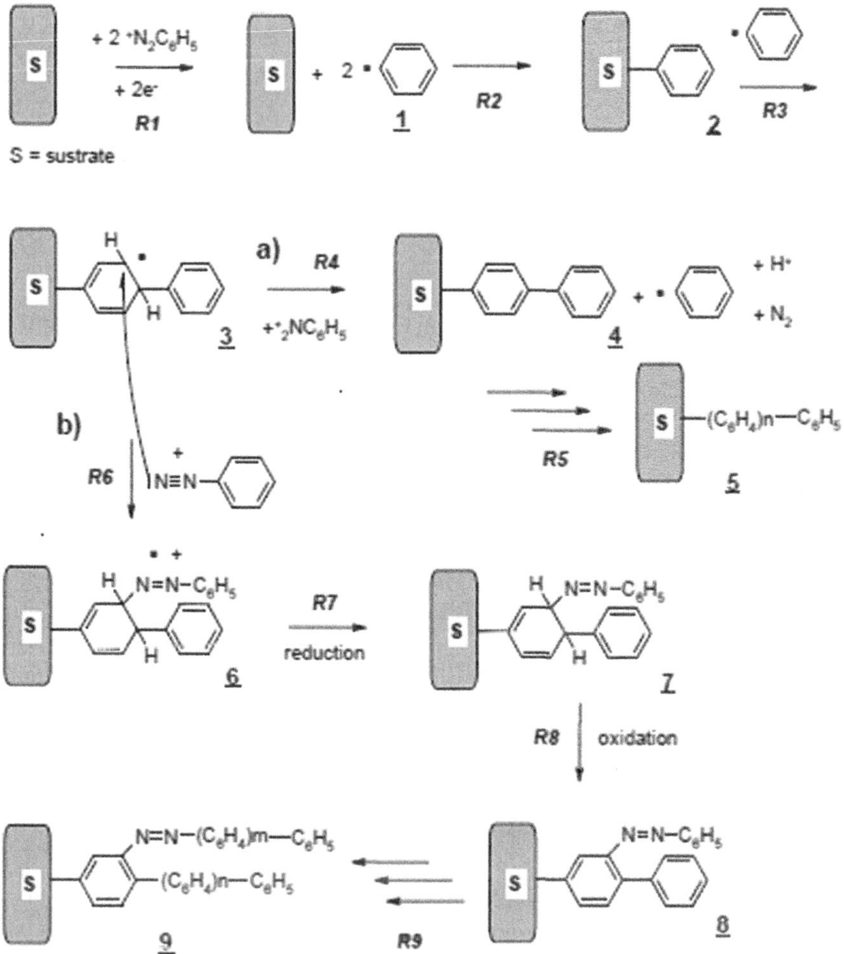

Scheme 2 The grafting mechanism of diazonium salts. Reused with permission from reference [82]. Copyright (2020) American Chemical Society

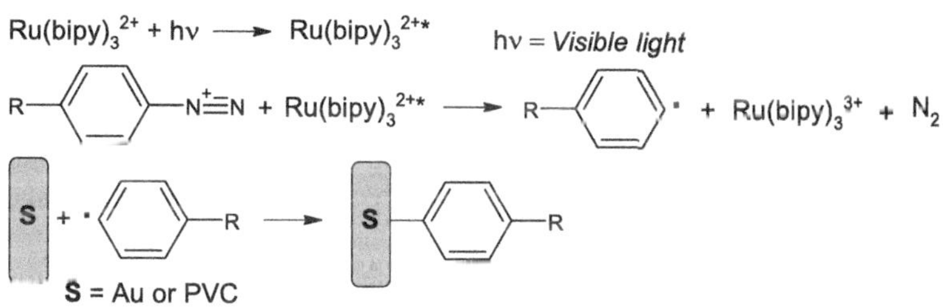

Scheme 3 Photoredox grafting mechanism of diazonium salts. From reference [33]. Copyright American Chemical Society

Scheme 4 Electrografting of a diazonium salt on SiH

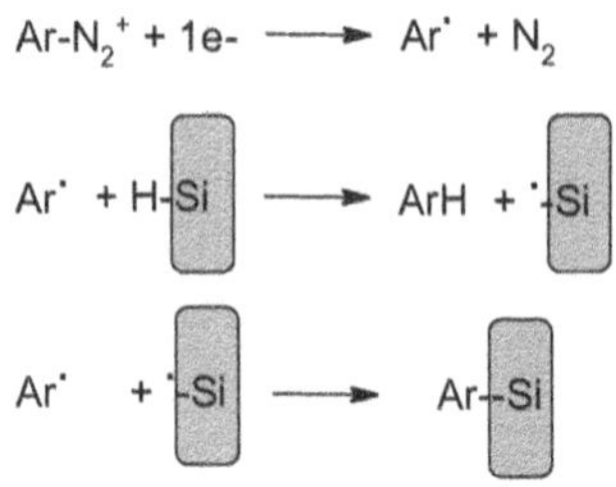

$Ru(bipy)_3{}^{2+}$ is regenerated. Under 456 nm light, in absence of sensitizer, azosulfones are grafted on gold, these compounds present perfectly reversible voltammograms, therefore the reaction cannot be related to an electron transfer, it implies an energy transfer via the $^{1}n\pi^*$ singlet excited state [17]; a similar mechanism is probably involved in the photografting of diazonium salts under UV light [34].

The possible involvement of an aryl cation stemming from the heterolytic dediazonation of the diazonium salt ($ArN_2{}^+ = Ar^+ + N_2$) is examined (see Chaps. 3 and 6). Experiments performed under conditions where the only carbocation is produced indicate that it can react with the surface but the films are thinner than when the radical is involved. The aryl radical is the dominating species in the electrografting reaction [83].

On hydrogenated Si, the mechanism involves a hydrogen atom abstraction as shown in Scheme 4. A diazonium salt is reduced by one electron to an aryl radical that abstracts a hydrogen atom from SiH, coupling of the Si and aryl radicals provides the modified surface [84–86].

On polymers, the radical abstracts a hydrogen atom from the surface of a polymer or add on double bonds; but these hypotheses await an experimental confirmation [87].

Kinetic studies concerning the grafting of diazonium salts and other compounds are rather scarce. The spontaneous grafting of 4-nitrobenzenediazonium salts onto gold electrodes was studied via quartz crystal microbalance (Quartz Crystal Microbalance, QCM) from aqueous solutions of the salt; the experimental data were fitted with a Langmuir isotherm, however, the concentration used was that of the diazonium salt not that of the reacting radical, besides the measurements were performed for 10 min, a time-lapse sufficient for the formation of multilayers, therefore the kinetics include both the grafting on gold and on a previously grafted aryl group [88]. The same remarks apply to the grafting of Ag nanocubes [89]. To prevent the measurement from reflecting grafting on both the surface and the first grafted aryl group, initial rates are obtained by EQCM (Electrochemical Quartz Cristal Microbalance) [90, 91] and by Electrochemical Impedance Spectroscopy (EIS) [91]: the grafting rate at $t = 0$ s is estimated as 0.6–0.7 nmol s^{-1} cm^{-2} and depends more on the grafted compound than on the substrate. The calculated energy of the transition states is: 38.4 kcal mol^{-1} for 4-nitrobenzenediazonium on graphene oxide. These calculations indicate that the grafting reaction is barrierless, meaning that the grafting reaction will proceed through the adsorption of the diazonium cation, nitrogen bond cleavage, and grafting of the aryl radical (Chap. 6) [72].

7 The Structure of the Films, Monolayers, Multilayers, Multifunctional Films

At the beginning of this chapter, it was shown that due to their very high reactivity, aryl radicals are attached not only to the substrate surface but also to already grafted phenyl groups; therefore oligomeric films are obtained often referred to as multilayers. The mechanism of the formation of the oligomers on the material surface is given in Scheme 2. The formation of polyaryl film is considered sometimes as a drawback of this method. For example, one of the parameters that affect the performance of a sensor device is the rate of electron transfer faster for monolayers than for multilayers; the organization of the film on the transducer surface is also a parameter that must be taken into account (see Chaps. 18 and 19) but as shown below multilayers provide very interesting characteristics to the surface.

7.1 Monolayers

Since the discovery of the method there have been many attempts to obtain grafted monolayers with aryl diazonium salts and other compounds by using different approaches: (i) controlling the charge during the electrochemical reduction by chronoamperometry or cyclic voltammetry; [92, 93] (ii) controlling the immersion time and the concentration in solution during their spontaneous reduction; [94] (iii) adapting the nature of the solvent; [95] (iv) using diazonium salt of diarylsulfides or protected hydroxylamines; [96, 97] (v) using diazonium salts with bulky substituents such as 3,5-bis-t-butylbenzenediazonium salt, [98, 99] and diazonium salt of polypyridile Ru(II) complex, [100] triisopropylsilyl (TIPS)-protected ethynyl aryldiazonium salt, [101] calixarenes; [102] (vi) using a radical scavenger such as 2,2-diphenyl-1-picrylhydrazyl (DPPH) [103], chloranil and dichlone [104] or in the presence of ascorbic acid, [105] (vii) transferring Langmuir Blodgett monolayers of octadecylamine to a gold electrode maintained at a potential sufficiently positive to insure the grafting of the amine, in this case the monolayer presents a local order [106].

Monolayer films may serve as coupling agents [107] the faster electron transfer is favorable for sensors and biosensors, [108] (see Chaps. 18 and 19) molecular electronics [109] (see Chap. 12), optoelectronics [110] and coupling different materials on surface including CNT, polymers, and nanoparticles [6–9] More detailed information about the structure and the characterization of monolayers will be given in Chap. 5.

7.2 Multilayers

Scheme 2 shows the formation of multilayers, as long as electrons can transfer through the film, diazonium salts are reduced at the film solution interface leading to thicker and thicker films; the process stops when the electron transfer becomes too slow; the process is therefore self-limiting. This translates into the rapid decrease of the cathodic current in voltammogram and chronoamperometry as in Fig. 3 for an iron electrode biased at −0.8 V/SCE in a solution of 4-nitrobenzenediazonium; the current does not follow Cottrell's equation law and decreases much faster [111]. An exception to this standard behavior occurs with the (unsubstituted) benzenediazonium tetrafluoroborate (Fig. 3a) which leads rapidly to the formation of a micrometer thick conductive film.

Thick organic layers are obtained with aryl diazonium salts containing redox substituents, like nitro or anthraquinone groups when the potential of the electrode encompasses the reduction potential of these redox moieties. Under such conditions, nitrophenyl and anthraquinone redox groups serve as relays for the transfer of electrons to the film solution interface, permitting the reduction of diazonium cations at this interface and the growth of the film up to 90 nm (Fig. 4) [112].

Micrometer thick films are also obtained by reducing benzenediazonium on an iron surface, but the exact mechanism responsible for this high thickness is unknown. In spite of their thickness these films remain conductive and can be electrochemically derivatized by 4-nitro or 4-bromobenzenediazonium salts giving a second layer of nitro or bromophenyl groups, they can also be metalized by copper [111].

When chemical grafting is achieved in the presence of reductive agents the growth of the layer does not depend on the film conductivity as the radicals are formed in the bulk of the solution therefore the thickness of the film may achieve several hundreds of nm when iron powder was used as reducing agent [42].

These oligomeric films limit the electrochemical response to some redox groups inside the film; for example, Fig. 5a shows that for thicker and thicker layers of

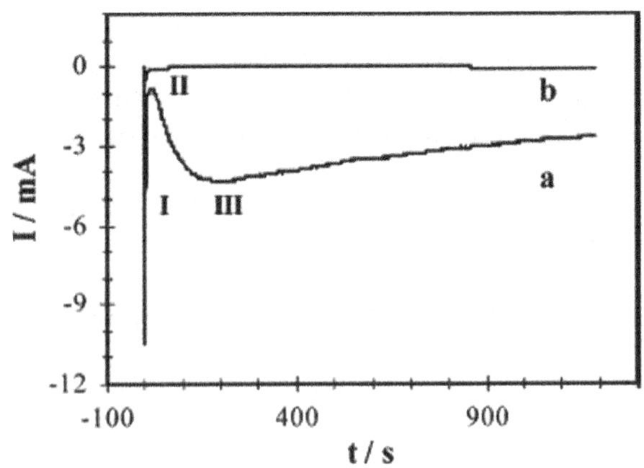

Fig. 3 Chronoamperometric curve of an iron plate ($s = 8$ cm^2), maintained at $E = -$ 0.8 V/SCE, in ACN + 0.1 M NBu_4BF_4 + 10 mM **a** $^+N_2C_6H_5\ BF_4^-$ and **b** $^+N_2C_6H_4NO_2\ BF_4^-$ solutions. Reused with permission from reference [111]. Copyright (2020) American Chemical Society

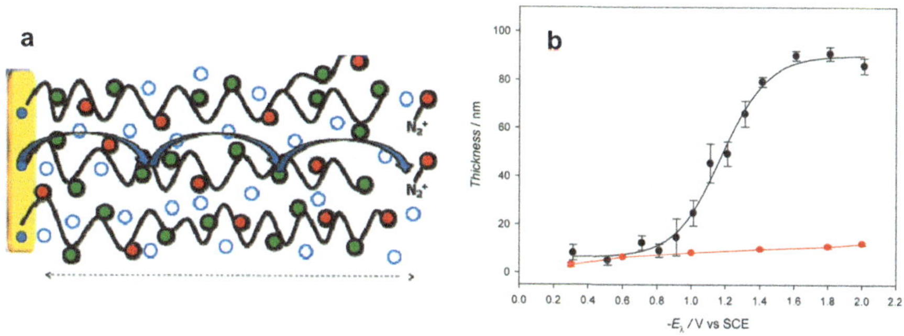

Fig. 4 **a** Schematic presentation of the electron transfer through redox groups. **b** Film thickness on gold surfaces, electrografted from 2 mM solutions of 4-nitro and 4-methylbenzenediazonium (black line and redline, respectively) in ACN + 0.1 M NBu_4BF_4 using 10 consecutive cyclic voltammetric cycles at a sweep rate of 1 V s^{-1} for varying values of the final potential. Comparison of these two curves shows that redox groups are necessary for obtaining very thick layers. Reused with permission from reference [112]. Copyright (2020) American Chemical Society

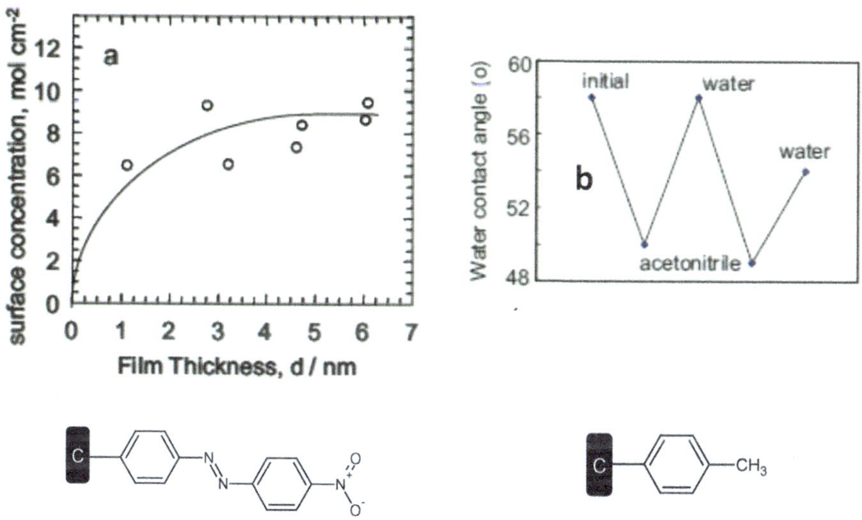

Fig. 5 **a** Surface concentration 4-nitroazobiphenyl groups as a function of the film thickness; **b** water contact angles on methylphenyl film after sonication (20 min) in the solvents indicated; experiments were performed in order left to right. Reused with permission from reference [115]. Copyright (2005) American Chemical Society

polynitrophenylene groups the number of groups that can be detected by electrochemistry reaches a limit, in other words for a film thicker than about 4 nm all the nitrophenyl groups inside the film cannot be detected by electrochemistry, most likely by lack of solvation. Along the same lines, with such multilayered films, only the outer part plays an active role in coupling reactions due to the reduced accessibility of the inner part of the film [113].

The oligomers that constitute the multilayers are flexible and their structure responds to changes in their environment, for example, their thickness, barrier properties, and wettability (Fig. 5b) are reversibly modified when exposed to different solvents and/or applied potentials. These switchable properties are proposed to originate in film swelling and shrinking which are possible because of the loosely packed multilayer structure [114, 115].

Spontaneous grafting of aryl diazonium salts on reducing substrate surface (carbon and coinage metals) also permits the formation of multilayered film and its thickness depends form the immersion time and the concentration of the diazonium salt.

Polyaryl films are highly beneficial for composites, [116] as shown in Chaps. 20 and 21; for example, the deformation of the oligomers under strain is a factor that permits to increase of the Interfacial Shear Strength (IFSS) of carbon fiber-epoxy composites [117]. Phenol–formaldehyde resins possess excellent properties but emit toxic substances during the decomposition of the crosslinking agent. Modified carbon black derivatized with 4-hydroxymethylphenyl groups permits the crosslinking of the resin. This method increases by 26% the flexural strength of the resin composite and 1100% for abrasive tools. This performance is certainly due to the strong bonding of aryl groups on the surface, but also the interaction of surface oligomers with the resin [118]. Thick organic layers are also necessary on coinage metal surfaces to protect them when they are exposed to corrosive environments [119, 120] or for the removal of heavy metal ions with complexing ligands [121].

8 Surface Functional Groups

During three decades the number of newly synthesized aryl diazonium salts has considerably increased with the aim of imparting specific properties to various surfaces. One must first recall that for building complex architectures two different ways are possible [122]. Either the complex molecule is synthetized, equipped with a diazonium group and grafted to the surface, or a simple aryl diazonium bearing a reactive 4-substituent is attached to the surface and this substituent is coupled to the desired complex molecule. Going through the literature, one finds that the second method is used overwhelmingly, mainly for sensors and biosensors (Chaps. 18 and 19). The surface postfunctionalization has been recently reviewed [123].

4-nitrobenzenediazonium salt is the most used diazonium salt due to the easy characterization of the modified surface due to: (i) the electrochemical activity of -NO_2 in aprotic and aqueous media which is used to detect and quantify the nitrophenyl groups on the material surface; [11, 12] (ii) its characteristic strong infrared signals at 1520 and 1340 cm^{-1} (iv) its characteristic XPS signal at 406 eV, but also its possible electrochemical or chemical reduction into –NH_2 group [22, 124] that permits to build further chemistry, by peptidic coupling for example [125].

4-carboxybenzenediazonium is used to form grafted organic films with hydrophilic properties; it also permits to design of surfaces with complexing properties of heavy metal ions [126, 127]. The main application is the coupling to a plethora of different species containing primary amine groups after the activation of carboxylic group with EDC (1-ethyl-3-(3-dimethylaminopropyl)carbodiimide hydrochloride)/NHS (N-hydroxysuccinimide). Carbon Nano Fiber electrode surfaces decorated with carboxyphenyl moieties are activated in the presence of EDC/NHS and linked with SARS CoV-2 nucleocapsid antigen [128]. This platform is used for the detection of virus antigens via swabbing. This electrochemical immunosensor has shown very good sensitivity, LOD 0.8 pg/mL without interferences of other virus antigens like influenza A and HCoV. The surface modification by carboxylic groups provides the basis for the construction of numerous sensing devices which will be detailed in Chap. 19 [129, 130].

4-aminophenyl groups grafted on surfaces are obtained from the 4-aminobenzenediazonium or from the reduction of the nitrophenyl group. 2D black phosphorous surface is modified with nitrophenyl groups that are reduced into amino groups; these groups serve for linking Au nanoparticles which in their turn enable catalytic reduction of 4-nitrophenol [131]. Multifunctional graphene-polyimide nanocomposites are prepared using aminophenyl functionalized graphene nanosheets; these composites present improved mechanical and electrical properties compared with those of pure polyimide due to the homogeneous dispersion of modified graphene sheets and their strong interfacial covalent bonds to the polyimide matrix [132]. Grafted aminophenyl groups (Surface-C_6H_4-NH_2) are transformed into phenyldiazonium groups (Surface-C_6H_4-N_2^+), these groups are reduced to tether two different materials, in this way reduced graphene oxide is tethered to a gold surface Au-C_6H_4-RGO [129].

Bisdiazonium salts are prepared directly [133] and used to prepare multilayered films bonded to a surface. For example, a variety of supporting electrode materials are modified by electrografting with a [2,2′-bipyridine]-5,5′-bis(diazonium) rhenium complex leading to a multilayered film with photocatalytic activity for the reduction of CO_2 [134]. Benzene-4,4′-bis(diazonium), and 1,1′-biphenyl-4,4′-bis-(diazonium) (but also p-diiodobenzene, p-diiodobiphenyl) react negatively charged SWCNT (by reaction with KC_4), the presence of two diazonium groups permits to bind some SWCNT together [135].

Azide substituted diazonium salt is grafted on carbon surface and reacted with alkynes by Cu(I)-catalyzed Huisgen 1,3-dipolar cycloaddition (alkyne-azide) click (CuAAC) reaction. This reaction is used to attach cobalt phtalocyanines to GC [136]. Click chemistry is also used to attach rhodium complex on the surface of CF-CNT (Carbon-Fiber-Carbon Nanotube), this assembly presents catalytic properties for the electrochemical reduction of NAD to NADH (Scheme 5) [137]. Conversely, the alkyne group can be attached to the surface and reacted with the azide group of a molecule in solution. A diazonium salt substituted 4-alkyne group protected by a bulky triisopropylsilyl group is covalently grafted on BDD (Boron Doped Diamond).

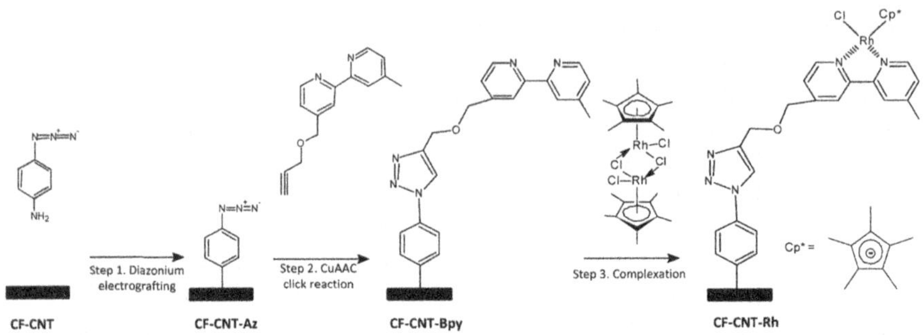

Scheme 5 Synthetic route followed for the functionalization of a carbon electrode surface with [Cp*RhIII(bpy)Cl]$^+$. Reused with permission from reference [137]. Copyright (2017) American Chemical Society

Deprotection provides an alkyne group, the copper (I) catalyst is generated electrochemically and this permits the click reaction with an azide substituted ferrocene (Scheme 6) [138, 139].

Aryl diazonium salts containing fluorine, chlorine, bromine, or iodine atoms are used to characterize the grafted layers on carbon and metal surfaces by XPS, Tof–SIMS, and RBS spectroscopies [22, 140, 141] Perfluoroalkyl groups were attached on copper nanostructures by spontaneous grafting of the corresponding diazonium salt in ACN, the resulting surface presents anti-icing properties with a considerably delayed freezing time [143].

Phtalimide moieties are grafted on carbon, BDD and gold surface through diazonium chemistry; using a protection-deprotection strategy, surfaces tethered hydroxylamine groups have been obtained. These hydroxylamine groups enable the attachment of aldehyde-containing horseradish peroxidase which retains its electroactivity [144].

Porphyrins containing different metal ions are grafted on GC surface by reduction of their corresponding diazonium salt to obtain catalytic surfaces [143]. An

Scheme 6 Surface modification of BDD by grafting and click chemistry. Reused with permission from reference [139]. Copyright (2018) Elsevier

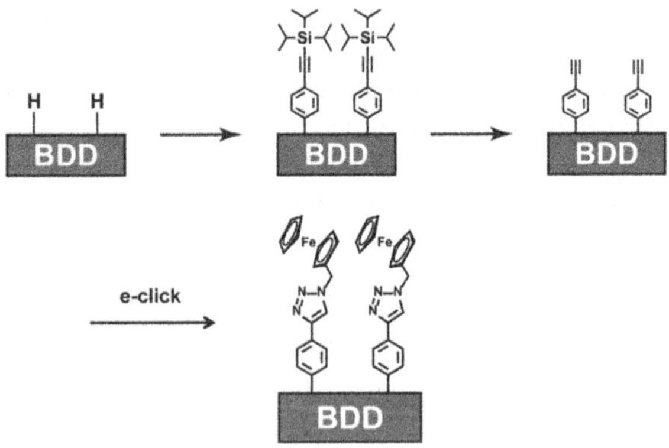

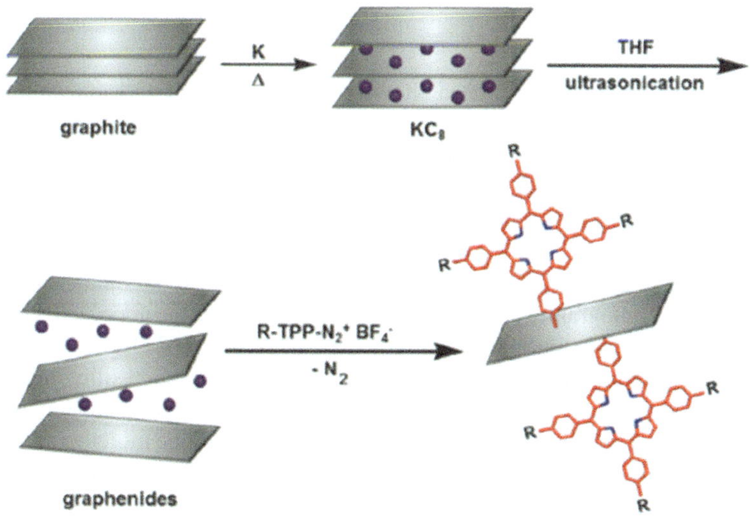

Scheme 7 Preparation of porphyrin-modified grapheme nanosheets. Reused with permission from reference [146]. Copyright (2017) American Chemical Society

iron porphyrin layer attached spontaneously to GC presents a catalytic effect for oxygen reduction in saturated oxygen solution of PBS (Phosphate-buffered-saline), and surface-bonded manganese porphyrins present electrocatalytic activity for Water Oxidation and Oxygen Reduction Reactions [144, 145]. Porphyrins are spontaneously grafted to graphene intercalated with KC8 (Scheme 7) [146]. Iron porphyrins electrografted on metal oxide surface (mesoporous antimony doped tin oxide and planar tin-doped indium oxide) show high stability at a wide pH range level and serve as catalyst for Oxygen Reduction Reaction [147]. Porphyrin-modified surfaces are also used in organic electronics [148] and non-linear optics [149].

Other metal complexes are attached to different surfaces by diazonium chemistry. Ruthenium complexes of bi and terpyridine are bonded to GC [150]. Copper complexes of nitriloacetic acid are used for binding histidine-tagged horseradish peroxidase to GC, this bonded enzyme remains active [151]. Rhodium complex with N,N or N,P ligands anchored to GC present catalytic properties for hydroamination; [152] manganese and rhenium complexes of bipyridines bonded to GC catalyze the reduction of CO_2 [153]. An heteroleptic polypyridyle Ru(II) complex is grafted on HOPG to give an organized, crystalline monolayer as shown by STM; this is a unique feature in diazonium chemistry [100].

Calixarenes are the subject of Chap. 13; specifically substituted diazonium salts for catalysis or polymerization are described in Chaps. 15 and 16, respectively.

8.1 *Mixed Layers*

For almost two decades it has been shown that two different aryl diazonium salts can be grafted simultaneously to give a bi-component modified surface and many published papers deal with the composition of mixed layers, their structure, and spatial orientation [154]. There exists three main strategies to prepare mixed layers: (i) simultaneous grafting of two diazonium salts from their mixture, [155, 156] (ii) successive grafting of different diazoniums has been achieved in several ways: using nanoshaving, [157] PDMS (poly(dimethylsiloxane) molds patterning [158] and nanosphere lithography [159, 160], protection-deprotection approach, [161] and (iii) grafting of multifunctional diazonium via calixarenes [162, 163].

It is then possible to design the properties of surface-bonded films in order to manage their interaction with their environment or with targeted biomolecules [154]. For example, a mixed layer containing phenyl phosporylcholine moieties and phenyl butyric acid chains has been prepared: short chains zwitterionic entities show anti-biofouling properties and low resistance transfer, the phenylbutyric acid groups serve for the construction of an immunosensor operating in blood [164].

9 New Substrates

Since the initial experiments on GC, [21, 22] the substrates derivatized by diazonium salts have steadily increased and this section focuses on new materials.

Carbon materials now include carbon powder, [165] carbon black, [166, 167] mesoporous carbon, [168] carbon cloth, [169] carbon felts, [170] carbon nanotubes, [171, 172] (see Chap. 7), HOPG and graphene [173] in its different forms (graphene oxide, reduced graphene oxide,...) (see Chap. 8), diamond [174].

Metals. In addition to gold widely used because of its stability and brightness (the modification of gold gratings is reported in Chap. 17), other metals are modified by diazonium salts: iron, [141] nickel, [175, 176], zinc, [177] platinum, [178]. Even coated by an oxide layer Al, [49, 179] Ti, [180] stainless steel, [181] are spontaneously derivatized as their open circuit potential is sufficient to reduce the diazonium salts. Metal oxides are also derivatized: ITO, [182, 183] wide band gap metal oxide surfaces (TiO_2, SnO_2, ZrO_2, ZnO, In_2O_3:Sn) [184–187] as well as semiconductors: Si, [86, 92] and polymers [3, 188].

Note that using $ArN^+{\equiv}N\ AuCl_4^-$ one obtains directly aryl surface-modified gold nanoparticles (see Chap. 9) [189].

Black Phosphorus is a layered two-dimensional semiconductor that exhibits favorable charge-carrier mobility, tunable bandgap, and highly anisotropic properties, but it presents a poor ambient stability that impedes its use in most applications, including electronics. Black Phosphorus nanosheets are easily oxidized under ambient conditions because of the existence of a lone electron pair in each phosphorus atom. The spontaneous covalent functionalization of exfoliated Black Phosphorus by reaction with aryl diazonium increases its stability up to three weeks of ambient exposure [190]. However, surface modification using diazonium salts produces mixed covalent and noncovalent functionalization of BP and induces oxidation; better results are obtained using aryliodonium salts which induce less oxidation (as could be expected from its more negative electrochemical reduction potential) by inhibiting bridged oxygen formation through attachment to surface oxygen and phosphorus sites [57]. Covalent azide ($Ar\text{-}N{=}N^+{=}N^-$) functionalization of Black Phosporus nanosheets is also reported, leading to significant enhancement of the ambient stability. A reaction mechanism is proposed involving the generation of nitrene as a reactive intermediate, which attacks the phosphorus atom bearing a lone electron pair and forms the P=N double bonds after 48 h reaction. Consequently, five-coordinate bonding of phosphorus atoms is achieved, completely passivating the reactive Black Phosphorus nanosheets and resulting in enhancement of its ambient stability [191].

MoS_2 is also a layered compound where a layer Mo atoms is bonded to two layers of sulfur atoms as shown in Fig. 6 [192]. It can be derivatized spontaneously, [193] by chemical reduction (KI, [192]; ferrocene and cobaltocene [194]); by lithium intercalation, chemical exfoliation, and subsequent quenching of the negative charges residing on the MoS_2 [195]); by electrochemistry [196] of diazonium salts, the carbon of the aryl group being bonded to a S atom [197, 198]. Mixed alkyl aryl layers are prepared by exfoliation of MoS_2 with BuLi (Li^+ intercalation) and reaction with alkyliodides and diazonium salts [199]. The kinetics of the reaction are measured but, as above, the values are not correct as the authors used the concentration of the diazonium salt to fit the Langmuir isotherm [200]. This method is used for: the fabrication of efficient gate dielectric in MoS_2 transistors, [196], sensors, [201], functionalized MoS_2/polyurethane sponge as a scavenger for oil in water [202], broadband optical limiters, [203] anticancer therapy [204].

MXenes are a family of two-dimensional layered compounds comprising a stack of "metal–carbon-metal" atom planes as shown in Fig. 7. These materials are exfoliated by insertion of Na^+ or Li^+, and the nanosheets are modified by diazonium chemistry. For example, Ti_3C_2 is exfoliated by Na^+ intercalation and reacted with phenylsulfonic acid diazonium salt [205]. Ti_3C_2Tx MXene is exfoliated and spontaneously modified by 4-nitrophenyl groups; the degree of modification of MXene is adjusted through the concentration of the diazonium salt solution. The work function of functionalized MXene is tunable by regulating the quantity of grafted diazonium

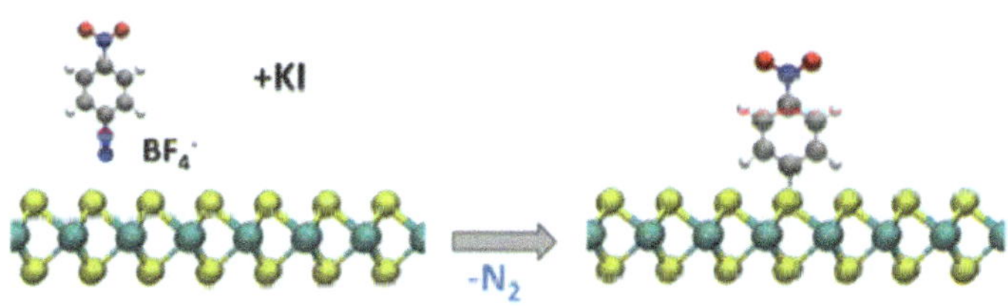

Fig. 6 The grafting of MoS_2 by diazonium salts. Reused with permission from reference [192]. Copyright (2021) American Chemical Society

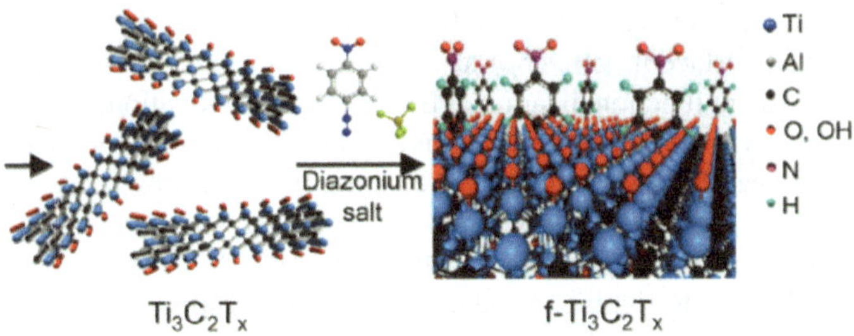

Fig. 7 Surface modification of Ti_3C_2Tx nanoshheets by 4-nitrobenzenediazonium salt. Reused with permission from reference [206]. Copyright (2021) American Chemical Society

surface groups, with an adjustable range of around 0.6 eV. This modified MXene permits the fabrication of field-effect-transistor devices [206]. Ti_3C_2Tx nanosheets are selectively modified with amidoxime functional groups by diazonium chemistry; this permits to improve significantly their stability in aqueous solution. They permit the efficient, rapid, and recyclable uranium extraction from aqueous solutions containing competitive metal ions [207]. Ti3C_2Tx MXene is covalently modified by reaction with the diazonium salt of 1-aminoanthraquinone for supercapacitor applications [208].

10 Indirect Grafting

Aryl radicals are very reactive species, when created in the vicinity of the electrode they react with the surface of many materials and create directly electrografted organic films with different thicknesses as shown previously in this chapter. The same species may undergo other reactions (Chap. 2) in solution such as (i) the attack of an arene to give biaryls (Gomberg Bachman reaction) [209, 210]; (ii) abstraction of a hydrogen or halogen atom (Sandmeyer reaction) from a molecule in solution or the solvent itself; [211] (iii) reaction with vinylic compounds to initiate the formation of vinyl radicals that in their turn attack already attached polyphenylene layer; the Surface Electroinitiated Emulsion Polymerization (SEEP) and Graftfast processes are described in Chap. 23 [212, 213].

This solution chemistry is in competition with the electrografting reaction, it will take its full extent if the last one is blocked. This is possible using 2,6-dimethylbenzendiazonium (2,6-DMBD) that shows a particular behavior during its electrochemical reduction by cyclic voltammetry as is shown in Fig. 8 [214]:

An irreversible cathodic wave is observed at $Ep = -0.22$ V/Ag/AgCl ($= -0.1$ V/SCE) as in the case of other aryldiazonium salts, but remains almost unchanged after 10 cycles. The behavior of 2,6-DMBD is very different from that of other aryldiazonium salts as the electrode surface is not passivated by the organic film as

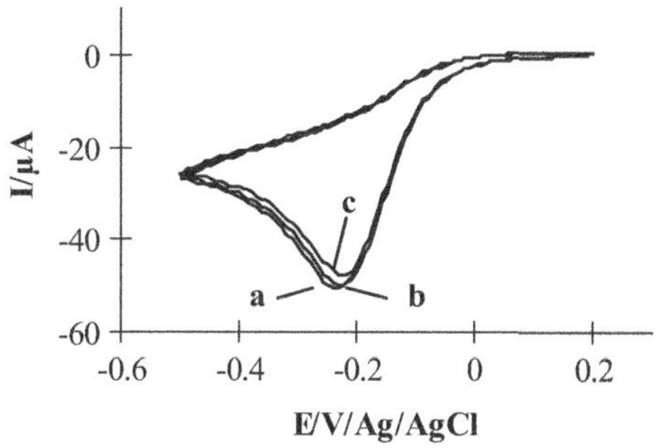

Fig. 8 Cyclic voltammetry in ACN + 0.1 M NBu4BF4 on a glassy carbon electrode ($d = 2$ mm) of **a** 4 mM 2, 6-dimethylbenzenediazonium tetrafluoroborate, **a** first, **b** second, and **c** 10th scan, Scan rate) 0.1 V s^{-1}. l. Reused with permission from reference [214]. Copyright (2014) American Chemical Society

described above (Fig. 1). After these voltammetric cycles, XPS and IRRAS measurements do not indicate the presence of an organic layer on the GC and copper surface and DFT calculations confirm a decrease of more than 50% of the bonding energy by comparison with a simple phenyl group. The absence of surface reaction of the 2,6-dimethylphenyl radical is due to the steric effects of two methyl groups in ortho positions to the diazonium group [214]. Formation of the 2,6-dimethylphenyl radical is also achieved chemically under sonication or by heating at 60 °C. This absence of surface reaction does not imply the absence of solution reactions as shown below. Indeed, new reactions are observed that permit the grafting of nearly any compound provided it can be solubilized in the same solution as the diazonium salt.

Iodide Abstraction. Direct electrografting of alkyl iodides on metals and glassy carbon [16] occurs at very negative potentials (−2.5 V/SCE for 1-iodohexane), at such potential, the alkyl radicals are reduced to their anion, therefore, decreasing the yield of the surface reaction as shown by the long electrolysis times [215, 216]. This problem can be circumvented by use of 2,6-dimethylphenyl radicals in the presence of alkyliodides. Figure 9 Aa shows the electrochemical reduction of 1-iodo-1*H*, 1*H*, 2*H*, 2*H*-perfluorooctane ($ICH_2CH_2C_6F_{13}$) along with an irreversible two-electron wave Ep ~ −1.95 V/Ag/AgCl [214]. Figure 9Ab presents the one-electron irreversible reduction wave of 2.6-DMBD. When these two compounds ($ICH_2CH_2C_6F_{13}$ and 2,6-DMBD) are mixed together in ACN one can see in Fig. 9B that the voltammogram on GC decreases and comes close to zero indicating a passivated surface after 8 scans at the potential of 2,6-DMBD.

GC plates modified by chronoamperometry in the same solution as that of Fig. 9B, present IRRAS, XPS, ellipsometry, and water contact angle in agreement presence of perfluoroalkyl groups on the GC plate without the signal of iodine atoms. The grafted film is very stable as it withstands 2 h in boiling toluene in a Soxhlet and a cathodic potential excursion until—1.8 V/Ag/AgCl by cyclic voltammetry. These results indicate that the perfluoroalkyliodide can be grafted at the potential of 2,6-DMBD [214]. A mechanism is proposed in Scheme 8. 2,6-DMBD is electrochemically reduced at

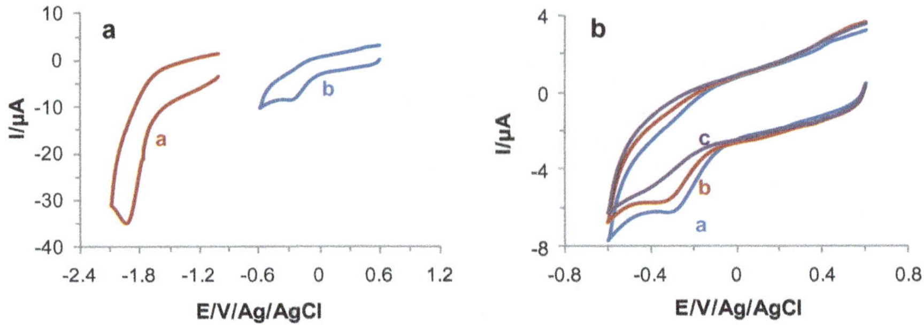

Fig. 9 Cyclic voltammetry on a *GC* electrode ($d = 2$ mm) in an ACN + 0.1 M NBu_4PF_6 solution of $ICH_2CH_2C_6F_{13}$ and 2,6-DMBD (1 mM each). **a** each species recorded separately: (a) 1, (b) 2,6-DMBD. **b** mixture of both species (a) 1st, (b) 2nd, and c) 8th scan. $v = 0.2$ V/s. Reused with permission from reference [214]. Copyright (2014) American Chemical Society

Scheme 8 Mechanism for the indirect grafting of alkyliodides on GC surface via the 2,6-DMBD

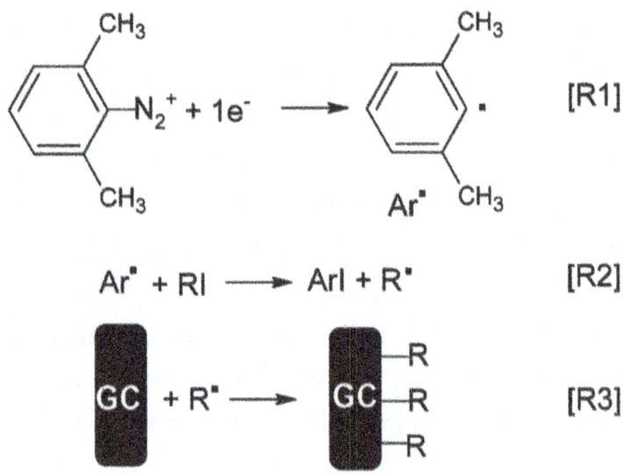

the electrode, the resulting 2,6-dimethylphenyl radical (2,6-DMPh·) does not react with the GC electrode due to its steric hindrance [R1]. 2,6-dimethyl radical abstract an iodine atom from perfluoroalkyliodide (RI) to give an alkyl radical [R2]. These very reactive species attach to the GC surface to give a perfluoroalkyl layer [R3].

There is a positive potential shift of about $\Delta Ep = 1.7$ V when comparing the indirect grafting of alkyliodides with the direct electrografting. Such a large catalytic effect, obtained by diverting the reactivity of aryl radicals, has never been observed even in the presence of metals such as Ag or Pd [217, 218]. This electrocatalysis effect is even more pronounced with alkyl bromides $\Delta Ep = 2.3$ V, because they are electrochemically reduced at potentials close to the background discharge, Ep ~ −2.6 V/SCE on Au [219].

As alkyliodides, aryliodides [220] are electrochemically reduced at negative potentials; their reduction involves almost a bielectronic transfer along a concerted process, the aryl radical is immediately transformed to its anion. This is the case with iodobenzene Ep = −2.23 V/SCE [221]. Nevertheless, modification of a GC electrode surface with aryl groups is possible by cyclic voltammetry but after 100 scans, which indicates a difficult reaction with a very low faradic yield [216]. Using 2,6-DMBD

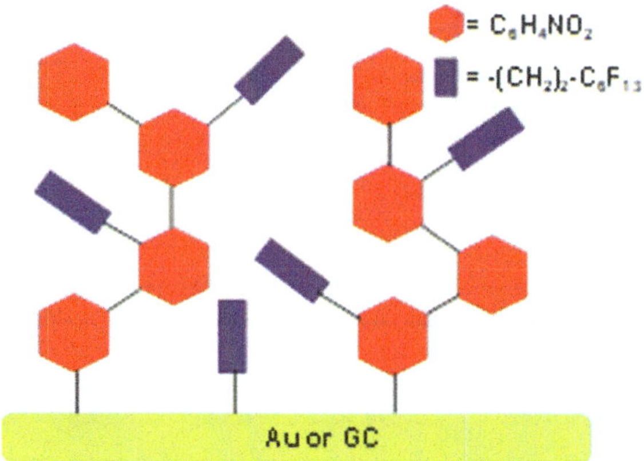

Fig. 10 Mixed layers obtained by simultaneous reduction of a diazonium salt and an alkyl halide. Reused with permission from reference [225]. Copyright (2016) American Chemical Society

as precursor (Ep = −0.22 V/SCE) of the 2,6-dimethylphenyl radical in the presence of iodobenzene in ACN solution it is possible, after few cyclic voltammograms, to graft aryl groups along the mechanism of Scheme 8 [222]. The same procedure is used to graft 4-iodonitrobenzene that cannot be directly electrografted because it is electrochemical reduced to nitrobenzene via ECE mechanism, [223] 4-iodoaniline and 5-iodo-2-pyridine (the corresponding diazonium salt is very unstable) [224].

Aryl radicals from diazonium salts (without steric hindrance) undergo two parallel reactions in the presence of alkyl iodides: (i) they abstract iodine atoms (leading to alkyl radicals that attack the surface) and (ii) attach to the surface leading to mixed alkyl/aryl layers with a very complex structure (Fig. 10). By tuning the ratio between the diazonium salt and the alkyl iodide, it is possible to obtain surfaces with controlled properties (thickness, water contact angle, …) [225].

Bromide Abstraction. Easily available alkylbromides with terminal functional groups provide a functionalized surface that can be further modified. Carboxylic groups can be attached to GC, metals, or polymer surfaces by indirect grafting of 6-bromohexanoic acid. Hexanoic groups attached to a gold surface can be further post-modified with aminoferrocene by peptidic coupling. They also are used to prepare layer-by-layer (LbL) assembly when the modified gold plate is immersed in poly-electrolyte solutions containing alternatively branched polyethyleneimine- (PEI) and poly(acrylic acid) (PAA) as is shown in Fig. 11 [219]. These LbL layers have shown a certain capacity to retain $Ru(NH_3)_6{}^{2+/3+}$ ions within the assembly (Fig. 11).

H atom Abstraction. H atom abstraction is also possible starting from 2,6-dimethylphenyl radicals; in this way the activation of C–H bonds are achieved from acetonitrile [211], nitromethane or methyl amine [226]. The mechanism is shown below for CH_3CN, CH_3NO_2, and CH_3NH_2, Scheme 9. As above [R1] is the formation of the 2,6-dimethylphenyl radical, this radical abstracts a hydrogen atom to give the three radicals that react with the surface [R4], the formation of oligomers from CH_3CN results from the reduction of the radical to its anion [R5], this anion reacts with the attached cyano group (Thorpe reaction) to give an amino group [R6]. Repetition of this reaction permits to obtain a 56 nm thick film on gold. The $\cdot CH_2NO_2$ and

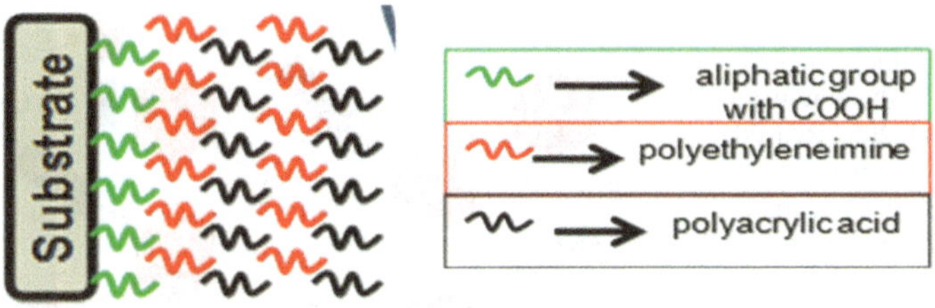

Fig. 11 Formation of layer-by-Layer assembly of PEI and PAA on a surface modified with alkylcarboxylate ions. Reused with permission from reference [219]. Copyright (2016) American Chemical Society

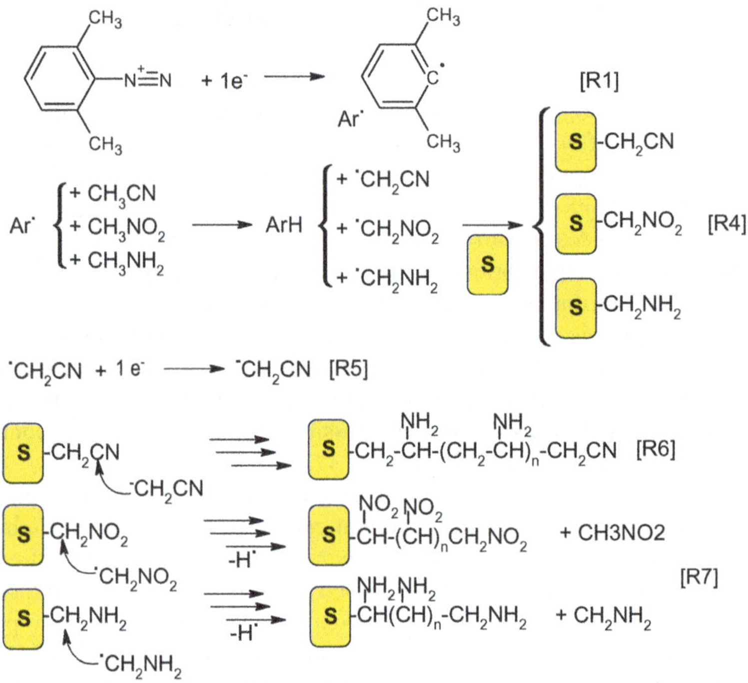

Scheme 9 Mechanism of the indirect grafting of CH_3CN, CH_3NO_2, CH_3NH_2 on gold surfaces

$\cdot CH_2NH_2$ radicals abstract an H atom from the first grafted group to give a grafted radical, coupling a grafted radical with $\cdot CH_2NO_2$ and $\cdot CH_2NH_2$ permits to obtaining (~6 nm) of oligomeric films [R7].

The above reactions enlarge the spectrum of surface modification reactions to a much broader spectrum of organic groups, in principle, any molecule with an H, I, Br atom can be grafted on a surface.

11 Conclusion

The surface chemistry of diazonium salts is now widely used and the reaction is now well described, however, some points remain to be precised or verified. For example, the exact path leading from the diazonium to its radical, this should permit to model of the concentration of the radical in the vicinity of the electrode and by consequence the rate of reaction on surfaces. The formation of *Surface*-N=N-aryl that has been observed in some cases should be further investigated to precise on which surfaces, with which substituents it is favored or disfavoured. Bis-diazonium salts are particularly interesting they are used to bind nanoobjects together, but the yield is still low, formation of real 2D structures of bonded nanoobjects seems an interesting objective. As soon as new materials are prepared (2D materials for example) they are modified by diazonium chemistry to provide additional properties, this will hopefully continue. The conditions under which the surfaces are modified are now well established and a number of activations are available, but new solvents are used; after ionic solvents, Deep Eutectic Solvents are now investigated [25].

In addition to the direct grafting an indirect grafting method was developed, it is based on atom abstraction, and furnishes a bunch of new radicals that react with surfaces. This should enlarge the scope of grafted entities.

Applications of diazonium chemistry are steadily increasing and seem to be limited by the sole imagination of the authors, but the ultimate achievement is industrial application, some are already on the market, but it is hoped that some more will become available.

References

1. Pujari SP, Scheres L, Marcelis ATM, Zuilhof H (2014) Angew Chem Int Ed 53:6322–6356. https://doi.org/10.1002/anie.201306709
2. Vericat C, Vela ME, Benitez G, Carro P, Salvarezza RC (2010) Chem Soc Rev 39:1805–1834. https://doi.org/10.1039/b907301a
3. Hetemi D, Pinson J (2017) Chem Soc Rev 46:5701–5713. https://doi.org/10.1039/c7cs00150a
4. Pinson J, Thiry D (eds) (2020) Surface modification of polymers: methods and applications. Wiley, Weinheim, Germany. ISBN: 978-3-527-81923-2
5. Mo F, Qiu D, Zhang L, Wang J (2021) Chem Rev 121:5741–5829. https://doi.org/10.1021/acs.chemrev.0c01030
6. Pinson J, Podvorica FI (2020) Curr Opin Electrochem 24:44–48
7. Berisha A, Chehimi MM, Pinson J, Podvorica FI (2016) Electrode surface modification using diazonium salts. In: Bard AJ, Zoski CG (eds) Electroanalytical chemistry, vol 26. CRC Press, Boca Raton, FL. ISBN:978-1-498763377-9
8. Chehimi MM (ed) (2012) Aryl Diazonium salts. New coupling agents in polymer and surface science. Wiley-VCH, Weinheim, Germany, ISBN: 978-3-527-32998-4
9. Mahouche-Chergui S, Gam-Derouich S, Mangeney C, Chehimi MM (2011) Chem Soc Rev 40:4143–4166. https://doi.org/10.1039/c0cs00179a
10. Gooding JJ, Ciampi S (2011) Chem Soc Rev 40:2704–2718. https://doi.org/10.1039/C0CS00139B

11. Bélanger D, Pinson J (2011) Chem Soc Rev 40:3995–4048. https://doi.org/10.1039/c0cs00149j
12. Pinson J, Podvorica F (2005) Chem Soc Rev 34:429–439. https://doi.org/10.1039/b406228k
13. Vase KH, Holm AH, Pedersen SU, Daasbjerg K (2005) Langmuir 21:8085–8089. https://doi.org/10.1021/la050933e
14. Koefoed L, Pedersen SU, Daasbjerg K (2016) ChemElectroChem 3:495–501. https://doi.org/10.1002/celc.201500512
15. Wang D, Buriak JM (2005) Surf Sci 590:154–161. https://doi.org/10.1016/j.susc.2005.06.018
16. Chehimi MM, Hallais G, Matrab T, Pinson J, Podvorica FI (2008) J Phys Chem C 112:18559–18565. https://doi.org/10.1021/jp807044j
17. Médard J, Decorse P, Mangeney C, Pinson J, Fagnoni M, Protti S (2020) Langmuir 36:2786–2793. https://doi.org/10.1021/acs.langmuir.9b03878
18. Adenier A, Chehimi MM, Gallardo I, Pinson J, Vilà N (2004) Langmuir 20:8243–8253. https://doi.org/10.1021/la049194c
19. Hernández-Muñoz LS, González-Fuentes MA, Díaz-Sánchez BR, Fragoso-Soriano R, Vázquez-López C, González FJ (2012) Electrochim Acta 63:287–294. https://doi.org/10.1016/j.electacta.2014.04.189
20. Li D, Luo Y, Onidas D, He L, Jin M, Gazeau F, Pinson J, Mangeney C (2021) Adv Colloid Interface Sci 294:102479
21. Delamar M, Hitmi R, Pinson J, Savéant J-M (1992) J Am Chem Soc 114:5883–5884. https://doi.org/10.1021/ja00040a074
22. Allongue P, Delamar M, Desbat B, Fagebaume O, Hitmi R, Pinson J, Savéant J-M (1997) J Am Chem Soc 119:201–207. https://doi.org/10.1021/ja963354s
23. Hetemi D, Combellas C, Kanoufi F, Pinson J, Podvorica FI (2016) Elec Comm 68:5–9. https://doi.org/10.1016/j.elecom.2016.04.001
24. Richard W, Evrard D, Busson B, Humbert C, Dalstein L, Tadjeddine A, Gros P (2018) Electrochim Acta 283:1640–1648. https://doi.org/10.1016/j.electacta.2018.07.073
25. Lenne Q, Andrieux V, Levanen G, Bergamini J-F, Nicolas P, Paquin L, Lagrost C, Yann R, Leroux YR (2021) Electrochimica Acta 369: 137672. https://doi.org/10.1016/j.electacta.2020.137672
26. Benedetto A, Balog M, Viel P, Le Derf F, Sallé M, Palacin S (2008) Electrochim Acta 53:7117–7122. https://doi.org/10.1016/j.electacta.2008.05.001
27. Lee L, Brooksby PA, Hapiot P, Downard AJ (2016) Langmuir 32:468–476. https://doi.org/10.1021/acs.langmuir.5b03233
28. Koefoed L, Vase KH, Stenlid JH, Brinck T, Yoshimura Y, Lund H, Pedersen SU, Daasbjerg K (2017) ChemElectroChem 4:3212–3221. https://doi.org/10.1002/celc.201700772
29. Romanczyk PP, Kurek SS (2020) Electrochim Acta 351:136404. https://doi.org/10.1016/j.electacta.2020.136404
30. Koefoed L, Pedersen SU, Daasbjerg K (2017) Langmuir 33:3217–3222. https://doi.org/10.1021/acs.langmuir.7b00300
31. Raviola C, Protti S (2020) Eur J Org Chem 33:5292–5304. https://doi.org/10.1002/ejoc.202000143
32. Hari DP, Hering T, Koenig B (2018) Arene functionalization by visible light photoredox catalysis in visible light photocatalysis in organic chemistry. In: Stephenson C, Tehshik Y, MacMillan DWC (eds). Wiley VCH, pp 253–281. https://doi.org/10.1002/9783527674145
33. Bouriga M, Chehimi MM, Combellas C, Decorse P, Kanoufi F, Deronzier A, Pinson J (2013) Chem Mater 25:90–97. https://doi.org/10.1021/cm3032994
34. Busson M, Berisha A, Combellas C, Kanoufi F, Pinson J (2011) Chem Commun 47:12631–12633. https://doi.org/10.1039/c1cc16241a
35. Verberne-Sutton SD, Quarels RD, Zhai X, Garno JC, Ragains JR (2014) J Am Chem Soc 136:14438–14444. https://doi.org/10.1021/ja505521k
36. Médard J, Combellas C, Kanoufi F, Pinson J, Chauvin J, Deronzier A (2018) J Phys Chem C 122:19722–19730. https://doi.org/10.1021/acs.jpcc.8b06541

37. Schroll P, Fehl C, Dankesreiter S, König B (2013) Org Biomol Chem 11:6510–6514. https://doi.org/10.1039/c3ob40990b
38. Garcia A, Hanifi N, Jousselme B, Jégou J, Palacin S, Viel P, Berthelot T (2013) Adv Funct Mater 23:3668–3674. https://doi.org/10.1002/adfm.201203544
39. Sergeeva NN, Chaika AN, Walls B, Murphy BE, Walshe K, Martin DP, Richards BDO, Jose G, Fleischer K, Aristov VU, Molodtsova OV, Shvets IV, Krasnikov SA (2018) Nanotechnology 29:275705. https://doi.org/10.1088/1361-6528/aabf11
40. González MCR, Brown A, Eyley S, Thielemans W, Mali KS De Feyter S (2020) Nanoscale12:18782. https://doi.org/10.1039/d0nr05244b
41. Zeb G, Viel P, Palacin S, Le XT (2015) RSC Adv 5:50298–50305. https://doi.org/10.1039/c5ra07875j
42. Mevellec V, Roussel S, Tessier L, Chancolon J, Mayne-L'Hermite M, Deniau G, Viel P, Palacin S (2007) Chem Mater 19:6323–6330. https://doi.org/10.1021/cm071371i
43. Enciso AE, Doni G, Nifosì R, Palazzesi F, Gonzalez R, Ellsworth AA, Coffer JL, Walker AV, Giovanni M. Pavan GM, Mohamed AM, Simanek EE (2017) Nanoscale 9:3128–3132. https://doi.org/10.1039/c6nr09679d
44. Xia Y, Martin C, Seibel J, Eyley S, Thielemans W, van der Auweraer M, Mali KS, De Feyter S (2020) Nanoscale 12:11916–11926. https://doi.org/10.1039/d0nr03309j
45. Adenier A, Cabet-Deliry E, Chaussé A, Griveau S, Mercier F, Pinson J, Vautrin-Ul (2005) Chem Mater 17:491–501. https://doi.org/10.1021/cm0490625
46. Hurley BL, McCreery RL (2004) J Electrochem Soc 151:B252-259. https://doi.org/10.1149/1.1687428
47. Combellas C, Delamar M, Kanoufi F, Pinson J, Podvorica F (2005) Chem Mater 17:3968–3975. https://doi.org/10.1021/cm050339q
48. Lehr J, Williamson BE, Downard AJ (2011) J Phys Chem C 115:6629–6634. https://doi.org/10.1021/jp111838r
49. Berisha A, Hazimeh H, Galtayries A, Decorse P, Kanoufi F, Combellas C, Pinson J, Podvorica FI (2016) RSC Adv 6:78369–78376. https://doi.org/10.1039/c6ra15313e
50. Squillace O, Perrault T, Gorczynska M, Caruana A, Bajorek A, Brotons G (2021) Colloids Surf B 197:111427. https://doi.org/10.1016/j.colsurfb.2020.111427
51. Delaporte N, Trudeau ML, Bélanger D, Zaghib K (2020) Materials 13:942. https://doi.org/10.3390/ma13040942
52. Luo Y, Xiao Y, Onidas D, Iannazzo L, Ethève-Quelquejeu M, Lamouri A, Félidj N, Mahouche-Chergui S, Brulé T, Gagey-Eilstein N, Gazeau F, Mangeney C (2020) Chem Commun 56:6822–6825. https://doi.org/10.1039/d0cc02842h
53. Lei Y, Huang Q, Gan D, Huang H, Chen J, Deng F, Liu M, Li X, Zhanga X, Wei Y (2020) J Environ Chem Engin 8:103780. https://doi.org/10.1016/j.jece.2020.103780
54. Griffete N, Herbst F, Pinson J, Ammar S, Mangeney C (2011) J Am Chem Soc 133:1646–1649. https://doi.org/10.1021/ja108928
55. Chenga C, Jia P, Xiao L, Geng J (2019) Carbon 145:668–676. https://doi.org/10.1016/j.carbon.2019.01.079
56. Mamane V, Mercier G, Shukor JA, Gleize J, Azizan A, Fort Y, Vigolo B (2014) Beilstein J Nanotechnol 5:537–545. https://doi.org/10.3762/bjnano.5.63
57. Van Druenen M, Davitt F, Collins T, Glynn C, O'Dwyer C, Holmes JD, Collin G (2018) Chem Mater 30:4667–4674. https://doi.org/10.1021/acs.chemmater.8b01306
58. Güell AG, Roodenko K, Yang F, Hinrichs K, Gensch M, Sanz F, Rappich J (2006) Mater Sci Eng B 134:273–276. https://doi.org/10.1016/j.mseb.2006.07.005
59. Hinge M, Gonçalves ES, Pedersen SU, Daasbjerg K (2010) Surf Coat Technol 205:820–827. https://doi.org/10.1016/j.surfcoat.2010.07.125
60. Gorp HV, Walke P, Teyssandier J, Hirsch BE, Uji-i H, Tahara K, Tobe Y, Van der Auweraer M, De Feyter S (2020) J Phys Chem C 124:1980–1990. https://doi.org/10.1021/acs.jpcc.9b09808
61. Schirowski M, Hauke F, Hirsch A (2019) Chem Eur J 25:12761–12768. https://doi.org/10.1002/chem.201902330

62. Laurentius L, Stoyanov SR, Gusarov S, Kovalenko A, Du R, Lopinski GP, McDermott MT (2011) ACS Nano 5:4219–4227. https://doi.org/10.1021/nn201110r
63. Barosi A, Berisha A, Mangeney C, Pinson J, Dhimane, Dalko PI (2021) Mater Adv 2:2358–2365. https://doi.org/10.1039/d1ma00022e
64. Berisha A, Combellas C, Kanoufi F, Médard J, Decorse P, Mangeney C, Kherbouche I, Seydou, M, Maurel F, Pinson J (2018) Langmuir 34:11264–11271. https://doi.org/10.1021/acs.langmuir.8b01584
65. Li H, Kopiec G, Müller F, Nyßen F, Shimizu K, Ceccato M, Daasbjerg K, Plumeré N (2020) J Am Chem Soc Au 142:8662–8671. https://doi.org/10.1021/jacsau.0c00108
66. Greenwood J, Phan TH, Fujita Y, Li Z, Ivasenko O, Vanderlinden W, Van Gorp H, Frederickx W, Lu G, Tahara K, Tobe Y, Uji-i H, Mertens SFL, De Feyter S (2015) ACS Nano 5:5520–5525. https://doi.org/10.1021/acsnano.5b01580
67. Sampathkumar K, Diez-Cabanes V, Kovaricek P, del Corro E, Bouša M, Hošek J, Kalbac M, Frank O (2019) J Phys Chem C 123:22397–22402. https://doi.org/10.1021/acs.jpcc.9b06516
68. Ossonon BD, Bélanger D (2017) Carbon 111:83–93. https://doi.org/10.1016/j.carbon.2016.09.063
69. Brymora K, Fouineau J, Eddarir A, Chau F, Yaacoub N, Grenèche J-M, Pinson J, Ammar S, Calvayrac F (2015) J Nanopart Res 17:438. https://doi.org/10.1007/s11051-015-3232-x
70. Betelu S, Tijunelyte I, Boubekeur-Lecaque L, Ignatiadis I, Ibrahim J, Gaboreau S, Berho C, Toury T. Guenin E, Lidgi-Guigui N, Félidj N, Rinnert E, Lamy de la Chapelle M (2016) J Phys Chem C 120:18158–18166. https://doi.org/10.1021/acs.jpcc.6b06486
71. Mesnage A, Lefèvre X, Jégou P, Deniau G, Palacin S (2012) Langmuir 28:11767–11778. https://doi.org/10.1021/la3011103
72. Berisha A (2019) J Chem 5126071. https://doi.org/10.1155/2019/5126071
73. Kong L, Enders A, Rahman TS, Dowben PA (2014) J Phys: Condens Matter 26:443001. https://doi.org/10.1088/0953-8984/26/44/443001
74. Gross AJ, Tanaka S, Colomies C, Giroud F, Nishina Y, Cosnier S, Tsujimura S, Holzinger M (2020) ChemElectroChem 7:4543–4549. https://doi.org/10.1002/celc.202000953
75. Le Comte A, Chhin D, Gagnon A, Retoux R, Brousse T, Bélanger D (2015) J Mater Chem A 3:6146–6156. https://doi.org/10.1039/c4ta05536e
76. Berisha A, Combellas C, Kanoufi F, Pinson J, Podvorica FI (2011) Electrochim Acta 56:10762–10766. https://doi.org/10.1016/j.electacta.2011.01.049
77. Morales-Martínez D, Lartundo-Rojas L, González FJ (2020) ChemElectroChem 7:4431–4439. https://doi.org/10.1002/celc.202001096
78. Mesnage A, Esnouf S, Jégou P, Deniau G, Palacin S (2010) Chem Mater 22:6229–6239. https://doi.org/10.1021/cm1014702
79. Schmidt G, Gallon S, Esnouf S, Bourgoin J-P, Chenevier P (2009) Chem Eur J 15:2101–2110. https://doi.org/10.1002/chem.200801801
80. Lohmann S-H, Trerayapiwat KJ, Niklas J, Poluektov OG, Sharifzadeh S, Ma X (2020) ACS Nano 14:17675–17682. https://doi.org/10.1021/acsnano.0c08782
81. Suehiro T, Masuda S, Tashiro T, Nakausa R, Taguchi M, Koike A, Rieker A (1986) Bull Chem Soc Jpn 59:1877–1886. https://doi.org/10.1246/bcsj.59.1877
82. Doppelt P, Hallais G, Pinson J, Podvorica F, Verneyre S (2007) Chem Mater 19:4570–4575. https://doi.org/10.1021/cm0700551
83. Berisha A, Combellas C, Kanoufi F, Decorse F, Oturan N, Médard J, Seydou M, Maurel F, Pinson J (2017) Langmuir 33:8730–8738. https://doi.org/10.1021/acs.langmuir.7b01371
84. Buriak JM, Sikder MdDH (2015) J Am Chem Soc 137:9730–9738. https://doi.org/10.1021/jacs.5b05738
85. Ait El Hadj F, Amiar A, Cherkaoui M, Chazalviel J-N, Ozanam F (2012) Electrochim Acta 70: 318–324. https://doi.org/10.1016/j.electacta.2012.03.072
86. Henry de Villeneuve C, Pinson J, Bernard MC, Allongue P (1997) J Phys Chem B 101:2415–2420. https://doi.org/10.1021/jp962581d
87. Chehimi MM, Lamouri A, Picot M, Pinson J (2014) J Mater Chem C 2:356–363. https://doi.org/10.1039/c3tc31492h

88. Jayasundara DR, Cullen RJ, Soldi L, Colavita PE (2011) Langmuir 27:13029–13036. https://doi.org/10.1021/la202862p
89. Han X, Lee HK, Lee YH, Hao W, Liu Y, Phang IY, Li S, Ling XY (2016) J Phys Chem Lett 7:1501–1506. https://doi.org/10.1021/acs.jpclett.6b00501
90. Bouden S, Pinson J, Vautrin-Ul C (2017) Electrochem Com 81:120–123. https://doi.org/10.1016/j.elecom.2017.06.007
91. Shkirskiy V, Levillain E, Gautier C (2021) ChemPhysChem 22:1074–1078. https://doi.org/10.1002/cphc.202100154
92. Allongue P, Henry de Villeneuve C, Cherouvrier G, Cortes R, Bernard, MC (2003) J Electroanal Chem 550–551:161–174. https://doi.org/10.1016/S0022-0728(03)00076-7
93. Anariba F, Viswanathan U, Bocian DF, McCreery RL (2006) Anal Chem 78:3104–3112. https://doi.org/10.1021/ac052042h
94. Podvorica FI, Kanoufi F, Pinson J, Combellas C (2009) Electrochim Acta 54:2164–2170. https://doi.org/10.1016/j.electacta.2008.10.017
95. Fontaine O, Ghilane J, Martin P, Lacroix J-C, Randriamahazaka H (2010) Langmuir 26:18542–18549. https://doi.org/10.1021/la103000u
96. Nielsen LT, Vase KH, Dong M, Besenbacher F, Pedersen SU, Daasbjerg K (2007) J Am Chem Soc 129:1888–1889. https://doi.org/10.1021/ja0682430
97. Yates ND, Dowsett MR, Bentley P, Dickenson-Fogg JA, Pratt A, Blanford CF, Fascione MA, Parkin A (2020) Langmuir 36:5654–5664. https://doi.org/10.1021/acs.langmuir.9b01254
98. Combellas C, Kanoufi F, Pinson J, Podvorica FI (2008) J Am Chem Soc 130:8576–8577. https://doi.org/10.1021/ja8018912
99. Gillan L, Teerinen T, Johansson L-S, Smolander M (2020) Sensor Int 2:100060. https://doi.org/10.1016/j.sintl.2020.100060
100. Nguyen VQ, Sun X, Lafolet F, Audibert JF, Miomandre F, Lemercier G, Loiseau F, Lacroix JC (2016) J Am Chem Soc 138:9381–9384. https://doi.org/10.1021/jacs.6b04827
101. Leroux YR, Hui F, Noël JM, Roux C, Hapiot P (2010) J Am Chem Soc 132:14039–14041. https://doi.org/10.1021/ja106971x
102. Troian-Gautier L, Martinez-Tong DE, Hubert J, Reniers F, Sferrazza M, Mattiuzzi A, Lagrost C, Jabin I (2016) J Phys Chem C 120:22936–22945. https://doi.org/10.1021/acs.jpcc.6b06143
103. Menanteau T, Levillain E, Breton T (2013) Chem Mater 25:2905–2909. https://doi.org/10.1021/cm401512c
104. López I, Dabos-Seignon S, Breton T (2019) Langmuir 35:11048–11055. https://doi.org/10.1021/acs.langmuir.9b01397
105. Rodríguez González MC, Brown A, Eyley S, Thielemans W, Mali KS, De Feyter S (2020) Nanoscale 12:18782–18789. https://doi.org/10.1039/D0NR05244B
106. Gabaji M, Médard J, Hemmerle A, Pinson J, Michel JP (2020) Langmuir 36:2534−2542. https://doi.org/10.1021/acs.langmuir.9b03601
107. Lee L, Leroux YR, Hapiot P, Downard AJ (2015) Langmuir 31:5071–5077. https://doi.org/10.1021/acs.langmuir.5b00730
108. Hetemi D, Noël V, Pinson J (2020) Biosensors 10:4. https://doi.org/10.3390/bios10010004
109. Sayed SY, Bayat A, Kondratenko M, Leroux Y, Hapiot P, McCreery RL (2013) J Am Chem Soc 135:12972–12975. https://doi.org/10.1021/ja4065443
110. Wang A, Ye J, Humphrey MG, Zhang C (2018) Adv Mater 30:1705704. https://doi.org/10.1002/adma.201705704
111. Adenier A, Combellas C, Kanoufi F, Pinson J, Podvorica FI (2006) Chem Mater 18:2021–2029. https://doi.org/10.1021/cm052065c
112. Ceccato M, Bousquet A, Hinge M, Pedersen SU, Daasbjerg K (2011) Chem Mater 23:1551–1557. https://doi.org/10.1021/cm1033244
113. Evrard D, Lambert F, Policar C, Balland V, Limoges B (2008) Chem Eur J 14:9286–9291. https://doi.org/10.1002/chem.200801168
114. Downard AJ (2009) Int J Nanotechnol 6:233–244. https://doi.org/10.1504/IJNT.2009.022916
115. Brooksby PA, Downard AJ (2005) J Phys Chem B 109:8791–8798. https://doi.org/10.1021/jp046095z

116. Sandomierski M, Voelkel A (2021) J Inorg Organomet Polym Mater 31:1–21. https://doi.org/10.1007/s10904-020-01725-0
117. Demir B, Henderson LC, Walsh TR (2017) ACS Appl Mater Interfaces 9:11846–11857. https://doi.org/10.1021/acsami.6b16041
118. Sandomierski M, Buchwald T, Strzemiecka B, Voelke A (2020) J Appl Polym Sci 137:48160. https://doi.org/10.1002/app.48160
119. Chaussé A, Chehimi MM, Karsi N, Pinson J, Podvorica F, Vautrin-Ul C (2002) Chem Mater 14:392–400. https://doi.org/10.1021/cm011212d
120. Bernard MC, Chaussé A, Cabet-Deliry E, Chehimi MM, Pinson J, Podvorica F, Vautrin-Ul C (2003) Chem Mater 15:3450–3462. https://doi.org/10.1021/cm034167d
121. Cui Y, Hu Z-J, Yang J-X, Gao H-W (2012) Microchim Acta 176:359–366. https://doi.org/10.1007/s00604-011-0725-x
122. Gooding JJ, Ciampi S (2011) Chem Soc Rev 40:2704–2718. https://doi.org/10.1039/C0CS00139B
123. Gautier C, López I, Breton T (2021) Mater Adv 2:2773–2810. https://doi.org/10.1039/d1ma00077b
124. Cougnon C, Gohier F, Bélanger D, Mauzeroll J (2009) Angew Chem Int Ed 48:4006–4008. https://doi.org/10.1002/anie.200900498
125. Corgier BP, Bellon S, Anger-Leroy M, Blum LJ, Marquette CA (2009) Langmuir 25:9619–9623. https://doi.org/10.1021/la900762s
126. Bouden S, Chaussé A, Dorbes S, El Tall O, Bellakhal N, Dachraoui M, Vautrin-Ul C (2013) Talanta 106:414–421. https://doi.org/10.1016/j.talanta.2013.01.021
127. Bouden S, Bellakhal N, Chaussé A, Dachraoui M, Vautrin-Ul C (2014) Electrochim Acta 125:149–155. https://doi.org/10.1016/j.electacta.2014.01.083
128. Eissa S, Zourob M (2021) Anal Chem 93:1826–1833. https://doi.org/10.1021/acs.analchem.0c04719
129. Qi M, Zhang Y, Cao C, Zhang M, Liu S, Liu G (2016) Anal Chem 88:9614–9621. https://doi.org/10.1021/acs.analchem.6b02353
130. Rather JA, Khudaish EA, Kannan P (2018) Analyst 143:1835–1845. https://doi.org/10.1039/C7AN02092A
131. Zhang L, Gao L-F, Li L, Hu C-X, Yang Q-Q, Zhu Z-Y, Peng R, Qiang Wang Q, Peng Y, Jin J, Zhang H-L (2018) Mater Chem Front 2:1700–1706. https://doi.org/10.1039/C8QM00237A
132. Ok-Kyung P, Jun-Yeon H, Munju G, Joong HL, Bon-Cheol K, Nam-Ho Y (2013) Macromolecules 46:3505–3511. https://doi.org/10.1021/ma400185j
133. Marshall N, Locklin JJ (2011) Langmuir 27:13367–13373. https://doi.org/10.1021/la2024617
134. Orchanian NM, Hong LE, Skrainka JA, Esterhuizen JA, Popov DA, Marinescu SC (2019) ACS Appl Energy Mater 2:110–123. https://doi.org/10.1021/acsaem.8b01745
135. Schirowski M, Abellán G, Nuin E, Pampel J, Dolle C, Wedler V, Fellinger TP, Spiecker E, Hauke F, Hirsch A (2018) J Am Chem Soc 140:3352–3360. https://doi.org/10.1021/jacs.7b12910
136. Mpeta LS, Sen P, Nyokong T (2020) J Electroanal Chem 860:113896. https://doi.org/10.1016/j.jelechem.2020.113896
137. Zhang L, Vilà N, Kohring G-W, Walcarius A, Etienne M (2017) ACS Catal 7:4386–4394. https://doi.org/10.1021/acscatal.7b00128
138. Leroux YR, Fei H, Noël J-M, Roux C, Hapiot P (2010) J Am Chem Soc 132:14039–14041. https://doi.org/10.1021/ja106971x
139. Yamamoto T, Akahori M, Natsui K, Saitoh T, Einaga Y (2018) Carbon 130:350–354. https://doi.org/10.1016/j.carbon.2017.12.098
140. Combellas C, Kanoufi F, Pinson J, Podvorica FI (2005) Langmuir 21:280–286. https://doi.org/10.1021/la0481061
141. Adenier A, Bernard MC, Chehimi MM, Cabet-Deliry E, Desbat B, Fagebaume O, Pinson J, Podvorica F (2001) J Am Chem Soc 123:4541–4549. https://doi.org/10.1021/ja003276f
142. Gam-Derouich S, Pinson J, Decorse P, Luo Y, Herbaut R, Royon L, Mangeney C (2018) Chem Comm 54:8983–8986. https://doi.org/10.1039/C8CC02601G

143. Dimé AKD, Bousfiha A, Devillers CH (2020) Curr Opin Electrochem 24:69–78. https://doi.org/10.1016/j.coelec.2020.07.004
144. Kudas Z, Atmaca U, Saruhan T, Celik M, Ekinci D (2020) Electroanalysis 32:1379–1390. https://doi.org/10.1002/elan.201900707
145. Marianov AN, Jiang Y (2019) ACS Sustain Chem Eng 7:3838–3848. https://doi.org/10.1021/acssuschemeng.8b04735
146. Dasler D, Schäfer RA, Minameyer MB, Hitzenberger JF, Hauke F, Drewello T, Hirsch A (2017) J Am Chem Soc 139:11760–11765. https://doi.org/10.1021/jacs.7b04122
147. Harris TGAA, Gotz R, Wrzolek P, Davis V, Knapp CE, Ly K, Hildebrandt P, Schwalbe M, Weidinger I, Zebger I, Fischer A (2018) J Mater Chem A 6:15200. https://doi.org/10.1039/c8ta02983k
148. Yao X, Sun X, Lafolet F, Lacroix J-C (2020) Nano Lett 20:6899–6907. https://doi.org/10.1021/acs.nanolett.0c03000
149. Wang A, Yu W, Huang Z, Zhou F, Song J, Song Y, Long L, Cifuentes MP, Humphrey MG, Zhang L, Shao J, Zhan C (2016) Sci Rep 6:23325. https://doi.org/10.1038/srep23325
150. Jousselme B, Bidan G, Billon M, Goyer C, Kervella Y, Guillerez S, Hamad EA, Goze-Bac C, Mevellec J-Y, Lefrant S (2008) J Electroanal Chem 621:277–285. https://doi.org/10.1016/j.jelechem.2008.01.026
151. Blankespoor R, Limoges B, Schoellhorn B, Syssa-Magale JL, Yazidi D (2005) Langmuir 21:3362–3375. https://doi.org/10.1021/la047139y
152. Tregubov AA, Vuong KQ, Luais E, Gooding JJ, Messerle BA (2013) J Am Chem Soc 135:16429–16437. https://doi.org/10.1021/ja405783gl
153. Sun C, Rotundo L, Garino C, Nencini L, Yoon SS, Gobetto R, Nervi C (2017) ChemPhysChem 18:3219–3229. https://doi.org/10.1002/cphc.201700739
154. Jiang C, Silva SM, Fan S, Wu Y, Alam MT, Liu G, Gooding JJ (2017) J Electroanal Chem 785:265–278. https://doi.org/10.1016/j.jelechem.2016.11.043
155. Liu G, Chockalingham M, Khor SM, Gui AL, Gooding JJ (2010) Electroanalysis 22:918–926. https://doi.org/10.1002/elan.200900539
156. Zhang L, Vilà N, Walcarius A, Etienne M (2018) ChemElectroChem 5:2208–2217. https://doi.org/10.1002/celc.201800258
157. Brooksby PA, Downard A (2005) Langmuir 21:1672–1675. https://doi.org/10.1021/la0468848
158. Downard A, Garret DJ, Tan ESQ (2006) Langmuir 22:10739–10746. https://doi.org/10.1021/la061148k
159. Corgier BP, Bélanger D (2010) Langmuir 26:5991–5997. https://doi.org/10.1021/la904521w
160. Nguyen V-Q, Schaming D, Martin P, Lacroix J-C (2019) Langmuir 35:15071–15077. https://doi.org/10.1021/acs.langmuir.9b02811
161. Leroux YR, Hui F, Noël J-M, Roux C, Downard AJ, Hapiot P (2011) Langmuir 27:11222–11228. https://doi.org/10.1021/la202250y
162. Mattiuzzi A, Jabin I, Mangeney C, Roux C, Reinaud O, Santos L, Bergamini J-F, Hapiot P, Lagrost C (2012) Nat Commun 3:1130. https://doi.org/10.1038/ncomms2121
163. Troian-Gautier L, Mattiuzzi A, Reinaud O, Lagrost C, Jabin I (2020) Org Biomol Chem 18:3624–3637. https://doi.org/10.1039/D0OB00070A
164. Jiang C, Alam MT, Silva SM, Taufik S, Fan S, Gooding JJ (2016) ACS Sensors 1:1432–1438. https://doi.org/10.1021/acssensors.6b00532
165. Delaporte N, Lajoie G, Collin-Martin S, Zaghib K (2020) Sci Rep 10:3812. https://doi.org/10.1038/s41598-020-60633-y18
166. Toupin M, Belanger D (2008) Langmuir 24:1910–1917. https://doi.org/10.1021/la702556n
167. Belmont JA, Amici RM, Galloway CP (1998) Reaction of carbon black with diazonium salts, resultant carbon black products, and their uses. United States, US5851280 A 1998-12-22
168. Li Z, Yan W, Dai S (2005) Langmuir 21:11999–12006. https://doi.org/10.1021/la051608u
169. Orchanian NM, Hong LE, Marinescu SC (2019) ACS Catal 9:9393–9397. https://doi.org/10.1021/acscatal.9b03134

170. Cao N, Guo J, Cai K, Xue Q, Zhu L, Shao Q, Gu X, Zang X (2020) Sep Purif Technol 251:117308. https://doi.org/10.1016/j.seppur.2020.117308
171. Bensghaier A, Mousli F, Lamouri A, Postnikov PS, Chehimi MM (2020) Chem Afr 3:535–569. https://doi.org/10.1007/s42250-020-00144-5
172. Unwin PR, Guell AG, Zhang G (2016) Acc Chem Res 49:2041–2048. https://doi.org/10.1021/acs.accounts.6b00301
173. Hirsch A, Englert JM, Hauke F (2013) Acc Chem Res 46:87–96. https://doi.org/10.1021/ar300116q
174. Raymakers J, Haenen K, Maes W (2020) J Mater Chem C 7:10134–10165. https://doi.org/10.1039/c9tc03381e
175. Jacques A, Devillers S, Delhalle J, Mekhalif Z (2013) Electrochim Acta 109:781–789. https://doi.org/10.1016/j.electacta.2013.07.178
176. Mesnage A, Lefevre X, Jegou P, Deniau G, Palacin S (2012) Langmuir 28:11767–11778. https://doi.org/10.1021/la3011103
177. Torrens M, Ortiz M, Turner APF, Beni V, O'Sullivan CK (2015) Chem Eur J 21:671–681. https://doi.org/10.1002/chem.201405121
178. Ghilane J, Delamar M, Guilloux-Viry M, Lagrost C, Mangeney C, Hapiot P (2005) Langmuir 21:6422–6429. https://doi.org/10.1021/la050401y
179. Atmane YA, Sicard L, Lamouri A, Pinson J, Sicard M, Masson C, Nowak S, Decorse P, Piquemal JY, Galtayries A, Mangeney C (2013) J Phys Chem C 117:26000–26006. https://doi.org/10.1021/jp406356s
180. Alageel O, Abdallah MN, Luo ZY, Del-Rio-Highsmith J, Cerruti M, Tamimi F (2015) Dent Mater 31:105–114. https://doi.org/10.1016/j.dental.2014.11.002
181. Le XT, Zeb G, Jegou P, Berthelot T (2012) Electrochim Acta 71:66–72. https://doi.org/10.1016/j.electacta.2012.03.076
182. Samanta S, Bakas I, Singh A, Aswal DK, Chehimi MM (2014) Langmuir 30:9397–9406. https://doi.org/10.1021/la501909r
183. Kim YS, Fournier S, Lau-Truong S, Decorse P, Devillers CH, Lucas D, Harris KD, Limoges B, Balland V (2018) ChemElectroChem 5:1625–1630. https://doi.org/10.1002/celc.201800418
184. Bangle R, Sampaio RN, Troian-Gautier L, Meyer GJ (2018) ACS Appl Mater Interfaces 10:3121–3132. https://doi.org/10.1021/acsami.7b16641
185. Lamberti F, Agnoli S, Brigo L, Granozzi G, Giomo M, Elvassore N (2013) ACS Appl Mater Interfaces 5:12887–12894. https://doi.org/10.1021/am403292x
186. Bui-Thi-Tuyet V, Cannizzo C, Legros C, Andrieux M, Chaussé A (2019) Surfaces Interfaces 15:110–116. https://doi.org/10.1016/j.surfin.2019.01.012
187. McNeill AR, Martinez-Gazoni R, Reeves RJ, Allen MW, Downard AJ (2021) ChemPhysChem 22:1344–1351. https://doi.org/10.1002/cphc.202100240
188. Buck E, Lee S, Stone LS, Cerruti M (2021) ACS Appl Mater Interfaces 13:7021–7036. https://doi.org/10.1021/acsami.0c16509
189. Mohamed AA, Neal SN, Atallah B, AlBab ND, Alawadhi HA, Pajouhafsar Y, Abdou HE, Workie B, Sahle-Demessie E, Han C, Monge M, Lopez-de-Luzuriaga JM, Reibenspies JH, Chehimi MM (2018) J Organomet Chem 877:1–11. https://doi.org/10.1016/j.jorganchem.2018.07.032
190. Ryder CR, Wood JD, Wells SA, Yang Y, Jariwala D, Marks TJ, Schatz GC, Hersam MC (2016) Nat Chem 8:597–602. https://doi.org/10.1038/nchem.2505
191. Liu G, Gao P, Zhang T, Zhu X, Zhang M, Chen M, Du P, Wang GW, Ji H, Yang J, Yang S (2019) Angew Chem Int Ed 58:1479–1483. https://doi.org/10.1002/anie.201813218
192. Daukiya L, Teyssandier J, Eyley S, El Kazzi S, Rodriguez G, Miriam C, Pradhan B, Thielemans W, Hofkens J, De Feyter S (2021) Nanoscale 13:2972–2981. https://doi.org/10.1039/d0nr07310e
193. Chu XS, Yousaf A, Li DO, Tang AA, Debnath A, Ma D, Green AA, Santos EJG, Wang QH (2018) Chem Mater 30:2112–2118. https://doi.org/10.1021/acs.chemmater.8b00173
194. Yan EX, Caban-Acevedo M, Papadantonakis KM, Brunschwig BS, Lewis NS (2020) ACS Mater Lett 2:133–139. https://doi.org/10.1021/acsmaterialslett.9b00241

195. Knirsch KC, Berner NC, Nerl HC, Cucinotta CS, Gholamvand Z, McEvoy N, Wang Z, Abramovic I, Vecera P, Halik M, Sanvito S, Duesberg GS, Nicolosi V, Hauke F, Hirsch A, Coleman JN, Backes C (2015) ACS Nano 9:6018–6030. https://doi.org/10.1021/acsnano.5b00965
196. Casademont H, Fillaud L, Lefevre X, Jousselme B, Derycke V (2016) J Phys Chem C 120:9506–9510. https://doi.org/10.1021/acs.jpcc.6b01630
197. Er E, Hou HL, Criado A, Langer J, Möller M, Erk N, Liz-Marzán LM, Prato M (2019) Chem Mater 31:5725–5734. https://doi.org/10.1021/acs.chemmater.9b01698
198. Park Y, Shin S, An Y, Ahn JG, Shin G, Ahn C, Bang J, Baik J, Kim Y, Jung J, Lim H (2020) ACS Appl Mater Interfaces 12:40870–40878. https://doi.org/10.1021/acsami.0c09096
199. Xin C, Cian B, Lloret V, Badylan NM, Wolff S, Gillen R, Stimpel-Lindner T, Maultzsch J, Duesberg G, Knirsch K, Hirsch A (2021) Angew Chem Int Ed 60:13484–13492. https://doi.org/10.1002/anie.202103353
200. Li DO, Chu XS, Wang QH (2019) Langmuir 35:5693–5701. https://doi.org/10.1021/acs.langmuir.8b04288
201. Lihter M, Graf M, Iveković D, Zhang M, Shen TH, Zhao Y, Macha M, Tileli V, Radenovic A (2021) ACS Appl Nano Mater 4:1076–1084. https://doi.org/10.1021/acsanm.0c02628
202. Yu T, Mathias D, Lu S, Xu W, Naushad M, Szunerits S, Boukherroub R (2020) Sep Purif Technol 238:116420. https://doi.org/10.1016/j.seppur.2019.116420
203. Jiang P, Zhang B, Liu Z, Chen Y (2019) Nanoscale 11:20449–20455. https://doi.org/10.1039/c9nr06604g
204. Liu M, Zhu H, Wang Y, Sevencan C, Li BL (2021) ACS Mater Lett 3:462–496. https://doi.org/10.1021/acsmaterialslett.1c00073
205. Wang H, Zhang J, Wua Y, Huang H, Lia G, Zhang X, Wang Z (2016) Appl Surf Sci 384:287–293. https://doi.org/10.1016/j.apsusc.2016.05.060
206. Jing H, Yeo H, Lyu B, Ryou J, Choi S, Park JH, Lee BH, Kim YH, Lee S (2021) ACS Nano 15:1388–1396. https://doi.org/10.1021/acsnano.0c08664
207. Zhang P, Wang L, Huang Z, Yu J, Li Z, Deng H, Yin T, Yuan L, Gibson JK, Mei L, Zheng L, Wang H, Chai Z, Shi W (2020) ACS Appl Mater Interfaces 12:15579–15587. https://doi.org/10.1021/acsami.0c00861
208. Boota M, Urbankowski P, Porzio W, Barba L, Osti NC, Bleuel M, Keum JK, Mamontov E (2020) ACS Appl Energy Mater 3:4127–4133. https://doi.org/10.1021/acsaem.0c00314
209. Gomberg M, Bachmann WE (1924) J Am Chem Soc 42:2339–2343. https://doi.org/10.1039/c7cs00150a
210. Amaya T, Jin Y, Tobisu M (2019) Tetrahedron Lett 39:151062. https://doi.org/10.1016/j.tetlet.2019.151062
211. Berisha A, Combellas C, Kanoufi F, Pinson J, Podvorica FI, Ustaze S (2010) Chem Mater 22:2962–2969. https://doi.org/10.1021/cm100295n
212. Deniau G, Azoulay L, Bougerolles L, Palacin S (2006) Chem Mater 18:5421–5428. https://doi.org/10.1021/cm060739t
213. Tessier L, Deniau G, Charleux B, Palacin S (2009) Chem Mater 21:4261–4274. https://doi.org/10.1021/cm901430t
214. Hetemi D, Kanoufi F, Combellas C, Pinson J, Podvorica FI (2014) Langmuir 30:13907–13913. https://doi.org/10.1021/la503833j
215. Lund H, Daasbjerg K, Lund T, Occhialini D, Pedersen SU (1997) Acta Chem Scandinavica 51:135–144. https://doi.org/10.3891/acta.chem.scand.51-0135
216. Koefoed L, Pedersen SU, Daasbjerg K (2017) Covalent modification of glassy carbon surfaces by electrochemical grafting of aryl iodides. Langmuir 33:3217–3222. https://doi.org/10.1021/acs.langmuir.7b00300
217. Jouikov V, Simonet J (2012) Electrochem Commun 15:93–96. https://doi.org/10.1016/j.elecom.2011.12.008
218. Huang Y-F, Wu DY, Wang A, Ren B, Rondinini S, Tian Z-Q, Amatore C (2010) J Am Chem Soc 132:17199–17210. https://doi.org/10.1021/ja106049c

219. Hetemi D, Medard J, Decorse P, Combellas C, Kanoufi F, Pinson J, Podvorica FI (2016) Langmuir 32:6335–6342. https://doi.org/10.1021/acs.langmuir.6b01557
220. Hetemi D, Combellas C, Kanoufi F, Podvorica FI (2021) ChemPhysChem 22:1–7. https://doi.org/10.1002/cphc.202100296
221. Pause L, Robert M, Saveant J-M (1999) J Am Chem Soc 121:7158–7159. https://doi.org/10.1021/ja991365q
222. Combellas C, Kanoufi F, Pinson J, Podvorica FI (2019) Elec Com 98:119–123. https://doi.org/10.1016/j.elecom.2018.12.005
223. Savéant J-M, Bethel D (eds) (1990) Advances in physical organic chemistry, vol 26. Academic Press, p 39. https://doi.org/10.1016/S0065-3160(08)60044-1
224. Smida H, Lebegue E, Bergamini JF, Barriere F, Lagrost C (2018) Bioelectrochemistry 120:157–165. https://doi.org/10.1016/j.bioelechem.2017.12.006
225. Hetemi D, Hazimeh H, Decorse P, Galtayries A, Combellas C, Kanoufi F, Pinson J, Podvorica FI (2015) Langmuir 31:5046–5415. https://doi.org/10.1021/acs.langmuir.5b00754
226. Medard J, Berisha A, Decorse P, Kanoufi F, Combellas C, Pinson J, Podvorica FI (2020) Elec Acta 345:136170. https://doi.org/10.1016/j.electacta.2020.136170

Structures, Stability, and Safety of Diazonium Salts

Victor D. Filimonov, Elena A. Krasnokutskaya, and Alexander A. Bondarev

Abstract This chapter provides data on the qualitative and quantitative relationships between the structures of diazonium salts $Ar(Het)N_2^+X^-$ and their stability and safety. The effects of diazonium cation structure and the nature of the anion on stability are discussed. Examples of stable and safe triazenes as surrogates of diazonium salts are given. Unconventional methods for producing diazonium salts under low acidity conditions are also discussed.

The disadvantage of aromatic diazonium salts $Ar(Het)N_2^+X^-$ (DS) is their low stability, a tendency to decomposition during storage, heating, mechanical and electrostatic effects, the action of light, impurities and a number of other factors, especially in a dry state (although these disadvantages are the second side of the coin is high reactivity of DSs and is associated with the diverse transformations of these compounds). Often this decomposition proceeds with explosions, and many examples of such incidents have been published (see, for example, [1–3]). For these reasons, DSs are traditionally referred to as unstable compounds. At the same time, many DSs are described, which are called "stable" in publications. But the criteria for "stability" are very vague, often limited to the indication that these substances "can be stored for hours, days, weeks" without measuring the rate and energy of their decomposition. Such definitions are important from the standpoint of ease of preparation and practical use, but not the safety of DSs, since DSs are known that are stable when stored in a dry state, but can explode on heating or shocks, for example, trifluoroacetates $ArN_2^+CF_3CO_2^-$ [4] or nitrates $ArN_2^+NO_2^-$ [5] (other examples of this kind will be given below). Thus, it is necessary to distinguish at least two aspects of assessing the applicability and hazard/safety of DSs—the ability to be stored unchanged (storage stability) and dangerously exothermic decomposition under certain physical and chemical influences (explosion hazards).

V. D. Filimonov (✉) · E. A. Krasnokutskaya
National Research Tomsk Polytechnic University, Tomsk, Russian Federation
e-mail: filimonov@tpu.ru

A. A. Bondarev
Altai State University, Barnaul, Russian Federation

M. M. Chehimi et al. (eds.), *Aryl Diazonium Salts and Related Compounds*,
Physical Chemistry in Action, https://doi.org/10.1007/978-3-031-04398-7_2

The physical reason underlying the instability of the domain structure is the low bond strength Ar(Het)-N_2^+ (the activation barrier is about 30 kcal/mol according to many experimental data and calculations) and the thermodynamic advantage of the elimination of molecular nitrogen with the formation of highly reactive intermediates of a cationic or free radical structure capable of further diverse transformations. However, this fundamental quantity is just one of many components that affect the energies of decomposition of DSs.

Some aromatic DSs with strong electron-donating substituents in the *para*- and *ortho*-positions (HO, MeO, NR_2) are characterized by turned dark after preparation as a result of more or less rapid transformation into colored structures of the quinone type (quinone diazides) [1, 6], which is probably not directly related to the safety problems of the DS.

In this review, we will use the concept of "stability of DS" mainly to assess their stability during storage and the hazards associated with thermal and detonation decomposition.

It is known that the stability of the DS Ar(Het)$N_2^+X^-$ is influenced by both the structure of the diazonium cation Ar(Het)N_2^+ and the nature of the counterion X^-.

1 Influence of the Nature of Counterions

It is accepted that DSs with NO_2^-, NO_3^-, ClO_4^-, MnO_4^-, CrO_4^{2-}, or picrate anions are especially unstable and explosive. Storage-stable DSs include salts with heavy metal anions ($ZnCl_4^{2-}$, $SbCl_6^-$, $FeCl_4^-$, SbF_6^-, $AuCl_4^-$, etc.) and some low nucleophilic inorganic anions (BF_4^-, PF_6^-) [1]. For example, 4-hexyloxybenzenediazonium hexafluorostibate 4-$C_6H_{13}C_6H_4N_2^+$ SbF_6^- persists at 40 °C in dioxane for 12 days and 1,2-dichloroethane 410 days [7]. A lot of DSs with organic anions are mentioned in publications; however, not all of them have been isolated and reliably identified. Among the relatively stable in the solid state and easily obtained DSs with organic counterions, the structure of which has been proven and which exhibit pronounced "diazonium" properties can be distinguished, arenediazonium *o*-benzenedisulfonimides **1** [8, 9], arenediazonium saccharinates **2** [10], arenediazonium bis(trifluoromethylsulfonyl) imides **3** [11, 12], arendiazonium 1,1,2,3,3-pentacycanopropenides **4** [13]. For example, DSs **1** can be stored in the dry state at room temperature unchanged for two months. They melted with decomposition, at 110 °C but none exploded [8]. 4-$MeOC_6H_4N_2^+$ saccharinate was stable when stored at 4 °C for a week [10]. The most widely presented, available, and easy-to-obtain DSs of this type are arenediazonium sulfonates Ar(Het)N_2^+ RSO_3^- (RSO_3^- = HSO_4 **5**, TsO **6** [14, 15], 4-$C_{12}H_{25}C_6H_4SO_3$ **7** [16], TfO **8** [17], camphorsulfonate **9** [18], naphthalene-1,5-disulfonates **10** (in mixture with naphthalene-1-sulfonates) [19] (Scheme 1).

Anions of a polymer nature also exhibit a stabilizing DSs effect. Thus, Merrington et al. treated arendiazonium chlorides sulforesin Amberlyst A-15 and obtained polymer supported DSs ArN_2^+ $^-O_3SCH_2$Polym through ion exchange, which did

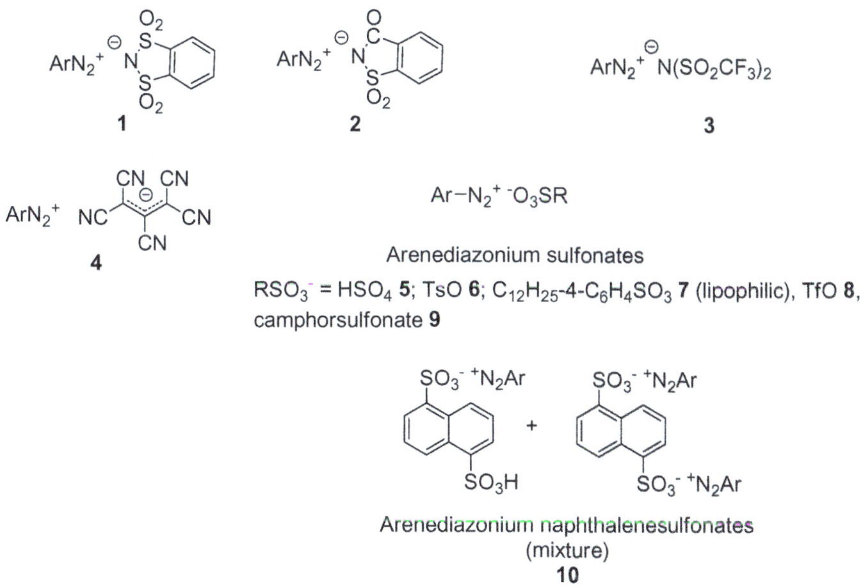

Scheme 1 Different diazonium salts

not lose activity in azo-coupling over an 8-day [20]. It has also been shown [21] that sulforesins KU-2 (Russian) or SPC-160H (South Korea) can serve as an effective proton source for one-pot diazotization-iodination of aromatic amines in water. The resulting polymer supported DSs ArN_2^+ $^-O_3SCH_2$Polym retain activity when stored for at least 9 days. Silica sulfuric acid HO_3S-SiO_2 also plays the role of an acid in the diazotization of amines, giving stable aryl diazonium salts supported on silica sulfuric acid (aryl diazonium silica sulfates), ArN_2^+ ^-O_3S-SiO_2, which begin to noticeably lose activity in azo-coupling reactions in the dry state between 6 and 24 h storage [22]. A similar type of the solid sulfonated derivatives belongs to the sulfonated cellulose, which provides diazotization of anilines by the action of $NaNO_2$, the resulting DSs ArN_2^+ ^-O_3S-Cellulose can be stored in a dessicator for 48 h without any loss of activity [23, 24].

Since the 1980s, it has been known that the stabilization of diazonium cations is provided by crown ethers due to the coordination of the diazonium group –$N{\equiv}N^+$ in the cavity of the crown ether [1, 25]. As a result of this coordination, the rate of thermal decomposition of the ArN_2^+ BF_4^- series in the presence of 21-Crown-7 slows down by more than two orders of magnitude in comparison with the decomposition in 1,2-dichloroethane (however, 12-crown-4, on the contrary, destabilizes the benzenediazonium cation) [26]. A similar stabilizing effect with respect to diphenylamine-4-diazonium chloride $PhNHC_6H_4$-4-N_2^+ Cl^- is provided by *p*-sulfonatocalix[4]arene, in the presence of which the rate of decomposition slows

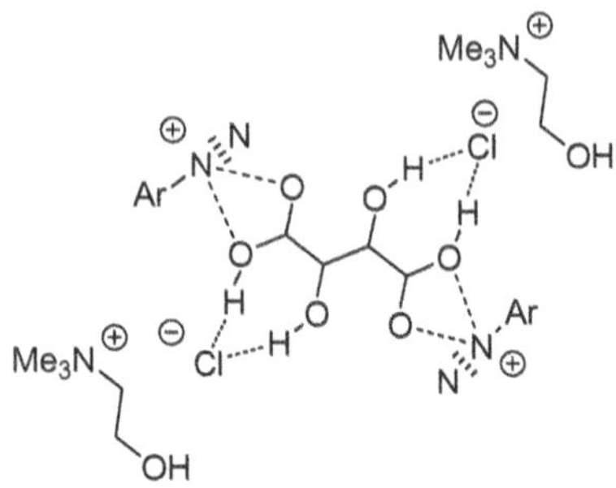

Fig. 1 Supposed structure of arenediazonium cations in DES [29a]

down about 10 times [10]. It was also shown that 4-NO_2-$C_6H_4N_2^+$ forms a 1:1 host–guest complex with cucurbit[7]uril, and this complex has higher thermal stability than the initial diazonium cation [28].

Recently, it was shown that diazotization of a number of anilines by the action of $NaNO_2$ occurs in a deep eutectic solvent (DES) consisting of tartaric acid and chlorocholine chloride (ChCl) diluted with ethanol [29a]. DS obtained under these conditions from 2-chloroaniline is stored in this DES at room temperature without loss of activity for 8 days. Replacing tartaric acid with malonic or oxalic acid reduces the yield of DS and, possibly, their stability. The authors [29a] explain the increased stability of DS in the DES by the coordination of two diazonium cations by four hydrogen bonds of the carboxyl groups of tartaric acid and the binding of two ChCl molecules with two hydroxyl groups (Fig. 1).

It should be noted, however, that it did not take into account that part of the tartaric acid should exist in the carboxylate form –COONa, and, in addition, the increased viscosity of DES was not taken into account, which should slow down the diffusion of DSs and their reactions with solution components, for example, EtOH. Quite recently an experimental and theoretical study of arenediazonium tetrafluoroborates reduction in new halide salts/polyol-based DESs has been published [29b].

Another variant of stabilization of diazonium cations was described by Doctorovich et al. [30], and it was shown that in the complexes [AlkN≡N–Fe(CN)$_5$] and [PhN≡N–Fe(CN)$_5$] the strengths of Alk(Ph)–N bonds are significantly higher than in free diazonium cations $[AlkN{\equiv}N]^+$ and $[PhN{\equiv}N]^+$ (more than 50 kcal/mol, depending on the calculation method). These complexes have not been experimentally studied, so it is difficult to say whether they are not varieties of the long-known and really stable "metal double diazonium salts", for example $(ArN_2^+)_2ZnCl_4^{2-}$ or $(ArN_2^+)_2CuCl_4^{2-}$ [1].

Few studies have been published in which comparative quantitative measurements of the effect of the nature of various anions on the stability of DSs were carried out. Thus, Bondarev et al. [31] determined the thermal decomposition energies of several DSs ArN_2^+ X^- with different counterions X^- = TfO, TsO, BF_4 by DSC/TGA and isothermal flow calorimetry methods at 70, 80, and 85 °C (Table 1). The same table shows the kinetic parameters of the decomposition reactions of DSs, calculated by approximating the Arrhenius equation (E_a—activation energy).

Table 1 Decomposition energies of diazonium salts according to DSC and isothermal flow calorimetry, and calculated kinetic parameters of decomposition reactions occurring at 25 °C [31]

DSs	DSC, $-\Delta H$, J/g (kJ/mol)	$-\Delta H^a$, kJ/mol	$P_{max}{}^a$, mW/g	$k_{298} \times 10^9$, s^{-1}	E_a, kJ/mol
$2\text{-}NO_2C_6H_4N_2^+\ TfO^-$	753.4 (225.3)	386.0	1.49	1.39	111.4
$3\text{-}NO_2C_6H_4N_2^+\ TfO^-$	840.4 (251.3)	230.0	14.15	2.90	159.7
$4\text{-}NO_2C_6H_4N_2^+\ TfO^-$	219.9 (65.7)	235.1	47.89	4.45	173.0
$4\text{-}MeOC_6H_4N_2^+\ TfO^-$	328.9 (93.5)	183.2	2.97	0.049	187.1
$4\text{-}NO_2C_6H_4N_2^+\ TsO^-$	24.5 (7.8) 323.0 (103.7)	232.4	34.82	48.91	131.7
$4\text{-}NO_2C_6H_4N_2^+\ BF_4^-$	229.2 (54.3)	156.0	31.63	18.75	140.3

[a] ΔH—integral enthalpy, P_{max}—maximum heat flow value at 80 °C

From the kinetic data, the following half-life times of decomposition reactions of 4-nitro derivatives of DS $4\text{-}NO_2C_6H_4N_2^+\ X^-$ at 85 °C were calculated, which depend on the nature of the anions and are for DS X = BF_4^- 2 h, 4 h for X = TfO^-, and 6 h for X = TsO^- [31]. At the same time, the calculation of kinetic data for normal conditions (25 °C), characterizing the stability of 4-nitrobenzenediazoniums during storage, shows a different order of the influence of anions on the half-life: X = BF_4^- 4.5 years, X = TsO^-, 5.0 years, and X = TfO^- 46 years [31]. Thus, the nature of the anion has a different effect on the stability of DS during thermal decomposition and during storage under normal conditions. It is very likely that storage stability depends primarily on the characteristics of the crystal lattice of solid DS, which change at temperatures close to melting.

It was also established also [31] that quantum-chemical calculations of the thermodynamics of the decomposition of a number of DSs by the B3LYP/aug-cc-pVDZ method according to the reactions (2–6) given in Table 2 give results close to the experimental values of the decomposition energies by isothermal flow calorimetry at different temperatures except for the case of $2\text{-}NO_2C_6H_4N_2^+\ {}^-OTf$.

The reason for the precipitation of $2\text{-}NO_2C_6H_4N_2^+\ {}^-OTf$ from the general dependence of the calculated and experimental decomposition energies is that its decomposition proceeds by a different mechanism [31]. The study of the decomposition products by GS-MS and LC-MS methods showed that for $2\text{-}NO_2C_6H_4N_2^+\ {}^-OTf$ there is no product of the substitution of the diazonium group by the triflate anion, i.e. reaction (1) is not realized, unlike reactions (2)–(6). The decomposition products of $2\text{-}NO_2C_6H_4N_2^+\ {}^-OTf$ are polymeric substances with repeating units $(NO_2C_6H_3)_n$, which can be formed from the intermediate 1-nitrocyclohexa-1,3-dien-5-yne **11** according to Scheme 2.

The alternative decomposition route does not go due to the limiting instability of 2-nitrobenzene-1-ylium cation **12**.

DSC methods are most often used in studies of the thermal decomposition of DSs. However, a comparison of the data in Tables 1 and 2 shows that DSC methods, in contrast to isothermal flow calorimetry, give lower values of decomposition energies,

Table 2 Experimental and calculated (B3LYP/aug-cc-pVDZ) values of the enthalpies of decomposition of diazonium salts [31]

Reaction	Calculated	Flow calorimetry		
	ΔH_{298}, kJ/mol	ΔH_{348}, kJ/mol	ΔH_{353}, kJ/mol	ΔH_{358}, kJ/mol
$2\text{-}NO_2C_6H_4N_2^+\ {}^-OTf \rightarrow 2\text{-}NO_2C_6H_4OTf + N_2$ (1)	−230	−414	−386	−396
$3\text{-}NO_2C_6H_4N_2^+\ {}^-OTf \rightarrow 3\text{-}NO_2C_6H_4OTf + N_2$ (2)	−243	−228	−230	−225
$4\text{-}NO_2C_6H_4N_2^+\ {}^-OTf \rightarrow 4\text{-}NO_2C_6H_4OTf + N_2$ (3)	−248	−200	−235	−250
$4\text{-}MeOC_6H_4N_2^+\ {}^-OTf \rightarrow 4\text{-}MeOC_6H_4OTf + N_2$ (4)	−201	−183	−183	−106
$4\text{-}NO_2C_6H_4N_2^+\ {}^-OTs \rightarrow 4\text{-}NO_2C_6H_4OTs + N_2$ (5)	−284	−253	−232	−231
$4\text{-}NO_2C_6H_4N_2^+\ {}^-OTs \rightarrow 4\text{-}NO_2C_6H_4F + BF_3 + N_2$ (6)	−188	−173	−156	−147

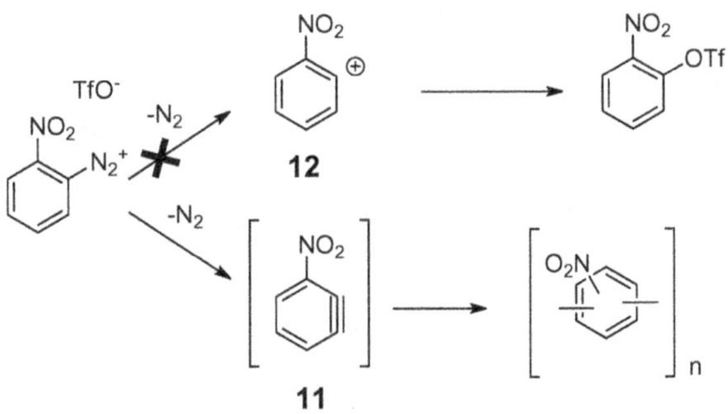

Scheme 2 Direction of thermal decomposition of $2\text{-}NO_2C_6H_4N_2^+\ {}^-OTf$

and also do not correlate with the calculated data and cannot be a measure of the DS stability during storage under normal conditions. This is probably due to that, in most DSC experiments, decomposition is preceded by melting, or by processes of crystal lattice rearrangement, and also by the processes of evaporation of decomposition products, which reduce the heat effect of the reactions. See also a recent article for a discussion of the possibilities and limitations of DSC methods for assessing thermally hazardous substances [32].

In addition to thermal decomposition, DS can explode on impact. It has been repeatedly observed that diazonium chlorides are more sensitive to detonation explosives than diazonium tetrafluoroborates [1]. In the work of Bondarchuk [33] in quantum chemical studies of the crystalline structures of benzenediazonium chloride (BDC) and tetrafluoroborate (BDT) at various pressures, it was shown that the stored energy content in BDC is by 1000 kJ/mol^{-1} higher than that of BDT, which predetermines the big explosion hazards of BDC.

Thon et al. [34] describes a method for stabilizing aqueous solutions of usually unstable arenediazonium chlorides by freezing them at −84 °C. At this temperature, frozen DSs can be stored for at least 21 days without noticeable loss of activity in a number of diazonium reactions.

Summarizing the above results of the effect of anions on the safety of DS, it is important to note the following. DSs are most dangerous in a dry isolated state, in which the effect of the anion on their properties is most clearly realized. While the chemical properties of DSs in solutions due to the destruction of crystal lattices are less dependent on the interactions of the diazonium cation with the anion. Traditionally, when explaining the chemical properties and reactivity of DSs, the diazonium cation is considered, implying the complete dissociation of DSs in solutions, although in many cases, especially for non-aqueous solutions, this has not been proven at all.

2 Influence of the Structure of Aryl and Heteroaryl Radicals

It is known that the effect of substituents in the aromatic nucleus of benzenediazoniums on the reactivity of DSs in dediazotation reactions does not obey the usual laws described by the Hammett equation. Only the use of dual substituent parameters (DSP), taking into account the induction σ_F and mesomeric σ^+_R effects of substituents, made it possible to find a correlation between the rate constants of dediazotation reactions and the nature of substituents in a limited number of *para*-substituted benzenediazonium cations $4\text{-}XC_6H_4N_2^+$ [1].

The processes of decomposition of DS are even more complex, they include dediazotation reactions, but are accompanied by other diverse transformations, therefore, it is difficult to expect simple relationships between the structure of the DS, stability, and decomposition energies. As a result of long-term observations, some empirical qualitative signs of the influence of the structure of DSs on the degree of their danger have been formed. For example, it was noticed that nitro groups in the ring increase the decomposition energy, while donor MeO groups decrease it. To some extent, experimental and calculated data on the thermal decomposition of arenediazonium triflates from Tables 1 and 2 help explain these effects.

At the same time, for arenediazonium sulfates $ArN_2^+\ HSO_4^-$, other dependences of stability on structure have been shown [35]. It turned out that sulfates of $2,6\text{-}Br_2\text{-}4\text{-}NO_2C_6H_2N_2^+$, $4\text{-}MeOC_6H_4N_2^+$, $4\text{-}MeC_6H_4N_2^+$ are stable as solids, but $2\text{-}CN\text{-}4\text{-}NO_2\text{-}C_6H_3N_2^+$ and $2,4\text{-}(NO_2)_2C_6H_3N_2^+$ are «extremely unstable and decompose rapidly into a mucus substance» [35].

It should be recognized that up to the present time no clear structural factors have been found that would make it possible to unambiguously predict the level of stability of aromatic DSs. The problem is aggravated by the fact that storage stability of aromatic solid DSs depends not only on their structure, but also on purity and production method. The presence of impurities, often uncontrolled by conventional analytical methods, can lead to both the acceleration of decomposition and its inhibition.

Probably the most extensive experimental data on the propensity to detonate and ignite various arenediazonium chlorides was done by Ullrich and Grewer [36] using drop-hammer tests. Detonation sensitivity is defined as impact energy (IE) and is measured in J. Sensitive explosives are characterized by IE 1–5 J, and the level of non-explosive substances is close to 50 J IE. For arenediazonium chlorides, IE is in the range 1–100 J [36]. The lowest detonation energy is possessed by nitro-substituted salts 1–2 J, and the detonation energy increases significantly in the series of *ortho*-, *meta*-, and *para*-nitrobenzenediazonium chlorides. Impact sensitivity decreases with increasing molecular weight of DSs, but compounds having two diazonium groups were found to be very sensitive to detonation, despite their high molecular weight. No relationship was found between detonation sensitivity and thermal stability. The results obtained cannot be explained by the structural parameters of the DSs or the electronic characteristics of the substituents. Only recently by Bondarchuk [37] found the parameter Ω, which satisfactorily links the results of quantum-chemical calculations of a wide range of crystalline arenediazonium chlorides with their detonation characteristics (Eqs. 7, 8).

$$\mathrm{IE} = 1.2955\Omega + 6.3113 \tag{7}$$

$$\Omega = l_1 q_{\mathrm{NN}} / l_2 \nu A^7 \tag{8}$$

$q_{\mathrm{NN}} = q_{\mathrm{N}(\alpha)} + q_{\mathrm{N}(\beta)}$—Bader charges of N($\alpha$) and N($\beta$) atoms of diazonium group; l_1—C–N bond length; l_2—N≡N bond length; ν—IR frequency of N≡N; A—electron affinity from the energies of the lowest unoccupied (E_{LUMO}) and the highest occupied (E_{HOMO}) molecular orbitals taken with the opposite sign.

It is known that DSs of the pyridine series in comparison with aromatic DSs are less stable. Pyridines with a diazonium group at positions 4 and 2 are especially unstable and little known, although pyridine-3-diazonium tetrafluoroborate is known and commercially available. Attempts to obtain pyridine diazonium salts stabilized with TsO^-, TfO^- or camphor-10-sulfonate anions through diazotization of aminopyridines led to the formation of only the corresponding iodopyridines [38], or pyridinyl tosylates, triflates, and camphor-10-sulfonates [39–41a, b] (Scheme 3).

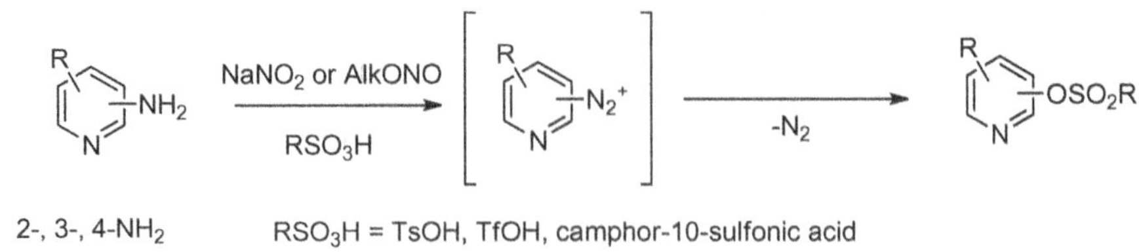

Scheme 3 Diazotization of aminopyridines leads to pyridinyl tosylates

In this way, aminopyridines are fundamentally different from anilines, which give relatively stable arenediazonium tosylates, triflates, and camphor-10-sulfonates [14, 15, 17, 18].

Attempts to diazotize aminopyridines by the action of $NaNO_2$ and TfOH in DMF led to the formation of N, N-dimethylaminopyridines [42], and the reaction of aminopyridines with $NaNO_2$ and H_3PO_4 in MeCN gives N-pyridinylacetamides [43].

Diazotization of N-oxides of aminopyridines and aminoquinolines in the presence of TsOH and KI also does not provide stable DS, but leads to N-oxides of iodopyridines [44].

The reason for the instability of pyridine diazonium salts is the thermodynamically favorable elimination of the nitrogen molecule, since pyridyl cations are more stable than phenyl cations [45]. According to DFT calculations of the bond strength of the aromatic nucleus with nitrogen in the diazonium cations ArN_2^+ fall in the following order: Ph > 3-Py > 4-Py > 2-Py [45].

In contrast to pyridinediazoniums, DSs of five-membered heteroarenes $HetN_2^+X^-$ (especially among pyrazole and imidazole derivatives) are widely known and often exhibit increased stability [1, 46]. An interesting example in this series is the recently obtained nitrogen-rich inner bis(diazonium) salts, the 4,4′-dinitro-5,5′-diazo-2,2′-bisimidazole **13** and the 4,4′-dinitro-5,5′-diazo-3,3′-bispyrazole **14** zwitterions [47] (Scheme 4).

These substances are high-energy materials and can be environmentally friendly primary explosive candidates. They combine high detonation sensitivity (IE = 1 J) with high thermal stability. So, the decomposition temperature for **13** is 199 °C, and for **14** 150 °C [47]. Other diazonium compounds of this kind are 4-diazo-3,5-dinitropyrazole zwitterion (DDNPz) [48] and zwitterionic diazonium tetrazolyl-1,2,3-triazolate [49], but they are less thermally stable. It is important to note that all these compounds provide additional example of the fact that the stability during storage of DSs does not at all correlate with their explosion hazards.

DSs of polymer nature are described. Several examples of diazotization of poly(4-aminostyrene) are known, for example, by the action of $NaNO_2$/HCl at 0 °C [50]. Bräse et al. [25] obtained a series of polymer-bound diazonium salts **15–23** by diazotization of polymer-bounded amines or ion exchange.

The Merrifield resin (chloromethylated polystyrene, cross-linked with 1–2% divinylbenzene) was used as a polymer matrix (Scheme 5). It turned out that most of the obtained polymeric DSs are relatively stable. The study of the kinetics of thermal

Scheme 4 4,4′-dinitro-5,5′-diazo-2,2′-bisimidazole **13** and the 4,4′-dinitro-5,5′-diazo-3,3′-bispyrazole **14** zwitterions

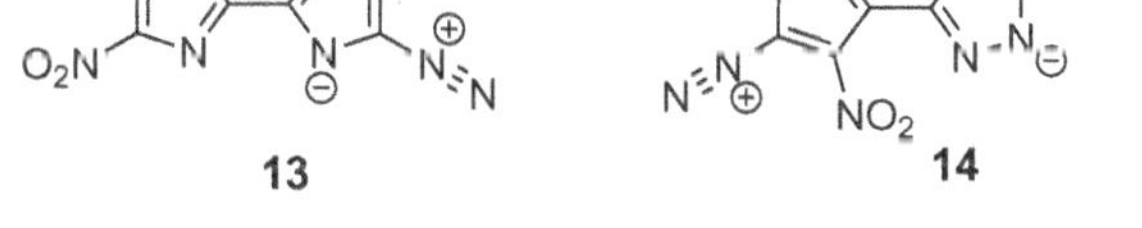

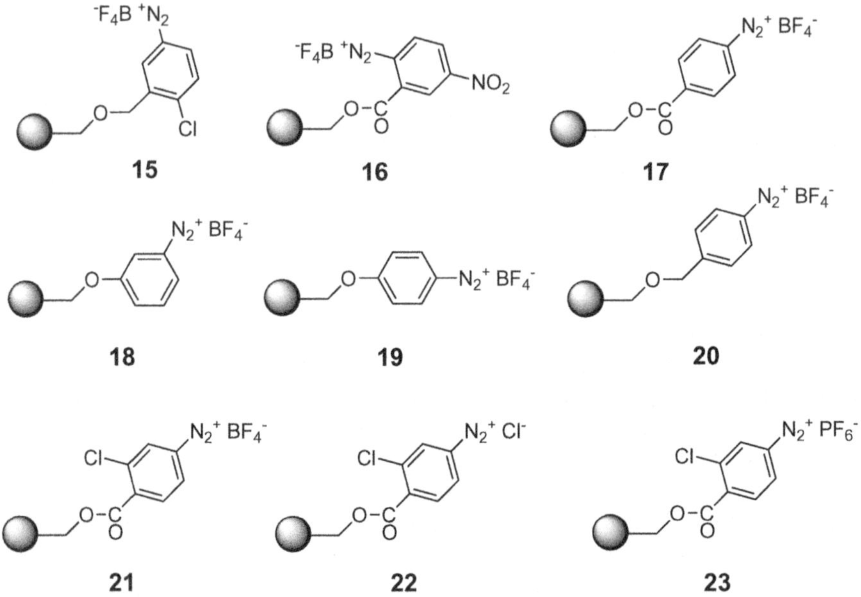

Scheme 5 Diazonium salts of Merrifield resins

decomposition of polymeric DSs **15–23** made it possible to find the activation energies of decomposition and to establish the half life-time at 20 °C, which is (in days): **20**—1.3, **22**—5, **17**—15, **15**—21, **21**—350, **16**—6100. For DS **19**, half life-time is highly dependent on the method of preparation. It is noteworthy that diazonium chloride **22** is much less stable than diazonium tetrafluoroborate **21**, which is similar in structure. This coincides with the known data for diazonium chlorides and tetrafluoroborates of monomeric structure and may indicate some similarity in the factors of interaction with counterions in DSs with monomeric and polymer structures in stabilizing diazonium cations. In general, the results obtained in these studies are comparable in their values with the above values of the decomposition energies and half life-time for arenediazonium triflates and tosylates [31].

A strategy similar to [25] was recently used to obtain polymer-bound arenediazonium chloride on Merrifields resin, Polym-$CH_2OCH_2C_6H_3$(2-Cl)-5-$N_2^+Cl^-$ [51]. Films of acrylic-type copolymers with covalently bound benzenediazonium fragments were prepared and described [52] (Scheme 6).

These films have been proposed as colorometric sensors for the determination of phenols by azo-coupling reactions. They have been shown to remain stable for at least 13 days [52]. Preparation of polyvalent reactive polymers based upon poly(N-(2-hydroxypropyl)methacrylamide) with diazonium reactive groups Polym-$OCOC_6H_4N_2^+CF_3CO_2^-$ and their application in the modification of the chimeric group B oncolytic virus enadenotucirev (EnAd) are described in [53]. The *para*-aminobenzoyl group grafted to cellulose can be diazotized by the action of

Scheme 6 Polymer-bound arenediazonium chloride

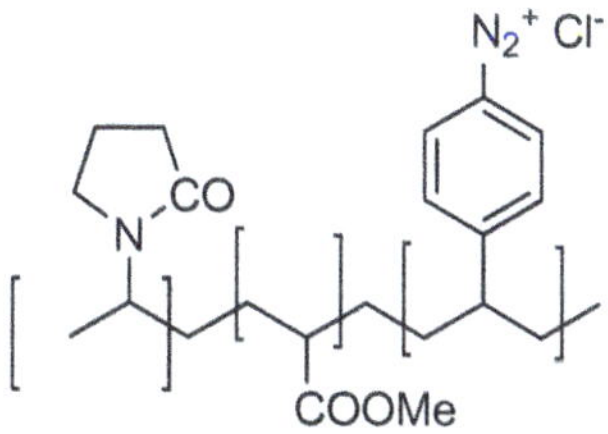

$NaNO_2/H_2SO_4$ [54]. The stability of polymeric DSs obtained in [53, 54] has not been determined.

Thus, there are the following ways and methods of stabilization of DSs in the solid state available for large-scale practical use. First, it is the selection of a stabilizing counterion, the set of which is quite diverse today. In terms of stabilizing effects and availability (cost), one can add to the long-known BF_4^- and relatively new aryl(alkyl) sulfonates RSO_3^-, including sulforesins, and $^-O_3S\text{-}SiO_2$. An additional advantage of arenediazonium sulfonates **6–9** is their good solubility, both in water and in many organic solvents. Another way to stabilize DSs is the stabilization of diazonium cations. In this respect, the strongest effect was shown by covalent immobilization of diazonium cations into polymer matrices.

It is also important to note that computational quantum-chemical methods are emerging that make it possible to predict to some extent the effect of the structure of a DS on the degree of danger during thermal decomposition [31] or detonation [37].

Nevertheless, despite the progress made, DSs were and remain potentially hazardous compounds, especially in the dry solid state. Formulated 12 important rules [2] that should be followed in laboratories to reduce the hazards in handling DSs. In this regard, it is appropriate to comment on the work [55], which shows that the ball milling treatment of some arenediazonium tetrafluoroborates and piezoelectric materials ensures the decomposition of DSs and the generation of aryl radicals capable of entering into useful transformations. For all the scientific importance of the obtained result for the chemistry of diazonium compounds, the conditions for carrying out this process, combining shock and electrostatic effects, are ideal for initiating explosions of DSs. A similar example of the use of hazardous shock effects is described Mukherjee et al. [56], in which arenediazonium tetrafluoroborates, when rubbed in ball-milling, were converted into various diaryl chalcogenides. Considering the importance and promising of free-solvent reactions for organic synthesis, it seems likely that an increasing number of similar works using DSs can be expected. However, it should always be borne in mind that several diazonium tetrafluoroborates are listed as hazardous in the *Bretherick's Handbook of Reactive Chemical Hazards* [57], and for most others, hazard characteristics have not yet been measured. Therefore, the importance of warnings in publications about the potential risks of experiments with dry DSs is becoming increasingly relevant [3].

Considering the above, the flow chemistry methods developed in recent years are especially relevant for DSs transformations, since they mitigate risks (see, for example, reviews [58–61]).

2.1 Triazenes as Safe Surrogates for Diazonium Salts

It has long been established that aryltriazenes ArN=N-NR_2 in acidic media are in equilibrium with DSs ArN_2^+ X^-, and therefore can enter into the same reactions as DSs. At the same time, triazenes are more stable and not prone to hazardous explosive decomposition. A recent comparative DSC study of the thermal stability of the series arenediazonium tetrafluoroborates ArN_2^+ BF_4^- and triazenes ArN=N-NR_2 [62] showed that in most cases triazenes decompose at higher temperatures, and some of them are stable up to 200 °C. Some of the data obtained are shown in Table 3.

This review does not aim to describe in detail the chemical properties and transformations of triazenes. However, it is pertinent to note that, in general, triazenes show a reactivity similar to DSs and can be considered as safe synthetic equivalents of DSs in traditional reactions of nucleophilic substitution of the diazonium

Table 3 Decomposition temperatures ($T_{decomp.}$) diazonium salts and triazenes [62]

Diazonium salt	$T_{decomp.}$, °C	Triazene	$T_{decomp.}$, °C
PhN_2^+ BF_4^-	110	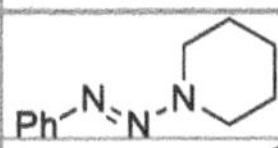	>200
2-$HOCOC_6H_4N_2^+$ BF_4^-	Explosive	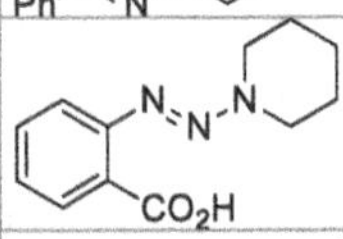	100
4-$ClC_6H_4N_2^+$ BF_4^-	140	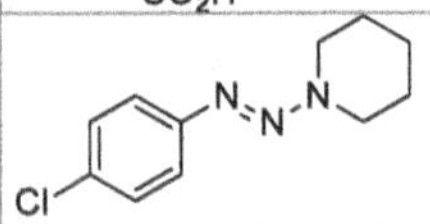	>200
		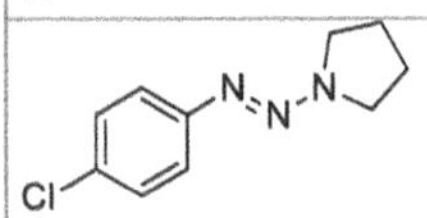	150
		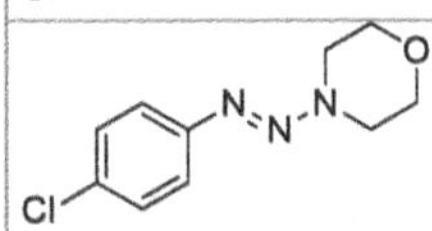	>200
2-$ClC_6H_4N_2^+$ BF_4^-	170	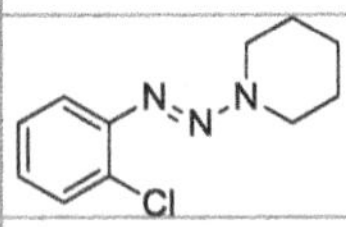	>200
4-$NO_2C_6H_4N_2^+$ BF_4^-	150 (explosive)	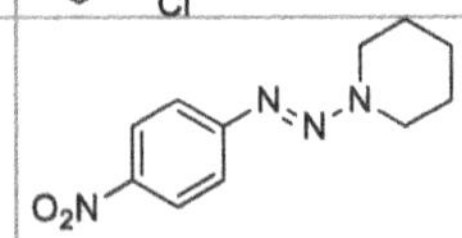	>200
4-$MeOC_6H_4N_2^+$ BF_4^-	140	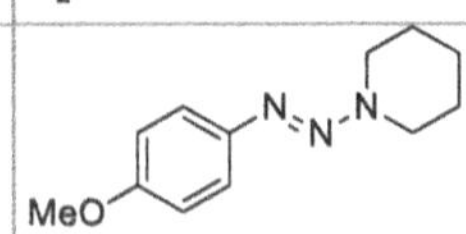	150

group, azo-coupling reactions, and for molecular grafting to solid surfaces (see, for example [63–65]). Of course, the use of triazenes does not allow one to completely get rid of DSs, since in most methods triazenes are obtained from DSs and amines. However, this eliminates the most dangerous isolation procedures and operations with dry forms of DSs. In a number of works by Bräse et al. [25, 66–68], the possibilities of synthetic use of polymer-bound triazenes obtained from polymer-bound DSs have been shown. Their use further minimizes the potential risks of working with DSs. More recently, reactions of some DSs with monomethylated lysyne (Kme) have produced triazenes selectively grafted to proteins [51].

2.2 Unconventional Methods for Synthesis of Diazonium Salts

The main methods for obtaining DS are described and analyzed in the book [1], which, together with the second book of *Zollinger, H. Diazo Chemistry II (Aliphatic, Inorganic and Organometallic Compounds), Vol. 2, VCH, Weinheim 1995* can be considered the bible of diazonium chemistry of the twentieth century. Since then, no comprehensive reviews of DS synthesis methods have been published. From the time of Peter Griess (1858) to the present, the vast majority of DSs are obtained through diazotization of aromatic or heterocyclic amines with compounds that generate NO^+—nitrosonium ion in acidic media. These reagents are most often salts of nitrous acid, alkyl nitrites, or nitrogene oxides.

All synthetic improvements in DSs production are aimed at increasing efficiency and safety, reducing the cost and amount of waste generated, and making diazotization processes green. An important trend is a decrease in the acidity of the medium during diazotization or the complete elimination of the use of strong acids.

In the last decade, TsOH has become widespread as an affordable medium strength acid. Several variants of using TsOH as an acid component of diazotization reactions have been proposed. Diazotization of a wide range of anilines by the action of "Resin NO_2^-" (see below) and TsOH in AcOH allowed for the first time to obtain and identify relatively stable arenediazonium tosylates [14]. Later arenediazonium tosylates are easily obtained by diazotization with *t*-BuONO/TsOH in EtOAc [15]. Aromatic amines and 5-aminouracil by the action of $NaNO_2$/TsOH in the presence of NaN_3 in water at room temperature form arenediazonium tosylates, are rapidly converted to aromatic azides [16]. TsOH in combination with $NaNO_2$ or *t*-BuONO and $CuBr_2$/NaBr or *t*-$Bu_4N^+Br^-$ leads to diazotization-bromination of anilines, and the use of $PhCH_2(Et)_3N^+Cl^-$ under the same conditions provides diazotization-chlorination [69]. TsOH is also convenient for performing diazotization by co-grinding amines with $NaNO_2$ [70], or *t*-BuONO [71] in aqueous pastes in free-solvent diazotization. Subsequent treatment of the resulting aqueous pastes with KI [70] or Cu_2X_2 (X = Cl, Br) [71] gives the corresponding aryl halides ArX (X = Cl, Br, I).

Free-solvent diazotization-iodination of anilines by co-grinding with $NaNO_2$ and KI has been shown to work even with an accessible and ecologically safe acidic reagent, $NaHSO_4$ [72]. A method of diazotization of a wide range of anilines with *t*-BuONO and catalytic quantities of salicylic acid in a $MeCN/H_2O$ mixture followed by in situ arylation of enol acetates without demanding thermal or photochemical activation has been described [73] (Scheme 7).

Kamble et al. [29] showed for the first time that diazotization of anilines occurs in deep-eutectic solvents (DES) with the formation of fairly stable DS (Fig. 1). Chlorocholine chloride (ChCl) with tartaric, malonic, or oxalic acids with the addition of ethanol was studied as DES. The ChCl-tartaric acid system showed the best results. DSs formed in DES are capable of azo-coupling with 2-naphthole, acetoacetanilide derivatives, and 3-methyl-1-phenyl pyrazolone [29].

One of the greenest diazotization methods is described by Tundo et al. [74]. Diazotization of some anilines was performed with $NaNO_2$ in a liquid CO_2/H_2O solvent at 65 bar. The resulting arenediazonium hydrogen carbonates ArN_2^+ HCO_3^- interact with unreacted amine, therefore the final product of this reaction is the triazenes ArN=N-NHAr, and when diazotization is carried out under the indicated conditions in the presence of KI, the final products are aryl iodides ArI [74] (Scheme 8).

There are known examples of diazotization in the presence of some NH-acids, the anions of which quite effectively stabilize the formed diazonium cations. These include *o*-benzenedisulfonimide **1**, in the presence of which the diazotization of anilines by the action of i-PnONO yields stable arenediazonium o-benzenedisulfonimides in 85–99% yields [8] (Scheme 9).

Importantly, according to X-ray data, these compounds **1** represent true DS [8], but not alternative triazenes as might be expected.

A similar function is performed by saccharin, in the presence of which and under the action of *t*-BuONO, anilines are converted into arenediazonium saccharinates **2**

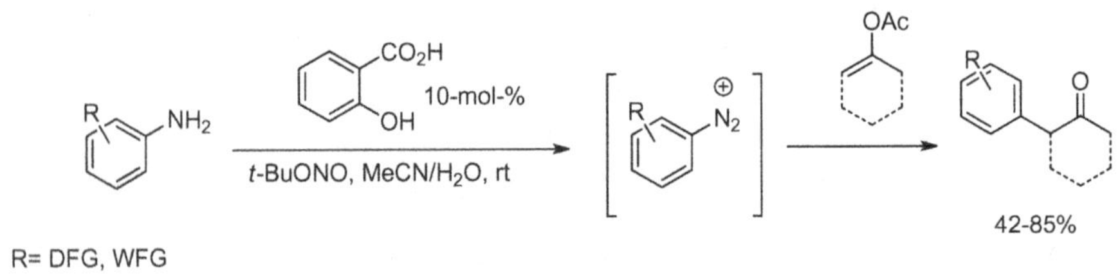

Scheme 7 Diazotization of anilines with *t*-BuONO and catalytic quantities of salicylic acid in a $MeCN/H_2O$ mixture followed by in situ arylation of enol acetates

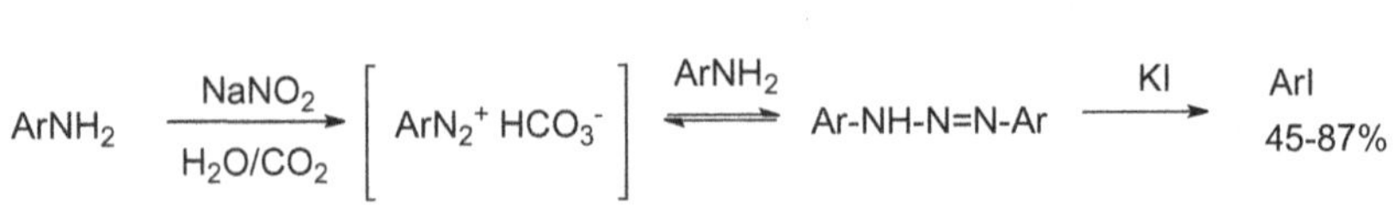

Scheme 8 Diazotization of anilines with $NaNO_2$ in a liquid CO_2/H_2O solvent at 65 bar

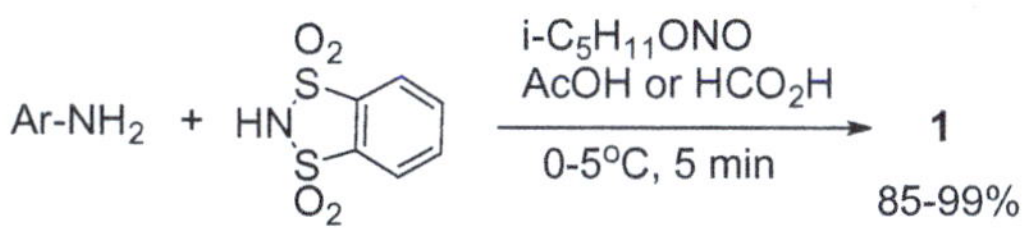

Ar= Ph, 2-MeC_6H_4, 3-MeC_6H_4, 4-MeC_6H_4, 2-$MeOC_6H_4$, 3-$MeOC_6H_4$, 4-$MeOC_6H_4$, 2-ClC_6H_4, 3-ClC_6H_4, 4-Cl-C_6H_4, 3-BrC_6H_4, 4-BrC_6H_4, 2-$NO_2C_6H_4$, 3-$NO_2C_6H_4$, 4-$NO_2C_6H_4$, 4-$Me_2NC_6H_4$, 2,6-$Me_2C_6H_3$, 2-MeO-5-ClC_6H_3, 2,6-$Cl_2C_6H_3$, 1-naphthyl

Scheme 9 Diazotization in the presence of some NH-acids

[10]. These DSs were not specially isolated but were used in situ in azo-coupling reactions. Preparatively isolated 4-methoxybenzenediazonium saccharinate was stable when stored at 4 °C for a week.

It was published that bis(trifluoromethylsulfonyl) amine $(CF_3SO_2)_2NH$ can also act as an acid component of diazotization of various anilines at low temperature (ice acetone bath) [11, 12]. The resulting arenediazonium bis(trifluoromethane) sulfonimides **3** were used without isolation in the azo-coupling reaction with 4-hydroxy benzaldehyde (Scheme 10).

4-Nitrobenzenediazonium bis(trifluoromethane) was preparatively isolated with a yield of 88% [11]. Another area under development is the replacement of corrosive liquid acids with solid acids. These include the aforementioned sulforesins [20, 21], silica sulfuric acid HO_3S-SiO_2 [22], sulfonated cellulose [23, 24], in the presence of which, upon the action of $NaNO_2$ on aromatic amines, polymer-bounded DSs are formed. The advantage of these acidic agents is also the relative stability of the resulting DSs, as discussed above, as well as the use of water instead of organic solvents as a medium. Clays such as Brønsted acids have also been proposed as acid components of diazotization reactions. In works [75, 76] it was found that under the action of two samples of local Indian kaolinite and bentonite clays or K-10 montmorillonite and $NaNO_2$ at room temperature in water takes place a rapid diazotization of a number of anilines. The formed DSs are fixed on these clays and can react in the azo-coupling reaction with phenols [75] or arylate arenes giving biaryles [76].

There are a few methods of diazotization without the use of Brønsted acids. A good example of such transformations is the diazotization of various anilines and heterocyclic amines by the action of *t*-BuONO in MeCN in the presence of B_2pin_2 [77]. It

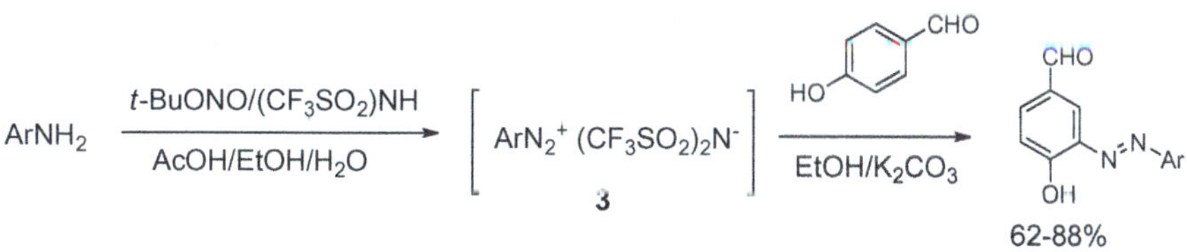

Scheme 10 Bis(trifluoromethylsulfonyl) amine $(CF_3SO_2)_2NH$ for the diazotization of various anilines

was shown that under these conditions the initial diazotization occurs, followed by the transformation of DSs into pinacol arylboronates Ar(Het)Bpin (Scheme 11).

Apparently, under the indicated conditions, B_2pin_2 plays the role of a Lewis acid during diazotization. Previously, the same Chines group showed that a similar reaction occurs in the presence of benzoyl peroxide at room temperature [78].

It was also shown that diazotization of some anilines can proceed under the action of *i*-PnONO in MeCN in the presence of dimethyldisulfide Me_2S_2 with the formation of aromatic methylsulfides ArSMe in moderate yields [79].

Diazotization-halogenation of aromatic and heterocyclic amines can take place with $NaNO_2$ and N-halosuccinimides in DMF under free-acid conditions [80] (Scheme 12).

The authors [80] believe that in the reaction between $NaNO_2$ and N-halosuccinimides nitrosyl halides ONO-X are formed, which provides further diazotization.

As for unconventional diazotizing agents, the following can be noted. A new nitrite containing ionic liquid derived from the O-nitrosation of N-methyl-N-hydroxybutylimidazolinium chloride (IL-ONO) provides diazotization of some anilines in HCl with subsequent azo-coupling reactions [81]. Anion-exchange resins saturated with nitrite ions from sodium nitrite are also relatively new diazotizing agents (Resin NO_2^-) [20, 21, 82]. This polymeric diazonium agent is capable of diazotizing a variety of anilines with donor and acceptor substituents in HCl, giving

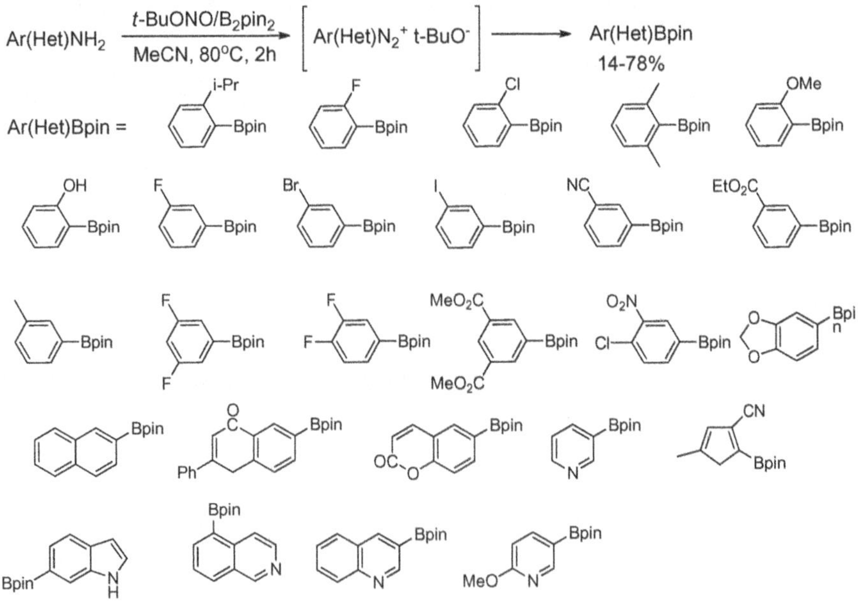

Scheme 11 Diazotization of various anilines and heterocyclic amines by the action of *t*-BuONO in MeCN in the presence of B_2pin_2

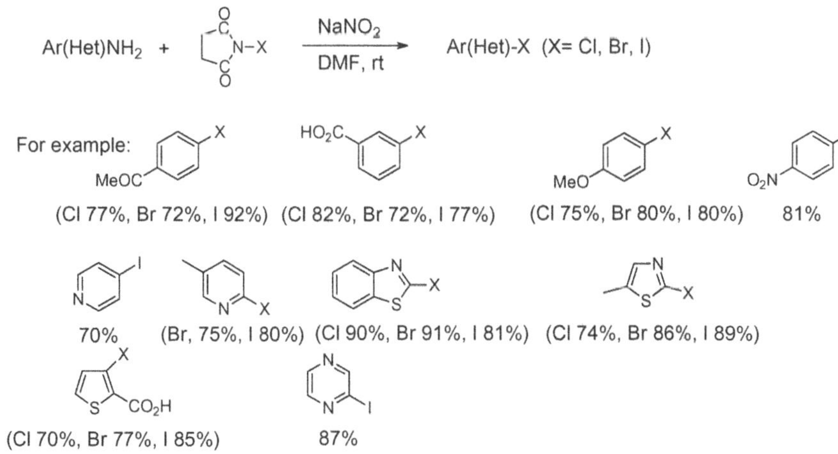

Scheme 12 Diazotization-halogenation of aromatic and heterocyclic amines

arenediazonium chlorides [20], and in water in the presence of TsOH, forming diazo nium tosylates ArN_2^+ TsO^- [21, 82]. The advantage of Resin NO_2^- is the absence of release of nitrogen oxides during diazotization [82], which often accompanies the diazotization of amines by the action of $NaNO_2$ or AlkONO. Polymeric diazotizing agent based on poly (4-vinylpyridine) cross-linked with divinyl benzene [P_4-VP]NO_2 proposed in the paper [83]. The authors [83] used [P_4-VP]NO_2 in combination with [P_4-VP]HCl for diazotization of anilines in acetonitrile. The resulting solutions of diazonium chlorides were used to azo-coupling reactions.

Diazotization of amines in solutions by the action of NO_2 has long been known [1], but Kaupp et al. proposed waste-free versions of such diazotization, acting on solid aromatic amines with gaseous or liquid diazotizing agents NO_2, NO, NOCl [84]. In this case, solid arene diazonium nitrates or chlorides are obtained in quantitative yields, without impurities and any waste. Using this approach, it was also possible to convert N-(4-nitrophenyl) acetamide, some nitrosoarenes, and 4-chloro-N-(4-methoxybenzylidene) aniline into DSs with 100% yields [84] (Scheme 13).

The obtained solid DSs when rubbed with primary or secondary aromatic amines give triazenes, enter into azo-coupling reactions with 2-naphthol and barbituric acids and form aryl iodides with KI. The yields of target products in these reactions are 98–100% [84]. With all the effectiveness of these transformations (room temperature, quantitative, or almost quantitative preparative yields, waste-free, and acids-free), they are dangerous, especially during grinding procedures.

A fundamentally new approach to the formation of diazonium compounds, which differs from those described above, has been demonstrated Guella et al. [85]. This work shows that benzenediazonium cation $C_6H_5N_2^+$ under chemical ionization can be produced by the reaction of phenylium ions $C_6H_5^+$ with N_2 in gas phase and is observed in MS experiments. Of course, today this result has only academic interest,

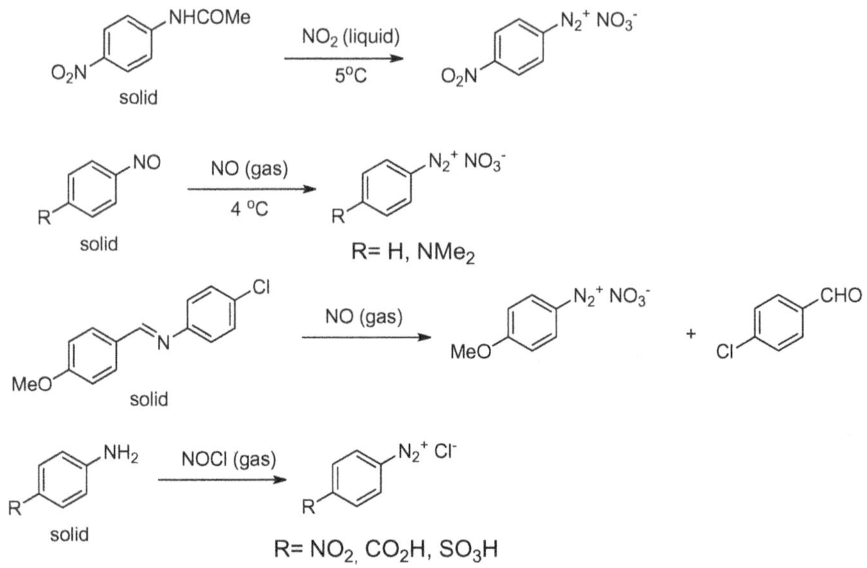

Scheme 13 Diazotization of amines in solutions by the action of liquid or gas reagents

but in the future, with the development of methods and equipments for a large-scale electronic ionization of organic molecules, it may also be of practical importance.

References

1. Zollinger H (1994) Diazo chemistry I (aromatic and heteroaromatic compounds), vol 1. VCH, Weinheim
2. Sheng M, Frurip D, Gorman D (2015) Reactive chemical hazards of diazonium salts. J Loss Prev Proc Ind 38:114–118. https://doi.org/10.1016/j.jlp.2015.09.004
3. Firch JD, Fairlamb IJS (2020) A need for caution in the preparation and application of synthetically versatile aryl diazonium tetrafluoroborate salts. Org Lett 22:7057–7059. https://doi.org/10.1021/acs.orglett.0c02685
4. Colas C, Goeldner M (1999) An efficient procedure for the synthesis of crystalline aryldiazonium trifluoroacetates—synthetic applications. Eur J Org Chem 1357–1366. https://doi.org/10.1002/(SICI)1099-0690(199906)1999:6<1357::AID-EJOC1357>3.0.CO;2-P
5. Zhang QZ, Zhang S, Liu Z, Zhang J (2001) Synthesis of solid arenediazonium nitrates under nonaqueous conditions. Synth Commun 31:1243–1247. https://doi.org/10.1081/SCC-100104011
6. Quintero B, Morales JJ, Quirós M, Martínez-Puentedura MI, Cabeza MDC (2000) Dediazoniation of p-hydroxybenzenediazonium ion in a neutral aqueous medium. Free Radical Biol Med 29:464–479. https://doi.org/10.1016/S0891-5849(00)00321-X
7. Müller U, Utterodt A, Mörke W, Deubzer B, Herzig C (2001) New insights about diazonium salts as cationic photoinitiators. J Photochem Photobiol A Chem 140(140):53–66. https://doi.org/10.1016/S1010-6030(01)00382-3

8. Barbero M, Crisma M, Degani I, Fochi R, Perrachino P (1998) New dry arenediazonium salts, stabilized to an exceptionally high degree by the anion of *o*-benzenedisulfonimide. Synthesis 1171–1175. https://doi.org/10.1055/s-1998-2132
9. Antenucci A, Barbero M, Dughero S, Ghigo G (2020) Copper catalysed Gomberg-Bachmann-Hey reactions of arenediazonium tetrafluoroborates and heteroarenediazonium *o*-benzenedisulfonimides. Synth Mech Aspects Tetrahedron 76:131632. https://doi.org/10.1016/j.tet.2020.131632
10. Khaligh NG, Johan MR, Ching JJ (2019) Saccharin: a cheap and mild acidic agent for the synthesis of azo dyes via telescoped dediazotization. Green Proc Synth 8:24–29. https://doi.org/10.1515/gps-2017-0133
11. Khaligh NG (2017) Telescopic synthesis of azo compounds via stable arenediazonium bis(trifluoromethane)sulfonimide salts by using tert-butyl nitrite. Dyes Pigm 139:556–560. https://doi.org/10.1016/j.dyepig.2016.11.058
12. Khaligh NG (2018) A facile and sustainable protocol to the preparation of aryl iodides using stable arenediazonium bis(trifluoromethylsulfonyl) imide salts via the telescopic process. Heteroat Chem 29:e21418. https://doi.org/10.1002/hc.21418
13. Kachanov AV, Slabko OYu, Kaminskii VA (2012) A method for the synthesis of stable aryldiazonium salts possessing a 1,1,2,3,3-pentacyanopropenide anion. Tetrahedron Lett 53:5807–5808. https://doi.org/10.1016/j.tetlet.2012.08.098
14. Filimonov VD, Trusova ME, Postnikov PS, Krasnokutskaya EA, Lee YM, Hwang HY, Kim H, Chi K-W (2008) Unusually stable, versatile, and pure arenediazonium tosylates: their preparation, structures, and synthetic applicability. Org Lett 10:3961–3964. https://doi.org/10.1021/ol8013528
15. Michelač M, Siljanovska A, Košmrlj J (2021) A convenient approach to arenediazonium tosylates. Dyes Pigm 184:108726. https://doi.org/10.1016/j.dyepig.2020.108726
16. Kutonova KV, Trusova ME, Postnikov PS, Filimonov VD (2012) The first example of the copper free chloro and hydrodediazoniation of aromatic amines using sodium nitrite, CCl_4, and $CHCl_3$. Russ Chem Bull Int Ed 61:206–208. 1066-5285/12/6101-206
17. Filimonov VD, Krasnokutskaya EA, Kassanova AZh, Fedorova VA, Stankevich KS, Naumov NG, Bondarev AA, Kataeva VA (2019) Synthesis, structure, and synthetic potential of arenediazonium trifluoromethanesulfonates as stable and safe diazonium salts. Eur J Org Chem 665–674. https://doi.org/10.1002/ejoc.201800887
18. Vajpayee V, Moon ME, Lee S, Ravikumar S, Kim H, Ahn B, Choi S, Hong SH, Chi K-W (2013) Halogenation and DNA cleavage via thermally stable arenediazonium camphorsulfonate salts. Tetrahedron 69:3511–3517. https://doi.org/10.1016/j.tet.2013.02.016
19. Qiu J, Tang B, Ju B, Xu Y, Zhang S (2017) Stable diazonium salts of weakly basic amines. Convenient reagents for synthesis of disperse azo dyes. Dyes Pigm 136:63–69. https://doi.org/10.1016/j.dyepig.2016.08.026
20. Merrington J, James M, Bradley M (2002) Supported diazonium salts—convenient reagents for the combinatorial synthesis of azo dye. Chem Commun 140–141. https://doi.org/10.1039/b109799g
21. Filimonov VD, Semenischeva NI, Krasnokutskaya EA, Tretyakov AN, Hwang HY, Chi K-W (2008) Sulfonic acid based cation-exchange resin: a novel proton source for one-pot diazotization-iodination of aromatic amines in water. Synthesis 185–187. https://doi.org/10.1039/b109799g
22. Zarei A, Hajipour AR, Khazdooz L, Mirjalili BF, Chermahini AN (2009) Rapid and efficient diazotization and diazo coupling reactions on silica sulfuric acid under solvent-free conditions. Dyes Pigm 81:240–244. https://doi.org/10.1016/j.dyepig.2008.10.011
23. Nemati F, Elhampour A (2012) Cellulose sulphuric acid as a biodegradable catalyst for conversion of aryl amines into azides at room temperature under mild conditions. J Chem Sci 124:889–892. https://doi.org/10.1007/s12039-012-0261-1
24. Nemati F, Elhampour A (2012) Green and efficient diazotization-iodination of aryl amines using cellulose sulfuric acid as a biodegradable and recyclable proton source under solvent-free condition. Sci Iranica 19:1594–1596. https://doi.org/10.1016/j.scient.2012.10.015

25. Bräse S, Dahmen S, Popescu C, Schroen M, Wortmann F-J (2004) The structural influence in the stability of polymer-bound diazonium salts. Chem Eur J 10:5285–5296. https://doi.org/10.1002/chem.200400386
26. Kuokkanen T (1997) Effect of ring size on the complexation and decomposition of benzenediazonium ion in the presence of crown ethers in 1,2-dichloroethane and the gas phase. J Phys Org Chem 10:67–75. https://doi.org/10.1002/(SICI)1099-1395(199702)10:2%3c67::AID-POC859%3e3.0.CO;2-O
27. Zhang Y, Tu G, Cao W (2002) Inclusion complexation of diphenylamine-4-diazonium chloride and p-sulfonatocalix[4]arene. Supramol Chem 14:473–475. https://doi.org/10.1080/1061027021000002242
28. Xu H, Wang Q (2019) Cucurbit[7]uril/CuCl promoting decomposition of 4-nitrobenzenediazonium in aqueous solution. Chin Chem Lett 30:337–339. https://doi.org/10.1016/j.cclet.2018.03.014
29. (a) Kamble SS, Shankarling GS (2019) Room temperature diazotization and coupling reaction using DES-Ethanol system: a green approach towards the synthesis of monoazo pigments. Chem Commun 55:5970–5973. https://doi.org/10.1039/c9cc01114e. (b) Antenucci A, Bonono M, Ghigo G, Gontrani L, Barolo C, Dughera S (2021) How do arenediazonium salts behave in deep eutectic solvents? A combined experimental and computational approach. J Mol Liq 339:116743. https://doi.org/10.1016/j.molliq.2021.116743
30. Doctorovich F, Escola N, Trapani C, Estrin DA, Lebrero MCG, Turjanski AG (2000) Stabilization of aliphatic and aromatic diazonium ions by coordination: an experimental and theoretical study. Organometallics 19:3810–3817. https://doi.org/10.1021/om000004c
31. Bondarev AA, Naumov EV, Kassanova AZh, Stankevich KS, Krasnokutskaya EA, Filimonov VD (2019) First study of the thermal and storage stability of arenediazonium triflates comparing to 4-nitrobenzenediazonium tosylate and tetrafluoroborate by calorimetric methods. Org Proc Res Div 23:2405–2415. https://doi.org/10.1021/acs.oprd.9b00307
32. Green SP, Wheelhouse KM, Payne AD, Hallett JP, Miller PW, Bull JA (2020) On the use of differential scanning calorimetry for thermal hazard assessment of new chemistry: avoiding explosive mistakes. Angew Chem Int Ed 59:15798–15802. https://doi.org/10.1002/anie.202007028
33. Bondarchuk SV (2017) Impact sensitivity of crystalline phenyl diazonium salts: a first-principles study of solid-state properties determining the phenomenon. Int J Quantum Chem 117:e25430. https://doi.org/10.1002/qua.25430
34. Thon D, Fürst MCD, Altmann L-M, Heinrich MR (2018) Frozen aryldiazonium chlorides in radical reactions with alkenes and arenes. Tetrahedron 74:5289–5294. https://doi.org/10.1016/j.tet.2018.05.089
35. Qiu J, Tang B, Ju B, Zhang S, Jin X (2020) Clean synthesis of disperse azo dyes based on peculiar stable 2,6-dibromo-4-nitrophenyl diazonium sulfate. Dyes Pigm 173:107920. https://doi.org/10.1016/j.dyepig.2019.107920
36. Ullrich R, Grewer Th (1993) Decomposition of aromatic diazonium compounds. Thermochim Acta 225:201–211. https://doi.org/10.1016/0040-6031(93)80188-G
37. Bondarchuk SV (2019) Impact sensitivity of aryl diazonium chlorides: limitations of molecular and solid-state approach. J Mol Graph Model 89:114–121. https://doi.org/10.1016/j.jmgm.2019.03.008
38. Krasnokuskaya EA, Semenischeva NI, Filimonov VD, Knochel P (2007) A new, one-step and effective protocol for iodination of aromatic and heterocyclic compounds via aprotic diazotization of amines. Synthesis 81–84. https://doi.org/10.1055/s-2007-990950
39. Tretyakov AN, Krasnokutskaya EA, Gorlushko DA, Ogorodnikov VD, Filimonov VD (2011) A new one-pot solvent-free synthesis of pyridinyl tosylates via diazotization of aminopyridines. Tetrahedron Lett 52:85–87. https://doi.org/10.1016/j.tetlet.2010.10.163
40. Krasnokutskaya EA, Kassanova AZh, Estaeva MT, Filimonov VD (2014) A new synthesis of pyridinyl trifluoromethanesulfonates via one-pot diazotization of aminopyridines in the presence of trifluoromethanesulfonic acid. Tetrahedron Lett 55:3771–3773. https://doi.org/10.1016/j.tetlet.2014.05.052

41. (a) Kassanova AZh, Krasnokutskaya EA, Beisembai PS, Filimonov VD (2021) A novel convenient synthesis of pyridinyl and quinolinyl triflates and tosylates via one-pot diazotization of aminopyridines and aminoquinolines in solution. Synthesis 48:256–262. https://doi.org/10.1055/s-0035-1560392. (b) Sanzhiev AN, Krasnokutskaya EA, Erin KD, Filimonov VD (2021) Diazotization of aminopyridines in the presence of camphorsulfonic acid. Synthesis and some properties of pyridinyl camphorsulfonates. Russ J Org Chem 57:922–929. https://doi.org/10.1134/S1070428021060063
42. Sanzhiev AN, Potapova MI, Krasnokutskaya EA, Filimonov VD (2020) A novel one-pot synthesis of N, N-dimethylaminopyridines by diazotization of aminopyridines in dimethylformamide in the presence of trifluoromethanesulfonic acid. Russ J Org Chem 56:1023–1028. https://doi.org/10.1134/S1070428020060093
43. Chudinov AA, Dovbnya RS, Krasnokutskaya EA, Ogorodnikov VD, Filimonova IL (2016) A new transformation of aminopyridines upon diazotization in acetonitrile with the formation of N-pyridinylacetamides. Russ Chem Bull Int Ed 65:2312–2314. https://doi.org/10.1007/s11172-016-1583-9
44. Krasnokutskaya EA, Chudinov AA, Filimonov VD, New A (2018) Simple, and general synthesis of N-oxides of Iodopyridines and Iodoquinolines via the diazotization-iodination of heterocyclic amino N-oxides in the presence of *p*-toluenesulfonic acid in water. Synthesis 50:1368–1372. https://doi.org/10.1055/s-0036-1591738
45. Breton GW (2018) DFT study of *ortho*, *meta* and *para*-pyridyl cations. Pyridynium found? Comput Theor Chem 1133:51–57. https://doi.org/10.1016/j.comptc.2018.04.013
46. Ledenyova IV, Didenko VV, Shikhaliev KhS (2014) Chemistry of pyrazole-3(5)-diazonium salts. Chem Heterocyl Compd 50:1214–1243. https://doi.org/10.1007/s10593-014-1585-1
47. Tang Y, Imler GH, Parrish DA, Shreeve KM (2019) $C_6N_{10}O_4$: thermally stable nitrogen-rich inner bis(diazonium) zwitterions. Org Lett 21:8201–8204. https://doi.org/10.1021/acs.orglett.9b02875
48. Du Y, Zhang J, Peng P, Su H, Li S, Pang S (2017) Synthesis and characterization of three pyrazolate inner diazonium salts: green, powerful and stable primary explosives. New J Chem 41:9244–9249. https://doi.org/10.1039/c7nj00876g
49. Klapötke TM, Krumm B, Pflüger C (2016) Isolation of a moderately stable but sensitive zwitterionic diazonium tetrazolyl-1,2,3-triazolate. J Org Chem 81:6123–6127. https://doi.org/10.1021/acs.joc.6b01098
50. Li X, Wang X, Ye G, Xia W, Wang X (2010) Polystyrene-based diazonium salt as adhesive: a new approach for enzyme immobilization on polymeric supports. Polymer 51:860–867. https://doi.org/10.1016/j.polymer.2009.12.036
51. Nwajiobi O, Mahesh S, Streety X, Raj M (2021) Selective triazenation reaction (STaR) of secondary amines for tagging monomethyl lysine post-translational modifications. Angew Chem Int Ed 60:7344–7352. https://doi.org/10.1002/anie.202013997
52. Bustamante SE, Vallejos S, Pascual-Portal BS, Muños A, Meńdia A, Rivas BL, Carcia FC, Carcia JM (2019) Polymer films containing chemically anchored diazonium salts with long-term stability as colorimetric sensors. J Hazard Mater 365:725–732. https://doi.org/10.1016/j.jhazmat.2018.11.066
53. Francini N, Cochrane D, Illingworth S, Purdie L, Mantovani G, Fisher K, Seymour LW, Spain SG, Alexander C (2019) Polyvalent diazonium polymers provide efficient protection of oncolytic adenovirus enadenotucirev from neutralizing antibodies while maintaining biological activity in vitro and in vivo. Bioconjug Chem 30:1244–1257. https://doi.org/10.1021/acs.bioconjchem.9b00189
54. Singhaphan P, Unob F (2021) Thread-based platform for nitrite detection based on a modified Griess assay. Sens Actuator B Chem 327:128938. https://doi.org/10.1016/j.snb.2020.128938
55. Kubota K, Pang Y, Miura A, Ito H (2019) Redox reactions of small organic molecules using ball milling and piezoelectric materials. Science 366:1500–1504. https://doi.org/10.1126/science.aay8224
56. Mukherjee N, Chatterje T, Ranu BC (2013) Reaction under ball-milling: solvent-, ligand-, and metal-free synthesis of unsymmetrical diaryl chalcogenides. J Org Chem 78.11110–11114. https://doi.org/10.1021/jo402071b

57. Bretherick L (2017) In: Urben PG (ed) Handbook of reactive chemical hazards, 8th edn. Elsevier. ISBN: 978-008101059-4; 978-008100971-0
58. Oger M, Le Grognec E, Felpin F-X (2015) Handling diazonium salts in flow for organic and material chemistry. Org Chem Front 2:590–614. https://doi.org/10.1039/c5qo00037h
59. Deadman BJ, Collins SG, Maguire AR (2015) Taming hazardous chemistry in flow: the continuous processing of diazo and diazonium compounds. Chem Eur J 21:2298–2308. https://doi.org/10.1002/chem.201404348
60. Hu T, Baxendale IR, Baumann M (2016) Exploring flow procedures for diazonium formation. Molecules 21:918. https://doi.org/10.3390/molecules21070918
61. Movsisyan M, Delbeke EIP, Berton JKET, Battilocchio C, Ley SV, Stevens CV (2016) Taming hazardous chemistry by continuous flow technology. Chem Soc Rev 45:4892–4928. https://doi.org/10.1039/c5cs00902b
62. Schotten C, Leprevost SK, Yong LM, Hughes CE, Harris KDM, Browne DL (2020) Comparison of the thermal stabilities of diazonium salts and their corresponding triazenes. Org Proc Res Div 24:2336–2341. https://doi.org/10.1021/acs.oprd.0c00162
63. Schotten C, Aldmairi AH, Sagatov Y, Shepherd M, Browne DL (2016) Protected diazonium salts: a continuous-flow preparation of triazenes including the anticancer compounds dacarbazine and mitozolomide. J Flow Chem 6:218–225. https://doi.org/10.1556/1846.2016.00025
64. Lozynskyi A, Sabadakh O, Luchkevich E, Taras T, Vynnytska R, Karpenko O, Novikov V, Lesyk R (2018) The application of anthraquinone-based triazenes as equivalents of diazonium salts in reaction with methylene active compounds. Phosphorus Sulfur Silicon 193:409–414. https://doi.org/10.1080/10426507.2018.1452236
65. Hudson JL, Jian H, Leonard AD, Stephenson JJ, Tour JM (2006) Triazenes as a stable diazonium source for use in functionalizing carbon nanotubes in aqueous suspensions. Chem Mater 18:2766–2770. https://doi.org/10.1021/cm060020l
66. Wippert NA, Jung N, Bräse S (2019) Synthesis of arylamides via ritter-type cleavage of solid-supported aryltriazenes. ACS Comb Sci 21:568–572. https://doi.org/10.1021/acscombsci.9b00096
67. Avemaria F, Zimmermann V, Bräse S (2004) Synthesis of aryl azides via post-cleavage modification of polymer-bound triazenes. Synlett 1163−1166. https://doi.org/10.1055/s-2004-82298
68. Bräse S (2004) The virtue of the multifunctional triazene linkers in the efficient solid-phase synthesis of heterocycle libraries. Acc Chem Res 37:805–816. https://doi.org/10.1021/ar0200145
69. Lee YM, Moon ME, Vajpayee V, Filimonov VD, Chi K-W (2010) Efficient and economic halogenation of aryl amines via arenediazonium tosylate salts. Tetrahedron 66:7418–7422. https://doi.org/10.1016/j.tet.2010.07.005
70. Gorlushko DA, Filimonov VD, Krasnokutskaya EA, Semenischeva NI, Go BS, Hwang HY, Chi K-W (2008) Iodination of aryl amines in a water-paste form via stable aryl diazonium tosylates. Tetrahedron Lett 49:1080–1082. https://doi.org/10.1134/S1070428008080253
71. Moon ME, Choi Y, Lee YM, Vajpayee V, Trusova ME, Filimonov VD, Chi K-W (2010) An expeditious and environmentally benign preparation of aryl halides from aryl amines by solvent-free grinding. Tetrahedron Lett 51:6769–6771. https://doi.org/10.1016/j.tetlet.2010.10.099
72. Gorlushko DA, Filimonov VD, Semenishcheva NI, Krasnokuskaya EA, Tret'yakov AN, Go BS, Hwang HY, Cha EH, Chi K-W (2008) A simple and efficient procedure for diazotization–iodination of aromatic amines in aqueous pastes by the action of sodium nitrite and sodium hydrogen sulfate. Russ J Org Chem 44:1243–1244. https://doi.org/10.1016/j.tetlet.2010.10.099
73. Felipi-Blanco D, Gonzales-Gomes JC (2018) Salicylic acid-catalyzed arylation of enol acetates with anilines. Adv Synth Catal 360:2773–2778. https://doi.org/10.1002/adsc.201800427
74. Tundo P, Loris A, Selva M (2007) Formation and reaction of diazonium salts in a CO_2/H_2O system. Green Chem 9:777–779. https://doi.org/10.1039/b700207f

75. Bahulyan D, John L, Lalithambika M (2003) Modified clays as efficient acid-base catalyst systems for diazotization and diazocoupling reactions. Synth Commun 33:863–869. https://doi.org/10.1081/SCC-120016343
76. Pandey G, Török B (2017) K-10 montmorillonite-catalyzed solid phase diazotizations: environmentally benign coupling of diazonium salts with aromatic hydrocarbons to biaryls. Green Chem 19:5390–5395. https://doi.org/10.1039/c7gc02804k
77. Qiu D, Jin L, Zheng Z, Meng H, Mo F, Wang X, Zhang Y, Wang J (2013) Synthesis of pinacol arylboronates from aromatic amines: a metal-free transformation. J Org Chem 78:1923–1933. https://doi.org/10.1021/jo3018878
78. Mo F, Jiang Y, Qiu D, Zhang Y, Wang J (2010) Direct conversion of arylamines to pinacol boronates: a metal-free borylation process. Angew Chem Int Ed 49:1846–1849. https://doi.org/10.1002/anie.200905824
79. Allaire FS, Lyga JW (2001) Nonaqueous diazotization of arylamines in the presence of dimethyldisulide; The convenient synthesis of arylmethylsulfides from anilines. Synth Commun 31:1857–1861. https://doi.org/10.1081/SCC-100104335
80. Mukhopadhyay S, Batra S (2018) Direct transformation of arylamines to aryl halides via sodium nitrite and N-halosuccinimide. Chem Eur J 24:14622–14626. https://doi.org/10.1002/chem.201803347
81. Valizadeh H, Shomali A (2012) A new nitrite ionic liquid (IL-ONO) as a nitrosonium source for the efficient diazotization of aniline derivatives and in-situ synthesis of azo dyes. Dyes Pigm 92:1138–1143. https://doi.org/10.1016/j.dyepig.2010.11.010
82. Trusova ME, Krasnokutskaya EA, Choi Y, Chi K-W, Filimonov VD (2011) A green procedure for the diazotization–iodination of aromatic amines under aqueous, strong-acid-free conditions. Synthesis 2154–2158. https://doi.org/10.1055/s-0030-1260046
83. Zarchi MAK, Karimi M (2012) Diazotization of anilines and diazo coupling with a coupling component mediated by a polymer-supported sodium nitrite and a polymeric acid. J Appl Polym Sci 123:2762–2767. https://doi.org/10.1002/app.34402
84. Kaupp G, Herrmann A, Schmeyers J (2002) Waste-free chemistry of diazonium salts and benign separation of coupling products in solid salt reactions. Chem Eur J 8:1395–1406. https://doi.org/10.1002/1521-3765(20020315)8:6%3c1395::AID-CHEM1395%3e3.0.CO;2-L
85. Guella G, Ascenzi D, Franceschi P, Tosi P (2005) Gas-phase synthesis and detection of the benzenediazonium ion, $C_6H_5N_2^+$. A joint atmospheric pressure chemical ionization and guided ion beam experiment. Rapid Commun Mass Spectrom 19:1951–1955. https://doi.org/10.1002/rcm.2009

Kinetics and Mechanisms of Aryldiazonium Ions in Aqueous Solutions

Carlos Bravo-Díaz and Elisa González-Romero

Dedicated to Prof. Laurence S. Romsted (Rutgers, the State University of New Jersey, USA) for introducing us to the very complex—yet fascinating—chemistry of aryldiazonium ions, for his continuous support, encouragement, extremely fruitful discussions, and, most importantly, for his friendship.

Abstract In aqueous acid solution and in mixed alcohol-water solvents ($[H_3O^+] > 10^{-2}$ M), in the dark and in the absence of reductants, the spontaneous decomposition of aryldiazonium, ArN_2^+, salts proceeds through borderline S_N1 ($D_N + A_N$) -S_N2 mechanisms. The rate constant values depend strongly on the nature of the substituents attached to the aromatic ring of ArN_2^+ and, for those with electron-withdrawing substituents, on solution composition. The product distribution is proportional to the composition of the solvation shell of the *ipso* carbon, which reflects the composition of the water/cosolvent mixture. However, upon decreasing moderately the acidity, reactions involving the formation of diazohydroxides, ArN_2OH, diazoethers, ArN_2OR, and diazoates, ArN_2O^-, become competitive and may even be the main decomposition pathway. The stability of ArN_2OH, ArN_2OR, and ArN_2O^- species (which may coexist with ArN_2^+ in solution) is intimately related to the *Z-E* (*syn-anti, cis-trans*) isomerization of the O-adducts, so that they may undergo further reactions when they are components of a Lewis acid-base equilibrium, or undergo homolytic scission to produce homolytic reduction products. In this book chapter, we aim to provide the reader with a practical and (hopefully) useful view of the complex chemistry of ArN_2^+ in aqueous and mixed alcohol-water solutions, mainly covering the kinetics and mechanisms of the reactions. In a last section, we introduce some analytical methods for the determination of diazonium salts and their degradation products.

C. Bravo-Díaz (✉)
Physical Chemistry Department, Universidad de Vigo, Vigo, Spain
e-mail: cbravo@uvigo.es

E. González-Romero
Analytical and Food Chemistry Department, Universidad de Vigo, Vigo, Spain

M. M. Chehimi et al. (eds.), *Aryl Diazonium Salts and Related Compounds*, Physical Chemistry in Action, https://doi.org/10.1007/978-3-031-04398-7_3

1 Introduction

Aryldiazonium, Ar-N_2^+, and ions played a major role in organic chemistry since their discovery in 1858 by Peter Griess [1, 2]. From a historical perspective, the chemistry of aryldiazonium salts has been developed, in most instances, in aqueous media, and classical applications of arenediazonium salts include the Sandmeyer, Balz-Schieman, Gomberg-Bachman, Pschorr, and Meerwein reactions [3–5]. In the last decades, there has been renewed interest in their use as reactants in solvents other than water and alcohols, in catalytic cross-coupling reactions, [6, 7] in the chemical-biological field through modifications of amino acids, peptides, and proteins, [8] with functionalization of surfaces [9–12] and as chemical probes for probing interfacial concentrations of nucleophiles in a variety of colloidal systems [13, 14].

The utility of aryl diazonium salts in the solid-state may be, however, diminished by their potential instability, which is often governed by the nature of the counterion, [3, 15] and during the second half of the twentieth century, some chemists focused in developing synthetic methods to prepare more stable aryldiazonium salts, mostly in the form of tetrafluoroborates, hexafluorophosphates, tosylates, and disulfonamides (Chap. 2) [15, 16]. The tetrafluoroborates, in particular, can often be easily prepared and isolated as crystalline solids that can be stored in the dark, in the absence of water, at low temperatures for months to years without significant decomposition (Chap. 2) [3].

When in solution, aryldiazonium, ArN_2^+, ions may behave as Lewis acids, thus reacting with Lewis bases (nucleophiles Nu^- or NuH followed by loss of a proton) to yield covalently bonded adducts, ArN_2-Nu, at the electrophilic reactive center (the β-nitrogen of ArN_2^+) [3, 4]. Typical examples of the ArN_2-Nu adducts are the azo dyes, which are formed through a C-coupling mechanism [3, 17]. The formation of azo dyes takes place when ArN_2^+ ions react with aromatic substrates containing strong electron donors such as hydroxy or amino groups [17]. Covalently bonded adducts can also be formed with other nucleophiles leading to O-, S-, N-, and P-adducts [3, 4, 18].

This book chapter will focus exclusively on the reactivity of aryldiazonium ions in aqueous solution, with particular emphasis on the formation of O-coupling adducts of the type ArN = NO-Nu because H_2O and alcohol/water mixtures, together with H_2O/CH_3CN mixtures (in CH_3CN, the corresponding acetamides are formed [19]) are, by far, the most important solvents for diazonium salts, and where a variety of nucleophiles can be present or dissolved depending on experimental conditions [3]. Details on reactions in other solvents can be found elsewhere [3, 18]. The synthesis of ArN_2^+, their classical reactions such as the Sandmeyer reaction, and their recent applications will not be covered as this material can be found in other chapters of the present book and in specialized books or reviews [3, 4, 17, 20, 21]. In addition, the chemistry of alkanediazonium, RN_2^+, ions will not be covered here because they are of much less practical application than the aromatic ones because of their much lower stability [20, 22].

Furthermore, the present book chapter is not intended to be exhaustive in covering all literature but is rather aimed to provide the reader a broad view of the complex chemistry of ArN_2^+ in aqueous solution, with special emphasis on the kinetics and mechanism of the reactions involved. Indeed, this solution chemistry has important consequences in the grafting of diazonium salts. In the next sections, we will discuss the various mechanisms in some detail.

2 A Brief Survey of the Chemistry of Arenediazonium Ions in Aqueous Solutions

One of the most important—but mechanistically complicated—reaction is the deceptively spontaneous decomposition of ground state $Ar\text{-}N_2^+$ in aqueous solution and in alcohol/water mixtures. Part of the complexity of the reaction arises from the fact that water and aliphatic and aromatic alcohols (ROH, ArOH) may also act as a neutral nucleophiles in addition to act as a solvents, and, depending on the pH, can even ionize so that species such as OH^- and ArO^- may be present in solution and act as potential nucleophiles [18, 23–25].

The reaction of ArN_2^+ with water was first reported by Griess in the late nineteenth century and its "apparent" complexity made it to be the origin of some scientific controversy between E. Bamberger and A. Hantzsch for almost two decades (1894–1912) because of the structures of the products [3, 4, 26, 27]. In aqueous acid solution (pH < 2), in the absence of reductants, and in the dark, the mechanism of the spontaneous decomposition of ArN_2^+ is still a matter of debate, as there is not yet clear evidence if the reaction proceeds through S_N1 ($D_N + A_N$) or S_N2 mechanisms, Scheme 1, leading to the formation of heterolytic products [19, 28–30]. A decrease in the acidity of the solution modifies the mechanism of their spontaneous decomposition so that between pH ~ 4 and ~9, diazohydroxides, ArN_2OH, and diazoethers, ArN = N-OR, are formed from the reaction of ArN_2^+ with H_2O and ROH, respectively, which in this case act as a nucleophiles) [3, 18, 26, 27]. When phenoxide ArO^- ions are present, ArN_2^+ may also react with them and diazoethers Ar-N = N-OAr´ may be formed [18]. Both ArN_2OH and ArN = N-OR or (Ar-N = N-OAr´) may undergo homolytic decomposition, leading to the formation of reduction products and, in some instances, tarry materials [4].

In aqueous alkaline solutions (pH > 9), ArN_2^+ function as Lewis acids, reacting at the terminal nitrogen with OH^-, and diazohydroxides ArN_2OH are formed which can further react with OH^- to finally form diazoates ArN_2O^-, which are quite stable and can be isolated [3, 18].

Spontaneous dediazoniations, thus, may involve a variety of mechanisms and the following example should be sufficient to illustrate how apparently minor modifications in the experimental conditions may lead to major changes in the dediazoniation kinetics and in the product distributions. In aqueous acid solution, in the dark and in the absence of reductants, dediazoniation of 4-nitrobenzenediazonium (PNBD)

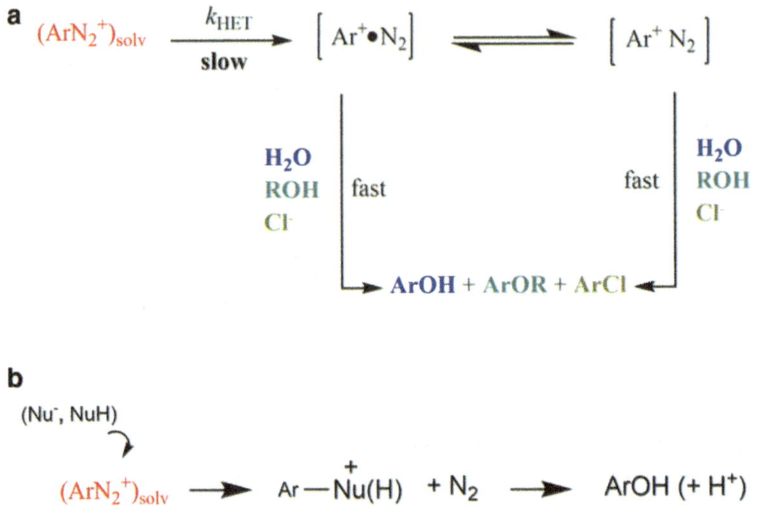

Scheme 1 **a** Ionic or heterolytic S_N1 ($D_N + A_N$) dediazoniation mechanism in the presence of various nucleophiles illustrating the formation of an intimate ion•molecule or a solvent separated ion-molecule pair that traps all nucleophiles available in their solvation shell **b** S_N2 dediazoniation mechanism. In both cases, heterolytic products are formed. The selectivity of ArN_2^+ against nucleophiles is very low, an argument commonly employed as an evidence for the S_N1 mechanism against the S_N2 one [3, 31, 32].

follows the S_N1 mechanism depicted in Scheme 1, with a half-life $t_{1/2} = 1383$ min [33]. However, addition of small amounts of MeOH leads to non-first-order kinetics compatible with a radical mechanism, so that $t_{1/2} \approx 6$ min at 10% (v:v) MeOH, Fig. 1. Interestingly enough, a further increase in [MeOH] makes the reaction to progressively revert to a first-order behavior, slowing down the rate of dediazoniation so that at percentages of MeOH higher than 90%, clean first-order kinetics are obtained with $t_{1/2} \approx 77$ min. In the absence of MeOH, 4-nitrophenol (product from heterolytic cleavage) is formed in quantitative yield; however, at only 20% MeOH, the reduction product nitrobenzene (product from homolytic cleavage) is obtained in more than 90% and the yield of 4-nitrophenol is less than 10%. Upon increasing the MeOH content of the solution, the formation of 4-nitroanisole (obtained from the heterolytic reaction between PNBD and MeOH, Scheme 1) becomes competitive and, at 99% MeOH, both nitrobenzene (60%) and 4-nitroanisole (40%) are obtained [33]. The reduction product 4,4-dinitrobiphenyl (obtained from the coupling reaction of two Ar· radicals) is detected only at MeOH percentages in the range 0.5–15%.

A major feature that needs to be considered when dealing with reactions where ArN_2ONu adducts are formed is the geometrical *Z*-*E* (*syn*-*anti*) isomerism, Scheme 2, leading to the *Z*-(*anti*) and *E*-(*syn*) forms. Isomerization is possible in reactions of ArN_2^+ with nucleophiles Nu^- where the nitrogen-nitrogen double-bond is retained, i.e., R–N = N-R' (R and R' can be alkyl or aryl residues). Their formation was recognized experimentally as early as 1938 by Hartley [34] after Hantzsch and Werner [35]

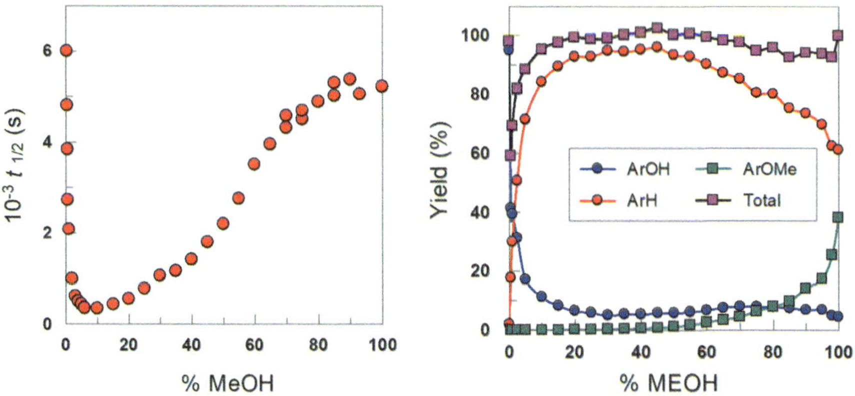

Fig. 1 Variations of the half-life (left) and of the product distribution (right) for the spontaneous decomposition of 4-nitrobenzenediazonium as a function of the percentage of MeOH (v:v) in solution. Up to four products can be detected depending on experimental conditions, 4-nitrophenol (ArOH), 4-nitroanisole (ArOMe), nitrobenzene (ArH), and 4,4-dinitrobiphenyl (only detected in the range 0.5–15%). Adapted with permission from ref. [33], copyright 2006 Wiley–VCH Verlag

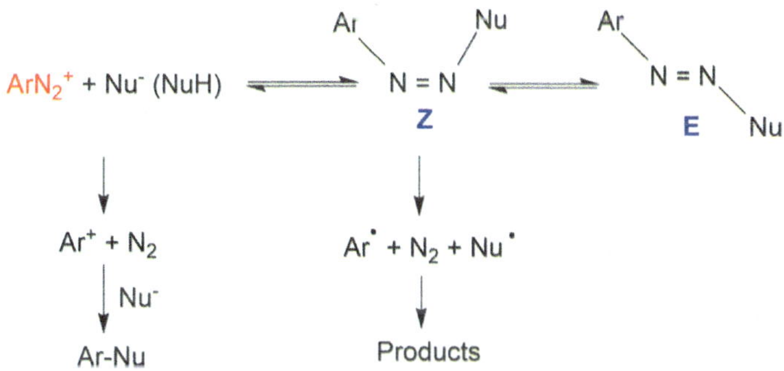

Scheme 2 *Z (syn)-E (anti)* isomerism following the formation of Ar-N = N-Nu adducts which may undergo homolytic scission. The competitive spontaneous decomposition of ArN_2^+ through the formation of a highly reactive aryl cation (S_N1 mechanism, Scheme 1) is also shown for illustrative purposes. Dediazoniations that follow homolytic mechanisms are always (as far as they are known today) faster than the heterolytic ones. Adapted with permission from ref. [18], copyright 2011 J. Wiley & Sons

put forward the hypothesis of their existence [3] and was investigated by employing a variety of nucleophiles such as Nu = O^- (diazoates), Nu = RO^- (diazoethers), Nu = SO_3^- (sulfonates), Nu = $-SO_2$-Ar (sulfones), Nu = $CONH_2$ (carboxamides) and Nu = S-R, S-Ar (diazothioethers), Nu = CN^- (cyanides) and Nu = R^1-NH-R^2 (triazenes). ArN_2OH adducts are components of a Lewis acid-base equilibrium and react with OH^- ions to give diazoates, ArN_2O^-.

In most instances, the *Z*-isomer is the kinetically controlled product, but the *E*-isomers are thermodynamically more stable. *Z*-isomers are formed in reactions with

OH^-, CN^-, RO^-, CN^- and SO_3^{2-}, whereas *E*-compounds are formed with reactants such as phenols and tertiary amines. Further details on the *Z-E* isomerism can be found elsewhere [3, 4, 18].

O-adducts may also be formed when reacting with nucleophiles Nu^- other than OH^-, but their formation, identification, and isolation may be very difficult because their stability depends on the leaving ability of the nucleophile involved in the reaction. If the reactant Nu^- is a poor leaving group (even if it is a good nucleophile), such as ascorbate or gallate—3,4,5-trihydroxybenzoate- ions, some stabilization may occur by conversion to the thermodynamically stable isomer (*Z-E* isomerization) [36, 37]. If Nu^- is a good nucleophile and good leaving groups, such as halide or acetate ions, the equilibrium lies largely on the side of reactants so that formation of O-adduct is negligible and ArN_2^+ undergoes spontaneous decomposition through the heterolytic mechanism [3, 23, 38]. In some instances, isomerization is not possible and the adduct splits homolytically to finally give reduction products [33, 39, 40].

3 Kinetics and Mechanisms of Spontaneous Decomposition of ArN_2^+ in Aqueous Solution and in Alcohol-Water Mixtures

3.1 Dediazoniations Under Acidic (pH < 2) Conditions

It is generally accepted that in aqueous acid (pH < 3) solution, in the absence of reductants and in the dark, the spontaneous hydrolysis of aryldiazonium salts afford phenols through bordeline S_N1-S_N2 mechanisms, Scheme 1 [3, 18, 26, 31]. A variety of studies including N_2 incorporation and rearrangement substituent effects, nucleophile and solvent effects have led to the widely accepted mechanism involving the rate-determining loss of N_2 and formation of an extremely reactive aryl cation intermediate, Ar^+, that reacts with nucleophiles in its solvation shell in a time faster than that required for cation diffusion out of the solvent cage [3, 13, 19, 26, 28]. However, recent reports, mostly based on theoretical calculations, suggest that the formation of a free phenyl cation is not an obligatory intermediate and that a borderline S_N1-S_N2 mechanism is compatible with experimental results [19, 28, 29].

Table 1 collects some values for the half-lives for the spontaneous decomposition of various arenediazonium ions in different aqueous solvents. As shown, ArN_2^+ decomposes spontaneously in aqueous acid and alcohol/water mixtures ($[H_3O^+] < 10^{-2}$ M), with have-lives ranging from minutes to hours, depending on the substituents in the aromatic ring. Solvolytic studies in alcohol-water (ROH/H_2O) or acetonitrile-water (ACN/H_2O) mixtures show that, in aqueous acid solution, aryldiazonium ions undergo preferential solvation by water, [23, 40, 41] results confirmed later by classical molecular dynamics calculations of solvent distribution around ArN_2^+ for

Table 1 Values of the half-lives and activation energies for the heterolytic (HET) and homolytic (HOM) dediazoniation processes of some X-ArN_2^+ in different solvents

X-ArN_2^+	Solvent (v:v)	$t_{1/2}$ (HET) s	$t_{1/2}$ (HOM) s	E_a (HET) kJ mol^{-1}	E_a (HOM) kJ mol^{-1}
X = 4-Me	H_2O	3450	–	110	–
	70–30 EtOH-H_2O	359	35	116	77
	90–10 BuOH-H_2O	1000	144	103	74
	75–25 BuOH	1200	4	–	76
	35–65 TFE-H_2O	616	–		
	70–30 TFE-H_2O	539	–		
X = 4-NO_2	H_2O	ca 83,000	–	119	–
	80–20 MeOH-H_2O	ca 16,000	1375	–	–
	95–5 MeOH-H_2O	3648	1824	–	–
X = 4-Br	H_2O	ca 180,000	–	~100	–
	25-75MeOH-H_2O	–	2390	~100	71
	75–25 MeOH-H_2O	–	2980	~100	74

Values compiled from Refs. [23, 33, 41, 43–45]

water/solvent mixtures [19]. The bulk product distribution reflects the relative nucleophile distribution in the solvation shell and the relative concentration of nucleophiles in solution [19, 23, 33, 40, 42].

3.2 *Kinetics and Mechanisms of Dediazoniation at Moderate Acidities (pH 3-8)*

The apparently “simple” decrease of the acidity of the solution (while keeping other experimental conditions constant) leads to a completely different—and much more complex—kinetic behavior than that observed under acidic conditions, deserving a specific section for the description of the mechanistic and practical consequences of the reactions involved. Production of higher amounts of homolytic products in alcohols than in water would be anticipated based on the nucleophilicity of the solvents. However, this is not the case. Dediazoniations in alcohols usually yield three or four products (including reduction ones) and diazo tars seldom form [4]. However, in weakly acidic or alkaline aqueous solutions, tarry materials may be formed and dediazoniations produce a complex mixture of other products [3, 4, 27].

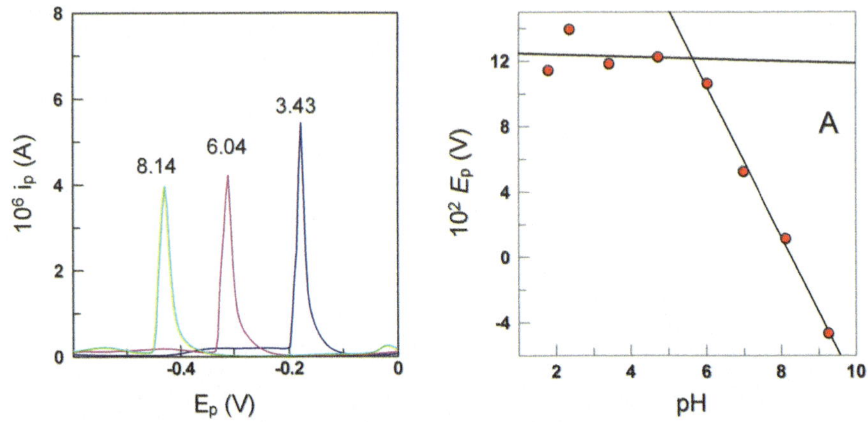

Fig. 2 Left. Illustrative variation of the first (irreversible) reduction peak of benzenediazonium ions with pHs. Right, variation of peak potential (E_p) of the first reduction peak of 4-methylbenzenediazonium ion with pH. Details on calculations can be found elsewhere [53]. Figures reproduced with permission from reference [53]. Copyright 2014 J. Wiley & Sons

At moderate acidities, the nucleophilic character of the solvent is reinforced, and the reactions of ArN_2^+ with H_2O and ROH (as nucleophiles) lead to the formation of diazohydroxides, ArN_2OH, and diazoethers, ArN_2OR, becoming competitive so that the observed rates of the reactions are much faster than under acidic (pH < 2) conditions and lead to the formation of a variety of products, including reduction products that in many instances are difficult to isolate or identify, suggesting that mechanisms other than the described S_N1-S_N2 mechanisms (Scheme 1) are operating and becoming increasingly important.

The formation of diazohydroxides ArN_2OH in aqueous solution was demonstrated on the basis of the variations of the peak currents and peak potentials (E_p) of the first reduction peak of a number of aryldiazonium ions with acidity, Fig. 2 [39, 46–51]. The dependence of Ep on pH is typical of acid-base systems and results were interpreted according to Scheme 3 [52].

Some values for the equilibrium constants K_R corresponding to the formation of hydroxides are shown in Table 2, indicating that effects of the nature of the substituents in the aromatic ring of ArN_2^+ are very small, suggesting an increase in the electron density on the atom adjacent to the ring during the reaction, neutralizing the positive charge on the diazonium group [54].

Scheme 3 Nucleophilic reactions of X-ArN_2^+ with H_2O and ROH to give a diazohydroxide, X-ArN_2OH, and diazoether, X-ArN_2OR, adducts. X stands for any substituent in the aromatic ring of the aryldiazonium ion

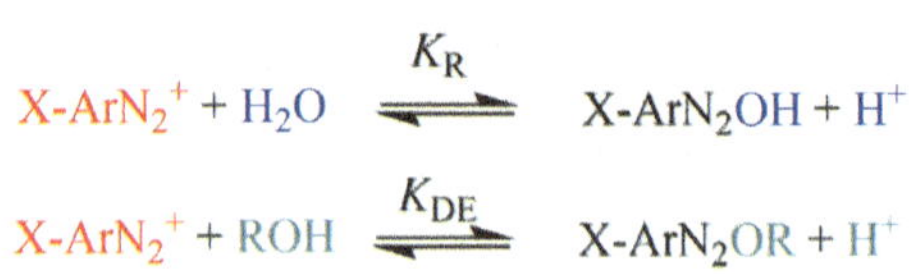

Table 2 Equilibrium constants for the formation of diazohydroxides and diazoethers (K_R and K_{DE}, respectively) defined according to Scheme 2C

$X\text{-}ArN_2^+$	X = H	X = 4-Me	X = 4-MeO	X = 4-Br	X = 4-NO_2
pK_R	5.70	5.72	5.80	5.20	5.24
pK_{DE}		3.6–5.3[a]		3.5[b]	4.18[c]

Values from: (a) ref. [40, 55], (b) ref. [45], (c) ref. [33]

A large body of solvolytic studies shows that the formation of O-diazo adducts with neutral nucleophiles (including the solvent) is quite common [3, 18, 33, 40, 45]. The studies permitted the recognition of the dual role of H_2O and ROH molecules as nucleophiles. On one side, H_2O and ROH act as solvents solvating the diazonium ions and allowing them to undergo thermal heterolytic decomposition, Scheme 1. On the other hand, H_2O and ROH molecules can act as nucleophiles reacting directly with arenediazonium ions to yield O-coupling adducts, Scheme 3, in the highly unstable *Z* configuration, i.e., *Z*-diazoethers. The *Z*-diazoethers then undergo either homolytic fragmentation, initiating radical processes leading to the formation of quantitative amounts of the reduction product, or isomerize to finally yield the *E*-isomer [18].

In solution, ArN_2^+, ArN_2OH, and/or ArN_2OR adducts may coexist in moderate acid solutions. The variation of the relative concentrations of these species as a function of acidity is shown in Fig. 3.

A product and kinetic analyses of dediazoniations in alcohol/water mixtures show that sigmoidal variations in both the observed rate constant k_{obs} and product yields with acidity for a number of arenediazonium ions [33, 40, 43, 45] as illustrated in Fig. 4. At relatively high acidities (–log([HCl] < 2), only products from ionic

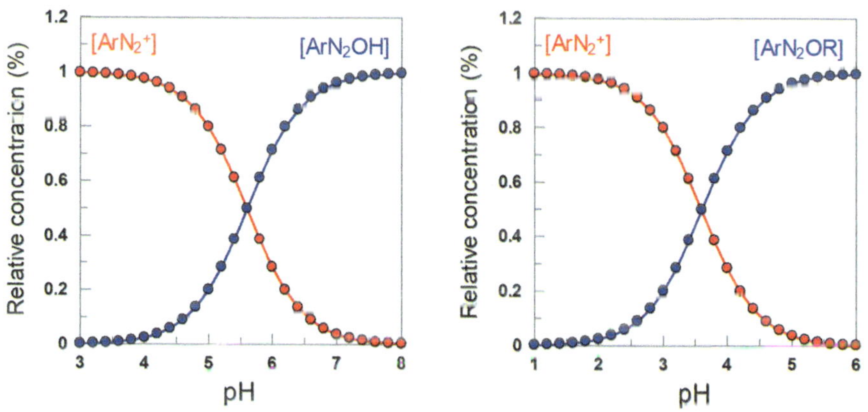

Fig. 3 Variations of the percentages of ArN_2^+, diazohydroxides ArN_2OH (left), and ArN_2OR (right) in aqueous solution and in ethanol/water mixtures, respectively, upon changing the solution acidity. Average values of $pK_R = 5.6$ and $pK_{DE} = 3.6$ were employed in calculations. The large difference between pK_R and pK_{DE} values makes that the formation of ArN_2OH in alcohol/water mixtures to be negligible compared to that of ArN_2OR

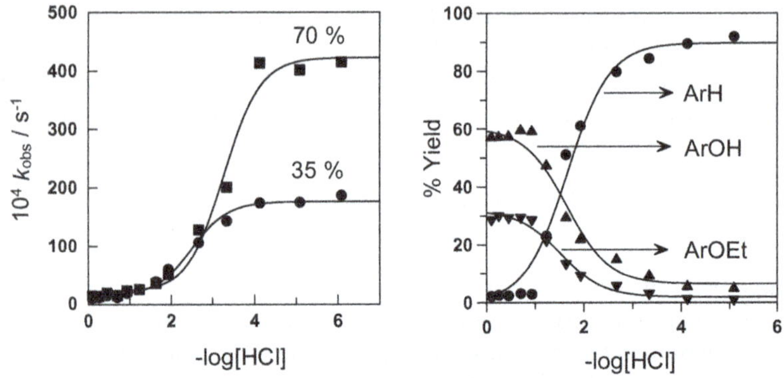

Fig. 4 Left: Effects of acidity on k_{obs} for the solvolytic decomposition of 4-methylbenzenediazonium, 4-Me-ArN_2^+ in 35 and 70% EtOH-H_2O mixtures. Right: Effects of acidity on the dediazoniation product distribution obtained in a 70% EtOH-H_2O mixture. The main solvolytic products were cresol, ArOH, methylphenetole, ArOEt, and toluene, ArH. Note the quantitative formation of the homolytic (reduction) product ArH at moderate acidities (pH = 3 −5). [4-Me-ArN_2^+] ~1 × 10^{-4} M, T = 60 °C. Reproduced with permission from reference [18] Copyright 2010 John Wiley & Sons

mechanisms (phenols, ArOH, ethers, ArOEt, halo derivatives, ArCl) are detected. But, upon decreasing the acidity, the homolytic product ArH becomes predominant at the expense of the heterolytic ones [40].

Sigmoidal variations of k_{obs} with acidity are usually observed in reactions of acid-base pairs where both forms are attainable and show different reactivities. Thus, in aqueous solutions or in alcohol/water mixtures at moderate acidities, two competing mechanisms are operating, Scheme 4, and Eq. 1 can be derived.

$$k_{obs} = \frac{k_{HET}[H^+] + k_{HOM}K_{DE}[ROH]}{K_{DE}[ROH] + [H^+]} \tag{1}$$

Two limiting situations can be considered, (i) When $[H^+] \gg K_{DE}[ROH]$, $k_{obs} \approx k_{HET}$, i.e., the reaction proceeds wholly through the ionic $D_N + A_N$ mechanism, and

Scheme 4 Competitive heterolytic (S_N1) and homolytic (O-coupling) mechanisms for the spontaneous dediazoniation of ArN_2^+ with water (X = H) and alcohols (X = R) under moderately acidic conditions

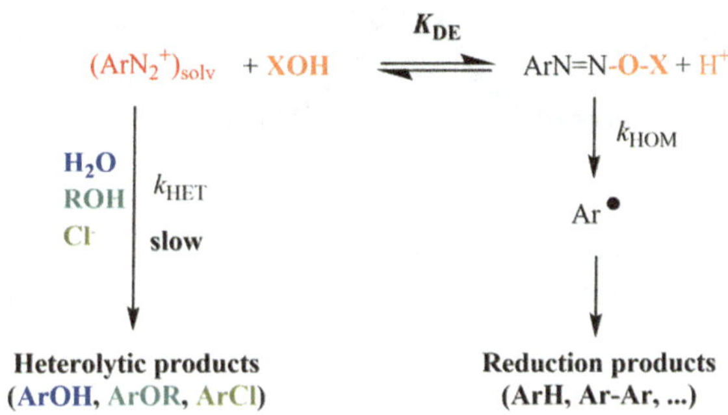

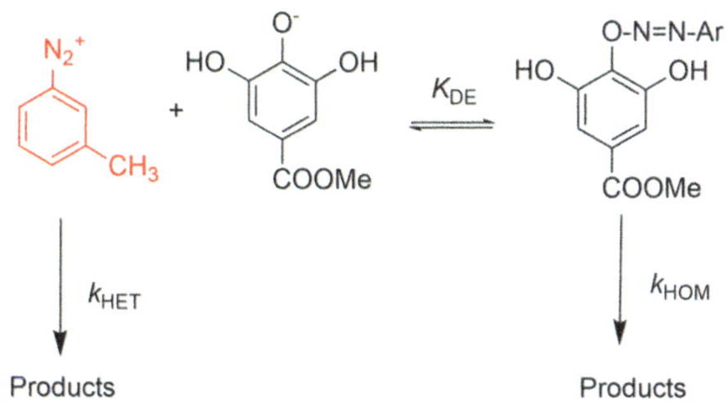

Scheme 5 Proposed reaction mechanism between 3-methylbenzenediazonium ions and methyl gallate, MG. The reaction takes place through the ionized form of MG (pKa = 8.03) and only tiny amounts of the ionized polyphenol are necessary for the reaction to proceed through the formation of the diazo ether at rates (k_{HOM}) much higher than those of the spontaneous decomposition (k_{HET}) [61].

only heterolytic products are obtained. On the other hand, when $[H^+] \ll K_{DE}[ROH]$, $k_{obs} \approx k_{HOM}$, i.e., the reaction proceeds wholly through the O-diazo ether and formation of homolytic reduction products is favored. Some representative values for the average pK' value, as well as those for k_{HET} and k_{HOM} are listed in Table 2.

Results in the presence of molecules bearing OH groups in their molecular structure such as ascorbic acid (vitamin C, VC) [24, 36, 56–59] are also consistent with the formation of *O*-adducts, which were detected experimentally [36, 58, 59] and unambiguously identified by ^{1}H and ^{13}C NMR spectroscopy as 3-*O*-arenediazo ascorbic acids. [56] O-adducts are also formed in reactions of ArN_2^+ with trihydric phenols such as gallic acid (GA, 3,4,5-trihydroxybenzoic acid) and methyl gallate (MG, methyl 3,4,5-trihydroxybenzoate) and catecholics such as hydroxytyrosol [25, 60–63]. For these reactions, a mechanism comprising consecutive, reversible steps, have been proposed. The first one involves a fast, bimolecular step, in which ArN_2^+ ions react with GA^{2-} to form a *Z*-diazoether complex, which is subsequently isomerized to the ionized form of the much more stable *E*-diazoether in a second, slow, unimolecular step (Scheme 5) [25, 60, 61].

3.3 *Dediazoniations in Alkaline (pH > 9) Aqueous Solutions*

Aromatic diazonium ions behave as Lewis acids, as they can react with OH^- ions to form diazohydroxides, ArN_2OH. The diazohydroxides then became Brönsted acids and may lose a proton to yield diazoates, Scheme 6.

Reports by Wittwer and Zollinger [64] showed that the variation of the degree of neutralization with acidity is different from that expected for a dibasic acid because the equilibrium constant for the formation of the diazoate is higher than that for the formation of the diazohydroxide, that is, $K_2 > K_1$. This behavior is in contrast with

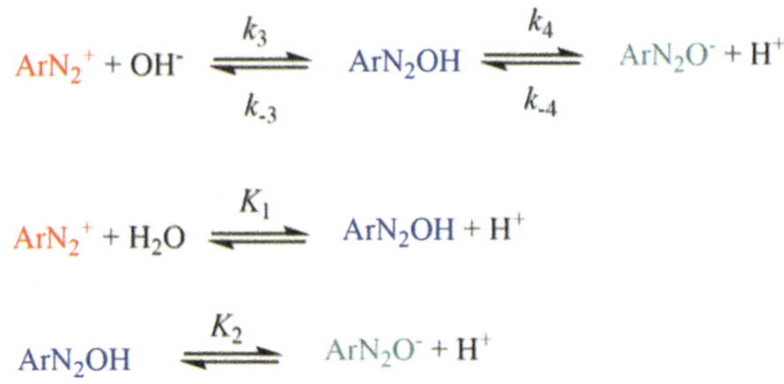

Scheme 6 Equilibria showing the formation of diazoates [18]

that obtained for typical dibasic acids, where $K_1 > K_2$. Wittwer and Zollinger [64] estimated, from neutralization curves, that $K_2 > 10^3\ K_1$ and for this reason, it was not possible to separate, at that time, the individual values but only could be determined their product K_1K_2. Individual values of K_1 and K_2 were first determined by Lewis and Suhr [65] by taking advantage of the fact that rate constant for the addition of OH^- ions to ArN_2^+ ions (k_3 in Scheme 6) is slower than the rate of deprotonation of the diazohydroxide (k_4). The general conclusion that can be drawn is that K_2 is greater than K_1 by 3–5 orders of magnitude, depending on the nature of the substituent in the aromatic ring.

By using the corresponding equilibrium constants, Eq. 2 has been derived, which shows that $(pK_1 + pK_2)/2 = pH_m$ equals the pH of the solution in which $[ArN_2^+] = [ArN_2O^-]$. The ratio $[ArN_2^+]$ / $[ArN_2O^-]$ changes 100 fold when the pH is changed by one [1] unit so that at $pH = pH_m + 1$, the reaction mixture only contains 1% of ArN_2^+ and 99% diazotate. Values for the rate constants, pK_1, pK_2 for a series of arenediazonium ions, and the variations of the relative amounts of ArN_2^+, ArN_2OH, and ArN_2O^- can be found elsewhere [3, 18].

$$pH = \frac{pK_1 + pK_2}{2} + \log \frac{[ArN_2O^-]}{[ArN_2^+]} \tag{2}$$

The relative concentrations of the diazonium and the diazoate ions (the major equilibrium forms) are shown in Fig. 5. The equilibrium concentrations of the diazohydroxide and the diazoanhydride are usually very small at all pH values, with a maximum at $pH = pKm$.

4 Analytical Chemistry of Diazonium Salts and Their Degradation Products

Regarding to analytical chemistry of these compounds and the chemistry around them, the identification and the quantification of the ArN_2^+ itself and all by-products generated in dediazoniation reactions will be the main challenge to get sufficient valuable data on these processes and to offer a convincing choice between the mechanistic

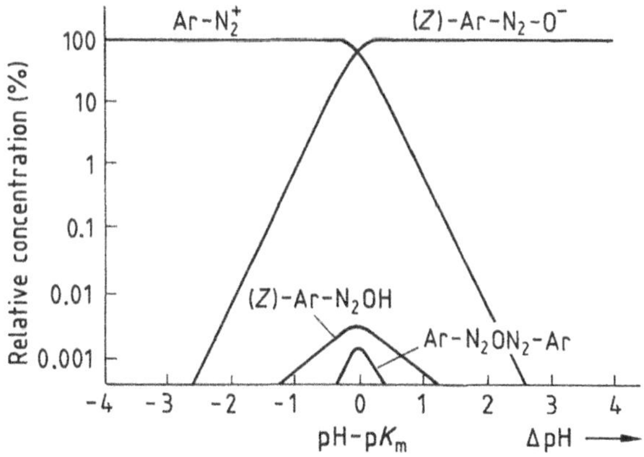

Fig. 5 Variation of the relative concentrations of various species in equilibrium with ArN_2^+ as a function of acidity. Reproduced with permission from ref. [3] Copyright 1994 J. Wiley & Sons

alternatives, in particular during surface reactions. The compounds to be identified and quantified are aromatic derivatives (typically, phenolics, haloarenes, benzene derivatives, biaryls, aromatic ethers, etc.), as well as the possible transient and stables intermediates, depending on the structural nature of ArN_2^+ and the reaction media composition.

As a starting point, some of the analytical techniques are able to monitor only one component at a time, and some others are limited to narrow ranges of useful experimental conditions, once the adjustable parameters to describe the measured property in any particular solution composition have been optimized [66]. As experts in the field, the readers know well that there is not only one technique to be used. The combination of several techniques is necessary in order to get insight into complete knowledge of what is going on in the dediazoniation reactions, how is the disappearance of the ion and its rate, which are the products formed and their reaction rates or the stable/transient intermediates that can be detected and monitored. On these bases, the possible degradation pathway in any matrix or composition of the reaction media (aqueous or non-aqueous solutions, colloidal and macromolecular systems, presence of different electrolytes, etc.) can be proposed. To achieve these goals, valuable and quality data that help to work out a real mechanism interpretation of dediazoniation reaction will be needed. Therefore, it is easy to understand that the use of several techniques that complement each other to solve the problem is the best way to validate the results obtained. Furthermore, the analytical signal of the measured property must be correlated with the concentration of the analyte (calibration) to obtain minimum uncertainty in concentration level, and then, high-quality kinetic data can be achieved. Keep in mind that this section is not intended to be exhaustive. The proposal is to address a brief description of most popular instrumental analytical techniques used for monitoring ArN_2^+ in kinetics studies up to date, with some references as representatives.

The UV–Vis spectra of ions and molecules in solution were the first instrumental methodology used to obtain both qualitative and quantitative chemical information and were widely applied to obtain kinetic data [23, 24, 33, 34, 38, 57, 58, 67, 68].

This technique is simple to use and relatively rapid, limited to transparent media and by the *cutoff point* of solvents. UV–Vis spectrophotometers with stopped-flow devices for monitoring very fast reactions are extremely valuable [69]. However, the UV region is not selective enough and most of the products involved in the dediazoniation reaction absorb in the same region, being impossible to monitor the reactive loss and/or products formation at the same time. For solving this situation, the absorption band can be shifted to the visible region, more selective, by a derivatization reaction that quantitatively transforms the unreacted ArN_2^+ into a compound with highly conjugated double bonds and heteroatoms. The coupling reaction with naphthols and naphthylamines to form azo dyes based on Griess reaction [7] provide the essential tool to monitor the degradation of ArN_2^+ by indirect monitoring the azo dye formed, but the experimental conditions must be chosen to ensure that azo dye formation is much faster than dediazoniation [17, 70]. In contrast with UV–Vis spectrophotometric technique, scant attention has been paid to the use of fluorescence for detection and monitoring dediazoniation. The ArN_2^+ derivatives with one aromatic ring system show a poor fluorescence, but enough for dediazoniations following up [70]. This technique is more sensitive (at least by two orders of magnitude) for polyaromatic molecules with rigid planar structures as naphthalene N_2^+ derivatives and much more selective than UV–Vis spectrophotometry.

Nuclear magnetic resonance, NMR, and Fourier transformed infrared, FT-IR, spectroscopies, together with Mass spectrometry by electrospray ionization, EIS-MS, are techniques that support all kinetics experiments, giving information about purity of standards, screening quantification, and structural elucidation and confirmation. They determine whether a synthetic reactant or product is the compound expected or planned. NMR and MS are remarkably close to each other in relation to the information obtained, but the advantage of MS is that it only needs small amount of sample (ng contrary to μg in NMR) [19, 30]. MS disadvantages are to be a destructive analytical technique and limited in solving structural isomerism, where NMR is still needed. The ^{13}C NMR or ^{1}H NMR techniques are additionally used in chemical analysis of dediazoniations by monitoring the chemical shift of the products in the spectra [71, 72]. Electron paramagnetic resonance, EPR, or Electron spin resonance, ESR, spectroscopy is another spectroscopic technique used to study radical chemical species in dediazoniation reactions, providing a valuable measure of how well the homolytic mechanism is going on and which are the radical intermediates found [70]. The principles of EPR are similar those to NMR, but give higher time resolution and sensitivity than NMR.

The most effective solution to prevent interferences is to use a separation technique previously to the detection. High-pressure liquid chromatography, HPLC, is a versatile technique widely used in kinetic studies of ArN_2^+ for obtaining kinetic data of products formation, choosing the chromatographic mode by reverse–phase and UV or diode array detectors [23, 24, 33, 40, 57, 58, 73]. However, this technique is time-consuming, tedious, and reactive samples cannot be obviously injected. The derivatization protocol mentioned above takes advantage of quenching instantaneously the dediazoniation reaction and preventing side reactions of highly reactive ArN_2^+ or its intermediates with the solvents of the mobile phase or other parts of the chromatographic system. The latter would ultimately lead to erroneous identification and quantification of dediazoniation products. At that point, the coupling reactions provide the best performance in the HPLC system and HPLC column and avoid pitfalls that can impact in the HPLC analysis [74]. Gas chromatography with flame-ionization detector, GC-FID, is one of the fastest and most useful separation techniques available [37, 72, 75], but limited to organic compounds that are volatile and not thermally labile. This is why HPLC has been gaining followers up to date. The most sophisticated detection system for both qualitative and sensitive quantitative analysis in GC or HPLC systems is Mass Spectrometry, MS, which provides retention-time data and peak area measurements, but structural information obtained from the fragmentation patterns can be used to positively identify the mixture components and to determine the products of the reactions with higher sensitivity and selectivity [76, 77]. However, this methodology requires several steps for sample preparation, expensive instrumentation, and trained professionals to operate the equipment.

The investigation of dediazoniations by electroanalysis showed that a variety of polarographic and voltammetric peaks can be chosen to obtain kinetic information of ArN_2^+ [46–48, 51, 78]. In this context, the selective detection is possible through the electroreduction on drop mercury electrode, DME, of the ArN_2^+ and products formed by differential pulse polarography, DPP, and the electrooxidation on carbon electrodes of some products (i.e., phenolics) by DP voltammetry [39, 47]. An interesting statement of these techniques, in comparison with all others reported herein, is that the reaction of 4-hexadecylbenzenediazonium ions, 16-ArN_2^+, is used as probe to better understand the colloidal interfaces and the distribution of antioxidants in very complex matrix as emulsions, can be directly monitored by linear sweep voltammetry, LSV [78, 79]. The advantages of these techniques rely on short analysis time, low cost, and simple sample preparation steps (unique dilution in electrolyte). Moreover, electroanalysis stands out for being useful for simultaneously detecting ArN_2^+ loss, transient intermediates as diazoethers [36], and products formed by dediazoniation with high sensitivity (detection limits lower than those obtained by the aforementioned techniques, except MS) and with similar selectivity to HPLC techniques in some dediazoniations [39, 46]. Furthermore, the electroanalysis can be applied not only in transparent solutions but also in both opaque solutions and suspensions.

5 Conclusions and Future Perspectives

The unique multimodal reactivity of ArN_2^+, while offering multiple options for the preparation of new compounds (Pd-catalyzed reactions, thermal or photosensitized radical chain reactions, and azo-coupling reactions) and materials, can also be fine tuned via the appropriate choice of experimental conditions and reaction partners. This overview shows a great potential for the formation of a wide variety of azo compounds and their use in chemistry. However, it should also be mentioned that protocols involving the use of arenediazonium salts have the drawbacks of ArN_2^+

being unstable in solution, highlighting the need for a careful control of experimental conditions, since apparently minor changes may lead to substantial modifications in the kinetics and mechanisms of their spontaneous decomposition, leading to large modifications in their effective concentrations in solution.

In this review, we focused on the main kinetic and mechanistic consequences of changing experimental conditions in the spontaneous decomposition of aryldiazonium salts to draw the attention of scientists interested in future applications of ArN_2^+ in solution. Indeed, the various applications of aryldiazonium salts, which are readily obtained from inexpensive anilines, will continue to attract attention of chemists, but the spontaneous reaction of diazonium salts in various solvents will need to be considered anyway as it may be competitive with the other processes involved.

Acknowledgments This book chapter was prepared during a sabbatical leave of CBD supported by the University of Vigo. We thank all colleagues for helpful discussions and, especially, to all students who participated with enthusiasm for years in the aryldiazonium project, making important contributions to this work. Financial support from Ministerio de Ciencia e Innovación (Spain), Xunta de Galicia and Universidad de Vigo is also acknowledged.

Credit authorship contribution statement. C. B-D. Sections 1–3 and 5: conceptualization, visualization, writing, review & editing.

E. G-R. Section 4: conceptualization, visualization, writing, review & editing.

References

1. Griess JP (1864) Philos Trans R Soc. London. https://doi.org/10.1098/rstl.1864.0018
2. Griess P (1858) Liebigs Ann Chem 106:123
3. Zollinger H (1994) Diazo chemistry I: aromatic and heteroaromatic compounds
4. Saunders KH, Allen RLM (1985) Aromatic Diazo compounds. Baltimore, MD, USA, E Arnold
5. Mo F, Qiu D, Zhang Y, Wang J (2018) Acc Chem Res 51:496–506. https://doi.org/10.1021/acs.accounts.7b00566
6. Roglans A, Pla-Quintana A, Moreno-Mañas M (2006) Chem Rev 106:4622–4643. https://doi.org/10.1021/cr0509861
7. Kostas ID (2018) Suzuki–Miyaura cross—coupling reaction and potential applications. MDPI AG
8. Sengupta S, Chandrasekaran S (2019) Org Biomol Chem 17:8308–8329. https://doi.org/10.1039/C9OB01471C
9. Mohamed AA, Salmi Z, Dahoumane SA, Mekki A, Carbonnier B, Chehimi MM (2015) Adv Coll Interface Sci 225:16–36. https://doi.org/10.1016/j.cis.2015.07.011
10. Chehimi MM (2012) Aryl diazonium salts: new coupling agents in polymer and surface science. Wiley
11. Granozzi G, Alonso-Vante N (2019) Electrochemical surface science: basics and applications. Mdpi AG
12. Hetemi D, Noël V, Pinson J (2020) Biosensors 10. https://doi.org/10.3390/bios10010004
13. Dar AA, Bravo-Diaz C, Nazir N, Romsted LS (2017) Curr Opin Colloid Interface Sci 32:84–93. https://doi.org/10.1016/j.cocis.2017.09.001

14. Bravo-Díaz C, Romsted LS, Liu C, Losada-Barreiro S, Pastoriza-Gallego MJ, Gao X, Gu Q, Krishnan G, Sánchez-Paz V, Zhang Y, Ahmad-Dar A (2015) Langmuir 31:8961–8979. https://doi.org/10.1021/acs.langmuir.5b00112
15. Firth JD, Fairlamb IJS (2020) Org Lett 22:7057–7059. https://doi.org/10.1021/acs.orglett.0c02685
16. Trusova ME, Kutonova KV, Kurtukov VV, Filimonov VD, Postnikov PS (2016) Res-Efficient Technol 2:36–42. https://doi.org/10.1016/j.reffit.2016.01.001
17. Zollinger H (1991) Color chemistry. VCH
18. Bravo Díaz C (2011) Diazohydroxides, diazoethers and related species. In: Rappoport Z, Liebman JF (eds) The chemistry of hydroxylamines, oximes and hydroxamic acids. Wiley, Chichester, UK
19. Cruz GN, Lima FS, Dias LG, el Seoud OA, Horinek D, Chaimovich H, Cuccovia IM (2015) J Org Chem 80:8637–8642. https://doi.org/10.1021/acs.joc.5b01289
20. Zollinger H (1995) Diazo chemistry II. In: Aliphatic, inorganic and organometallic compounds. Weinheim, Germany, VCH
21. Zollinger H (1983) Dediazoniations of arenediazonium ions and related compounds. In: Patai S, Rappoport Z (eds) The chemistry of triple bonded functional groups. Wiley
22. Moss RA (1974) Acc Chem Res 7:421–427. https://doi.org/10.1021/ar50084a005
23. Pazo-Llorente R, Bravo-Díaz C, González-Romero E (2003) Eur J Org Chem 2003:3421. https://doi.org/10.1002/ejoc.200300183
24. Costas-Costas U, Bravo-Díaz C, González-Romero E (2003) Langmuir 19:5197–5203. https://doi.org/10.1021/la026922s
25. Losada-Barreiro S, Sánchez-Paz V, Pastoriza-Gallego MJ, Bravo-Diaz C (2008) Helv Chim Acta 91:21–34. https://doi.org/10.1002/hlca.200890009
26. Hegarty AF (1978) Kinetics and mechanisms of reactions involving diazonium and diazo groups. In: Patai S (ed) The chemistry of diazonium and diazo compounds. Wiley, NY
27. Galli C (1988) Chem Rev 88:765. https://doi.org/10.1021/cr00087a004
28. Cuccovia IM, da Silva MA, Ferraz HM, Pliego Jr JR, Riveros JM, Chaimovich H (2000) J Chem Soc Perkin Tans 2:1896. https://doi.org/10.1039/b003079l
29. García Martínez A, de la Moya Cerero S, Osío Barcina J, Moreno Jiménez F, Lora Maroto B (2013) Eur J Org Chem 6098–6107. https://doi.org/10.1002/ejoc.201300834
30. Ussing BR, Singleton DA (2005) J Am Chem Soc 127:2888. https://doi.org/10.1021/ja043918p
31. Bravo-Diaz C (2009) Mini-Rev Org Chem 6:105–113. https://doi.org/10.2174/157019309788167693
32. Bentley TW, Ryu ZH (1994) J Chem Soc Perkin Trans 2:761. https://doi.org/10.1039/P29940002531
33. Pazo-Llorente R, Maskill H, Bravo-Díaz C, González-Romero E (2006) Eur J Org Chem 2006:2201. https://doi.org/10.1002/ejoc.200500946
34. Hartley GS (1938) J Chem Soc 633. https://doi.org/10.1039/JR9380000633
35. Hantzsch A, Werner A (1890) Ber Dtsch Chem Ges 23:11. 443.webvpn.fjmu.edu.cn/https://doi.org/10.1007/978-3-642-99003-8_13
36. Costas-Costas U, Gonzalez-Romero E, Bravo-Díaz C (2001) Helv Chim Acta 84:632–648. https://doi.org/10.1002/1522-2675(20010321)84;3%3c632::AID-HLCA632%3e3.0.CO;2 0
37. Hanson P, Jones JR, Taylor AB, Walton PH, Timms AW (2002) J Chem Soc Perkin Trans 2:1135
38. Canning PSJ, Mccrudden K, Maskill H, Sexton B (1999) J Chem Soc, Perkin Trans 2(12):2735. https://doi.org/10.1039/A905567C
39. González-Romero E, Malvido-Hermelo B, Bravo-Díaz C (2002) Langmuir 18:46. https://doi.org/10.1021/la010938l
40. Pazo-Llorente R, Bravo-Díaz C, González-Romero E (2004) Eur J Org Chem 2004:3221. https://doi.org/10.1002/ejoc.200400170
41. Fernandez-Alonso A, Bravo-Diaz C (2010) J Phys Org Chem 23:938. https://doi.org/10.1002/poc.1730

42. Pazo-Llorente R, Bravo-Diaz C, Gonzalez-Romero E (2003) Langmuir 19:9142. https://doi.org/10.1021/la034879i
43. Fernández-Alonso A, Bravo-Diaz C (2010) Helv Chim Acta 93:877. https://doi.org/10.1002/hlca.200900322
44. Crossley ML, Kienle RH, Benbrook CH (1940) J Am Chem Soc 62:1400–1404. https://doi.org/10.1021/ja01863a019
45. Fernández-Alonso A, Bravo-Diaz C (2008) Org Biomol Chem 6:4004–4011. https://doi.org/10.1039/B809521C
46. González-Romero E, Fernández-Calvar MB, Bravo-Díaz C (2002) Langmuir 18:10311. https://doi.org/10.1021/la026312s
47. Bravo-Díaz C, González-Romero E (2003) Electroanalysis 15:303–311. 1040-0397/03/0402-0303
48. Bravo-Díaz C, González-Romero E (2003) Electrochemical behavior of arenediazonium ions. New trends and applications. In: Current Topics in Electrochemistry. Research Trends, Trivandrum, India
49. Pastoriza-Gallego MJ, Losada-Barreiro S, Bravo Díaz C (2012) J Phys Org Chem 25:908–915. https://doi.org/10.1002/poc.2949
50. Fry AJ (1978) Electrochemistry of the diazo and diazonium groups. In: Patai S (ed) The chemistry of Diazo and Diazonium Groups. Wiley, NY
51. Viertler H, Pardini VL, Vargas RR (1994) The electrochemistry of triple bond. In: Patai S (ed) The chemistry of triple-bonded functional groups, supplement C. Wiley, NY
52. Zuman P (1969) Physical organic polarography. In: Zuman P, Perrin CL (eds) Organic polarography. Wiley, NY
53. Sienkiewicz A, Szymulaa M, Narkiewicz-Michaleka J, Bravo-Díaz C (2014) J Phys Org Chem 27:284–289. https://doi.org/10.1002/poc.3194
54. Lowry TH, Richardson KS (1987) Mechanism and theory in organic chemistry. Harper-Collins Pub, New York
55. Fernández-Alonso A, Pastoriza-Gallego MJ, Bravo-Diaz C (2010) Org Biomol Chem 8:5304–5312. https://doi.org/10.1039/c0ob00143k
56. Doyle MP, Nesloney CL, Shanklin MS, Marsh CA, Brown KC (1989) J Org Chem 54:3785–3789. https://doi.org/10.1021/jo00277a009
57. Costas Costas U, Bravo-Díaz C, González-Romero E (2005) Langmuir 21:10983–10991. https://doi.org/10.1021/la051564p
58. Costas-Costas U, Bravo-Díaz C, González-Romero E (2004) Langmuir 20:1631–1638
59. Pastoriza-Gallego MJ, Fernández-Alonso A, Losada-Barreiro S, Sánchez-Paz V, Bravo-Diaz C (2008) J Phys Org Chem 21:524–530. https://doi.org/10.1002/poc.1289
60. Losada-Barreiro S, Sánchez-Paz V, Bravo-Díaz C (2007) Helv Chim Acta 90:1559–1573. https://doi.org/10.1002/hlca.200790163
61. Losada-Barreiro S, Bravo-Diaz C (2009) Helv Chim Acta 92:2009–2023. https://doi.org/10.1002/hlca.200900080
62. Jaszczuk K, Dudzik A, Losada-Barreiro S, Szymula M, Narkiewicz-Michalek J, Bravo-Díaz C (2016) J Phys Org Chem 29:586–593
63. Dudzik A, Jaszczuk K, Losada-Barreiro S, Bravo-Díaz C (2017) New J Chem 41:2534–2542. https://doi.org/10.1039/C6NJ03670H
64. Wittwer R, Zollinger H (1954) Helv Chim Acta 37:1954. https://doi.org/10.1002/hlca.19540370707
65. Lewis ES, Surh H (1958) J Am Chem Soc 80:1367. https://doi.org/10.1021/ja01539a023
66. Brown KC, Doyle MP (1988) J Org Chem 53:3255–3261. https://doi.org/10.1021/jo00249a021
67. Zollinger H (2003) Color chemistry. In: Syntheses, properties, and applications of organic dyes and pigments, 3rd revised edn. Wiley-VCH Verlag, Zürich. https://doi.org/10.1002/anie.200385122 Is this book the vsame as ref 17
68. García-Meijide MC, Bravo-Díaz C, Romsted LS (1998) Int J Chem Kin 30:31–39. https://doi.org/10.1002/(SICI)1097-4601(1998)30:1<31::AID-KIN4>3.0.CO;2-V

69. Quintero B, Morales JJ, Quirós M, Martínez-Puentedura MI, Cabeza MC (2000) Free Radic Biol Med 29:464–479. https://doi.org/10.1016/s0891-5849(00)00321-x
70. Laali KK, Gettwert VJ (2001) J Fluor Chem 107:31–34. https://doi.org/10.1016/S0022-1139(00)00337-7
71. Chaudhuri A, Loughlin JA, Romsted LS, Yao J (1993) J Am Chem Soc 115:8351–8361. https://doi.org/10.1021/ja00071a050
72. Bravo-Díaz C, González-Romero E (2003) J Chromatogr A 989:221–229. https://doi.org/10.1016/S0021-9673(03)00170-5
73. Yasui S, Nakamura K, Ohno A (1984) J Org Chem 49:878–882. https://doi.org/10.1021/jo00179a024
74. Hanson P, Hammond RC, Goodacre PR, Purcell J, Timms AW (1994) J Chem Soc Perkin Trans 1(2):691–696. https://doi.org/10.1039/P2994000069
75. Pazo-Llorente R, Bravo-Díaz C, González-Romero E (2001) Fresenius J Anal Chem 369:582–586. https://doi.org/10.1007/s002160000694
76. Gunaseelan K, Romsted LS, González-Romero E, Bravo-Díaz C (2004) Langmuir 20:3047–3055. https://doi.org/10.1021/la0354279
77. Hanson P, Hammond RC, Goodacre PR, Purcell J, Timms AW (1994) J Chem Soc, Perkin Trans 1. 2:691–696. https://doi.org/10.1039/P2994000069
78. Gunaseelan K, Romsted LS, Pastoriza Gallego M-J, González-Romero E, Bravo-Díaz C (2006) Adv Colloid Interface Sci 123–126:303–311. https://doi.org/10.1016/j.cis.2006.05.007
79. Gunaseelan K, Romsted LS, González-Romero E, Bravo-Díaz C (2004) Langmuir 20:3047–3055. https://doi.org/10.1021/la0354279
80. Scaiano JC, Kim-Thuan N, Leigh WJ (1984) J Photochem 24:79–86. https://doi.org/10.1016/0047-2670(84)80009-X

Iodonium Salts as Reagents for Surface Modification: From Preparation to Reactivity in Surface-Assisted Transformations

Olga Guselnikova, Natalia S. Soldatova, and Pavel S. Postnikov

Abstract The chapter is dedicated to the modern trends in surface modification by iodonium salts as an alternative to the common diazonium chemistry. The iodonium salts with higher reduction potential commonly are not prone to the spontaneous reaction with surfaces that provide unique opportunities for the precise control of the grafting process. This chapter will include general information about the structure, reactivity, and preparation of iodonium salts, as well as a full overview of surface functionalization using iodonium salts.

1 Introduction

Surface chemistry drives the never-ending quest for new functional materials with controlled properties in modern materials science and technology [1–3]. Indeed, the diazonium chemistry can be considered one of the most investigated areas with tons of contributions in the functionalization of different materials. From 1992 [4], aromatic diazonium salts (DSs) occupied a special place in the covalent modification of surfaces due to the high reactivity, availability, and covalent nature of bonding of aryl groups to the surface. Nevertheless, the advantages of DSs sometimes become drawbacks. The high reactivity of DSs due to the low reduction potential and the tendency to spontaneously react with the surface sufficiently hamper the fine-tuning of surface features—possibility of patterning and control of layer thickness.

O. Guselnikova · N. S. Soldatova · P. S. Postnikov (✉)
Research School of Chemistry and Applied Biomedical Science, Tomsk Polytechnic University, 634050 Tomsk, Russian Federation
e-mail: postnikov@tpu.ru

O. Guselnikova
JST-ERATO Yamauchi Materials Space-Tectonics Project and International Center for Materials Nanoarchitectonics (WPI-MANA), National Institute for Materials Science (NIMS), 1-1 Namiki, Tsukuba 305-0044, Ibaraki, Japan

P. S. Postnikov
Department of Solid State Engineering, University of Chemistry and Technology, 16628 Prague, Czech Republic

M. M. Chehimi et al. (eds.), *Aryl Diazonium Salts and Related Compounds*, Physical Chemistry in Action, https://doi.org/10.1007/978-3-031-04398-7_4

From such a perspective, the iodonium salts (ISs) (diaryl- or arylalkynyl) are the better alternative. The reduction potential of ISs is higher than for appropriate diazonium derivatives that provide fine control for the surface modification and structure of grafted layers. Especially last years, such reactivity of ISs deserved special attention due to the discovery of novel stimuli for the homolysis of C–I bond.

In this chapter, we summarized the accumulated knowledge about the reactivity of ISs in surface functionalization and highlighted the ongoing trends in the application of such compounds in the design of smart materials.

2 Iodonium Salts: Preparation and Reactivity

ISs are typical λ^3-iodanes consisting of iodonium cation ($R_1R_2I^+$) and various anions (halides, borates, sulfonates, etc.) [5–8]. Commonly, ISs are utilized for vinylation, arylation, and alkynylation of a wide scope of nucleophiles (*C*, *N*, *S*, *P*, etc.) [7–12]. ISs are also applied for photocationic polymerization [13–15]; in chemistry of halogen bonds [16–20], catalysis [21, 22], and study of biological activity [23]. The high reactivity of ISs can be explained by the properties of iodoarenes as perfect leaving groups (similarly to N_2 in diazonium salts). Thus, the iodobenzene is hypernucleofuge according to the range of reports [24–26]. Release of iodobenzene leads to the homolysis of C–I bond via the reductive process with the generation of aryl radicals, which further attack the surface.

ISs were firstly prepared via condensation of iodosobenzene and iodoxybenzene in alkaline solution by Hartmann and Meyer in 1894 [27]. After this, Willgerodt made a huge contribution to the synthesis of new ISs [28]. In the 1950s, F. Beringer et al. became the founder of modern preparation approaches based on utilization of hypervalent iodine (III) and (V) compounds (HIC(III) or HIC(V)) in reactions with *C*-ligand source, but only G.F. Koser et al. provided a fresh breath to ISs's chemistry in 1980s.

The first approach comes down to the usage of prepared HIC(III) (di(acetoxy)iodobenzene (PIDA), [bis(trifluoroacetoxy)iodo]benzene (PIFA), Koser's reagent and iodosobenzene) and rarely HIC(V). Among the presented, HIC(III) PIDA and iodosobenzene are less active in electrophilic substitution and required the addition of strong acids (TfOH or $BF_3{\cdot}Et_2O$) (Fig. 1). Alkynes, alkenes, and arenes can participate in the reaction as a source of the second *C*-ligand. Additionally, corresponding boronic acids, trialkylsilyl, and trialkylstannanes are utilized for specific regioselectivity purposes or increased reactivity in reaction with HIC(III). These derivatives exclusively provide the synthetic pathway for the preparation of ISs bearing electron-withdrawing substituents in both aryl rings.

The choice of an appropriate synthetic scheme is an important question, which materials scientists should consider before application (Fig. 1). For instance, the anion in the structure of ISs determines solubility and reactivity. The most easy-to-use ISs contain weak nucleophilic anions, which increase solubility—triflates, trifluoroacetates, and tetrafluoroborates. Another significant issue is the choice

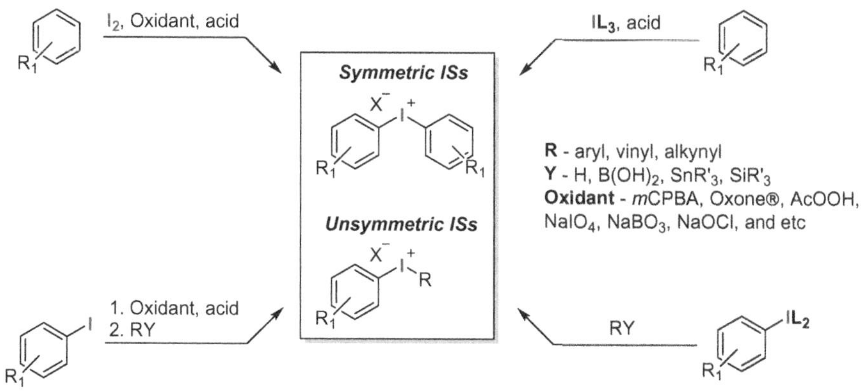

Fig. 1 The main synthetic routes to the preparation of ISs

between symmetric or unsymmetric ISs. In the case of unsymmetric ISs, the selectivity of C–I bond homolysis significantly depends on the electronic and steric effects of substituents. However, for radical processes, the selectivity control is still challengeable; therefore, symmetric ISs are preferable. Despite the clear way of homolysis of symmetric ISs, in contrast to unsymmetric ones, their preparation is often complicated, required specific reagents, and is low yielded, especially in case electron-withdrawing substituents as $-NO_2$ or CF_3-groups.

Nevertheless, ISs can be easily prepared from available substrates. In contrast to the high explosibility of DSs, ISs advantage from the explosion safety and perfect storage stability. These features provide promising opportunities for the handling of ISs in surface chemistry.

3 Surface Modification via Decomposition of Iodonium Salts

3.1 *Iodonium Salts as Modification Agent: General Insight*

ISs as modification agents draw the attention of materials scientists because they could facilitate the wettability variation [29], unique electronic properties [30], sensing [31], and new centers for more advanced functionalization [32]. The exciting performance is delivered through the generation of radicals via the activation of IS and further surface attack. The primary step here is the activation of ISs (Fig. 2), which can be performed via surface-assisted [33] electrochemical [34] and chemical [35] reduction of ISs; light-induced or plasmon-assisted decomposition [36] of IS.

The closest well-studied analogue of ISs as modification agent is diazonium salts (DSs). Modern materials scientists are faced with a choice between ISs and DSs in

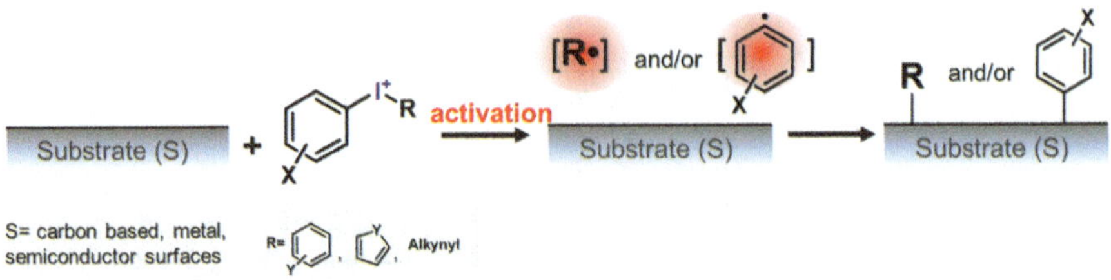

Fig. 2 Principal scheme of surface functionalization using ISs

each special case of application [37]. Therefore, for a better understanding of the ISs grafting details, mechanism, and application, through the whole chapter, we have to compare them from the point of view of (i) structure of resulting layers; (ii) activation methods; and (iii) reported modified substrates (Table 1).

DSs and ISs both graft to the surface with the formation of covalent bonds, which were observed by Raman spectroscopy for DSs grafted on gold nanoparticles [38]. Daasbjerg's group first evidenced the formation of the covalent bonds by ISs using ToF SIMS measurements on glassy carbon electrode (GCE): except the signals from organic layers, in Time-of-Flight Secondary Ion Mass spectra they found $[C_2–C_6H_4NO_2]^-$, $[C_2O–C_6H_4NO_2]^-$, $[O–C_6H_4NO_2]^-$ and other fragments confirming the strong and covalent attachment ISs-derived radicals [34]. DS- and IS-derived layers demonstrated excellent stability toward vigorous ultrasonication [39, 40].

The primary difference between ISs in comparison with DSs is in more negative reduction potential [37, 41]. The reduction potentials of ISs, DSs, and rarely used stable sulfonium salts (SSs) were compared during the modification of the gold surface by exploiting the potential gradient method] [42]. It was found that reduction potentials are −0.32 V versus SCE for DSs, −0.88 V versus SCE for ISs, and −1.39 V versus SCE for SSs. In general, DSs are easier to reduce than the corresponding ISs which, are more easily reduced than SSs. This divergence leads to the improved control of the surface grafting—the surface coverage can be varied by different modification times, the concentration of ISs, or the number of cyclic voltammetry scans [35, 43, 44]. Commonly, ISs produce close to monolayer structures, and only a few examples of additional attacks of grafted aryl layer are reported [37, 45], while DSs produce multilayer structures with thickness up to 10 nm [46, 47]. The azo moieties can also explain the increased thickness of these films in film structure due to the mechanism of diazonium group reduction [48], ISs are deprived of this feature.

The ISs have a critical feature in comparison with DSs: the iodonium center contains two substituents (aryl, vinyl, or alkynyl), which can be represented by identical functional groups (symmetric salts) or different (unsymmetric). Unsymmetric ISs are especially attractive because they are easier to prepare and inherently less wasteful [49]. The variety of available functional groups is much larger for ISs than for DSs: in contrast to the solely aromatic structure of DSs, one part of unsymmetric ISs can consist of alkynyl and heterocyclic as well [34, 35, 39, 50–53]. Therefore, unsymmetric ISs provide the possibility to prepare mixed layers on the surface from

Table 1 Comparison of basic parameters for DSs and ISs modification

Diazonium salts (DSs)	Iodonium salts (ISs)
Structure of layers	
Covalent nature of grafting—stability	
Poor control of grafting density	Better control of grafting
Mixed layers can be formed only using a two DS molecules	Mixed layers are possible via single IS molecule
Only aromatic	Aromatic, alkynyl, alkyl, heterocyclic
Multilayer	Close to monolayer
–N=N– bonds are common	Only polyphenylene
Activation	
Spontaneous reaction is common	Spontaneous reaction is not common [a]
Electrochemical	Electrochemical (more negative potentials)
Light[b]	Light (w and w/o photo activator)
Plasmon excitation	Plasmon excitation
Chemical activators are common [c]	Chemical activators are only in case of carbon [d]
Substrates	
Carbon-based materials	Carbon-based materials
Wide range of metals and metal oxides	Au, Pt
Semiconductors (BP, Si, GaS, MXene, TMDCs)	Semiconductors (ITO, FTO, BP)
Polymers	Surface modification of polymers is a still unknown
Patterning (electro, photo, AFM nanoshaving, soft lithography)	Patterning (photo, electro)

[a]Black phosphorous
[b]Photoredox catalyst is required
[c]For both substrate and DSs
[d]Activation of surface (not ISs)

single-molecule, simultaneously raising questions about the selectivity of modification. In the case of electrochemical modification, EWGs mostly attach to the surface [39, 50] while in the case of plasmon-induced grafting—the opposite trend is observed [51]. Simultaneously, for nucleophilic attack of carbon-based materials, no significant selectivity was observed [52]; therefore, this question is still under consideration. Diazonium modification also can provide multi-component layers using the mixture of different DSs, the recent achievements are accumulated by Gooding in [54]. Commonly, the surface ratio is dominated by the species with a more positive reduction potential.

The list of substrates modified by DSs is extremely large and includes planar and nanostructured surfaces, novel layered materials, and colloidal nanoparticles consisting of carbon allotropes, metal, metal oxides, semiconductors, polymers, and composites. However, iodonium modification is relatively recently known, and mostly planar surfaces have been studied (carbon, gold, and platinum, semiconductors), while only a few nanoobjects—nanofilm coated optical fiber and gold nanorods—up to now have been modified by ISs.

3.2 Electrochemical Modification

3.2.1 Carbon-Based Substrates

Firstly, Kim Daasbjerg et al., in this breakthrough study, performed grafting of symmetric IS bearing 3-nitrophenyl substituents by applying the negative potentials [53]. Further, in 2007 the same group started to explore the mechanistic aspects, firstly utilizing a range of unsymmetric ISs [34]. Comparing DSs with ISs concludes that surface coverages are up to ~25% lower for ISs than for DSs procedure. The further focus moved to the understanding of mechanistic aspects of ISs modification, where Daasbjerg proposed to use a range of unsymmetric ISs to cover substituents going over electron-withdrawing groups (EWG)—for example, $-NO_2$—to electron-donating (EDG)—for example, –OMe groups.

They studied the electrochemical grafting of ISs **1**, **2**, **3**, where two possible mechanisms can be realized (Fig. 3a). While pathway A leads to a substituted aryl radical formation, pathway B produces the corresponding phenyl radical (Fig. 3b). Authors demonstrate that whereas the first route dominates in the case of EWGs, a mechanistic crossover takes place when going to EDGs.

In further work [50], authors used quantum chemistry for mechanistic study—they offer a theory that the first single-electron uptake rules the energetics of the electrochemical reduction. Therefore, they assume that the electronic effects of substituents determine selectivity, EWGs in aryl ring led to the weakening of $C_{Ar}-I$; thus, C_{Ar} groups are prevailing on the surface (Fig. 3b). They also found a unique relation between experimental reduction potentials and theoretical vertical electron affinities, even embodying the electrochemical reduction process of grafted moieties.

Graphene occupies a special place among other carbon-based materials due to its unique optical and physic-chemical properties [55, 56]. Chemical functionalization is required to adapt graphene's properties to many applications [57]. Diazonium modification is often applied for these purposes - spontaneous dissociation or electrochemical grafting leads to the covalent bonding of the resulting radical with a carbon atom in the graphene lattice [58, 59]. However, most covalent functionalization approaches are spontaneous or defect-driven and are not suitable for applications requiring

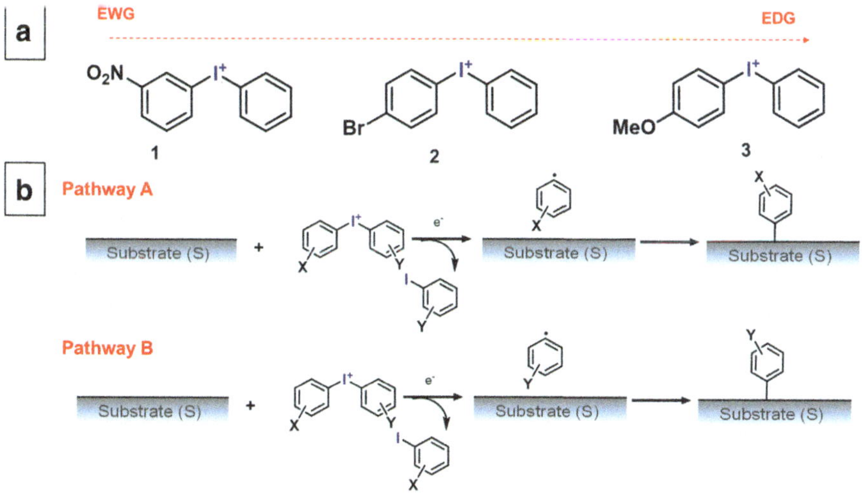

Fig. 3 **a** Structure of ISs **1–3**, and **b** possible mechanistic pathways for unsymmetric ISs grafting

directed assembly of molecules on graphene [43]. Therefore, surely ISs with higher reduction potentials logically came to this substrate. Firstly, Chan electrochemically functionalized epitaxial graphene on SiC(0001) with trifluoromethylphenyl-derived symmetric IS [43]. This IS was chosen due to the presence of unreactive marker element fluorine. Shifts in the F 1 s and C 1 s core-level positions observed in XPS indicate a redistribution of electronic states toward the $CF_3C_6H_4$–graphene covalent bond. The rehybridization of sp^2 to sp^3 states was also observed as an increase in C σ-bonding orbitals' density with respect to the C π-bonding orbitals. The same controlled character of modification was observed for 4-nitrophenyl ISs, where –NO_2 offers the advantage of being probed directly using electrochemical methods to quantify the grafting density. Stevenson showed first-time STM images with an atomic resolution of covalently modified graphene [43, 60].

Electron transfer kinetic studies have been performed using graphene samples with high and densities of single point defects (vacancies). There is no evidence for preferential attachment at the defect sites [61]. Stevenson also showed that diaryliodonium chemistry is a versatile method to selectively modify and induce changes in electronic and magnetic properties of graphene: 4-nitrophenyl radicals attachment gives rise to ferromagnetic properties [45].

The most recent research on comparing DS and IS modification of graphene was performed by De Feyter group [37]. They used highly ordered pyrolytic graphite (HOGP) and 4-nitro-containing DS and IS as a model substrate. They revealed that electrochemical modification using IS proceeds efficiently, leading to faster blocking of HOGP surfaces compared to DS. Raman spectroscopy clearly showed that the grafting efficiency is higher in the case of the IS. This observation was supported by molecular-resolution scanning tunneling microscopy measurements. XPS data show more enhanced electrochemical reduction of the nitro group to amino for DS rather than IS.

3.2.2 Metals and Semiconductors

Apart from carbon-based materials, gold and platinum surfaces were modified by ISs for the selective patterning of organic molecules [62, 63]. DSs spontaneously react with most metal surfaces [64], while ISs do not have this disadvantage. Control experiments were carried out without potential, and electrolytes failed to generate an organic film deposition. ISs can assemble on platinum and gold selectively. The local electrografting was imaged by XPS and scanning electrochemical microscopy, as the organic patterns block the electron transfer to aqueous redox probes. Other authors further utilized this crucial advantage of ISs for more complicated patterning [36].

Indium tin oxide (ITO) and fluorine-doped tin oxide (FTO) are essential materials for optoelectronics, photovoltaics, and other electro-optical and insulating applications [65], which were electrochemically modified by di(4-nitrophenyl)-iodonium IS [44, 66]. In contrast to the gold or carbon surface, where the formation of Au–C or C–C bonds was observed, XPS data reveal that electrochemically generated radicals do not bond with indium or tin; grafting occurs either through bonding to surface hydroxyl groups or through strong physisorption on ITO. As a result, electrochemical impedance spectroscopy showed that the electrochemical reduction gives a better isolation effect than an FTO electrode with a compact blocking layer on the TiO_2 surface.

In general, electrochemical approaches for IS grafting are well-studied and understood; it allows quite a precise tuning of the layer thickness via the variation of IS concentration and the number of cycles. The main disadvantage is the complicated experimental setup, a limited number of substrates, and the impossibility of selective grafting through the surface on the micro and nano level. Therefore, some alternative methods were appearing for the next 5 years.

3.3 Nucleophilic Substitution—A Straight Way for Modification of Carbon Allotropes

Further, electrochemical methods for carbon allotropes functionalization were superseded by nucleophilic substitution reaction (Fig. 4). Hirsch et al. firstly demonstrated the interaction of single-wall carbon nanotube (SWCNT) and synthetic spherical graphite with ISs in the presence of molten potassium heated at 150 °C for 8 h in the inert atmosphere [33]. Further, such investigations focused on graphene modification via the activation by solvated NaK_3 or KC_8 [67, 68]. In these cases, after treatment by a reducing agent, the graphene surface became negatively charged,

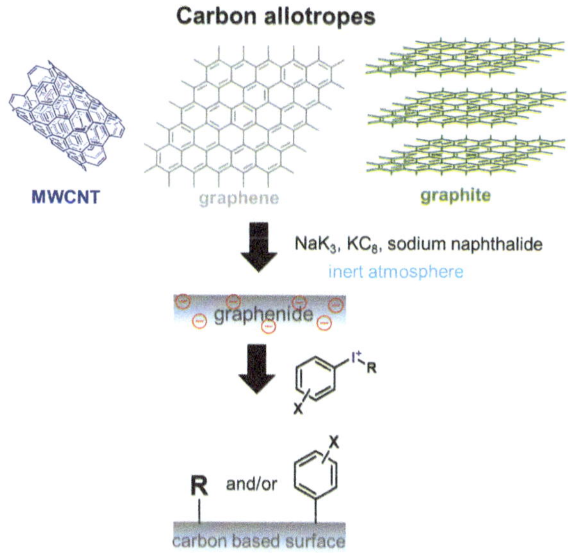

Fig. 4 Modification of carbon allotropes via nucleophilic substitution

forming "graphenides." These graphenides can actively interact with suitable electrophiles such as ISs. Hirsch proposed that due to the reductive activation, the Fermi level of graphene is shifted, and the reactivity toward radical reactions increases.

A similar approach was used by Swager but with a milder reducing agent (sodium naphthalide) at room temperature in a tolerant way to a variety of substituents (including EDG and EWG arenes, esters, and heteroarenes) [52]. However, it still requires an inert atmosphere and glovebox operation mode. Swager et al. performed a systematic study focused on the functionalization of carbon nanotubes (SWCNTs, DWCNTs, and MWCNTs, respectively), graphene [52], and graphite using both symmetric and unsymmetric ISs. The grafting selectivity was relatively low; the only preferential transfer of phenyl groups containing EWGs was demonstrated. The authors also compared the efficiencies of functionalization for ISs and DSs. ISs possess clear advantages over DSs due to the higher degree of covalent functionalization according to Raman, TGA, XPS, and UV-vis-NIR simultaneously without the need for stoichiometric or excess amount of the modification reagent. Slightly later, this method was adopted to design chemo resistive sensors with voltage-activated sensitivity to detect CO based on the SWCNTs [31]. SWCNTs were functionalized by pyridyl ligands for further coupling with iron porphyrins. Another exciting application was demonstrated for the evaluation of aryl group migration across graphene surfaces [69].

ISs can be used for surface chemistry in an alternative way. Thus, it was recently reported that iodoxoboroles could serve as a reagent for the formation of reactive aryne intermediates in mild conditions [26]. The aryne species can interact with unsaturated C=C bonds in graphene structure. Postnikov and Boukherroub implemented such an approach to modify reduced graphene oxide (rGO) via benzyne cycloaddition reaction [30]. Iodonium-induced treatment of rGO resulted in improved capacitance of electrode materials due to the low degree of functionalization conserving the conductivity and the increase of interlayer distance between agglomerated graphene layers providing enhanced transport property for electrolyte ions in the material.

3.4 Iodonium Salts—Spontaneous Modification?

DSs with relatively low reduction potentials are well known as reagents for modification of metals, metal oxides, and carbon surfaces [59], while ISs with the higher reduction potential need additional stimuli for the generation of radical and further surface attack [37]. Daasbjerg et al. in one of the first contributions claimed that IS grafting on GCE may occur spontaneously—the surface coverage extracted from cyclic voltammograms was $\approx 0.3 \times 10^{10}$ mol $\times$ cm^{-2} in contrast to electrochemical $\approx 4 \times 10^{10}$ mol $\times$ cm^{-2} [34]. However, this value was within the deviation of measurement and was not confirmed by further studies.

In 2018 Collins [70] and in 2020 Sofer [71] demonstrated the applicability of ISs for the covalent functionalization of few-layer black phosphorus (BP) (Fig. 5a), which recently has gained popularity due to its high carrier mobility and tunable bandgap, however, suffer from the tendency to oxidation and degradation [72]. They demonstrated that ISs could interact with exfoliated few layers of BP at room temperature without any additional activation or stimuli. Therefore, this can be considered the first example of spontaneous modification by ISs, confirmed by the detailed XPS and FTIR analysis (Fig. 5b). Unfortunately, the authors did not demonstrate stability toward heating or ultrasound as was done to confirm covalent grafting with DSs [40]. Collins and Sofer proposed that grafting can occur via arylation of both *O*- and *P*-nucleophiles and explained this reactivity by the excellent leaving ability of aryl iodide (Fig. 5a).

However, it should be noted that this explanation could not be considered as straightforward as other authors used the same ISs or with similar reactivity [29, 43, 51, 52] for the functionalization of gold and carbon surface, and no signs of spontaneous modification were detected. Therefore, probably this reactivity originates from the nature of BP and its unique electronic structure. This assumption was further confirmed by study of reactivity ISs in the interaction with typical 2D films of 1T' (metallic) and 2H (semiconducting)-MoTe2 with different electronic structure [73]. While the functionalization of the metallic MoTe2 occurs spontaneously, the semiconducting MoTe2 requires activation by light with the formation of covalent Te-C bond (confirmed by X-ray photon spectroscopy). However, the establishment of exact mechanism of spontaneous modification using ISs deserved further detailed studies.

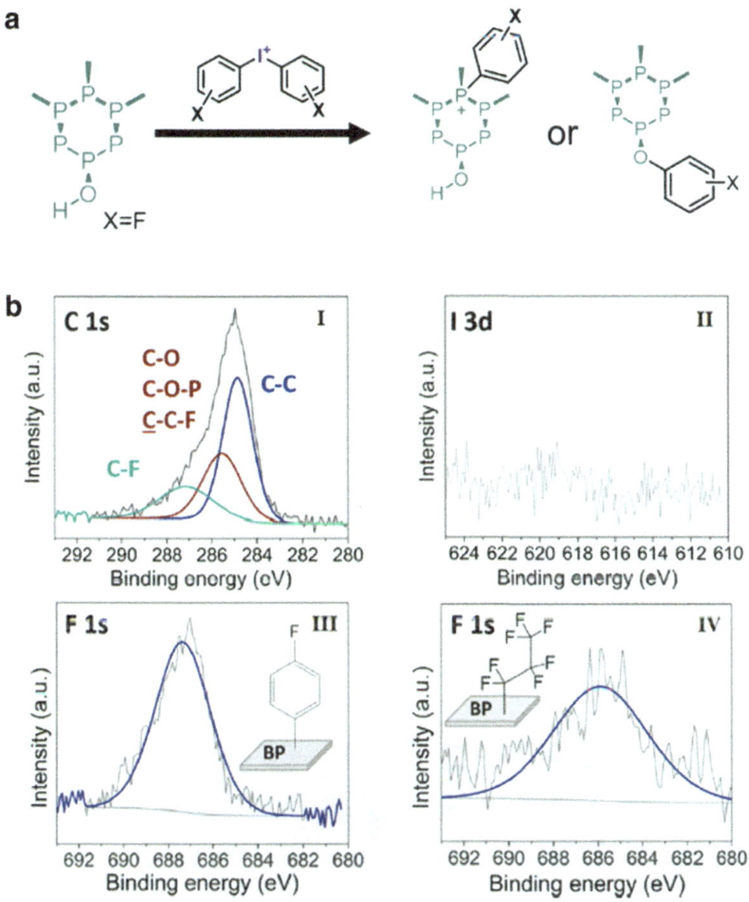

Fig. 5 **a** Aryl functionalization of BP using ISs through the *P* site and the *O* sites, **b** XPS analysis of functionalized BP. The C 1 s (I) displays three components due to the presence of C–C; C–O, C–O–P, and C–C–F; and C–F. (II) The I_{3d} and (III) F 1 s core-level spectra and (IV) F1s for functionalized BP. (Adapted with permission from [70]. Copyright (2018) American Chemical Society)

3.5 *Light as a New Stimulus for ISs Grafting*

ISs were applied for the pattering of the platinum surface using electrochemical modification [62], while photolithography approaches are commonly overperforming due to the possibility of multiple patterning from a single mask, permitting production at the same time a large number of patterned surfaces as in the microelectronic industry [74, 75]. This inspired Pinson and Deronzier to develop a photolithography procedure for gold surfaces (Figs. 6a, 7a) [36]. Also, the photoinduced reduction of ISs with dyes has been demonstrated since the end of the 1970s, [76] therefore

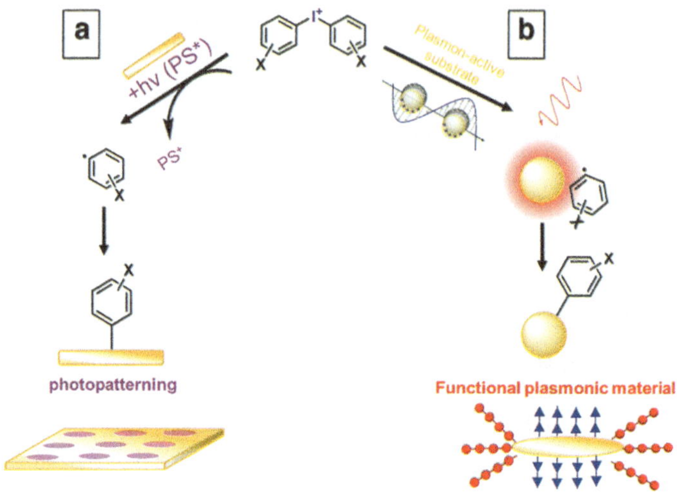

Fig. 6 UV light (**a**) and plasmon-induced (**b**) grafting of ISs onto the gold surfaces

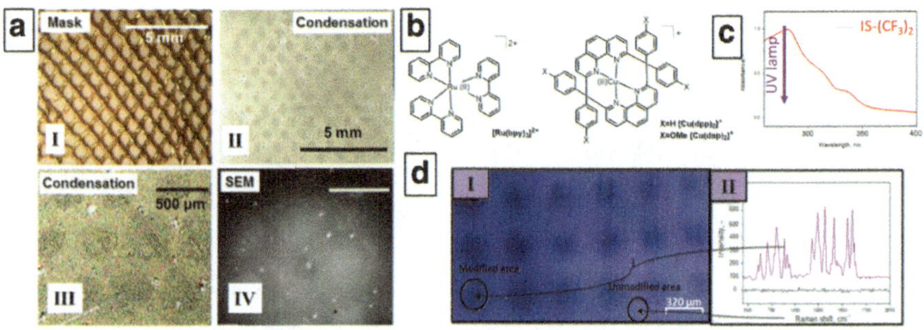

Fig. 7 **a** Patterning on a Au surface by photografting of ISs. (I) Optical image of mask. (II, III) Condensation image of patterned Au surface. (IV) SEM images of the patterned Au surface, **b** Chemical structures of PSs, ; **c** UV-vis spectrum of IS-$(CF_3)_2$, **d** (I) Confocal image of the patterned Au film by IS-$(CF_3)_2$ and UV light, (II) Raman spectra in the modified region and out of the modified region. (Adapted with permission from [36]. Copyright (2018) American Chemical Society)(Adapted with permission from [29]. Copyright (2019) Elsevier B.V.)

main principles were clear. Authors performed pattering of gold film using photosensitizes (PSs) $[Ru(bpy)_3]_2^+$; $[Cu(dpp)_2]^+$; $[Cu(dap)_2]^+$ (Fig. 7b) and 4-nitrophenyl-2,4,6-trimethylphenyliodonium salt. PS upon irradiation with visible light undergoes the excited state and forms a radical from ISs, which further reacts according to the previously described mechanisms (Fig. 6).

Further, Guselnikova et al. demonstrated that 3,5-bis(trifluoromethyl)phenyl)iodonium salt (IS-$(CF_3)_2$) can be grafted to the gold through activation by UV light without the addition of any PSs [29]. Because the maximum absorption of symmetric IS-$(CF_3)_2$ is at 250–300 nm and well-overlapped with the lamp emission wavelength (254 nm), the UV-irradiation is generating aryl radicals from IS-$(CF_3)_2$ (Fig. 7c). The attachment of fluorinated species leads to surface hydrophobicity with a water contact angle of 105°. The developed method was applied for simple one-step surface micrometer-sized patterning of the gold film with organic functional groups (Fig. 7d). The proposed strategy offers advantages such as strong covalent bonding of aryl radical, easiness of experimental procedure combined with the possibility to control surface architecture.

The light is an extremely attractive activator ISs grafting process; when light interacts with plasmon-active material (noble metal nanoparticles, nanofilms, etc.)—the excitation of plasmon resonance occurs [77]. A plasmon is commonly referred to as collective oscillations of an electron gas in metals, excited by an external electromagnetic wave at the metal–dielectric interface [78]. This process allows plasmonic material to interact with the surrounding media and initiate the chemical transformations [77, 79]. Consistently, Postnikov et al. recently demonstrated that interaction of gold-coated optical fiber with IS-$(CF_3)_2$ salt under laser irradiation led to the grafting of generated radicals to the gold surface (Fig. 8a) [80]. The grafting of fluorinated layers resulted in surface hydrophobicity, water repellence, and antifouling properties (Fig. 8b). The irradiation with the wavelengths out of a maximum of plasmon resonance does not give any attachment confirming the plasmonic nature of the process. They showed that utilization of 1 mM solution leads to the not closely packed layer, which is still available for further grafting providing the possibility to prepare mixed layers using another IS. The demonstrated method can be used for a wide range of ISs to obtain the tailored surface properties.

Further, Postnikov et al. moved to the study of the interaction between unsymmetric ISs and gold-coated fiber under plasmon excitation (Fig. 8c) [51]. Obviously, in the case of symmetric ISs, the regioselectivity of homolytic decomposition is not an issue. However, plasmon-induced modification using unsymmetric ISs is attractive due to the lower price and versatile synthetic procedures [49, 81]. Moreover, the decomposition of unsymmetric ISs under plasmon gives additional mechanistic insights into the interaction between plasmon energy and organic molecules [82]. Authors demonstrated unexpected regioselectivity of modification under plasmon excitation leading to the selective transfer of the EDG-bearing radical to the surface in contrast to the grafting of EWG-bearing radicals in case of electrochemical functionalization (Fig. 8c) [34, 50]. They provide three possible reaction mechanisms: initial excitation of electrons to LUMO orbital; hot electrons; and plasmonic heating [83]. Plasmonic heating was excluded by revealing the mixed layer formation under common heating compared to selective grafting under plasmon excitation. Further estimation of the mechanism was performed by quantum-chemical modeling and TDDFT calculations. The authors found that intramolecular excitation of an electron occurs followed by the regioselective cleavage of the C–I bond with the formation of EDG-bearing radicals, which cannot be explained by the hot electron excitation or thermal effects. Utilization of the excited plasmon energy leads to the change of the reaction way during the decomposition of ISs and radical generation.

Similar to ISs, DSs also can be activated by plasmon excitation, J.-Ch. Lacroix and Mangeney demonstrated that excitation of a plasmon in the place of its concentration

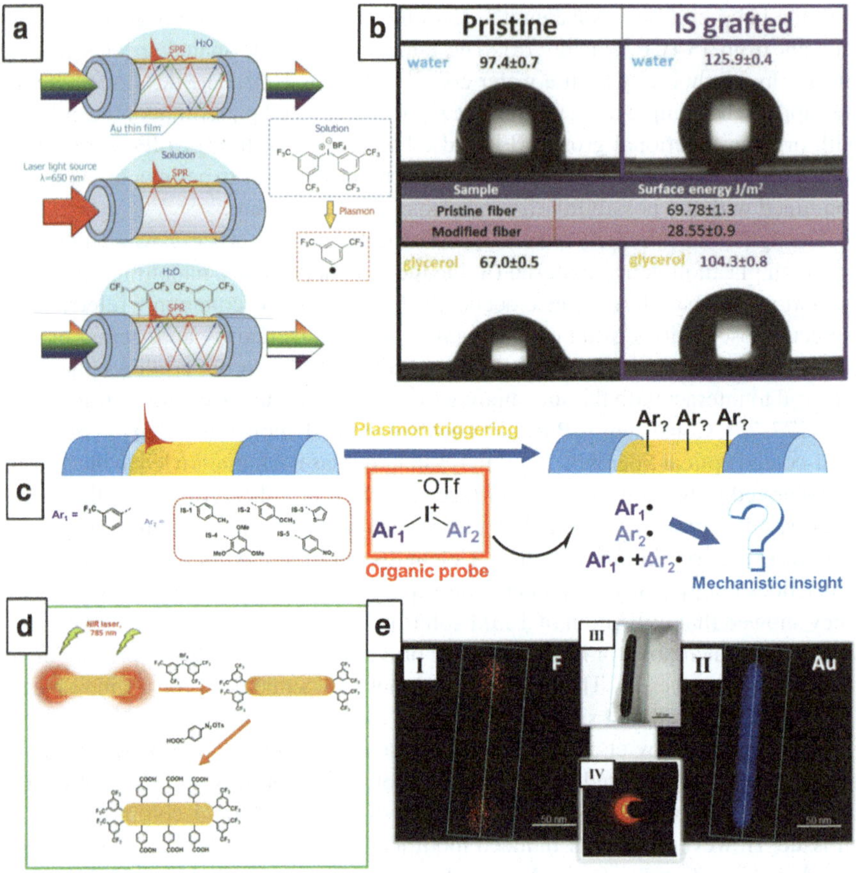

Fig. 8 **a** Schematic representation of plasmon-induced IS-$(CF_3)_2$ grafting of gold-coated optical fiber: excitation of SPR; plasmon-assisted initiation of cleavage of C–I bond and their grafting to the Au surface, **b** Water and glycerol contact angle (in °) measured on pristine and modified Au-coated fibers with values of surface energy, **c** Utilization of unsymmetrical ISs as organic probe for plasmon-induced reactions mechanisms. ; **d** plasmon-assisted grafting of AuNR tips, followed by the diazonium-based grafting of AuNR sides, **e** STEM/EDX mapping on the surface of AuNR after the plasmon-assisted grafting to the nanorod tip: (I and II) the spatial distribution of fluorine and gold, (II) the corresponding STEM image of the nanoparticle, and (IV) the numerically calculated distribution of EF near the nanoparticle tip under the illumination of a 2900 nm light (Reprinted with permission from [80]. Copyright (2018) WILEY-VCH Verlag GmbH & Co)(Reprinted with permission from [51]. Copyright (2020) American Chemical Society)(Reprinted with permission from [32]. Copyright (2020) Royal Society of Chemistry)

in anisotropic NP—edges of nano triangles or tips of nanorods in the presence of DSs leads to enhanced grafting [84]. However, again DSs spontaneously react with the gold surface; therefore, despite the advantages (fast reaction time, some extent of thickness control, and site-selective mode) [85] the degree of control is low and cannot be applied to prepare precisely modified structures.

However, further Lyutakov et al. combined plasmon-induced activation of ISs with the diazonium modification to prepare amphiphilic gold nanorods (AuNRs) with fluorinated species on the tips and carboxylated species on the other part for surface-enhanced Raman spectroscopy (SERS) biosensing (Fig. 8d) [32]. Firstly, the solution of ISs-$(CF_3)_2$ was mixed with AuNRs and irradiated with 785 nm laser wavelength to selectively excite longitudinal plasmon bands, further unoccupied part of NRs was grafted with 4-carboxyphenyldiazonium salt (Fig. 8e). Authors confirm the selective grafting by optical and STEM/EDX measurements on three types of AuNRs with various aspect ratios. The amphiphilic nature of AuNRs was used in SERS sensing: 785 nm laser was used for the selective sensing of oleophilic squalene—via interactions with fluorinated groups while 532 nm laser was used for hydrophilic catechol sensing—via interaction with carboxylic groups.

4 Conclusion and Outlook

In summary, ISs are universal modification agents for the carefully controlled grafting of organic functional groups, which can overperform other modification agent classes. However, their study has started relatively recently, and not so many substrates were involved. For example, the question of grafting of IS remains open as different activation methods lead to the contractionary results. Therefore, we believe that current challenges in the field are functionalization of layered semiconductor materials with the understanding of mechanisms of grafting and general finding of new stimuli of activation to cover such substrates as polymers, metal oxides, and composites. Moreover, we envision the active utilization of ISs for the design of sensors, optoelectronic, fillers, medical devices, and energy storage due to the perfect control of grafting and a wide range of functional groups.

References

1. Sosa-Vargas L, Kim E, Attias A-J (2017) Mater Horiz 4:570–583. https://doi.org/10.1039/C7MH00127D
2. Xu Z, Liu Y, Ren F et al (2016) Coord Chem Rev 320–321:153–180. https://doi.org/10.1016/j.ccr.2016.03.002
3. Georgakilas V, Tiwari JN, Kemp KC et al (2016) Chem Rev 116:5464–5519. https://doi.org/10.1021/acs.chemrev.5b00620
4. Delamar M, Hitmi R, Pinson J, Saveant JM (1992) J Am Chem Soc 114:5883–5884. https://doi.org/10.1021/ja00040a074
5. Yusubov MS, Yusubova RY, Nemykin VN et al (2012) Eur J Org Chem 2012:5935–5942. https://doi.org/10.1002/ejoc.201201064
6. Wirth T (eds) (2016) Hypervalent Iodine chemistry. Springer, Cham, pp 1–316. https://doi.org/10.1007/978-3-319-33733-3
7. Yusubov MS, Maskaev AV, Zhdankin VV (2011) ARKIVOC 2011:370–409. https://doi.org/10.3998/ark.5550190.0012.107

8. Merritt E, Olofsson B (2009) Angew Chem Int Ed 48:9052–9070. https://doi.org/10.1002/anie.200904689
9. Fañanás-Mastral M (2017) Synthesis (Stuttg) 49:1905–1930. https://doi.org/10.1055/s-0036-1589483
10. Aradi K, Tóth BL, Tolnai GL, Novák Z (2016) Synlett 27:1456–1485. https://doi.org/10.1055/s-0035-1561369
11. Hari DP, Nicolai S, Waser J (2018) In: PATAI'S chemistry of functional groups. Wiley, Chichester, UK, pp 1–58
12. Brand JP, Waser J (2012) Chem Soc Rev 41:4165. https://doi.org/10.1039/c2cs35034c
13. Crivello JV (2005) In: Initiators—poly-reactions—optical activity. Springer, Berlin, pp 1–48
14. Celiker T, Kaya K, Koyuncu S, Yagci Y (2020). Macromolecules. https://doi.org/10.1021/acs.macromol.0c00694
15. Villotte S, Gigmes D, Dumur F, Lalevée J (2020) Molecules 25:22–25. https://doi.org/10.3390/molecules25010149
16. Heinen F, Engelage E, Cramer CJ, Huber SM (2020) J Am Chem Soc 142:8633–8640. https://doi.org/10.1021/jacs.9b13309
17. Cavallo G, Murray JS, Politzer P et al (2017) IUCrJ 4:411–419. https://doi.org/10.1107/S2052252517004262
18. Soldatova NS, Postnikov PS, Suslonov VV et al (2020) Org Chem Front 7:2230–2242. https://doi.org/10.1039/D0QO00678E
19. Soldatova NS, Suslonov VV, Kissler TY et al (2020) Curr Comput-Aided Drug Des 10:230. https://doi.org/10.3390/cryst10030230
20. Aliyarova IS, Ivanov DM, Soldatova NS et al (2021) Cryst Growth Des 21:1136–1147. https://doi.org/10.1021/acs.cgd.0c01463
21. Heinen F, Engelage E, Dreger A et al (2018) Angew Chem Int Ed 57:3830–3833. https://doi.org/10.1002/anie.201713012
22. Zhang Y, Han J, Liu ZJ (2015) RSC Adv 5:25485–25488. https://doi.org/10.1039/c5ra00209e
23. Goldstein EJC, Citron DM, Warren Y et al (2004) Antimicrob Agents Chemother 48:2766–2770. https://doi.org/10.1128/AAC.48.7.2766-2770.2004
24. Okuyama T, Takino T, Sueda T, Ochiai M (1995) J Am Chem Soc 117:3360–3367. https://doi.org/10.1021/ja00117a006
25. Boye AC, Meyer D, Ingison CK et al (2003) Org Lett 5:2157–2159. https://doi.org/10.1021/ol034616f
26. Yoshimura A, Fuchs JM, Middleton KR et al (2017) Chem-A Eur J 23:16738–16742. https://doi.org/10.1002/chem.201704393
27. Hartmann C, Meyer V (1894) Berichte der Dtsch Chem Gesellschaft 27:1592–1599. https://doi.org/10.1002/cber.18940270286
28. Willgerodt C (1897) Berichte der Dtsch Chem Gesellschaft 30:56–58. https://doi.org/10.1002/cber.18970300110
29. Guselnikova O, Miliutina E, Elashnikov R et al (2019) Prog Org Coat 136:105211. https://doi.org/10.1016/j.porgcoat.2019.105211
30. Sviridova E, Li M, Barras A et al (2021) Electrochim Acta 369:137667. https://doi.org/10.1016/j.electacta.2020.137667
31. Savagatrup S, Schroeder V, He X et al (2017) Angew Chem-Int Ed 56:14066–14070. https://doi.org/10.1002/anie.201707491
32. Olshtrem A, Guselnikova O, Postnikov P et al (2020) Nanoscale 12:14581–14588. https://doi.org/10.1039/d0nr02934c
33. Hof F, Schäfer RA, Weiss C et al (2014) Chem-A Eur J 20:16644–16651. https://doi.org/10.1002/chem.201404662
34. Vase KH, Holm AH, Norrman K et al (2007) Langmuir 23:3786–3793. https://doi.org/10.1021/la0629227
35. Weissmann M, Baranton S, Coutanceau C (2010) Langmuir 26:15002–15009. https://doi.org/10.1021/la1024313

36. Médard J, Combellas C, Kanoufi F et al (2018) J Phys Chem C 122:19722–19730. https://doi.org/10.1021/acs.jpcc.8b06541
37. Steeno R, Rodríguez González MC, Eyley S et al (2020) Chem Mater 32:5246–5255. https://doi.org/10.1021/acs.chemmater.0c01393
38. Laurentius L, Stoyanov SR, Gusarov S et al (2011) ACS Nano 5:4219–4227. https://doi.org/10.1021/nn201110r
39. Sommerfeldt A, Pedersen SU, Daasbjerg K (2018) Electrochim Acta 261:356–364. https://doi.org/10.1016/j.electacta.2017.12.052
40. Civit L, Fragoso A, O'Sullivan CK (2010) Electrochem commun 12:1045–1048. https://doi.org/10.1016/j.elecom.2010.05.020
41. Datsenko S, Ignat'ev N, Barthen P, et al (1998) Zeitschrift für Anorg und Allg Chemie 624:1669–1673. https://doi.org/10.1002/(SICI)1521-3749(199810)624:10%3c1669::AID-ZAAC1669%3e3.0.CO;2-P
42. Koefoed L, Pedersen SU, Daasbjerg K (2016) ChemElectroChem 3:495–501. https://doi.org/10.1002/celc.201500512
43. Chan CK, Beechem TE, Ohta T et al (2013) J Phys Chem C 117:12038–12044. https://doi.org/10.1021/jp311519j
44. Charlton MR, Suhr KJ, Holliday BJ, Stevenson KJ (2015) Langmuir 31:695–702. https://doi.org/10.1021/la503522c
45. Gearba RI, Kim M, Mueller KM et al (2016) Adv Mater Interfaces 3:1–10. https://doi.org/10.1002/admi.201600196
46. Kariuki JK, McDermott MT (2001) Langmuir 17:5947–5951. https://doi.org/10.1021/la010415d
47. Guselnikova O, Postnikov P, Elashnikov R et al (2017) Colloids Surf A Physicochem Eng Asp 516:274–285. https://doi.org/10.1016/j.colsurfa.2016.12.040
48. Menanteau T, Dias M, Levillain E et al (2016) J Phys Chem C 120:4423–4429. https://doi.org/10.1021/acs.jpcc.5b12565
49. Bugaenko DI, Yurovskaya MA, Karchava AV (2018) Org Lett 20:6389–6393. https://doi.org/10.1021/acs.orglett.8b02676
50. Florini N, Michelazzi M, Parenti F et al (2013) J Electroanal Chem 710:41–47. https://doi.org/10.1016/j.jelechem.2013.01.023
51. Miliutina E, Guselnikova O, Soldatova NS et al (2020) J Phys Chem Lett 11:5770–5776. https://doi.org/10.1021/acs.jpclett.0c01350
52. He M, Swager TM (2016) Chem Mater 28:8542–8549. https://doi.org/10.1021/acs.chemmater.6b03078
53. Vase KH, Holm AH, Pedersen SU, Daasbjerg K (2005) Langmuir 21:8085–8089. https://doi.org/10.1021/la050933e
54. Jiang C, Moraes Silva S, Fan S et al (2017) J Electroanal Chem 785:265–278. https://doi.org/10.1016/j.jelechem.2016.11.043
55. Song S, Shen H, Wang Y et al (2020) Colloids Surf B Biointerfaces 185:110596. https://doi.org/10.1016/j.colsurfb.2019.110596
56. Li Q, Lu J, Gupta P, Qiu M (2019) Adv Opt Mater 7:1900595. https://doi.org/10.1002/adom.201900595
57. Nandanapalli KR, Mudusu D, Lee S (2019) Carbon N Y 152:954–985. https://doi.org/10.1016/j.carbon.2019.06.081
58. Lonkar SP, Deshmukh YS, Abdala AA (2015) Nano Res 8:1039–1074. https://doi.org/10.1007/s12274-014-0622-9
59. Mohamed AA, Salmi Z, Dahoumane SA et al (2015) Adv Colloid Interface Sci 225:16–36. https://doi.org/10.1016/j.cis.2015.07.011
60. Stevenson KJ, Veneman PA, Gearba RI et al (2014) Faraday Discuss 172:273–291. https://doi.org/10.1039/c4fd00038b
61. Gearba RI, Mueller KM, Veneman PA et al (2015) J Electroanal Chem 753:9–15. https://doi.org/10.1016/j.jelechem.2015.05.009

62. Dirk SM, Pylypenko S, Howell SW et al (2005) Langmuir 21:10899–10901. https://doi.org/10.1021/la052311z
63. Matrab T, Combellas C, Kanoufi F (2008) Electrochem commun 10:1230–1234. https://doi.org/10.1016/j.elecom.2008.06.006
64. Mesnage A, Lefèvre X, Jégou P et al (2012) Langmuir 28:11767–11778. https://doi.org/10.1021/la3011103
65. Azani M, Hassanpour A, Torres T (2020) Adv Energy Mater 10:2002536. https://doi.org/10.1002/aenm.202002536
66. Christiansen CD, Sørensen LA, Lund T (2018) J Electroanal Chem 809:44–51. https://doi.org/10.1016/j.jelechem.2017.12.050
67. Abellán G, Schirowski M, Edelthalhammer KF et al (2017) J Am Chem Soc 139:5175–5182. https://doi.org/10.1021/jacs.7b00704
68. Schäfer RA, Weber K, Koleśnik-Gray M et al (2016) Angew Chem-Int Ed 55:14858–14862. https://doi.org/10.1002/anie.201607427
69. He M, Swager TM (2020) J Am Chem Soc 142:17876–17880. https://doi.org/10.1021/jacs.0c05965
70. Van Druenen M, Davitt F, Collins T et al (2018) Chem Mater 30:4667–4674. https://doi.org/10.1021/acs.chemmater.8b01306
71. van Druenen M, Collins T, Davitt F et al (2020) Chem-A Eur J 26:17581–17587. https://doi.org/10.1002/chem.202003895
72. Li B, Lai C, Zeng G et al (2019) Small 15:1804565. https://doi.org/10.1002/smll.201804565
73. Guselnikova O, Fraser JP, Soldatova N, Sviridova E, Ivanov A, Rodriguez R, Ganin AY, Postnikov P (2022) The covalent functionalization of few-layered MoTe2 thin films with iodonium salts. Materials Today Chem 24:100846. https://doi.org/10.1016/j.mtchem.2022.100846
74. Neto AI, Levkin PA, Mano JF (2018) Mater Horiz 5:379–393. https://doi.org/10.1039/C7MH00877E
75. Lane AP, Maher MJ, Willson CG, Ellison CJ (2016) ACS Macro Lett 5:460–465. https://doi.org/10.1021/acsmacrolett.6b00075
76. Crivello JV, Lam JHW (1978) J Polym Sci Polym Chem Ed 16:2441–2451. https://doi.org/10.1002/pol.1978.170161004
77. Gellé A, Jin T, de la Garza L et al (2020) Chem Rev 120:986–1041. https://doi.org/10.1021/acs.chemrev.9b00187
78. Adleman JR, Boyd DA, Goodwin DG, Psaltis D (2009) Nano Lett 9:4417–4423. https://doi.org/10.1021/nl902711n
79. Guselnikova O, Audran G, Joly J-P et al (2021). Chem Sci. https://doi.org/10.1039/D0SC06470J
80. Miliutina E, Guselnikova O, Bainova P et al (2018) Adv Mater Interfaces 5:1800725. https://doi.org/10.1002/admi.201800725
81. Fontanesi C, Bortolotti CA, Vanossi D, Marcaccio M (2011) J Phys Chem A 115:11715–11722. https://doi.org/10.1021/jp2032115
82. Kazuma E, Kim Y (2019) Angew Chem Int Ed 58:4800–4808. https://doi.org/10.1002/anie.201811234
83. Kazuma E, Jung J, Ueba H et al (2018) Science 360:521–526. https://doi.org/10.1126/science.aao0872
84. Nguyen M, Kherbouche I, Gam-Derouich S et al (2017) Chem Commun 53:11364–11367. https://doi.org/10.1039/C7CC05974D
85. Kherbouche I, Luo Y, Félidj N, Mangeney C (2020) Chem Mater 32:5442–5454. https://doi.org/10.1021/acs.chemmater.0c00921

Control of the Aryl Layer Growth

Tony Breton and Christelle Gautier

Abstract Electroreduction of diazonium salts for the grafting of organic entities onto a wide variety of substrates has been extensively developed to generate modified surfaces for a large range of applications. The major issue with this approach lies in the lack of control of the grafting mechanism, which leads, in most of cases, to disorganized and loosely packed multilayers. Over the past decade, several strategies have been proposed to control the layer growth, and ideally stop at the monolayer stage. This chapter describes and compares the most relevant approaches, whether playing on the steric or kinetic aspects of the grafting technique.

1 Introduction

The full understanding of the grafting mechanisms has not been completed despite the significant amount of work carried out for almost 30 years. It is admitted that the grafting occurs thanks to phenyl radicals coupling, resulting from the nitrogen-carbon bond-breaking according to a concerted mechanism [1–5]. The production of highly reactive radicals at the electrode-solution interface is responsible for the very high efficiency of electrode functionalization. However, this major advantage becomes a weakness when a fine control of the layer formation is needed. Indeed, the radicals generated close to the electrode are grafted on the surface, but also on already immobilized aromatic structures in an uncontrolled way [6]. From a kinetic point of view, it is established that the formation of the first layer (adsorption on the substrate) is considerably faster than the subsequent polymeric growth which leads to dendritic formations [7]. Depending on the diazotized structures, the substrate, and the experimental conditions, layers ranging from monolayer to hundreds of nanometers can be obtained. In addition to this thickness variability, the chemical structure of the layers is not homogeneous because of the presence of nitrogen functions within the layer. In particular, the systematic presence of azo links has been demonstrated by XPS, [8] IR [9], and TOF–SIMS [6] analyses. Ideally, one should be able to adjust the layer

T. Breton (✉) · C. Gautier
Univ Angers, CNRS, MOLTECH-Anjou, SFR MATRIX, F-49000 Angers, France
e-mail: tony.breton@univ-angers.fr

M. M. Chehimi et al. (eds.), *Aryl Diazonium Salts and Related Compounds*, Physical Chemistry in Action, https://doi.org/10.1007/978-3-031-04398-7_5

thickness and structure to the targeted application. Conceptually, the control of the layer deposition can be considered from different angles. The most intuitive method consists in modulating the experimental parameters (diazonium concentration, deposition time, and grafting potential) to ensure that one of these parameters becomes a limiting factor of the growth. The second strategy aims to exploit the steric hindrance of diazonium salts to avoid, or at least limit, the grafting on aromatic rings already attached to the surface. The last approach, more generalist, aims to locally limit the concentration of the diazonium salts at the electrode/solution interface to lower the production rate of reactive radicals. Each of these approaches has been the subject of different developments, which are detailed below.

2 How to Characterize the Growth of Organic Layers?

The grafting control, and thus the possibility of accessing a molecular monolayer, involves the fine characterization of the deposited layer. Indeed, distinguishing a "pure"monolayer from a multilayer of a few nanometers is not easy, especially in the case of a disorganized radical growth. When the grafted entity is electroactive, its surface coverage can be determined by integrating the charge exchanged during the redox process by cyclic voltammetry. This method is reliable for very thin layers, but it has been shown that for thicker organic layers, a fraction of the redox sites is not electroactive [10]. The electrode/redox unit distance and/or the lack of supporting electrolyte permeability of the layer may explain this partial activity. A quantitative analysis can also be obtained by quartz crystal microbalance, which also offers the possibility of an in situ monitoring. The main limitation of this technique is that electrografting is often coupled to a significant species physisorption. These species, which can easily be removed by ultrasonic treatments, can influence, in a variable proportion, the mass increase measurement [11]. The use of XPS provides unambiguous information on the atomic composition of the surface. In most cases, the probed thickness is much greater than the organic layer one, but it is possible to estimate the surface coverage by exploiting the R/C ratio (with R and C the atomic percentages of a characteristic element and the carbon, respectively) [5]. The common drawback of the three aforementioned characterization techniques is that they cannot confirm the homogeneous distribution of the groups attached to the surface: is it a compact film or do we generate clusters and/or oligomeric structures? A more reliable indication of whether or not we get a real monolayer can be obtained by the thickness estimation of the film. For this purpose, an atomic force microscope (AFM) is commonly exploited to remove a part of the organic layer and then perform a profilometry of the scratch thus created. This "scratching technique" allows to measure layer thicknesses down to less than one nanometer [7]. However, this type of measurement can only be achieved on substrates whose roughness is very low compared to the film thickness. For that purpose, films of pyrolyzed photoresist resin, a glassy carbon-like surface with a low roughness (typically less than 0.5 nm), are commonly used [12]. Ellipsometry can also be exploited to measure layer thicknesses, but this method involves estimating the refractive index of the grafted material, which can generate approximations [13]. To characterize more finely the distribution of species on the surface, the scanning tunneling microscopy (STM) can be used. Its lateral resolution is much greater than that of AFM, which makes it possible to distinguish the molecular objects and thus obtain the finest characterization [14–16].

3 Impact of Grafting Conditions on the Layer Growth

To assess the impact of the grafting conditions, the nature of the substrate has to be taken into account. Indeed, the coupling of species with the surface requires both a good electrochemical reactivity to generate radicals at the interface and a good chemical reactivity allowing the coupling with the radical. Very early on, studies on highly oriented pyrolytic graphite (HOPG) showed the limits of the approach: edge sites, rich in defects, generate rapid grafting and extensive polymerization, while flat basal plans, made mainly of sp^2 carbon, are poorly reactive (Fig. 1) [17]. We observe this lack of reactivity with all similar carbon structures: bodies of carbon nanotubes, [18] graphene sheets, [18], etc. Therefore, the production of layers with controlled thickness can only be considered on substrates presenting a homogeneous reactivity.

The concentration usually used for the grafting is on the order of one millimolar, but it has been shown that increasing this concentration leads to a significant increase in surface concentration [8]. The generation of radicals being an electrochemical process, it is possible, at a given concentration, to play on two experimental parameters to regulate the electron flow and act on the layer formation: the applied potential and the electrolysis time. Studies carried out with 4-nitrobenzene diazonium have

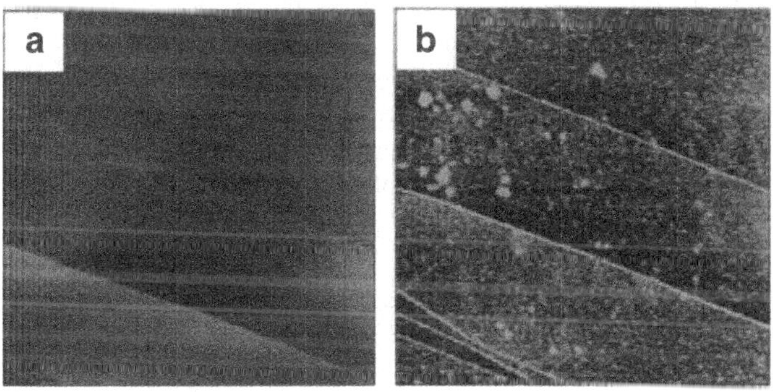

Fig. 1 Topographic in situ AFM images ($5 \times 5\ \mu m^2$) of a HOPG substrate collected in 0.5 mM 4-Diazo-N, N-diethylaniline fluoroborate (0.1 M Bu_4NBF_4/CH_3CN): **a** open circuit; **b** following one cycle between −0.1 and −0.9 V versus Ag/AgCl. Adapted with permission from reference [17]. Copyright (1999) American Chemical Society

shown that, at fixed potential, the surface coverage increases with time until a steady-state value [19]. This maximum only depends on the applied potential and is reached after about ten minutes. Typically, using the 4-nitrobenzene diazonium, known as the benchmark for the diazonium grafting, electrochemical surface coverage reaches 5, 12, and 19 × 10^{-10} mol cm^{-2} for respective potentials of −0.5, −0.8, and −1 V versus Fc^{+}/Fc. As modeled by Savéant et al. [20] this potential dependence is explained by the fact that, during a progressive passivation, the electron transfer constant decrease is inversely proportional to the driving force. Except in the case where an electron relay is incorporated in the organic layer, [21] self-limiting films are observed for all diazoniated structures. It can also be noted that the method chosen for the electrochemical grafting (chronoamperometry vs cyclic voltammetry) does not influence the structuring of the layers even if it has been observed that voltammetric cycling can generate, with certain diazoniums, a more pronounced layer growth [22]. Furthermore, all the studies carried out agree that the density of the layers formed by reduction of diazonium salts is relatively low. The few works linking the electrochemical surface coverage to the thickness measured by atomic force microscopy (AFM) show surface concentration per slice of monolayer of, at most, 50% of that expected for a compact layer [23, 24]. It thus appears obvious that the structural control of the layers is not limited to the thickness control but also involves the concept of compactness. However, these two aspects should be linked if we assume that the absence (or limitation) of the polymerization allows more complete reactivity on the substrate and therefore higher compactness.

4 Limiting the Grafting to a Monolayer

Two very different approaches have been developed to produce ultrathin layers. The first one consists in forming a thick layer from precursors having easily cleavable bonds allowing the degradation of the outer part of the multilayer formed. The second one consists in stopping the layer growth by limiting the quantity of grafted species. These two approaches will be developed in this paragraph and the latter will be divided into three parts: modulation of the deposition conditions, blocking of the aryl reactive positions, and local control of the diazonium concentration.

4.1 Degradation of an Existing Multilayer

This control strategy of the thickness of diazonium-based grafted layers is the first to be developed in this chapter because (i) it is historically the first that has emerged and (ii) it is based on an approach that is very different from all the others since it involves two distinct steps: the formation of a multilayer followed by its degradation to obtain a monolayer or a near-monolayer (Fig. 2).

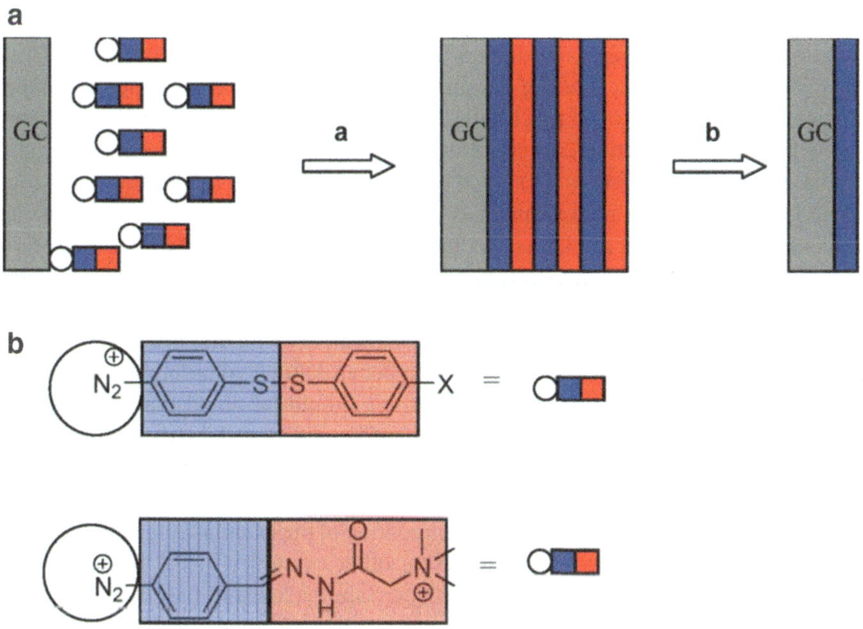

Fig. 2 **a** Formation of the organic layer (through grafting of aryl diazonium salts); **b** degradation (through electrochemical or chemical processes). Reprinted with permission from reference [25]. Copyright (2009) American Chemical Society

Two options, based on similar principles, have been developed. The first one consisted of the immobilization of diazoniums bearing electrochemically cleavable disulfide functions [26] and the second one consisted in the use of diazoniums bearing hydrazone functions hydrolyzable into aldehyde [25]. These cleavage reactions, whether electrochemically or chemically generated, were able to lead to the formation of monolayers of thiophenolates in the first case and aldehyde in the second.

The first approach is based on the initial hypothesis that radical attack is only possible on the outer ring of already grafted molecules (in red on the top diazonium molecule on the Fig. 2) because of the steric constraints [26]. Thus, electroreduction of 4-[(4-chlorophenyl)disulfanyl] benzenediazonium led to a first step to a chlorinated organic multilayer of 3 nm thickness (determined by AFM). Electroreduction of the formed layer, resulting in the cleavage of the S–S bonds, led to the disappearance of the chlorine signal and the decrease of sulfur atomic concentration (obtained by XPS analysis), as well as a decrease of the thickness to 1.5 nm. The surface coverage of $4.0\ 10^{-10}$ mol cm^{-2}, determined by integration of the charge obtained by cyclic voltammetry, confirms the small amount of immobilized species (of the order of the monolayer) after degradation of the initial multilayer.

The second approach, developed 2 years later by the same team, was based on the electroreduction of the benzaldehyde Girard's reagent T hydrazone diazonium salt (bottom molecule on Fig. 2) [25]. This bulky group, located on para to the diazonium function, allows minimizing multilayers formation thanks to its blocking effect. Thus, the radicals formed after grafting of a first molecules layer will probably have more chance to form dimers or to react with the outer part of the first layer (presumably through radical abstraction and coupling reactions), than to graft on the inner part of the grafted layer. The studies showed that the effectiveness of the hydrazone toward radical attack was more marked in aqueous medium than in organic one, as evidenced by the higher thicknesses obtained in DMSO (before and after hydrolysis of the protective group). It appears, however, by comparison with the direct grafting of formylbenzenediazonium salt, that the approach allows to limit the immobilized species quantity since it allows the formation of layers with aldehyde functionality of two times lower thicknesses, presenting a better defined voltammetric signal and giving access to a lower surface. Moreover, it should be noted that the identical hydrazone and aldehyde surface coverages (before and after degradation of the film by acid hydrolysis) confirmed the efficiency of the cleavage and the robustness of the link between the surface and the grafted species.

This original method, involving the degradation of the external part of a multilayer to lead to a thinner layer, does not present other examples in the literature. The strategies developed in the rest of the chapter are exclusively based on the direct formation of thin layers, thanks to the control of the parameters responsible for the film growth.

4.2 Modulation of the Deposition Conditions

As explained previously, the limitation of the reduction potential, whether in cyclic voltammetry or chronoamperometry, should allow the preparation of a near-monolayer. The theoretical value calculated for a close-packed monolayer coverage on a perfectly flat surface by phenyl entities is 12 10^{-10} mol cm^{-2}, [5] however, a value of 6.7 10^{-10} mol cm^{-2} was also proposed by the same authors for nitrophenyl entities [27]. For surfaces that are not atomically flat, it is necessary to consider a roughness factor, between 1.5 and 2.5 for commonly used vitreous carbon surfaces [28]. On monocrystalline doped diamond surface, monolayer films (estimated by AFM) were obtained from 4-nitrobenzene diazonium at low driving force (i.e., −0.2 V vs Ag/AgCl) [22]. XPS angle-resolved analysis showed that, unlike thicker layers, generated at more cathodic potentials, the nitro groups were oriented outwardly of the layer, arguing for an interface organization. Via this strategy, a surface coverage of 2.5×10^{-10} mol cm^{-2} was obtained on PPF for the grafting of 4-nitrophenyl and 4-nitroazobene diazonium at controlled potential, with thicknesses consistent with the formation of monolayers [29]. This value only corresponds to about 20% of a compact monolayer on a flat surface, which is more in the range of a sub-monolayer than a monolayer. Also on PPF, by exploiting cyclic voltammetry, a work carried out on the diazotized derivatives of biphenyl, stilbene, and terphenyl has shown

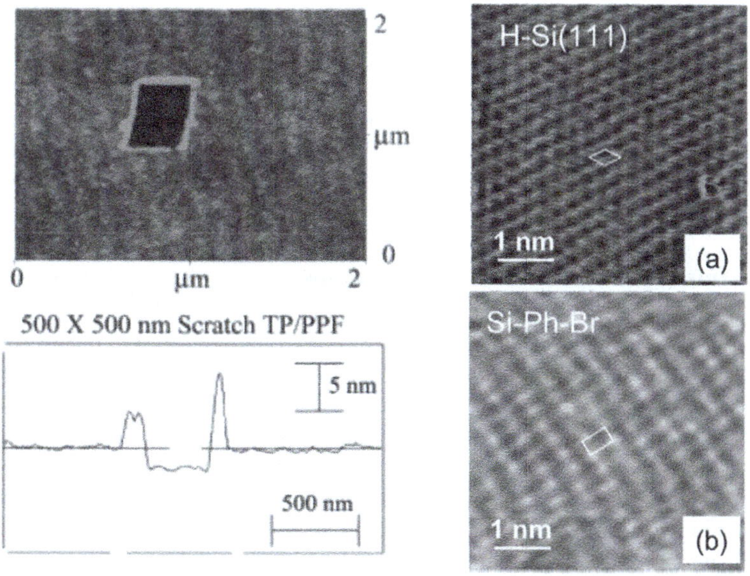

Fig. 3 Upper Left: tapping mode AFM image of a terphenyl monolayer showing a 500 nm × 500 nm trench in the monolayer formed with contact mode AFM. Lower left: line profile across the trench in upper image. Right: 5 nm × 5 nm high-resolution STM images of a H/Si(111) **a** and Br/Si **b** modified surfaces. Reprinted with permission from reference [7]. Copyright (2003) American Chemical Society. Reprinted with permission from reference [30]. Copyright (2003) Elsevier

that thicknesses close to monolayers (estimated by AFM) are obtained via a single voltammetric cycle up to −0.6 V versus Ag^{+}/Ag (Fig. 3) [7].

Repetitive cycles in this potential range led to the formation of multilayers. By limiting the charge exchanged during electrochemical grafting, it has been shown that it is possible to generate monolayers on hydrogenated silicon (111) [30]. The charge required, consumed in chronoamperometric mode, depends on the potential applied and on the substituent beared by the diazonium salt. Associated surface concentrations of 5 and 6.7 × 10^{-10} mol cm^{-2} were determined by Rutherford backscattering and STM imaging, respectively. Surprisingly, the imaging showed the presence of an organized monolayer over the whole surface of the silicon (Fig. 3). To explain this unexpected coverage, the authors suggested a possible preorganization at the electrode-solution interface via a π-stacking or a grafting of aryl radicals made easier by the presence of already grafted entities.

4.3 Blocking of Reactive Positions

Since it is known that the mechanism of multilayer formation is based on successive reactions between the radicals formed during the reduction of diazonium cations and the species already immobilized, one of the first ways used to limit the film growth was to block the reactive positions of the diazonium salts in order to inhibit the grafting of radicals as soon as a first layer is formed on the surface. The work carried out with this objective can be classified into two categories: the introduction of stable groups, which remain permanently bound to the layer, and the introduction of groups that can then be cleaved after the grafting procedure.

4.3.1 Introduction of Non-cleavable Groups

Most diazonium salts used to modify surfaces are functionalized in position 4 (para to the diazonium function), the most accessible position to the external environment. In this case, the film growth takes place via a radical attack on positions 3 and 5 of the already immobilized aryl groups. The electroreduction of a diazonium substituted on these two positions by bulky tert-butyl groups on copper, gold, and silicon electrodes allowed to limit the radical attacks on the first layer formed, driving to a near-monolayer formation [31]. These findings are supported both by the evolution of cyclic voltammograms, which show the behavior of a non-blocked surface, and the thinness of the layers measured by ellipsometry (between 1 nm on gold and 1.6 nm on SiH). This work highlights the importance of the nature of the bulky group since studies carried out under the same conditions with 3,5-trifluoromethyl benzenediazonium led to the formation of thick layers on copper (16.7 nm). The hypothesis put forward by the authors is that the trifluoromethyl groups are less bulky and probably insufficient to prevent the film from growing. To further understand the effect of steric hindrance on the film's growth, investigations have been carried out on a wide range of diazonium salts substituted in different positions [32]. First observation: the reduction of diazonium salts substituted in positions 2 and 6 (ortho to the diazonium function) by methyl groups or in position 2 by an ethyl group (slightly more bulky) does not lead to any modification of the surface as shown by the cyclic voltammogram recorded in a $K_3[Fe(CN)_6]$ solution, which is almost identical to the one recorded on a bare electrode. On the other hand, when the diazonium is substituted by only one methyl group in ortho, a film growth is observed on the surface. This experimental observation was validated by computational results showing that the thermodynamics driving force for phenyl attachment is greatly reduced for 2,6-dimethylphenyl, but only mildly affected for 2-methylphenyl. Secondly, the presence of two tert-butyl groups in positions 3 and 5 is essential to inhibit the growth of the layer. Indeed, the presence of the same group in position 4 or of two methyl groups in positions 3 and 5 led to the formation of multilayers. These first results, obtained during the reduction of 3,5-bis-tert-butyl benzenediazonium and demonstrating the inhibition of the aryl radicals attack on the immobilized aryl groups, inspired other studies where analytical methods such as AFM, STM, or Raman spectroscopy were able to bring additional information. The comparison of the grafting on graphite surfaces (HOPG or graphene) of the aforementioned diazonium with that of a well-known commercially available molecule, 4-nitrobenzene diazonium, has allowed to highlight notable differences in the structure of the layers obtained from the two precursors [15]. More than simply concluding that monolayers are formed in one case (with 3,5-bis-tert-butyl benzenediazonium) and multilayers in the other (with 4-nitrobenzene diazonium), it was shown that surface modification with sterically hindered diazoniums leads to a much higher grafting density. AFM measurements show five times lower layer thicknesses with 3,5-bis-tert-butyl benzenediazonium (0.67 nm vs 3.6 nm), Raman spectroscopy measurements indicate a greater extent of covalent functionalization (by a larger intensity ratio of the D- and G-peaks), and STM measurements show a density of the adsorbed species higher for monolayers (Fig. 4).

This increased grafting density for monolayers is explained by a larger proportion of radicals derived from 3,5-bis-tert-butyl benzenediazonium in solution capable of a direct grafting on the surface because the tert-butyl moieties prevent further radical attachment onto pregrafted species.

In the same vein, aryl diazonium salts substituted in 3,4,5-positions by electron-donating or electro-withdrawing groups were reduced on graphite surface. The results highlighted the impact of the electronic effect on the grafting efficiency [16]. It appeared that whatever the electronic effect of the substituents located in meta and para of the diazonium function (positions sensitive to radical attacks), the film's growth stopped at the monolayer stage. The control of the layer is thus linked, in this case, exclusively to the physical blocking of the reactive positions, as shown by the low thicknesses measured by AFM (0.5–0.7 nm) for all samples. However, STM and Raman spectroscopy analyses revealed a significant variability of the grafting density depending on the electronic properties of the substituents. Indeed, the organic layers formed from compounds substituted with three methoxy groups show a higher grafting density than those obtained from compounds substituted with

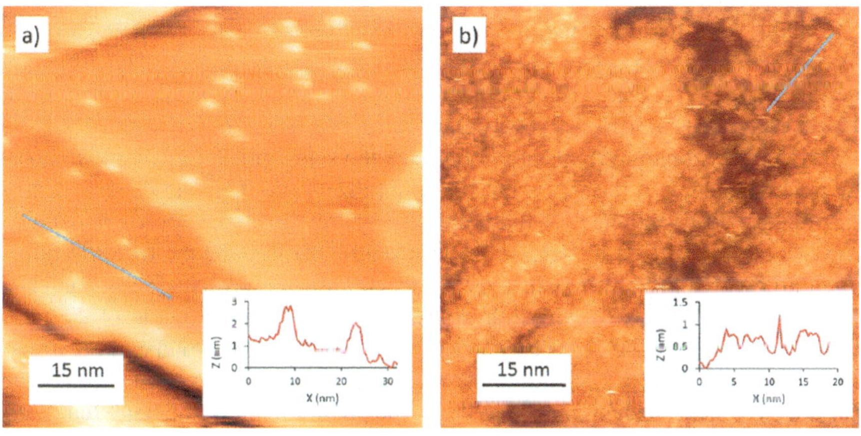

Fig. 4 STM images of clusters after grafting from a 1 mM solution of **a** 4-nitrobenzene diazonium and **b** 3,5-bis-tert-butyl benzenediazonium on CVD graphene on Cu. Reprinted with permission from reference [15]. Copyright (2015) American Chemical Society

methyl groups, which are themselves denser than the layers with ester or nitro functionalities. This increasing grafting density for the electron-donating groups could be related to a better match in the energy levels of molecular orbitals between reactants since SOMO energy levels of the electron-rich radicals are relatively close to the Fermi level of graphite (−4.7 eV). These very interesting and instructive fundamental studies have allowed us to better target the parameters governing the formation of thin and dense layers from substituted diazonium salts. Even if these substituents are not likely to have any applicative aims, a study showed that it was possible to take advantage of the obtaining of dense monolayers to form, using scanning probe nanolithography, nanocorrals with different sizes, shapes, and orientations [33]. These nanocorrals were then used to laterally confine molecular self-assemblies of 10,12-pentacosadiynoic acid.

To combine the approach based on the crowding of attackable positions with the presence of a functional group possessing photophysical properties, a heteroleptic Ru(II) complex involving 5-substituted-2,2′-bipyridine-based bulky ligands was deposited on carbon and ITO electrodes [34]. Thicknesses of the layers grown on ITO, estimated by AFM, either by imaging a scratch in the films or by measuring the depth of the holes left after removal of well-organized nanospheres, have demonstrated the formation of monolayers, whereas a multilayer formation was reported for the non-sterically hindered complex analog. In addition to validating the formation of monolayers, STM measurements revealed a short-range self-organization of the molecules on HOPG (Fig. 5).

From a general point of view, all the studies carried out on aryldiazonium-based layers in which the 3,4 or 5 positions are protected from further radical attack, either because they are all substituted or thanks to bulky groups that protect the three positions, have shown that this approach is relevant to obtain monolayers with high grafting density.

Calix[4]arene-tetra-diazonium salts are hindered structures with a particular design that could also lead to the formation of thin organic layers. The developed approach relies on the large rim grafting thanks to the diazonium reduction route and the small rim is decorated with post-functionalizable groups (Fig. 6) [35–37].

The four diazonium groups are oriented in the same direction, which facilitates the preorganization of the species to be immobilized, and the methylene bridges at the ortho positions and the post-funtionalizable groups at the para position avoid the polyaryl formation. The different methods used to characterize the formed organic layers allowed us to confirm, whatever the appending arms present on the small rim, the formation of layers (i) whose thickness corresponds to the height of the tetra-diazonium precursor (between 1.09 and 1.3 nm), and (ii) rather compact, because of the strong blocking effect regarding the answer of the monolayer toward dopamine. The added value of such systems also lies in the possibility of introducing functional entities in the para position of the diazonium functions. The concept has been validated thanks to the immobilization of ferrocenemethylamine on appending carboxyl through peptide coupling. Depending on whether the four appending groups are identical or not, the structure of the initial macrocycle plays a major role in the density of functional groups after chemical coupling. It should be noted that the

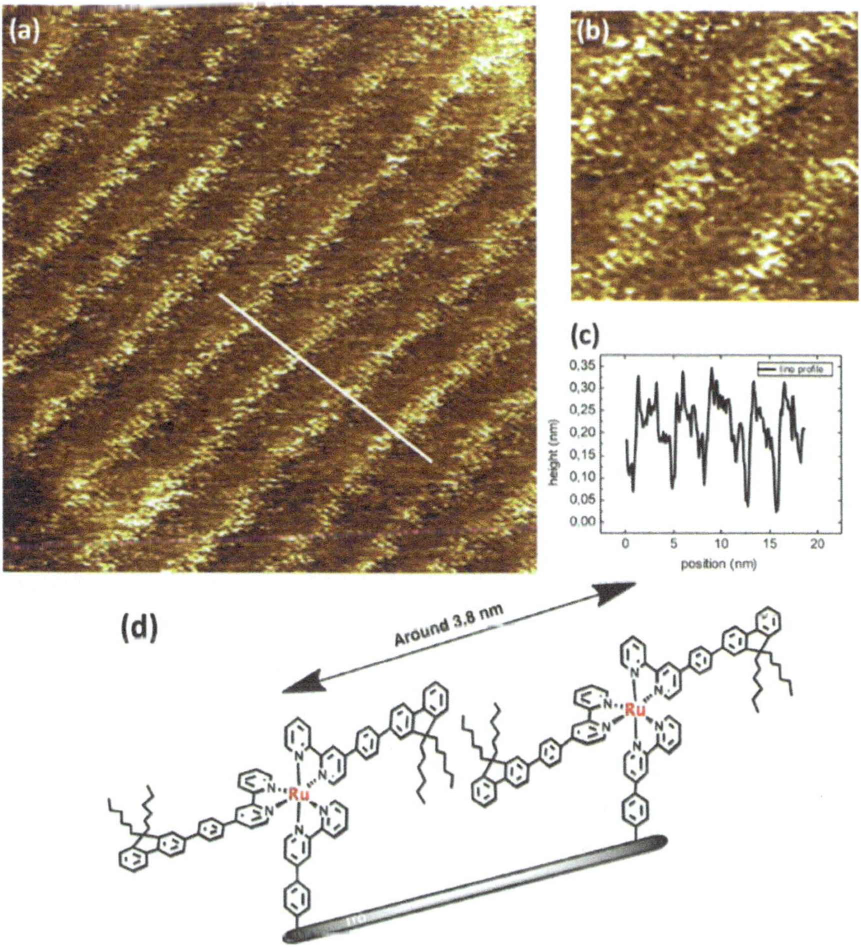

Fig. 5 **a** STM image of Ru(II) complex grafted on HOPG: 33 × 30 nm^2. **b** STM image, 11 × 11 nm^2. **c** Line profile of monolayer molecular stripes. **d** Proposed scheme for the [Ru(II)] complexes in supramolecular side-by-side organization. Reprinted with permission from reference [34]. Copyright (2016) American Chemical Society

post-functionalization of substrates initially modified with calix[4]arenes exhibiting one or four appending COOH groups at the small rim led to a twofold increase in Fc surface coverage. This result indicates that the COOH functions, when born by the same macrocycle, cannot all be post-functionalized and that two ferrocenyl moieties at most could be attached to each calixarene subunit [35]. Indeed, the value obtained after post-functionalization of immobilized tetra-carboxyl calix[4]arenes ($1.6\ 10^{-10}$ mol cm^{-2}) is similar to the one obtained after post-functionalization of analogs exhibiting four appending ethynyl groups through CuAAC reaction with ferrocenyl methyl azide moieties ($1.3\ 10^{-10}$ mol cm^{-2}) [37]. An interesting

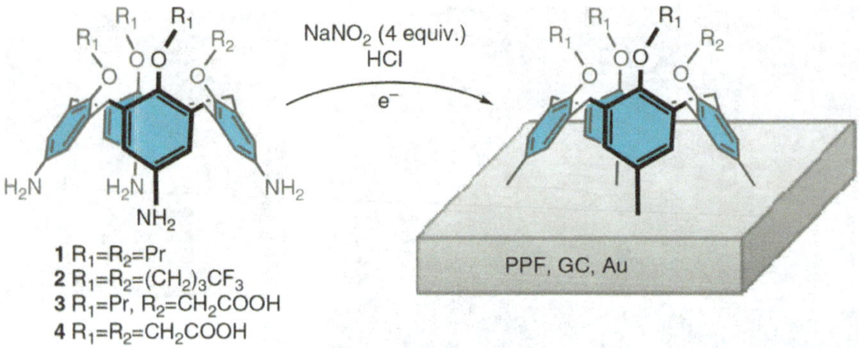

Fig. 6 Structures of the calix[4]tetraanilines used in this work. The modification procedure relies on the in situ diazotation of the calix[4]tetraanilines in the presence of $NaNO_2$ in acidic aqueous medium followed by the electrochemical reduction of the corresponding diazonium salts. Reprinted with permission from reference [35]. Copyright (2012) Nature Research

option to circumvent this problem and maximize the post-functionalization yield, was the formation of mixed monolayers of post-functionalizable calix[4]arenes and non-functionalizable ones used as diluent, aiming to favor the accessibility of COOH groups toward molecules in solution [38]. Although they were obtained in a single step (by simultaneous co-reduction of the two types of differently substituted calixarene), the mixed layers were obtained with controlled proportions for each of the two compounds. This result is rarely possible when a mixture of two diazoniums is reduced since the proportion of the two species on the surface is often governed by the reduction potential of each compound. In this particular case, the two constituent entities in the starting grafting solution exhibit very close reductive potentials because they share a common scaffold.

These studies carried out on the grafting of calix[4]arene-tetra-diazonium cations have shown that it is possible to limit the amount of species immobilized on the surface and to introduce entities of interest by post-functionalizing the organic layer formed. The strategy presented in the following paragraph has been thought to be able to also fulfill these two objectives.

4.3.2 Introduction of Cleavable Bulky Groups

This approach consists in grafting on a substrate an aryldiazonium salt having in para position a function protected by a cleavable bulky group. This very bulky protective group allows to avoid radical attacks on already grafted aryl layers (on para and ortho positions) and thus to limit the grafting of the film to the monolayer stage. The deprotection of the immobilized functions then leads to a reactive platform allowing the anchoring of functional entities by various post-functionalization reactions (Huisgen 1,3 dipolar cycloaddition, Sonogashira, Glaser, or amide coupling) [39].

This method was first described in 2010 [40]. In this early work, triisopropylsilyl (TIPS)-protected ethynyl aryldiazonium salt has been electroreduced on carbon surface and the resulting thin film was deprotected by immersion in a TBAF solution to provide an ethynyl reactive platform able to react with azidomethyl ferrocene by Cu(I)-catalyzed Huisgen 1,3-dipolar cycloaddition. After deprotection (and before post-functionalization), the layer thickness of 0.65 nm was measured by AFM scratching, and the voltammogram recorded in a ferrocene millimolar solution which was almost indiscernible from that obtained on a freshly polished GC electrode, confirm that the obtained layer was thin and corresponded to a near-monolayer. In addition, the post-functionalization reaction of the platform thus obtained with azidomethyl ferrocene allowed to highlight a better accessibility of the acetylene moieties in comparison with those immobilized from the direct reduction of the unprotected diazonium Eth-ArN^{2+} (Γ_{Fc} = 4.4 10^{-10} mol cm^{-2} for a monolayer, against 2.2 10^{-10} mol cm^{-2} for a layer of 6 nm thickness) (Fig. 7).

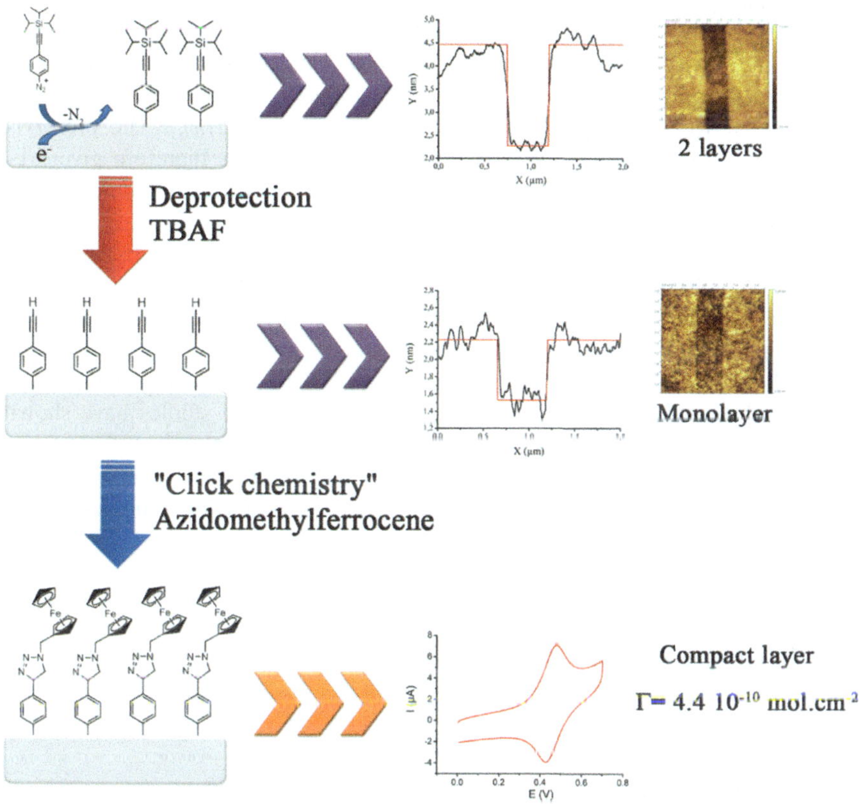

Fig. 7 Principles of the different modification steps of the method. Reprinted with permission from reference [40]. Copyright (2010) American Chemical Society

Regarding the "introduction of cleavable bulky groups" approach, the method developed in this study (reduction of TIPS-protected ethynyl aryldiazonium salt and deprotection) remains widely used (approximately 4/5 of the studies). It has been used to immobilize a wide variety of molecules (ferrocene, [41, 42] porphyrin, [43] azobenzene, [44, 45] copper complexes, [46, 47] tetrathiafulavlene, [48] fluorene [49] and peptides [50]). Most of these studies focus on exploiting the properties or the reactivity of post-grafted species and not on studying the initial monolayer platform.

A more fundamental study looked at the impact of the protecting groups' size (trimethylsilyl (TMS), triethylsilyl (TES), and TIPS) on the ethynyl layer properties [51]. First, it is important to note that the TMS groups are bulky enough to prevent radicals attack on the already grafted aryl layer, since thicknesses of the order of the monolayer (0.6 nm), identical to those obtained for more bulky groups, were measured by AFM scratching after deprotection of the ethynyl functions. A notable difference was however identified from the point of view of the grafting density, as evidenced by the modification of the electrochemical response of different redox probes in solution on the modified surfaces. The most important blocking of electronic transfer, after deprotection, for the layer obtained from TMS-protected ethynyl aryldiazonium salt made it possible to demonstrate a smaller distance d between immobilized species ($d_{TMS} < d_{TES} < d_{TIPS}$). The preserved accessibility of the reactive functions after deprotection has therefore given rise to increasing ferrocene surface coverages (immobilized by post-functionalization) with a decreasing protective group size.

The relatively large pinholes left by the cleavage of TIPS groups have been exploited to immobilize other species either by reduction of diazonium [52] or by oxidation of amine [53, 54] and thus form mixed nitrophenyl/ferrocene layers.

A few rare examples have shown that this "protection-deprotection" method was also applicable to the grafting of diazonium salts bearing protected alcohol, [55] amine, [53] or carboxy functions [56, 57]. In general, these studies have shown, like the previous ones (TIPS-protected ethynyl), that the introduction of protecting cleavable groups is a good option to lead to obtaining thin reactive layers with good accessibility toward post-graftable species in solution, due to the pinholes formed by the removal of bulky groups after deprotection.

4.4 Lowering the Local Concentration of Reactive Radicals

4.4.1 Use of Ionic Liquids

The use of ionic liquids as a solvent for electrochemical reactions was extended to the grafting of diazonium salts during the 2000s. Carbon nanotubes, [58] then glassy carbon electrodes, [59] were modified by exploiting ionic liquids based on imidazolium backbone. It has been shown that the grafting of 4-nitrobenzene diazonium in a ionic liquid (1-ethyl-3-methylimidazolium bis (trifluoromethylsulfonyl) imide) led to a limitation of the surface coverage to about 1.7×10^{-10} mol cm^{-2} [60]. This value, determined by voltammetric integration, is in agreement with the presence of a sub-monolayer. A systematic study of the grafting, involving several voltammetric cycles in three different ionic liquids, demonstrated the impact of their viscosity (η) on the film formation [61]. An exponential decrease in the surface coverage was observed as the viscosity increased, with an order of magnitude loss when ranging from 0.34 to 100 cP (Fig. 8). This effect, not correlated with the ionic liquid structure, resulted in a linear dependence of the logarithm of the surface coverage as a function of the logarithm of the viscosity. The addition of co-solvent (acetonitrile) to ionic liquids allowed a wide range of viscosities to be obtained and surface coverage to be modulated between 2 and 20×10^{-10} mol cm^{-2}. For viscosities higher than 100 cP, a limiting value of approximately 2×10^{-10} mol cm^{-2} was estimated. These results were supported by AFM scratching on modified gold electrodes, which showed thickness values on the order of 1 nm, consistent with a monolayer-like structure (Fig. 8). Comparison of the topographies between layers generated in acetonitrile and ionic liquids did not show any significant difference. Controlling the grafting by simply increasing the viscosity is not easily understandable because, if it is obvious that the viscosity tends to lower the diffusion of the diazotized species toward the electrode, it should also decrease the diffusion of the reactive species produced, and therefore increase the grafting yield. It is possible that the observed control results from the evolution of several kinetic and/or mechanistic factors due to the particular environment represented by ionic liquids.

By using protic ionic liquids, this approach can be extended to the grafting of diazonium salts in situ generated. Studies carried out using 4-nitroaniline and 4-chloroaniline in the presence of a ionic liquid composed of 2-methoxypyridine, trifluoroacetic acid (ratio 1:2) and sodium nitrite have shown an efficient functionalization of glassy carbon [62]. The layers prepared by cyclic voltammetry exhibited

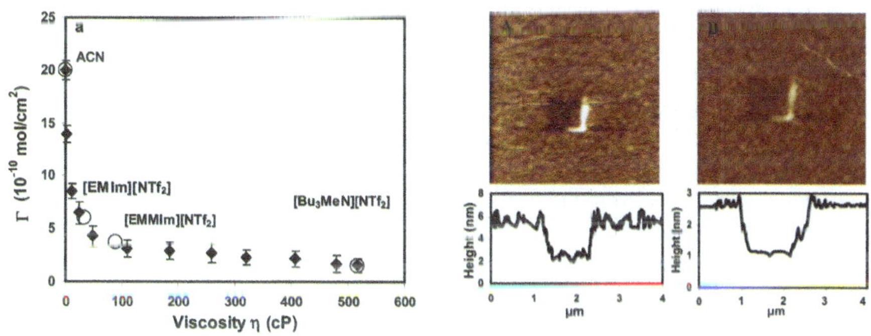

Fig. 8 Upper right. Tapping mode AFM images recorded after scratch experiment with AFM tip in contact mode. Image size 4 μm^2 × 4 μm^2. Lower right. Cross-section of the generated scratch. Left. Evolution of surface concentration of the attached NP groups as function of the viscosity of the grafting media. (◆) grafting in [Bu_3MeN][NTf_2] with the addition of various amount of co-solvent ACN. (O) grafting in pure ionic liquids and ACN. Reprinted with permission from reference [61]. Copyright (2010) American Chemical Society

similar composition compared to that obtained from grafting in usual solvents, which seems to mean that the grafting mechanism is identical. The electrochemical surface concentration was close to 20×10^{-10} mol cm^{-2}, which is in good agreement with a compact monolayer. This strategy, applied to a chronoamperometric grafting on a PPF surface led to layers having an average thickness of 1.5 nm, or 1 to 2 layers of nitrobenzene entities. The relevance of the technique lies in the fact that increasing the grafting time did not lead to a thickness increase, which allows very good control at fixed potential. The factors involved in self-limiting grafting are not clearly identified, but the large size of the ions of the ionic liquid at the interface could slow down or even stop the formation of multilayers. Concurring results were obtained by using an acidic Bronsted ionic liquid (1-butyl-3-methylimidazolium hydrogensulfate) to graft 4-nitrobenzene diazonium in situ generated from the corresponding aniline [63]. On PPF carbon, the surface coverage reached a maximum of some 10^{-10} mol cm^{-2} whatever the charge exchanged for the grafting, which is consistent with a self-limited process. Voltammetry carried out in the presence of an electrochemical probe showed that the layers are more compact than those prepared in a conventional aqueous medium while their average thickness is limited to only 0.5 nm (vs 2.4 nm in aqueous medium). AFM analyses also demonstrated lower roughness and better homogeneity.

4.4.2 Use of Diazonium Precursors

Limiting the concentration of reactive species can also be envisaged by limiting the production of the corresponding diazonium salts at the electrode-solution interface. This ingenious strategy was exploited by using a precursor, transformed into diazonium, via a coupled electrochemical/chemical mechanism [64]. The first step in the sequence was to generate protons in the diffusion layer by electro-oxidation of N, N'-diphenylhydrazine. This proton is able to trigger the cleavage of a triazene precursor, leading to the corresponding diazonium. The second step aims to electro-reduce the diazonium obtained to generate the radical, which will be grafted to the electrode (Fig. 9). This technique implements a sequence composed of an oxidation pulse (+ 0.9 V vs SCE for 0.5 s) and then a negative pulse (−0.66 V vs SCE for 0.5 s). The major advantage of this approach is the possibility to repeat the sequence in order to modulate the surface covered in an extremely fine way. Experiments carried out with a ferrocene derivative showed a perfectly defined electrochemical system whatever the surface concentration. A surface coverage of 3×10^{-10} mol cm^{-2} and a thickness around 1.5 nm (measured by ellipsometry) were reached after 15 cycles, which is consistent with a monolayer formation (Fig. 9). Once again, the film density was found relatively low (about 20% of that expected for a compact monolayer), but the lateral resolution of the AFM was not sufficient to perceive pinholes.

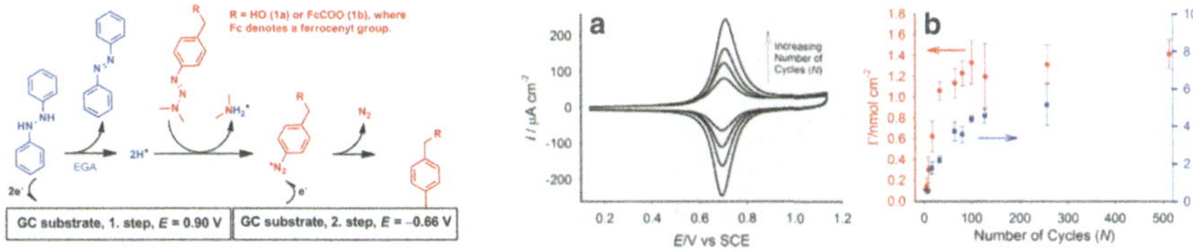

Fig. 9 Left. Mechanism suggested for electrografting of aryl groups on carbon surfaces by exploiting the reaction of aryltriazenes with an electrogenerated acid (EGA). Right. **a** Cyclic voltammograms on GC electrodes (obtained using N = 8, 16, 32, and 64 electrografting cycles). **b** Surface coverage of electrochemically active ferrocene units (red, left axis) and film thickness measured by ellipsometry (blue, right axis) as functions of N. Reprinted with permission from reference [64]. Copyright (2011) American Chemical Society

4.4.3 Use of Redox Mediators

Still, with the idea of controlling the concentration of active species at the electrode-solution interface to avoid polymerization, the reverse strategy was studied: locally lowering the aryl radical concentration via a secondary electrochemical reaction. To do this, a simple approach was implemented by adding 2,2-diphenylpicrilhydrazyl (DPPH) to the 4-nitrobenzene diazonium deposition solution [65]. The first studies showed that an increasing concentration of DPPH led to obtaining layers of decreasing surface coverage, with a limit of 6×10^{-10} mol cm^{-2} on glassy carbon, value corroborated by QCM measurements. These results were then confirmed by AFM scratching analyzes on a PPF surface, showing a thickness of 0.9 nm, consistent with the presence of a monolayer (Fig. 10) [24]. Initially explained by the trapping of excess aryl radicals produced in the diffusion layer, it was later shown that the inhibition is in fact due to the presence of the reduced form of DPPH, which diffuses in solution and reacts with approaching diazonium cations according to a cross-redox reaction [66]. Aryl radicals produced far from the interface can no longer attack the already grafter rings and react with the solvent or other species to provide by-products, thereby limiting the polyaryl layer growth (Fig. 11).

The proposed mechanism, validated by modeling, [66] is of a pure diffusion type and is based on the concentration balance between the diazonium salt and the redox mediator. The rather high functionalization density can be explained by the progressive plugging of the holes under prolonged electrolysis or multiple cycling, even at electrochemically less active carbon sites. This technique, initially developed in the context of the electrochemical grafting, has been extended to the spontaneous functionalization on carbonaceous surfaces [67]. A surface coverage of 6 $\times$ 10^{-10} mol cm^{-2}, equivalent to a monolayer, was obtained for a concentration of redox inhibitor ten times lower than when electrochemical induction is applied.

This approach has been exploited in the context of the development of carbon powder supercapacitors and has allowed to significantly increase in the proportion of electroactive species in the film [68]. The efficiency of the technique lies in the kinetics of the cross redox reactions, and therefore in the difference between the reduction potential of the diazonium considered and the redox potential of the

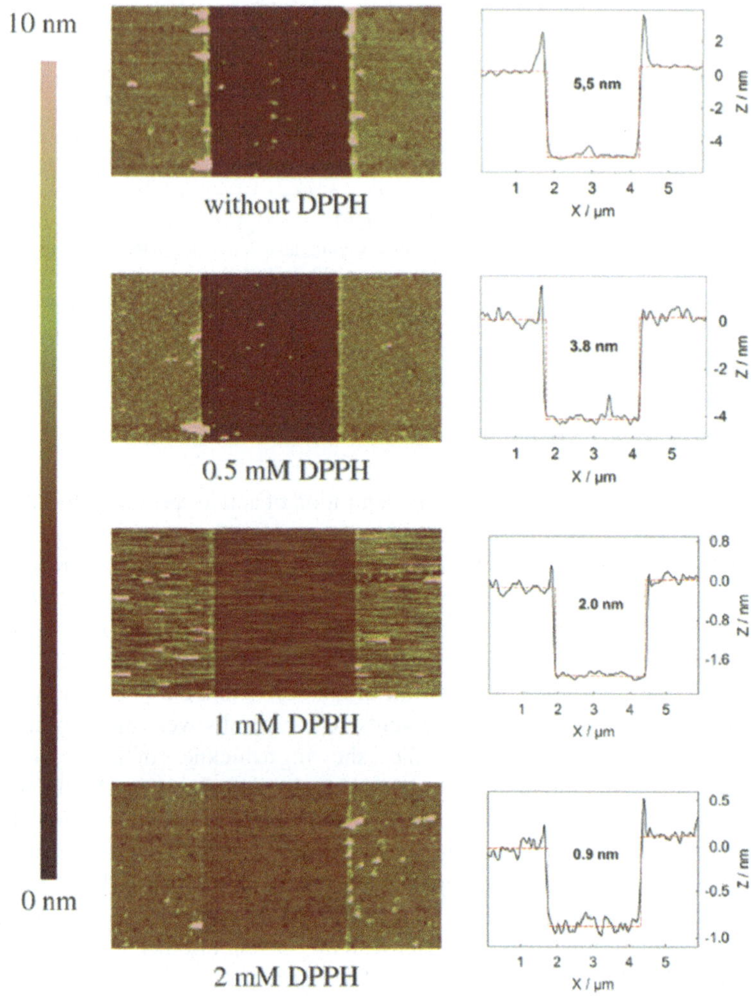

Fig. 10 AFM topography images (6 μm × 3 μm) and the corresponding depth profiles of nitrophenyl PPF modified with 1 mM of 4-nitrobenzene diazonium and increasing DPPH concentrations. Reprinted with permission from reference [24]. Copyright (2015) Royal Society of Chemistry

mediator used. To control the grafting of diazonium characterized by low reduction potential, it is possible to replace DPPH by quinones (chloranil or dichlone), which have shown similar control efficiency by limiting the thickness of the layers to 0.6–0.9 nm [69]. It is also possible to avoid the redox mediator addition by exploiting the reduction of atmospheric dioxygen into superoxide ions at a fixed potential around −0.8 V versus $Ag/AgNO_3$ [70]. This approach leads to the formation of ultrafine layers (approximately 0.6 nm) with surface coverages of 4–5 × 10^{-10} and

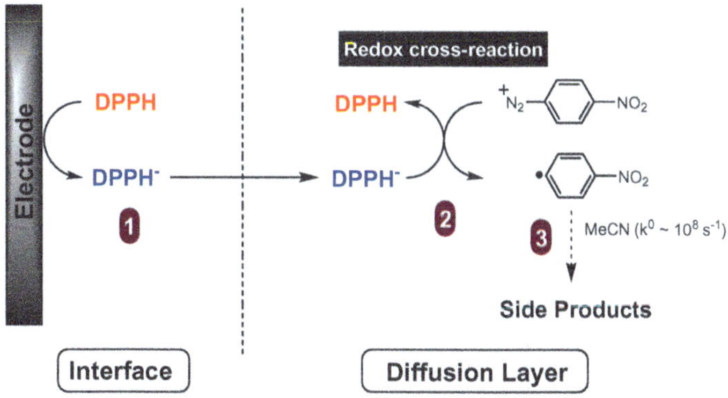

Fig. 11 Mechanism of the grafting inhibition by redox cross-reaction. Reprinted with permission from reference [66]. Copyright (2018) Wiley–VCH GmbH

3×10^{-10} mol cm^{-2} using nitrobenzene diazonium and 1-antraquinone diazonium, respectively.

Surprisingly, when this strategy is used, whether by the spontaneous or electro-induced route, the presence of azo bridges in the layer (revealed by the presence of an XPS peak at 400 eV) is higher than that determined on thicker films [24, 67]. Those results highlight the coexistence of two film formation mechanisms and seem to show that azo bridges are preferentially localized at the electrode-organic film interface (up to 50% of the linked aryl groups). The technique has been exploited on glassy carbon within the framework of the development of catalytic surfaces to produce reactive platforms dedicated to the immobilization of TEMPO groups via peptide coupling [71] or click chemistry (Fig. 12) [72]. The advantages provided by thin layers have been clearly demonstrated in terms of electronic transfer kinetic and interfacial reactivity. The approach has also shown its relevance to HOPG substrates by offering the possibility of modulating the thickness of the nitrophenyl films from the multilayer stage to the sub-monolayer one (thickness varying from 4 to 0.8 nm)

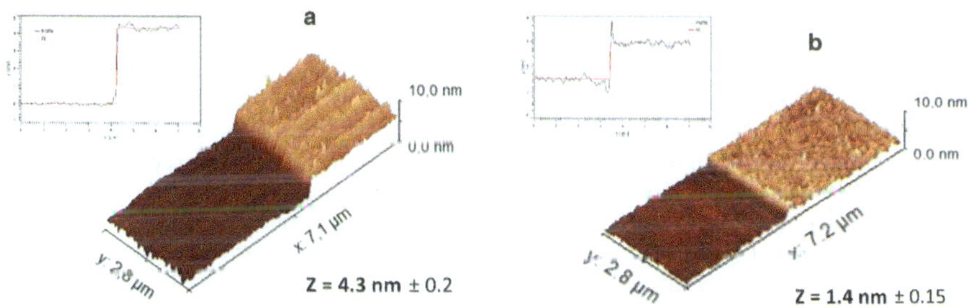

Fig. 12 AFM images of a PPF modified by 10 cycles between 0.4 and −0.6 V (vs. $Ag/AgNO_3$ 0.01 M) in a 1 mM 4-azidobenzene diazonium solution. **a** Without DPPH in the electrolytic medium and **b** in presence of 2 mM of DPPH. Reprinted with permission from reference [72]. Copyright (2020) Elsevier

[73, 74]. For the thinnest layers, AFM images showed an overall coverage of about 70–80% of the surface.

5 Concluding Remarks

Among the different strategies developed to control grafting and, if possible, stop at the monolayer stage, none combines all the advantages in terms of efficiency, versatility, and simplicity. It is obvious that controlling the potential or the charge exchanged is not a viable solution in the sense that many studies are necessary before identifying the conditions allowing a good control. Using sterically hindered diazotized structures is an attractive route because a strict limitation at the monolayer stage can be obtained. This strategy, which suffers from synthetic limitations, finds its full interest thanks to the use of the protection-deprotection approach, which allows the preparation of reactive monolayers for further "on-surface" chemistry. The fact that the surface compactness obtained is low due to the size of the protecting groups can however be detrimental when compact layers are targeted. The use of redox mediator is a technique involving a much simpler experimental approach, applicable to all types of diazonium salts and apparently producing more compact layers. On the other hand, the process being only diffusion-controlled is not likely to strictly stop at the monolayer stage depending on the experimental conditions. The approach consisting of the use of ionic liquids as a solvent is extremely promising thanks to a very simple implementation and a wide application range. However, data are still lacking to well understand the control phenomenon in itself and particularly its ability to strictly inhibit the attack of immobilized aryl groups to produce real monolayers.

References

1. Downard AJ, Roddick AD, Bond AM (1995) Covalent modification of carbon electrodes for voltammetric differentiation of dopamine and ascorbic acid. Anal Chim Acta 317:303–310. https://doi.org/10.1016/0003-2670(95)00397-5
2. Downard AJ, Roddick AD (1995) Protein adsorption at glassy carbon electrodes: the effect of covalently bound surface groups. Electroanalysis 7:376–378. https://doi.org/10.1002/elan.1140070414
3. Delamar M, Hitmi R, Pinson J, Saveant JM (1992) Covalent modification of carbon surfaces by grafting of functionalized aryl radicals produced from electrochemical reduction of diazonium salts. J Am Chem Soc 114:5883–5884. https://doi.org/10.1021/ja00040a074
4. de Villeneuve CH, Pinson J, Bernard MC, Allongue P (1997) Electrochemical formation of close-packed phenyl layers on Si(111). J Phys Chem B 101:2415–2420. https://doi.org/10.1021/jp962581d
5. Liu Y-C, McCreery RL (1995) Reactions of organic monolayers on carbon surfaces observed with unenhanced Raman spectroscopy. J Am Chem Soc 117:11254–11259. https://doi.org/10.1021/ja00150a024

6. Doppelt P, Hallais G, Pinson J et al (2007) Surface modification of conducting substrates. Existence of Azo bonds in the structure of organic layers obtained from diazonium salts. Chem Mater 19:4570–4575. https://doi.org/10.1021/cm0700551
7. Anariba F, DuVall SH, McCreery RL (2003) Mono- and multilayer formation by diazonium reduction on carbon surfaces monitored with atomic force microscopy “Scratching.” Anal Chem 8
8. Saby C, Ortiz B, Champagne GY, Belanger D (1997) Electrochemical modification of glassy carbon electrode using aromatic diazonium salts. 1. Blocking effect of 4-nitrophenyl and 4-carboxyphenyl groups. Langmuir 9
9. Itoh T, McCreery RL (2002) In situ raman spectroelectrochemistry of electron transfer between glassy carbon and a chemisorbed nitroazobenzene monolayer. J Am Chem Soc 124:10894–10902. https://doi.org/10.1021/ja020398u
10. Ceccato M, Nielsen LT, Iruthayaraj J et al (2010) Nitrophenyl groups in diazonium-generated multilayered films: which are electrochemically responsive? 10
11. Ghodbane O, Chamoulaud G, Bélanger D (2004) Chemical reactivity of 4-bromophenyl modified glassy carbon electrode. Electrochem Commun 6:254–258. https://doi.org/10.1016/j.elecom.2003.12.013
12. Ranganathan S, McCreery RL Electroanalytical performance of carbon films with near-atomic flatness. 8
13. Adenier A, Combellas C, Kanoufi F et al (2006) Formation of polyphenylene films on metal electrodes by electrochemical reduction of benzenediazonium salts. Chem Mater 18:2021–2029. https://doi.org/10.1021/cm052065c
14. Steeno R, Rodríguez González MC, Eyley S et al (2020) Covalent functionalization of carbon surfaces: diaryliodonium versus aryldiazonium chemistry. Chem Mater 32:5246–5255. https://doi.org/10.1021/acs.chemmater.0c01393
15. Greenwood J, Phan TH, Fujita Y et al (2015) Covalent Modification of graphene and graphite using diazonium chemistry: tunable grafting and nanomanipulation. ACS Nano 9:5520–5535. https://doi.org/10.1021/acsnano.5b01580
16. Tahara K, Kubo Y, Lindner B et al (2019) Steric and electronic effects of electrochemically generated aryl radicals on grafting of the graphite surface. Langmuir 35:2089–2098. https://doi.org/10.1021/acs.langmuir.8b03339
17. Kariuki JK, McDermott MT (1999) Nucleation and growth of functionalized aryl films on graphite electrodes. Langmuir 7
18. Mohamed AA (2015) Functionalization of nanomaterials with aryldiazonium salts. Adv Colloid Interface Sci 21
19. Downard AJ (2000) Potential dependence of self limited films formed by reduction of aryldiazonium salts at glassy carbon electrodes. Langmuir 16:9680–9682. https://doi.org/10.1021/la000866i
20. Amatore C, Saviant JM, Tessier D (1983) Kinetics of electron transfer to organic molecules at solid electrodes in organic media. J Electroanal Chem 9
21. Ceccato M, Bousquet A, Hinge M et al (2011) Using a Mediating effect in the electroreduction of aryldiazonium salts to prepare conducting organic films of high thickness. Chem Mater 23:1551–1557. https://doi.org/10.1021/cm1033244
22. Uetsuka H, Shin D, Tokuda N et al (2007) Electrochemical grafting of boron-doped single-crystalline chemical vapor deposition diamond with nitrophenyl molecules. Langmuir 23:3466–3472. https://doi.org/10.1021/la063241e
23. Brooksby PA, Downard AJ (2004) Electrochemical and atomic force microscopy study of carbon surface modification via diazonium reduction in aqueous and acetonitrile solutions. Langmuir 20:5038–5045. https://doi.org/10.1021/la049616i
24. Menanteau T, Levillain E, Downard AJ, Breton T (2015) Evidence of monolayer formation via diazonium grafting with a radical scavenger: electrochemical, AFM and XPS monitoring. Phys Chem Chem Phys 17:13137–13142. https://doi.org/10.1039/C5CP01401H
25. Malmos K, Dong M, Pillai S et al (2009) Using a hydrazone-protected benzenediazonium salt to introduce a near-monolayer of benzaldehyde on glassy carbon surfaces. J Am Chem Soc 131:4928–4936. https://doi.org/10.1021/ja809816x

26. Nielsen LT, Vase KH, Dong M et al (2007) Electrochemical approach for constructing a monolayer of thiophenolates from grafted multilayers of diaryl disulfides. J Am Chem Soc 2
27. DuVall SH, McCreery RL (1999) Control of catechol and hydroquinone electron-transfer kinetics on native and modified glassy carbon electrodes. Anal Chem 71:4594–4602. https://doi.org/10.1021/ac990399d
28. Pontikos NM, McCreery RL (1992) Microstructural and morphological changes induced in glassy carbon electrodes by laser irradiation. J Electroanal Chem 324:229–242. https://doi.org/10.1016/0022-0728(92)80048-9
29. Brooksby PA, Downard AJ (2005) Multilayer nitroazobenzene films covalently attached to carbon. an AFM and electrochemical study. J Phys Chem B 109:8791–8798. https://doi.org/10.1021/jp046095z
30. Allongue P, Henry de Villeneuve C, Cherouvrier G et al (2003) Phenyl layers on H-Si(111) by electrochemical reduction of diazonium salts: monolayer versus multilayer formation. J Electroanal Chem 550–551:161–174. https://doi.org/10.1016/S0022-0728(03)00076-7
31. Combellas C, Kanoufi F, Pinson J, Podvorica FI (2008) Sterically hindered diazonium salts for the grafting of a monolayer on metals. J Am Chem Soc 130:8576–8577. https://doi.org/10.1021/ja8018912
32. Combellas C, Jiang D, Kanoufi F et al (2009) Steric effects in the reaction of aryl radicals on surfaces. Langmuir 25:286–293. https://doi.org/10.1021/la8025792
33. Verstraete L, Greenwood J, Hirsch BE, De Feyter S (2016) Self-assembly under confinement: nanocorrals for understanding fundamentals of 2D crystallization. ACS Nano 10:10706–10715. https://doi.org/10.1021/acsnano.6b05954
34. Nguyen VQ, Sun X, Lafolet F et al (2016) Unprecedented Self-organized monolayer of a Ru(II) complex by diazonium electroreduction. J Am Chem Soc 138:9381–9384. https://doi.org/10.1021/jacs.6b04827
35. Mattiuzzi A, Jabin I, Mangeney C et al (2012) Electrografting of calix[4]arenediazonium salts to form versatile robust platforms for spatially controlled surface functionalization. Nat Commun 3.https://doi.org/10.1038/ncomms2121
36. Troian-Gautier L, Martínez-Tong DE, Hubert J et al (2016) Controlled modification of polymer surfaces through grafting of calix[4]arene-tetradiazoate salts. J Phys Chem C 120:22936–22945. https://doi.org/10.1021/acs.jpcc.6b06143
37. Buttress JP, Day DP, Courtney JM et al (2016) "Janus" calixarenes: double-sided molecular linkers for facile, multianchor point, multifunctional, surface modification. 8
38. Santos L, Mattiuzzi A, Jabin I et al (2014) One-pot electrografting of mixed monolayers with controlled composition. J Phys Chem C 118:15919–15928. https://doi.org/10.1021/jp5052003
39. Gautier C, López I, Breton T (2021) A post-functionalization toolbox for diazonium (electro)-grafted surfaces: review of the coupling methods. Mater Adv. https://doi.org/10.1039/D1MA00077B
40. Leroux YR, Fei H, Noël J-M et al (2010) Efficient covalent modification of a carbon surface: use of a silyl protecting group to form an active monolayer. J Am Chem Soc 132:14039–14041. https://doi.org/10.1021/ja106971x
41. Gietter AA, Pupillo RC, Yap GP et al (2013) On-surface cross-coupling methods for the construction of modified electrode assemblies with tailored morphologies. Chem Sci 4:437–443
42. Sayed SY, Bayat A, Kondratenko M et al (2013) Bilayer molecular electronics: all-carbon electronic junctions containing molecular bilayers made with "click" chemistry. J Am Chem Soc 135:12972–12975. https://doi.org/10.1021/ja4065443
43. Wei P-J, Yu G-Q, Naruta Y, Liu J-G (2014) Covalent grafting of carbon nanotubes with a biomimetic heme model compound to enhance oxygen reduction reactions. Angew Chem Int Ed 53:6659–6663. https://doi.org/10.1002/anie.201403133
44. Natsui K, Yamamoto T, Akahori M, Einaga Y (2015) Photochromism-induced amplification of critical current density in superconducting boron-doped diamond with an azobenzene molecular layer. ACS Appl Mater Interfaces 7:887–894. https://doi.org/10.1021/am5074613

45. Leroux YR, Hapiot P (2013) Photo-modulation of the permeation in azobenzene derivatives monolayer films electrografted on carbon substrates. Electrochem Commun 33:107–110. https://doi.org/10.1016/j.elecom.2013.04.032
46. Wang R-C (2015) A copper complex covalently grafted on carbon nanotubes and reduced graphene oxide promotes oxygen reduction reaction activity and catalyst stability. RSC Adv 7
47. Xi Y-T, Wei P-J, Wang R-C, Liu J-G (2015) Bio-inspired multinuclear copper complexes covalently immobilized on reduced graphene oxide as efficient electrocatalysts for the oxygen reduction reaction. Chem Commun 51:7455–7458. https://doi.org/10.1039/C5CC00963D
48. Jalkh J, Leroux YR, Vacher A et al (2016) Tetrathiafulvalene-tetracyanoquinodimethane charge-transfer complexes wired to carbon surfaces: tuning of the degree of charge transfer. J Phys Chem C 120:28021–28030. https://doi.org/10.1021/acs.jpcc.6b09459
49. Jalkh J, Thiery S, Bergamini J-F et al (2017) Influence of fluorene and spirobifluorene regioisomerism on the structure, organization, and permeation properties of monolayers. J Phys Chem C 121:14228–14237. https://doi.org/10.1021/acs.jpcc.7b04193
50. Matsubara T, Ujie M, Yamamoto T et al (2016) Highly sensitive detection of influenza virus by boron-doped diamond electrode terminated with sialic acid-mimic peptide. Proc Natl Acad Sci USA 113:8981–8984. https://doi.org/10.1073/pnas.1603609113
51. Leroux YR, Hapiot P (2013) Nanostructured monolayers on carbon substrates prepared by electrografting of protected aryldiazonium salts. Chem Mater 25:489–495. https://doi.org/10.1021/cm303844y
52. Lee L, Brooksby PA, Leroux YR et al (2013) Mixed monolayer organic films via sequential electrografting from aryldiazonium ion and arylhydrazine solutions. Langmuir 29:3133–3139. https://doi.org/10.1021/la400303x
53. Lee L, Leroux YR, Hapiot P, Downard AJ (2015) Amine-terminated monolayers on carbon: preparation, characterization, and coupling reactions. Langmuir 31:5071–5077. https://doi.org/10.1021/acs.langmuir.5b00730
54. Lee L, Gunby NR, Crittenden DL, Downard AJ (2016) Multifunctional and stable monolayers on carbon: a simple and reliable method for backfilling sparse layers grafted from protected aryldiazonium ions. Langmuir 32:2626–2637. https://doi.org/10.1021/acs.langmuir.5b04546
55. Liu W, Tilley TD (2015) Sterically controlled functionalization of carbon surfaces with $-C_6H_4CH_2X$ ($X = OSO_2Me$ or N_3) groups for surface attachment of redox-active molecules. Langmuir 31:1189–1195. https://doi.org/10.1021/la503796z
56. Lee L, Ma H, Brooksby PA et al (2014) Covalently Anchored carboxyphenyl monolayer via aryldiazonium ion grafting: a well-defined reactive tether layer for on-surface chemistry. Langmuir 30:7104–7111. https://doi.org/10.1021/la5013632
57. Price BK, Hudson JL, Tour JM (2005) Green chemical functionalization of single-walled carbon nanotubes in ionic liquids. J Am Chem Soc 127:14867–14870. https://doi.org/10.1021/ja053998c
58. Actis P, Caulliez G, Shul G et al (2008) Functionalization of glassy carbon with diazonium salts in ionic liquids. Langmuir 24:6327–6333. https://doi.org/10.1021/la703714a
59. Ghilane J, Martin P, Fontaine O et al (2008) Modification of carbon electrode in ionic liquid through the reduction of phenyl diazonium salt. Electrochemical evidence in ionic liquid. Electrochem Commun 10:1060–1063. https://doi.org/10.1016/j.elecom.2008.05.017
60. Fontaine O, Ghilane J, Martin P et al (2010) Ionic liquid viscosity effects on the functionalization of electrode material through the electroreduction of diazonium. Langmuir 26:18542–18549. https://doi.org/10.1021/la103000u
61. Shul G, Ruiz CAC, Rochefort D et al (2013) Electrochemical functionalization of glassy carbon electrode by reduction of diazonium cations in protic ionic liquid. Electrochim Acta 106:378–385. https://doi.org/10.1016/j.electacta.2013.05.082
62. Carvalho Padilha J, Noël J-M, Bergamini J-F et al (2016) Functionalization of carbon materials by reduction of diazonium cations produced in situ in a Brønstedt acidic ionic liquid. ChemElectroChem 3:572–580. https://doi.org/10.1002/celc.201500434
63. Kongsfelt M, Vinther J, Malmos K et al (2011) Combining aryltriazenes and electrogenerated acids to create well-defined aryl-tethered films and patterns on surfaces. J Am Chem Soc 133:3788–3791. https://doi.org/10.1021/ja111731d

64. Menanteau T, Levillain E, Breton T (2013) Electrografting via diazonium chemistry: from multilayer to monolayer using radical scavenger. Chem Mater 25:2905–2909. https://doi.org/10.1021/cm401512c
65. López I, Cesbron M, Levillain E, Breton T (2018) Diazonium grafting control through a redox cross-reaction: elucidation of the mechanism involved when using 2,2-diphenylpicrylhydrazyl as an inhibitor. ChemElectroChem 5:1197–1202. https://doi.org/10.1002/celc.201701331
66. Menanteau T, Levillain E, Breton T (2014) Spontaneous grafting of nitrophenyl groups on carbon: effect of radical scavenger on organic layer formation. Langmuir 30:7913–7918. https://doi.org/10.1021/la501437g
67. Menanteau T, Benoît C, Breton T, Cougnon C (2016) Enhancing the performance of a diazonium-modified carbon supercapacitor by controlling the grafting process. Electrochem Commun 63:70–73. https://doi.org/10.1016/j.elecom.2015.12.014
68. López I, Dabos-Seignon S, Breton T (2019) Use of selective redox cross-inhibitors for the control of organic layer formation obtained via diazonium salt reduction. Langmuir 35:11048–11055. https://doi.org/10.1021/acs.langmuir.9b01397
69. Pichereau L, López I, Cesbron M et al (2019) Controlled diazonium electrografting driven by overpotential reduction: a general strategy to prepare ultrathin layers. Chem Commun 55:455–457. https://doi.org/10.1039/C8CC08331B
70. Menanteau T, Dabos-Seignon S, Levillain E, Breton T (2017) Impact of the diazonium grafting control on the interfacial reactivity: monolayer versus multilayer. ChemElectroChem 4:278–282. https://doi.org/10.1002/celc.201600710
71. Cesbron M, Dabos-Seignon S, Gautier C, Breton T (2020) Enhanced electrocatalytic activity on TEMPO mixed film grafted by diazonium reduction. Electrochim Acta 345:136190. https://doi.org/10.1016/j.electacta.2020.136190
72. González MCR, Orive AG, Salvarezza RC, Creus AH (2016) Electrodeposition of gold nanoparticles on aryl diazonium monolayer functionalized HOPG surfaces. Phys Chem Chem Phys 18:1953–1960. https://doi.org/10.1039/C5CP06415E
73. González MCR, Carro P, Vázquez L, Creus AH (2016) Mapping nanometric electronic property changes induced by an aryl diazonium sub-monolayer on HOPG. Phys Chem Chem Phys 18:29218–29225. https://doi.org/10.1039/C6CP05910D

Grafting of Aryl Radicals onto Surfaces—A DFT Study

Avni Berisha and Mahamadou Seydou

Abstract The atomistic modeling aspects of the mechanism of diazonium salt grafting on different materials including metals, carbon-based materials, silicone and phosphorus-based materials are discussed in this chapter. These concern the dissociation, the grafting and spectroscopic signatures of diazonium salts. The aryl group is grafted favorably on different materials like metals, carbon-based nanomaterial's, silicone and phosphorus two-dimensional materials. In all cases the adsorption was found exothermic allowing to non-covalent, covalent or ionic bonds formation. The spectroscopic analysis allows to depict the characteristic of aryl grafting concerning the stretching modes N–N and C–N and bending C–N–N shifting toward low frequencies.

1 Introduction

The diazonium chemistry technique of grafting chemical systems onto a surface has emerged as an attractive approach for surface functionalization [1]. Numerous applications on metallic surfaces and carbon-based nanomaterials reinforce the technique's interest and expansion to other types of surfaces [2, 3]. The characterization of the grafted surface from aryldiazonium salts has been the concern of a vast number of experimental studies as presented in previous book Chapters. The relevance of theoretical studies that provide a molecular level explanation of the grafting reaction is vast.

A. Berisha
Chemistry Department of Natural Sciences Faculty, University of Prishtina, rr. "Nena Tereze" nr. 5, 10000 Prishtina, Kosovo
e-mail: avni.berisha@uni-pr.edu

Materials Science Nanochemistry Research Group, NanoAlb—Unit of Albanian Nanoscience and Nanotechnology, 1000 Tirana, Albania

M. Seydou (✉)
Université Paris Cité, CNRS, ITODYS, F-75013, Paris, France
e-mail: mahamadou.seydou@u-paris.fr

M. M. Chehimi et al. (eds.), *Aryl Diazonium Salts and Related Compounds*, Physical Chemistry in Action, https://doi.org/10.1007/978-3-031-04398-7_6

The main interest concerns the mechanism of grafting a system using diazonium through the dissociation of N_2^+ and the bonding of aryl moiety. In this context, atomistic simulations including electronic interaction details is an appropriate tool. In particular thermodynamic and kinetic of adsorption are investigated using standard Density Functional Theory (DFT) [4–9]. Large and rich discussions about the binding structure, stability, geometry, grafting energetics, interaction of aryldiazonium salt, etc. exist in literature. Electronic and spectroscopic properties were also investigated focusing on Infrared/Raman signature of the grafting in particular the carbon or nitrogen bonding on the host surface. The effect on the host and guest electronic, optical properties and transport properties were also investigated [9–11].

Density functional theory (DFT) is the most often utilized theoretical method for studying the adsorption mechanism of diazonium salts. This is most likely due to the excellent trade-off between simulation duration and precision of the findings. Thus, this technique has been used for a variety of functionalities, both molecular and periodic, which are more suited to metals and carbon-based nanomaterials [5, 8, 12]. Hartree–Fock and semi-empirical methods [13] were also used to determine the binding energies and simulate the vibration spectra of these functionalized compounds. Other methods such as Monte-Carlo simulations [14] in the grand canonical set and molecular dynamics [15] have been utilized to monitor grafting dynamics and/or the influence of temperature to a lesser extent.

2 The Adsorption/Grafting of the Diazonium Salts

2.1 Metallic Systems

Jiang et al. [16, 17] theoretically explored the binding phenyl radical onto metallic surfaces (Fe, Pd, Cu, Au and Ti) and showed that the binding of this radical onto metals is of chemical nature [16]. The most stable structures of phenyl radicals attachment onto these surfaces is presented in Fig. 1. As seen from the Fig. 1, the early transition metals prefer a flat-lying position onto the metal surface, whereas late transition metals prefer an upright position.

The highest found values for of the binding energy for the phenyl radical are: $\Delta E_{\text{Si-H(diamond lattice type)}} = 70$ kcal/mol, $\Delta E_{\text{Ti[0001],hcp}} = 106.7$ kcal/mol, $\Delta E_{\text{Fe[110],bcc}} = 41.4$ kcal/mol, $\Delta E_{\text{Cu[111],fcc}} = 27.9$ kcal/mol, $\Delta E_{\text{Au[111],fcc}} = 24.0$ kcal/mol, $\Delta E_{\text{Pd[111],fcc}} = 37.3$ kcal/mol.

Using the previous experimental findings on the grafting behavior of sterically hindered diazonium salts [18], where is evidenced that bulky groups at the 3,5-positions of the parent aryldiazonium salt hinder the layer, Combellas et al. [19] theoretically studied the binding energy and geometry of 2- methylphenyl, 2,6-dimethylphenyl and unsubstituted phenyl on the Cu(111) surface. The binding energy of 2 -methyl substitution is reduced slightly (9%) and the phenyl ring tilts toward the unsubstituted side, whereas the 2,6-dimethyl substitution reduces the binding

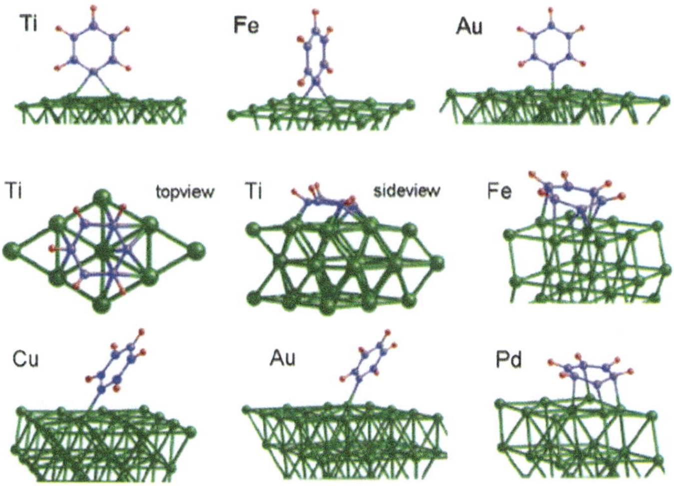

Fig. 1 Optimized structures of phenyl groups on metal surfaces for the upright and tilt or flat-lying configurations. Reproduced with permission from Ref. [16]. Copyright 2006 American Chemical Society

energy dramatically (>50%) and causes the Cu surface atom bounded to carbon upward the surface by as much as 0.6 Å. These results helped to corroborate their experimental finding. The upward move about 0.581 Å of the atom position (the C atom due to sp^2 to sp^3 hybridization change) on which the attached group (3,4,5-Tri-Methoxyphenyl) from diazonium is realized is evidenced for graphene surface through DFT calculation [20].

The formation of aryl-O bonds on iron oxide nanoparticles is shown via ab initio calculations and experimental measurements during the grafting of aryldiazonium salts. The binding energy of the phenyl group in the case of a carbon–iron bond is −59.15 kcal/mol, whereas for the carbon–oxygen bond the binding energy is −53.15 kcal/mol [21].

2.2 Au-C Bond Characterization

DFT calculations were used to show that during the electrografting of bromovaleric acid onto Au to produce Au–$(CH_2)_4$–COOH, a covalent bond is formed between the gold surface and the organic group. The observed Raman SERS spectroscopy Au–C band at 387 cm^{-1} identical to the calculated value, confirms this. The energy of Au-C(alkyl) bond is −37 kcal/mol, according to calculations performed on Au_{20} gold cluster. Furthermore, based on these findings, the authors compared the binding energy for thiols and demonstrated experimentally that the Au–$(CH_2)_4$–COOH can be reversibly transformed into Au–S–$(CH_2)_{17}$–CH_3 or Au–S–(CH_2)5–COOH when the grafted surface is simply transferred into thiol solution [22]. Médard et al. estimated that the Bond Dissociation Energy (BDE) values for the grafted carbon surface with the methylamino group ($C_{surface}$-C_{bond}) is -121.3 kcal/mol and for the grafted gold surface (Au surface-Cbond) −34.3 kcal/mol—a value close to that obtained for the valeric acid moiety grafted on gold: 36.9 kcal/mol. This indicates that the nitromethyl moiety's binding to the gold or carbon surface, as determined by BDE, forms a robust interface [23].

2.3 Evidence of Au–N Bond

Berisha et al. studied the mechanism of diazonium salt grafting through N or C [5]. Using DFT calculations and selective experimental production of carbocation and radicals, the binding energies for both hemolytic and heterolytic cleavages were explored. The results indicate that the homolytic cleavage is more favorable by 24.0 kcal/mol and that aryl radicals react with the surface of gold and the first grafted layer leading to nanometric thin films. The carbocations can also react on gold but the film growth is slower than with radicals and the obtained films were thinner. The diazonium cation can react directly with gold as it can be seen on Fig. 2. Moreover, the Au–(N=N–Ar) binding energy was found to be 7.9 kcal/mol, which is much weaker than the Au-Ar bond for which the binding energy was found to be 38.5 kcal/mol. Therefore, it appears that a direct cleavage of the Au–N bond catalyzed by the gold surface should also be considered. Further investigations should determine the relative importance of gold catalysis and carbocation reactions and exploit the use of carbocations to modify gold and other surfaces. The studies about the kinetic and the thermodynamic of diazonium taking into account the barrier of N_2 detachment and its diffusion on surface should be performed in order to elucidate the whole mechanism.

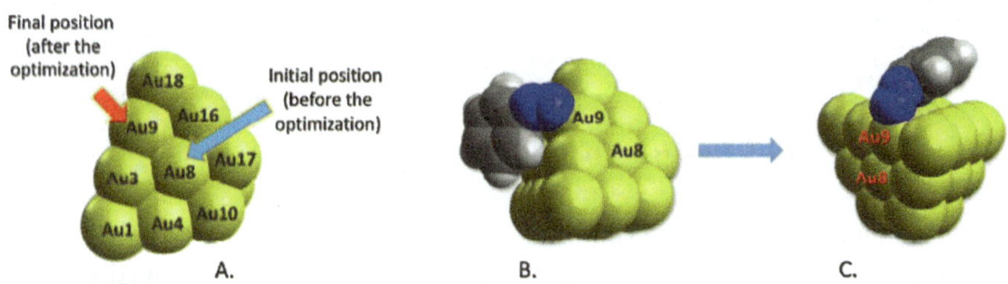

Fig. 2 Optimized structure for the Au19–N=N–Ph modified cluster. A. Initial and final positions of the aryl group in the Au-19 cluster. B,C. Two different views of the modified cluster. Reproduced with permission from Ref. [5]. Copyright 2017 American Chemical Society

2.4 Carbon Based Materials

In many materials, the adsorption of diazonium salts precedes the grafting reaction. Bensghaïer et al. studied the interaction of both: aryldiazonium salts (tetrafluoroborate salts of diazotized Azure A ($AA\text{–}N_2^+$), Neutral Red ($NR\text{–}N_2^+$) and CongoRed ($CR\text{–}N_2^+$) dyes and their corresponding radicals with the CNT. Relatively high binding energies between the dyes and the CNTs account for covalent bonding (33.46–44.22 kcal/mol with CNT-aryl distance 1.6 nm) presented in Fig. 3, according to quantum chemical calculations, though attachment by stacking also contributes to the formation of stable hybrids.

The adsorption of aryl diazonium salts is theoretically explored in case of other 2D nanomaterials like carbon (graphene oxide [4], graphydine [25]) or borophene [26]. The adsorption interaction of the benzenediazoniumtetrafluorborate onto borophene surface is clearly visible on Fig. 4, where the NonCovalent Interaction (NCI) surfaces are presented. In the case, the adsorption energy has a value of −24.03 kcal/mol, whereas the binding energy of the phenyl radical is 74.95 kcal/mol.

According to DFT calculations, aryl diazonium cations will bind more strongly to charged metallic SWCNTs species, resulting in preferential binding throughout the process as supported by experimental results [27].

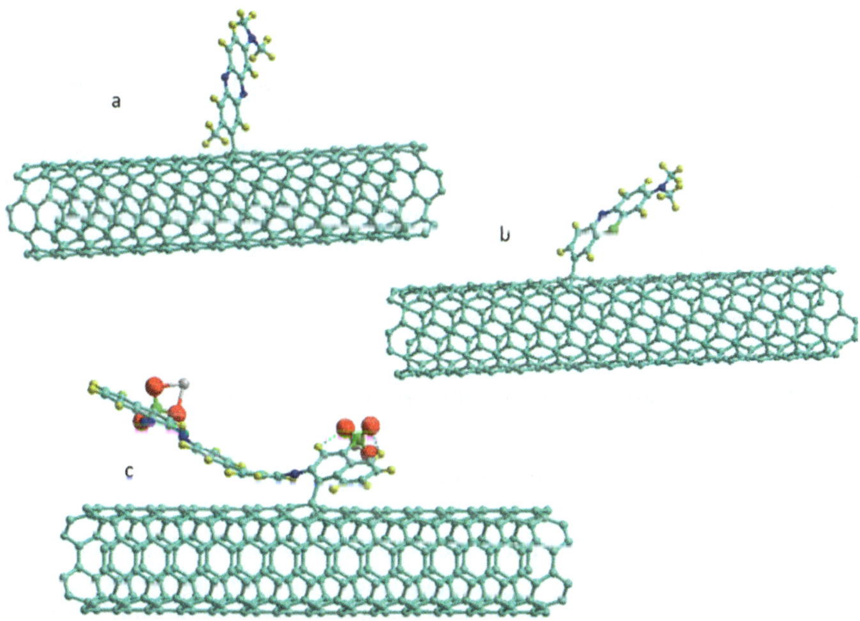

Fig. 3 Optimized structures of dye radicals grafted on (5, 5) CNT: AA• (**a**), NR• (**b**) and CR• (**c**). Carbon atoms are in cyan, oxygen in red, nitrogen in blue, sulfur in green, sodium in gray and hydrogen in yellow. Reproduced with permission from Ref. [24]. Copyright 2017 American Chemical Society

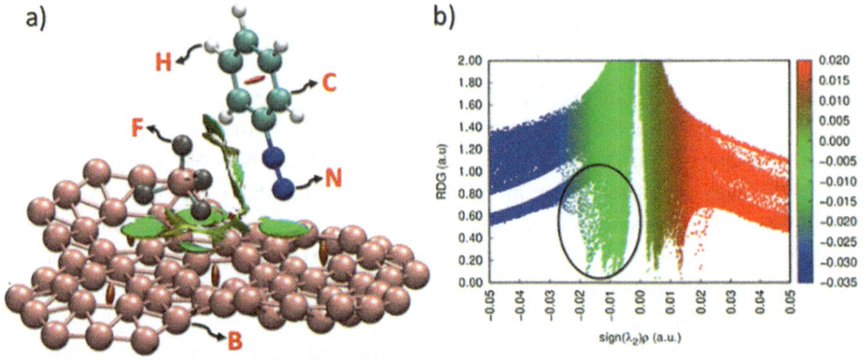

Fig. 4 **a** NonCovalent Interaction (NCI) surfaces and **b** the plot of RDG versus sign(λ)ρ for the interaction of benzenediazonium tetrafluoroborate/borophene. Reproduced with permission from Ref. [26]. Copyright 2021 Elsevier

The aryl group attaches to SWCNT covalently and favors a perpendicular orientation to the side wall, as presented in Fig. 5. With regard to rotation around the axis of the aryl-tube bond, a barrier of 8.07 kcal/mol was calculated, which is compatible with the equilibrium structure's thermodynamic stability at ambient temperature [6].

Adsorption energies in the case of graphene oxide lie in the range of −40 up to −60 kcal/mol in the case of vacuum and −22 up to −45 kcal/mol when the effect of the solvent is accounted. The range of adsorption energies is related to the type of aryl diazonium salt and adsorption sites on the graphene oxide surface [4]. Berisha investigated the adsorption energies of phenyl diazonium cation (PhN_2^+) on graphydine in vacuum, which ranged from −28.72 to −24.61 kcal/mol depending on the adsorption position. The adsorption energy (where the phenyldiazonium cation is flat into the central phenyl ring of graphyne with the diazonium group centered above the alkyne bond) was −14.85 kcal/mol in the presence of water as the solvent [25].

3 The Activation Energy of the Grafting Reaction

3.1 Carbon Based Nanomaterials

Ab initio calculations showed that paired phenyls attached onto the graphene are expected to remain grafted to the surface up to a few hundred degrees Celsius [28]. The binding energy for the grafted phenyl radical was −156.88 kcal/mol. The author suggests that there is no concerted bond-breaking/bond-re-forming mechanism in the diffusion case, Fig. 6.

When a phenyl group bonds to a graphene sheet, the sp^2 carbon is converted to an sp^3 carbon. Figure 7 displays the adsorption energy ($E_{ads.}$) of one phenyl group

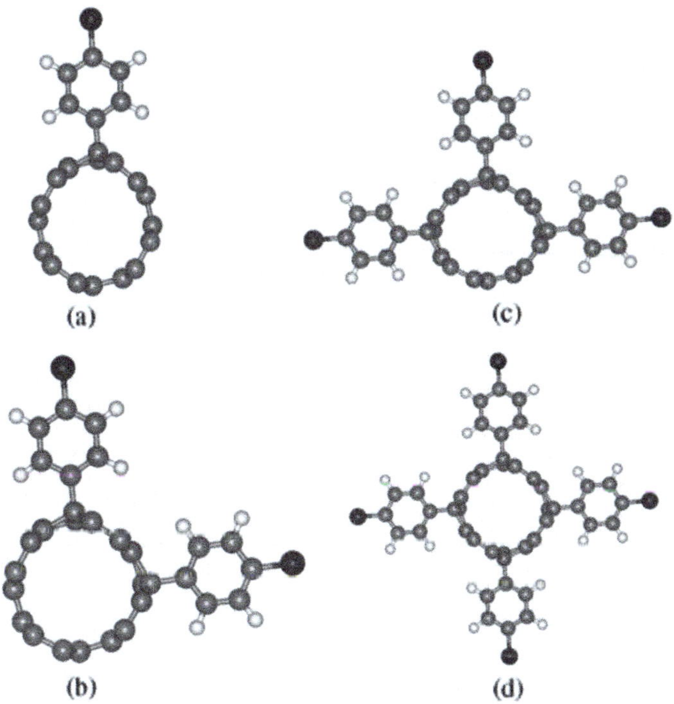

Fig. 5 Geometry obtained from fully optimization of the diazonium functionalized SWCNT at different coverages of the functional groups. The grey, black and white balls represent C, Cl and H atoms, respectively. Reproduced with permission from Ref. [6]. Copyright 2008 Taylor & Francis Group

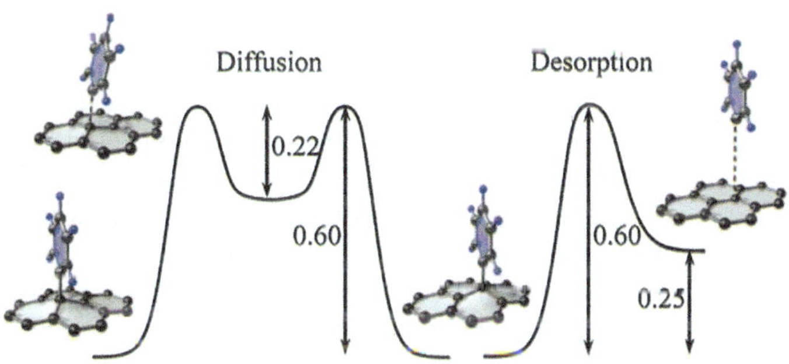

Fig. 6 Left barrier: energy profile and atomic configurations of initial, intermediate, and final states for phenyl diffusion. Right barrier: energy profile and atomic configurations of initial and final states for phenyl desorption. Energies are in electronvolts; carbon and hydrogen atoms are shown as gray and blue spheres, respectively. Graphene-phenyl distances at initial and transition states for diffusion (desorption) are 1.60 Å (1.60 Å) and 2.21 Å (2.24 Å). Reproduced with permission from Ref. [28] Copyright 2012 American Chemical Society

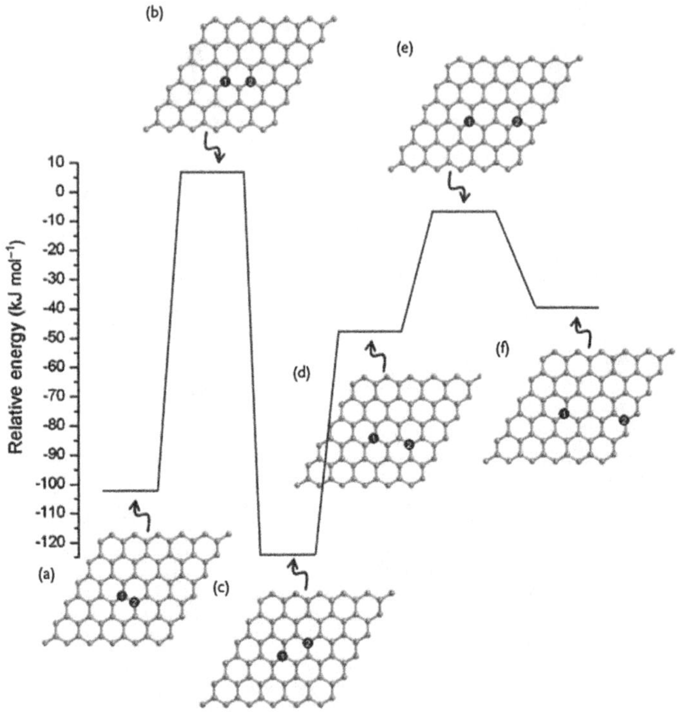

Fig. 7 Structures and energetics of different sites for the second attachment of phenyl on the graphene sheet. Black circles with the numbers 1 and 2 represent the site for the first and second phenyl attachments. Reproduced with permission from Ref. [29]. Copyright 2008 American Chemical Society

on a graphene sheet. The attachment of this group to the graphene sheet is not significant, meaning that the interaction between the aryl group and the graphene sheet is relatively weak, similarly to conventional physisorption (12 kcal/mol). This weak interaction suggests that a single attachment attack by the phenyl radical on the graphene sheet is challenging. The C–C bond that is produced in the optimal structure has a length of 1.60 Å, and the graphene carbon is pushed out of the surface by nearly 0.7 Å, creating strain in the graphene sheet by being changed into an sp^3 carbon. The initial phenyl group's attachment results in an unpaired electron on graphene. The unpaired electron is not concentrated on a single graphene carbon, but rather dispersed among a large number of carbon atoms in the vicinity of the transformed sp^3 carbon. Jiang et al. [29] showed that the binding energy is highly dependent on the surface coverage of the phenyl groups. The highest binding energy for a single phenyl groups onto the graphene is 29.4 kcal/mol. The authors found that the situation with regard to the binding energy onto the graphene-phenyl surfaces changes drastically, as shown in Fig. 7.

The electronegativity and placement of the halogen atoms on the grafted phenyl rings have a considerable impact on the graphene resistivity due to inductive and mesomeric effects [30]. The experimental results are in good accord with theoretical ab initio electron density estimates presented in Fig. 8.

Fine-tuning of the transport properties for the grafted graphene surfaces is evidenced by using non-equilibrium Green's function (NEGF) methods combined with density functional theory (DFT) calculations. The authors show that double-sided functionalized graphene is more energetically favorable than single-sided functionalized graphene, and that they display an unusual non-magnetic metallic state as well as a magnetic semiconducting state, respectively [21].

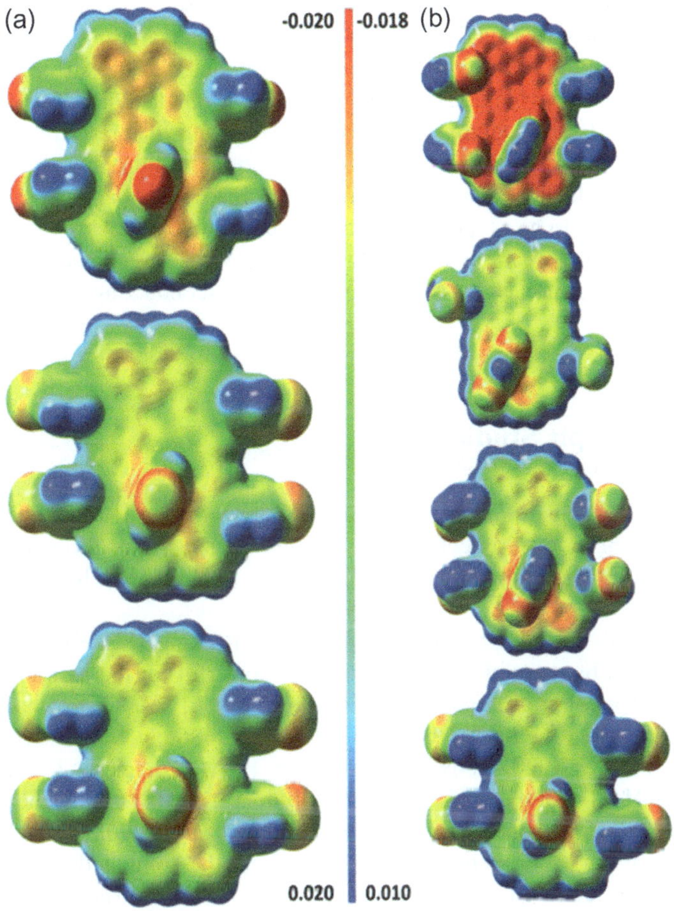

Fig. 8 Electron distribution on **a** p-fluorophenyl, p-chlorophenyl and p bromophenyl (from top), **b** o-chlorphenyl, 3,5 dichlorophenyl, m-chlorphenyl and p-chlorophenyl (from top) graphene sheets. Electron rich areas are in red, electron poor in blue. The scale is relative and limit values for the electron distribution are on both sides of the scale for both sets of images. Reproduced with permission from Ref. [30]. Copyright 2016 Royal Society of Chemistry

3.2 Phosphorus and Silicone Based Materials

According to DFT calculations, aryl groups' chemisorption onto exfoliated black phosphorus via the formation of P–C bonds onto a supercell of 16 phosphorus atoms is thermodynamically favorable, with two aryl moieties per supercell being more favorable than one aryl moiety. However, because phosphorus atoms adopt four-coordinate bonding, the introduction of new covalent bonds appears to cause significant lattice distortion. When the second aryl group is grafted (from −32.28 to −35.97 kcal/mol), the binding energy (for the phenyl moiety) increases [31]. The covalent modification of the few-Layer Black Phosphorus I also achieved by using aryl iodonium salts, offering the possibility to attach alkyl or aryl moieties [32]. The binding energy of the triazole moiety onto gold (modeled using Au13 gold cluster) is $\approx$56 kcal/mol (in solvent: water or acetonitrile) or $\approx$51 kcal/mol in vacuum [33].

A silicone surface containing nitrophenyl diazonium is theoretically investigated for spintronic applications. Different coverage rations of nitrophenyl groups were grafted onto the silicene surface, resulting in a decrease of its bandgap. The silicene system is predicted to be a bipolar magnetic semiconductor in the case of the grafted nitrophenyl group at a particular grafting ratio. The work investigated a tailor-made functionalization approach to realize magnetic semiconducting silicone surfaces [34].

4 Spectroscopic Studies

The use of DFT calculations to provide information with regard to direct peak assignment by comparing experimental versus theoretical spectra or through the use of Potential Energy Distribution (PED) [35] analysis has proven to be very important [32, 36].

Experimental Surface-Enhanced Raman Spectroscopy (SERS) measurements when supported by theoretical calculations are powerful for the assignment of the gold-carbon bond. For example: (a) Au20–Ph–NO_2 [36] (Au–C stretch assigned at 412 cm^{-1}); (b) Au20–Ph–CN (Au–C stretch, exp. at 488 cm^{-1} versus computed at 514 cm^{-1}) [37]; (c) Au20–Ph–NO_2 (M=Au: exp. 413 versus calc. at 424 cm^{-1}) [38]; (d) Au20–$(CH_2)_4$–COOH (at 387 cm^{-1}) [22]. Barosi et al. [10] studied the grafting of the picolinium-derived end-tethered polymer chains onto the gold surface, by close comparison of the experimental SERS spectra with the computed on (using Au20 gold cluster) presented in the Fig. 9 assigned to the Au–C stretching vibration at 410 cm^{-1}.

Minaev et al. [9]used DFT calculations at the B3LYP/6-31G(d,p) level to analyze the IR and Raman spectra of some aryl diazonium cations, X-$C_6H_4N_2^+$, functionalized with different chemical groups (X=Cl, Br, F, NO_2, NH_2, CH_3, CF_3, CN, OH, OCH_3, COOH, $COOC_2H_5$, etc.) in *ortho*-, *meta*- and *para*-positions. Calculations allowed the assignment of the characteristic bands of N–N and C-N bonds as shown

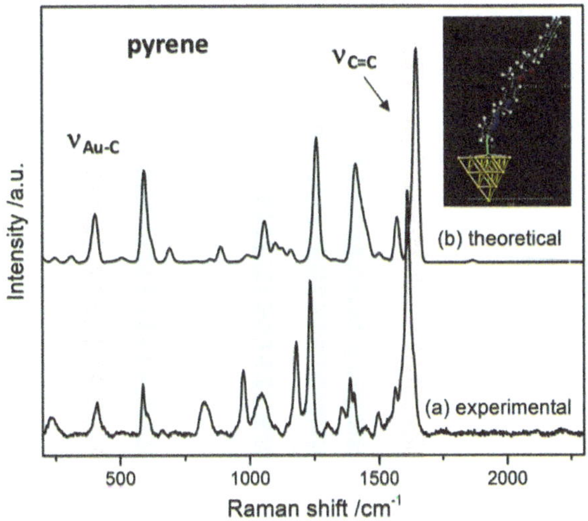

Fig. 9 SERS spectrum of Au20 grafted surface. The inset shows the optimized structure of grafted@Au20 used to compute the Raman spectrum (on the top) and experimental SERS spectrum of grafted@Au (on the bottom). Reproduced with permission from Ref. [10]. Copyright 2021 Royal Society of Chemistry

in Table 1. They proved the relationship between mesomeric effect on N–N and C-N vibrations modes depending on the substituent. Regarding the substation from donor to acceptor groups, the N–N bond order is well correlated with the frequencies decrease and the amplification intensity of the characteristic mode. For the radical cation, a distorted geometry was found leading to a red-shift of N–N frequency of 490 cm^{-1} relative to the neutral species. The binding character of the orbital describing the C–N bond demonstrates the absence of nitrogen loss during the Meerwein reaction. Therefore, the authors explain that the simple reduction of the cation is not sufficient to explain the detachment of N_2. A probable involvement of the triplet excited state of aryl diazonium in the reaction path is presented as an important factor. It is clear that the mechanism of N_2 elimination is not yet clearly understood and further studies are needed to clarify it.

More recently, Ramirez et al. [11] studied the optical spectra of carbon nanotubes functionalized by diazonium moieties and their chlorinated derivatives. Both ground and excited properties were finely described. The results indicated that adsorption of aryl diazonium on the CNT by π-stacking leads to a small redshifts of the optical spectra, while chemisorption causes a significant redshift and a strong increase in the oscillator strength of the lowest exciton known to be dark in pristine CNT as indicated in Fig. 10.

Table 1 Several vibrational bands of aryl diazonium cations X-C6H4N2 + calculated by DFT/B3LYP/6-31G** (Adapted from Ref. [9] Copyright 2009 Elsevier)

X	ν[N–N] (cm^{-1})	ν[C-N] (cm^{-1})	"Kekule" vibration (cm^{-1})
H	2349	1143	1386
o-Cl	2345	1162	1370
m-Cl	2352	1160	1380
p-Cl	2334	1162	1366
o-NO_2	2365	1171	1396
m-NO_2	2359	1164	1384
p-NO_2	2361	1134	1384
o-CH_3	2337	1117	1369
m-CH_3	2345	1143	1387
p-CH_3	2335	1157	1372
o-OH	2303	1126	1416
m-OH	2344	1134	1409
p-OH	2318	1176	1410
o-OCH_3	2330	1053	1384
m-OCH_3	2352	1120	1394
p-OCH_3	2312	1173	1383
o-$COOC_2H_5$	2335	1135	1373
p-$COOC_2H_5$	2322	1168	1374
p-F	2339	1159	1378
p-Br	2331	1162	1365
p-NH_2	2295	1188	1396
p-COOH	2353	1140	1374
p-SO_3H	2356	1138	1368
p-CN	2342	1144	1363
2,4,6-NO_2	2398	1130	1387
2,4,6-OH	2318	1254	1426
p-CF_3	2355	1139	1376
p-N'$(CH_3)_3$	2384	1133	1360

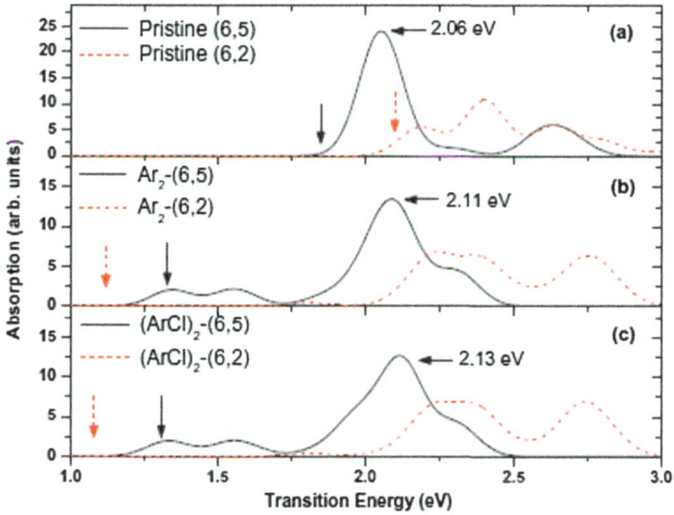

Fig. 10 Absorption spectra of (6, 5) and (6, 2) SWNTs and their functionalized derivatives calculated in vacuum using wB97XD/3-21G/ methodology. **a** Absorption spectra of pristine tubes; **b** Doubly arylated-SWNT hybrids; **c** Chlorinated derivatives of the doubly-arylated systems. Functionalization is realized via circumferential attachment of molecules to the different carbon rings at the opposite sides of the nanotube. Vertical arrows define the lowest-energy transition with the black color corresponding to the (6,5) and red to the (6,2) SWNT. Reprinted figure with permission from Ramirez et al. Reproduced with permission from Ref. [11]. Copyright 2012 Elsevier

References

1. Mahouche-Chergui S, Gam-Derouich S, Mangeney C, Chehimi MM (2011) Aryl diazonium salts: a new class of coupling agents for bonding polymers, biomacromolecules and nanoparticles to surfaces. Chem Soc Rev 40:4143–4166. https://doi.org/10.1039/C0CS00179A
2. Hari DP, König B (2013) The photocatalyzed Meerwein arylation: classic reaction of aryl diazonium salts in a new light. Angew Chem Int Ed 52:4734–4743. https://doi.org/10.1002/anie.201210276
3. Mo F, Dong G, Zhang Y, Wang J (2013) Recent applications of arene diazonium salts in organic synthesis. Org Biomol Chem 11:1582–1593. https://doi.org/10.1039/C3OB27366K
4. Berisha A (2019) Interactions between the aryldiazonium cations and graphene oxide: a DFT study. J Chem. https://doi.org/10.1155/2019/5126071
5. Berisha A, Combellas C, Kanoufi F, Decorse P, Oturan N, Médard J, Seydou M, Maurel F, Pinson J (2017) Some theoretical and experimental insights on the mechanistic routes leading to the spontaneous grafting of gold surfaces by diazonium salts. Langmuir 33:8730–8738. https://doi.org/10.1021/acs.langmuir.7b01371
6. Du AJ, Smith SC (2006) Structural and electronic properties of diazonium functionalized (4, 4) single walled carbon nanotube: an ab initio study. In: Molecular simulation. Taylor & Francis Group, pp 1213–1217
7. Li DO, Chu XS, Wang QH (2019) Reaction kinetics for the covalent functionalization of two-dimensional MoS2 by aryl diazonium salts. Langmuir 35:5693–5701. https://doi.org/10.1021/acs.langmuir.8b04288
8. Lo M, Seydou M, Bensghaïer A, Pires R, Gningue-Sall D, Aaron J-J, Mekhalif Z, Delhalle J, Chehimi MM (2020) Polypyrrole-wrapped carbon nanotube composite films coated on

diazonium-modified flexible ITO sheets for the electroanalysis of heavy metal ions. Sensors 20. https://doi.org/10.3390/s20030580
9. Minaev BF, Bondarchuk SV, Gîrţu MA (2009) DFT study of electronic properties, structure and spectra of aryl diazonium cations. J Mol Struct (Thoechem) 904:14–20. https://doi.org/10.1016/j.theochem.2009.02.022
10. Barosi A, Berisha A, Mangeney C, Pinson J, Dhimane H, Dalko PI (2021) Efficient construction of a redox responsive thin polymer layer on glassy carbon and gold surfaces for voltage-gated delivery applications. Mater Adv 2:2358–2365. https://doi.org/10.1039/d1ma00022e
11. Ramirez J, Mayo ML, Kilina S, Tretiak S (2013) Electronic structure and optical spectra of semiconducting carbon nanotubes functionalized by diazonium salts. Chem Phys 413:89–101. https://doi.org/10.1016/j.chemphys.2012.10.010
12. Bensghaïer A, Forro K, Seydou M, Lamouri A, Mičušik M, Omastová M, Beji M, Chehimi MM (2018) Dye diazonium-modified multiwalled carbon nanotubes: light harvesters for elastomeric optothermal actuators. Vacuum 155:178–184. https://doi.org/10.1016/j.vacuum.2018.06.018
13. Eldesoky AM, Nozha SG (2017) The adsorption and corrosion inhibition of 8-hydroxy-7-quinolinecarboxaldehyde derivatives on C-steel surface in hydrochloric acid. Chin J Chem Eng 25:1256–1265. https://doi.org/10.1016/j.cjche.2016.11.018
14. Oukhrib R, Abdellaoui Y, Berisha A, Abou Oualid H, Halili J, Jusufi K, Ait El Had M, Bourzi H, El Issami S, Asmary FA, Parmar VS, Len C (2021) DFT, Monte Carlo and molecular dynamics simulations for the prediction of corrosion inhibition efficiency of novel pyrazolylnucleosides on Cu(111) surface in acidic media. Sci Rep 11:3771. https://doi.org/10.1038/s41598-021-82927-5
15. Berisha A (2020) Experimental, Monte Carlo and molecular dynamic study on corrosion inhibition of mild steel by pyridine derivatives in aqueous perchloric acid. Electrochem 1:188–199. https://doi.org/10.3390/electrochem1020013
16. Jiang DE, Sumpter BG, Dai S (2006) Structure and bonding between an aryl group and metal surfaces. J Am Chem Soc 128:6030–6031. https://doi.org/10.1021/ja061439f
17. Tang Q, Jiang DE (2016) Computational insight into the covalent organic-inorganic interface. Chem Mater 28:5976–5988
18. Combellas C, Kanoufi F, Pinson J, Podvorica FI (2008) Sterically hindered diazonium salts for the grafting of a monolayer on metals. J Am Chem Soc 130:8576–8577. https://doi.org/10.1021/ja8018912
19. Combellas C, Jiang DE, Kanoufi F, Pinson J, Podvorica FI (2009) Steric effects in the reaction of aryl radicals on surfaces. Langmuir 25:286–293. https://doi.org/10.1021/la8025792
20. Ambrosio G, Drera G, Di Santo G, Petaccia L, Daukiya L, Brown A, Hirsch B, De Feyter S, Sangaletti L, Pagliara S (2020) Interface chemistry of graphene/Cu grafted by 3,4,5-tri-methoxyphenyl. Sci Rep 10:1–9. https://doi.org/10.1038/s41598-020-60831-8
21. Brymora K, Fouineau J, Eddarir A, Chau F, Yaacoub N, Grenèche JM, Pinson J, Ammar S, Calvayrac F (2015) Grafting of diazonium salts on oxides surface: formation of aryl-O bonds on iron oxide nanoparticles. J Nanopart Res 17:1–9. https://doi.org/10.1007/s11051-015-3232-x
22. Berisha A, Combellas C, Kanoufi F, Médard J, Decorse P, Mangeney C, Kherbouche I, Seydou M, Maurel F, Pinson J (2018) Alkyl-modified gold surfaces: characterization of the Au-C bond. Langmuir 34:11264–11271. https://doi.org/10.1021/acs.langmuir.8b01584
23. Médard J, Berisha A, Decorse P, Kanoufi F, Combellas C, Pinson J, Podvorica FI (2020) Electrografting of methylamine through C-H activation or oxidation to give highly aminated surfaces. Electrochim Acta 345:136170. https://doi.org/10.1016/j.electacta.2020.136170
24. Bensghaïer A, Lau Truong S, Seydou M, Lamouri A, Leroy E, Mičušik M, Forro K, Beji M, Pinson J, Omastová M, Chehimi MM (2017) Efficient covalent modification of multiwalled carbon nanotubes with diazotized dyes in water at room temperature. Langmuir 33:6677–6690. https://doi.org/10.1021/acs.langmuir.7b00711
25. Berisha A (2019) The influence of the grafted aryl groups on the solvation properties of the graphyne and graphdiyne—a MD study. Open Chem 17:703–710. https://doi.org/10.1515/chem-2019-0083

26. Berisha A (2021) First principles details into the grafting of aryl radicals onto the free-standing and borophene/Ag(1 1 1) surfaces. Chem Phys 544:111124. https://doi.org/10.1016/j.chemphys.2021.111124
27. Schirowski M, Tyborski C, Maultzsch J, Hauke F, Hirsch A, Goclon J (2019) Reductive diazotation of carbon nanotubes: an experimental and theoretical selectivity study. Chem Sci 10:706–717. https://doi.org/10.1039/c8sc03737j
28. Margine ER, Bocquet ML, Blase X (2008) Thermal stability of graphene and nanotube covalent functionalization. Nano Lett 8:3315–3319. https://doi.org/10.1021/nl801718f
29. Jiang DE, Sumpter BG, Dai S (2006) How do aryl groups attach to a graphene sheet? J Phys Chem B 110:23628–23632. https://doi.org/10.1021/jp065980
30. Bouša D, Pumera M, Sedmidubský D, Šturala J, Luxa J, Mazánek V, Sofer Z (2016) Fine tuning of graphene properties by modification with aryl halogens. Nanoscale 8:1493–1502. https://doi.org/10.1039/c5nr06295k
31. Ryder CR, Wood JD, Wells SA, Yang Y, Jariwala D, Marks TJ, Schatz GC, Hersam MC (2016) Covalent functionalization and passivation of exfoliated black phosphorus via aryl diazonium chemistry. Nat Chem 8:597–602. https://doi.org/10.1038/nchem.2505
32. Van Druenen M, Davitt F, Collins T, Glynn C, O'Dwyer C, Holmes JD, Collins G (2018) Covalent functionalization of few layer black phosphorus using iodonium salts and comparison to diazonium modified black phosphorus. Chem Mater 30:4667–4674. https://doi.org/10.1021/acs.chemmater.8b01306
33. Orqusha N, Phal S, Berisha A, Tesfalidet S (2020) Experimental and theoretical study of the covalent grafting of triazole layer onto the gold surface. Materials 13:2927. https://doi.org/10.3390/ma13132927
34. Dai J, Zeng XC (2015) Covalent nitrophenyl diazonium functionalized silicene for spintronics: a first-principles study. Phys Chem Chem Phys 17:17957–17961. https://doi.org/10.1039/c4cp04953e
35. Jamróz MH (2013) Vibrational energy distribution analysis (VEDA): scopes and limitations. Spectrochimica Acta Part A: Mol Biomol Spectrosc 114:220–230. https://doi.org/10.1016/j.saa.2013.05.096
36. Laurentius L, Stoyanov SR, Gusarov S, Kovalenko A, Du R, Lopinski GP, McDermott MT (2011) Diazonium-derived aryl films on gold nanoparticles: evidence for a carbon-gold covalent bond. ACS Nano 5:4219–4227. https://doi.org/10.1021/nn201110r
37. Mohamed AA, Neal SN, Atallah B, AlBab ND, Alawadhi HA, Pajouhafsar Y, Abdou HE, Workie B, Sahle-Demessie E, Han C, Monge M, Lopez-de-Luzuriaga JM, Reibenspies JH, Chehimi MM (2018) Synthesis of gold organometallics at the nanoscale. J Organomet Chem 877:1–11. https://doi.org/10.1016/j.jorganchem.2018.07.032
38. Ahmad R, Boubekeur-Lecaque L, Nguyen M, Lau-Truong S, Lamouri A, Decorse P, Galtayries A, Pinson J, Felidj N, Mangeney C (2014) Tailoring the surface chemistry of gold nanorods through Au-C/Ag-C covalent bonds using aryl diazonium salts. J Phys Chem C 118:19098–19105. https://doi.org/10.1021/jp504040d

Modification of sp^2 Carbon Allotropes with Diazonium Salts—Focus on Carbon Nanotubes Functionalization

Christine Vautrin-Ul

Abstract After a summary of the chronology of early studies on surface modification using diazonium salt presented for all types of sp^2 carbonaceous materials of macro, micro, and nano size, this chapter reviews the last decade's advances in the field of the carbon nanotube functionalization via diazonium salt chemistry. The different strategies of carbon nanotube functionalization based on aryl diazonium salt are then overviewed. The key parameters determining the selectivity of the diazonium salt functionalization reaction of single-wall carbon nanotubes (SWCNT) are discussed and more particularly the importance of their Fermi level as well as the substituents of the aryl diazonium. This chapter then focuses on the optical properties, including photoluminescence and electrical properties of the nanotubes specifically provided by the surface modification of CNTs via diazonium salt, and concludes with the presentation of some applications including catalytic platforms, nanocomposites, energy conversion, and (bio) chemical sensing.

1 Introduction

The electrochemical grafting of diazonium salts onto glassy carbon was first demonstrated in 1992 by Jean Pinson et al. [1]. Glassy carbon (GC) is a disordered amorphous carbon material that is non-graphitizing and composed largely of sp^2 hybridized carbon. Its structure was once thought to be an isotropic entanglement of layered graphitic ribbons [2], however it is now considered more likely to be comprised of either curved and folded graphitic sheets [3], or nanoonions [4]. In the same decade, the electrochemical grafting via a similar diazonium method was highlighted for other sp^2 ordered or disordered carbon materials: carbon fiber [5], Highly Oriented Pyrolytic Graphite (HOPG) [6], carbon black [7]Pyrolyzed Photoresists Film (PPF) [8] and from 2000–2010 has focused on the sp^2 carbon nanomaterials: carbon nanotubes SWCNTs (Single-Wall Carbon NanoTubes) [9], MWCNTs

C. Vautrin-Ul (✉)
Laboratoire Interfaces, Confinement, Materials and Nanostructures (ICMN)—UMR 7374, University of Orleans/CNRS, 2b rue de la Férollerie, 45 071, Cedex 2 Orléans, France
e-mail: christine.vautrin-ul@cnrs-orleans.fr

M. M. Chehimi et al. (eds.), *Aryl Diazonium Salts and Related Compounds*, Physical Chemistry in Action, https://doi.org/10.1007/978-3-031-04398-7_7

(Multiwall Carbon NanoTubes) [10] nanofibers [11], nano-horns [12], ordered mesoporous carbon [13], graphene [14] and more recently activated carbon (charcoal) [15], nanoonions [16], HSAG (High Surface Area nano-sized Graphite) [17], Graphene Quantum Dots (GQD) [18] and Carbon nanoDots (CD) [19].

A large number of functions can be grafted onto sp^2 carbon nanomaterials in aqueous, organic, or ionic liquid media [20]. Several grafting methods based on diazonium chemistry have been proposed: electrochemical or chemical reduction of diazonium salt generated or not *in-situ* are most used, grafting can also be induced by microwave or laser irradiation [21]. All these methods are not specific to the grafting of sp^2 carbon.

Diazonium chemistry is recognized as providing versatile building blocks in the chemist tool to design nanomaterials. Functionalization by diazonium grafting has had as booster effect on the development of applications of nanomaterials allowing the assembly of nanoobjects and the anchoring of nanocarbons on surfaces. Moreover, the intrinsic properties of nanocarbons such as their electrical conductivity, their chemical resistance, mechanical and optical properties as well as the properties of their surface induced by the diazonium functionalization have strengthened their potentiality for many high-tech applications such as: energy, chemical or biochemical sensors, or molecular electronics for example.

Among carbon nanomaterials, functionalization of carbon nanotubes via diazonium reaction has been extensively studied [21]for dispersing CNTs (Carbon Nano Tubes)in liquids, sorting CNTs by chirality, integrating CNTs into high-strength materials, attaching CNTs to specific locations on a semiconductor device and making CNT-based bioelectronic devices with biologically functional coatings [22, 23] and more recently [24], increasing the quantum yield of CNTs photoluminescence or the creation of optically active defect sites that can generate single photons on demand [25]. Several recent reviews have been written on the diazonium functionalization of nanomaterials [26] concerning particular functionalization of CNTs [20, 21].

This chapter will be dedicated essentially to the advances of the last ten years concerning the diazonium chemistry used for the functionalization of carbon nanotube materials chosen for their great importance in many applications.

2 Functionalization of the CNTs by Diazonium Chemistry. An Overview

2.1 *Electrochemical Grafting of CNT*

Grafting of CNTs was initially demonstrated by electrochemistry. However, this is not the easiest way to functionalize CNTs and more generally nanoobjects since it requires the preparation of robust and reproducible CNT electrodes. The group of

J. Tour [9] pioneered the functionalization of carbon nanotubes by electrochemical reduction of diazonium salt using CNT bulky paper electrode.

Chemical Vapor Deposition (CVD) is one of the most frequently used methods for the direct synthesis of vertically aligned films of CNTs (ACNTs) on different substrates [27]. ACNTs grown on conductive substrates are well-aligned, densely arrayed, and mutually separated, offering many important advantages over the entangled carbon nanotubes, such as high specific surface area and high electrical conductivity. It has been established that the electrografting of diazonium salts on ACNTs is a mild and efficient surface modification procedure that does not alter the ACNTs special structure. The ACNT modification permits to open of the tips and decorate the tubes or sensing applications [28].

Other approaches are based on CNT dispersion drop-casting or on electrophoretic deposition on several substrates such as gold or ITO (Indium Tin Oxide); they lead to CNT electrodes able that can be used for (bio) sensing application after functionalization via diazonium electroreduction [27].

Covalent anchoring of carbon nanotubes onto conductive substrates can be obtained by electroreduction of the diazonium salt generated *in-* or *ex-situ*. Although the precise orientation of CNT tethered on substrates by C–C bonds using diazonium salt reactions is argumentative [29, 30], it is widely agreed that the SWCNT interface is very stable in agreement with the known stability of tethered layers grafted from diazonium salts. Meanwhile, the assembly of CNT on the grafted aryldiazonium salt layer on an electrode surface greatly increased the electronic coupling between the electrode [31] and the outside biological environment, leading to biosensing applications [30].

Often, in order to have CNT electrodes that are easy to use, the sp^2 nanocarbons are incorporated into a formulation containing at least one binder and often conductive charge as carbon black to reach the percolation. This formulated paste can be used to prepare CNT electrodes: screen printed electrodes (SPE), carbon paste electrodes, or composite electrodes for example. However, the addition of a binder and often conductive charges influences the properties of these electrodes which cannot be attributed only to the presence of the nanocarbon. It should be noted that it is possible to buy commercial electrodes made from CNTs or other sp^2 nanocarbons (nanofibers, mesoporous carbons, graphene). Many publications deal with the chemical or biochemical sensing applications of commercial imprinted carbon nanotube electrodes via electrochemical grafting from diazonium salts [27].

2.2 *Other Methods of CNTs Functionalization via Diazonium Chemistry*

Other methods based on diazonium salts have been developed allowing direct grafting of the CNTs. The most studied is the chemical grafting in organic or aqueous solvent via the diazonium salt generated *ex-situ* or *in-situ* [26]. After dispersion in presence or not of a surfactant (Sodium Dodecyl Sulfate (SDS) for example) and sonication of the CNTs in the solvent; the functionalization can be carried out in aqueous or organic media using either a diazonium salt or an aniline derivative and a nitrite (usually isopentylnitrite in organic media or sodium nitrite in water) [32]. The reaction mixture is usually kept under stirring and under inert atmosphere and often heated (until 80 °C) during at least 12 h and sometimes even 48 h [20]. The functionalization via diazonium chemistry has been mostly studied for SWCNTs in comparison with MWCNTs. Due to their lower reactivity, their more difficult dispersion; the degree of functionalization is lower for MWCNTs than for SWCNTs in the same condition [10].

The most accepted proposed mechanism is based on the Gomberg–Bachmann mechanism, in which the reaction occurs via electron transfer from CNT into the diazonium species, releasing N_2 and forming a reactive aryl radical that reacts with CNTs to form arene bonds [10]. The degree of grafting obtained by this method leads to functionalized CNTs with a defect ratio, a functionalization degree and the solubility are very high. This is due to the very high reactivity of the radicals involved in this process. An illustration is the dispersion of SWCNTs in SDS solution in water in presence of diazonium salt, which leads to the functionalization with the aryl group of one carbon of the CNTs out of nine [10].

More lately, Schmidt et al. have studied the kinetic of the CNTs functionalization by the diazonium reaction in aqueous solution via absorption spectroscopy and Electron Spin Resonance Spectroscopy (ESR) [33, 34]. Their work has confirmed a free-radical chain mechanism initiated mainly by the diazo-anhydride homolytic decomposition followed by the electrophilic addition of the aryl radical on the SWCNT, this last step determines the selectivity. Different levels of selectivity and reactivity have been observed depending on the para substituents of the diazonium salts: electron-withdrawing groups ($-NO_2$ substituent for example) accelerate the reactions while electron-donating functions ($-OCH_3$ group for example) lead to slow reaction kinetic. In addition, if the electrophilicity of the aryl radical substituent is enhanced, the selectivity improves. However, it is observed that the tendency of the self-coupling (corresponding to the termination) increases with the electrophilicity of the substituent. It should be noted that the reaction rate changes a lot to turn to the ratio of metallic carbon nanotubes (m-CNTs) to semiconducting carbon nanotubes (sc-CNTs): diazonium salts functionalized more slowly than pure sc-CNT, whereas their coupling is accelerated on a mixture of sc-CNTs and m-CNTs.

The interaction of the diazonium salt with the CNTs surface is determined both by the nanotubes structure and the choice of the diazonium salt and its concentration as well as by the temperature of the functionalization reaction, the nature of the surfactant, and the pH of the aqueous medium [35, 36]. Some variants of these chemical approaches consist in adding a reductive agent such as iron powder [37] or hypophosphorous acid [38].

Rapid, economic, and environmentally friendly chemical approaches have been proposed for the functionalization of SWCNTs, they involve the room temperature grinding of CNT for several minutes with various aryldiazonium salts in the presence of potassium carbonate and various imidazolium-based ionic liquids as the solvent medium [39]; or the use of molten urea as green solvent [36].

Other approaches involved thermal functionalization [26], radiation-induced modification for example via microwave [21, 40] or via UV light [41] or visible light [42].

3 Properties of the CNT Induced by the Functionalization via Diazonium Chemistry

3.1 Solubilization—Application to the Separation of Metallic to Semiconducting SWCNT

The electronic properties of a nanotube change in correspondence to its structure [21]: single-walled carbon nanotubes (SWCNTs) vary from metallic to semiconducting, thus armchair nanotubes are metallic, while zigzag and chiral can be either metallic or semiconducting. While multi-walled carbon nanotubes (MWCNTs) are regarded as metallic conductors.

Pure SWCNTs represent a family of molecules with distinct properties; the differences in structure, especially size, strain, and symmetry, lead to variations in effective densities, electronic band structures (molecular orbitals), and dielectric properties as polarisability, spectroscopic features, and surface adsorption characteristics. Then these differences in the structure of SWCNTs obtained by the different elaboration methods lead to an alteration of their properties and consequently to their performances in the applicative devices [34]. For applications such as photovoltaic and photodetector devices highly pure sc-SWCNTs are essential because excitons generated from sc-SWCNT would be rapidly quenched in the vicinity of m-SWCNTs, even though m-SWCNTs are present in small quantities [43]. In order to exploit the exceptional properties of SWCNTs, it is necessary to apply time-consuming and costly separation methods such as centrifugation or electrophoresis based on the difference between the properties of the different sorts of CNTs such as solubility, density, or electrophoretic mobility. With this objective, diazonium chemistry can be used to promote selective covalent reactions able to give specific SWCNTs characteristics. The modification of SWCNTs makes the separation easier through the choice of the grafted functions, for example by centrifugation or electrophoresis via solvent/nonsolvent effects induced. Although the functionalization fundamentally alters the desired SWCNT properties, it is often possible to regenerate the original electronic structure using thermal or chemical treatments [34].

The selectivity for the reaction with the diazonium salt of the m-SWCNTs over sc-SWCNTs has been reported since 2003 [43]. Methods based on this selectivity have been used to separate m- and sc-SWCNTs in one-step process with more or less purity. More recently, Do et al. [44] demonstrated that the choice of the para-substituent of the aryldiazonium salts and consequently of the control of the electron transfer between SWCNTs and the diazonium salt, leads to the highly selective functionalization of m-SWCNT. The diazonium salts which have the highest electron-donating groups as phenyl substituents (hydroxy phenyl diazonium salt for example) functionalize only m-SWCNT [44].

The selectivity of the diazonium species with the nanotube surface is dependent not only on the nanotube structure but also on several experimental parameters as the substituent of the diazonium salt, its concentration, the reaction temperature, solution pH, and the nature of the surfactant… [35].

3.1.1 *Why Does the Diazonium Substituent Influence the Selectivity of the Functionalization?* [44]

The reaction between the diazonium salts and SWCNTs is performed by a charge transfer mechanism in which the diazonium compound extracts electrons from the SWCNTs to lead to the formation of a stable covalent aryl bond. The reactivity of SWCNT and diazonium species is driven by the difference between the redox potential of the nanotube and the diazonium salt. m-SWCNTs easily donate electrons to diazonium salts by means of the wide difference between the oxidation potential (EFermi) of SWCNTs and the reduction potential of the diazonium species [14]. Among sc-SWCNTs, large bandgap sc-SWCNTs, e.g.,(6, 5) SWNTs in Fig. 1a have lower oxidation potentials (valence band, V1) than the reduction potential (Eredox, low) of diazonium salts, consequently, these large bandgap sc-SWCNTs cannot give electrons to diazonium cations. Thus, there is no reaction between large bandgap sc-SWCNTs and diazonium salts. However, small bandgap sc-SWCNTs (e.g., (9, 8) SWNTs in Fig. 1a), possess higher oxidation potentials (valence band, V1) than the diazonium species reduction potential (Eredox, low), consequently, they can give electrons because of the appropriate gap between their redox potentials. Thus the selectivity of the functionalization is low between m-SWCNT over small bandgap sc-SWCNTs and leads to a low yield in their separation process. To avoid the functionalization between small bandgap sc-SWCNT and diazonium cations, it is necessary to increase the reduction potential of diazonium salts by choosing an electro-donating para-substituent on the diazonium phenyl group. To control the diazonium reduction potential via the use of the electron-donating hydroxy group as para-substituent of the diazonium phenyl ring, Do et al. showed that the diazonium moieties could be exclusively attached to m-SWCNT over sc-SWCNT. High purity separation of both m- and sc-SWCNT via density-induced separation has been shown by UV and photoluminescence in Fig. 1b.

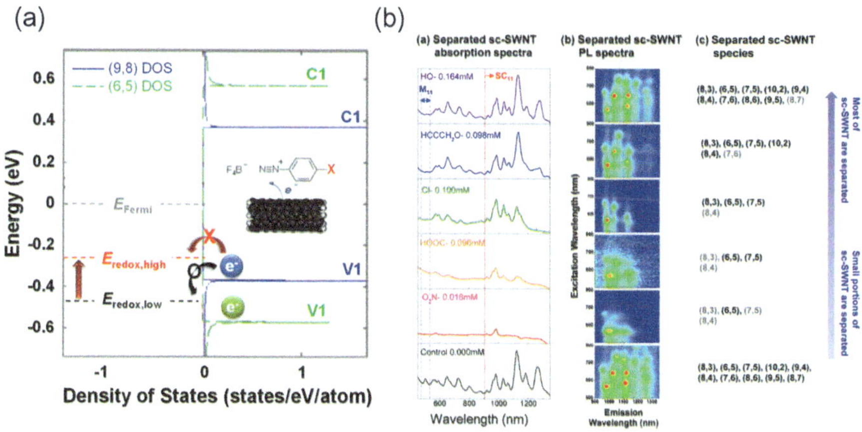

Fig. 1 **a** Schematic diagram of the mechanism of electron transfer between semiconducting SWCNT and diazonium salts with a withdrawing (E redox, low in black) and a electro-donating (E redox, high in red) para-substituent: valence level (V1) and conduction level (C1) in blue for sc-SWCNT with a small bandgap and green for sc-SWCNT with a large ban gap. **b** (a) UV−vis−NIR absorption spectra, (b) photoluminescence (PL) spectra, and (c) SWCNT assignments of separated sc-SWCNT using 4-nitrophenyl (O_2N- , 0.018 mM), 4-carboxyphenyl(HOOC− , 0.096 mM), 4-chlorophenyl (Cl − , 0.100 mM), 4-propargyloxyphenyl ($HCCCH_2O-$, 0.098 mM), and 4-hydroxyphenyl (HO− , 0.164 mM) diazonium functionalized SWCNT, along with that of unreacted original (0.000 mM) SWCNT: Reused with permission from Reference [44] Copyright © 2012, American Chemical Society

3.2 *Influence of the Surfactant on the Selectivity of the Functionalization via Diazonium Chemistry*

Strano's group [45] has shown the possibility to control the functionalization with different surfactants used to decorate SWCNTs. To understand and model the extent of diazonium functionalization of surfactant-decorated SWCNTs, they used experiments, molecular dynamics (MD) simulations, and equilibrium reaction modeling [45] They have shown that the adsorbed surfactant layer influences the functionalization of carbon nanotubes by diazonium cations in different routes: steric exclusion, electrostatic repulsion or attraction, direct chemical modification of the diazonium species. The use of anionic SDS and cationic cetyltrimethylammonium bromide as surfactant leads to the highest electrostatic effects, it follows that the accessibility of diazonium cations on the nanotube surface is affected by the difference in the charge of the surfactant. In the case of bile salt surfactants (except for sodium cholate), the surfactant backbone wraps tightly around the CNT surface and consequently leads to an exclusion effect and prevents the access of the diazonium. Conversely to the other bile surfactants, in the case of sodium cholate, the surfactant-diazonium coupling is able to form a reactive diazoester and consequently, it follows a high selectivity to the metallic CNTs and small bandgap sc-CNTs. Blanch et al. [35] examined the effect of several cationic, anionic, and neutral surfactants on the functionalization of CNTs via several diazonium. In this work, the preferential functionalization of sc-SWCNTs at neutral pH was observed only when the anionic surfactant SDS was used. Charge localization interactions with the surfactant are the hypothesis proposed by the authors to explain this behavior. On the whole, the charge transfer interaction between the surfactant and the CNT is the driving force guiding the selectivity of the functionalization, before the state of aggregation, the density of the adsorption layer, or the rate of dediazonation reaction (35).

3.3 Optical Properties Focus on Photoluminescence (PL)

Doping semiconductor SWCNT is a powerful and tunable means for modulating their optical properties. Single-photon emission in SWCNTs has been demonstrated recently at room temperature resulting from the ability to localize excitons at sp^3 centers introduced by low-level covalent functionalization [24, 25, 46, 47].

In recent years, controlled chemical functionalization has been studied to create optically active quantum defects in sc-SWCNT through covalent attachment of aryl functional groups which introduce scattered sp^3 defects in the sp^2 carbon lattice of the nanotube [46, 47]. Sidewall alkylation [48] and oxygen doping [49] can also produce fluorescent defects, however, these two approaches usually lack the same versatility as the diazonium chemistry and the introduced defects are less bright. The PL quantum yield can be improved significantly through diazonium chemistry method. [24, 49]. The doubly degenerate frontier orbitals can be split via accurate incorporation of aryl defects and thus a new, low-lying fluorescent state can be produced (E_{11}^-) [24]. These fluorescent defects are able to snare a mobile exciton (Fig. 2a), enabling the localized exciton to relax by radiation via E_{11}^- photoluminescence (PL) [25]. The defect PL occurs at red-shifted energies from the native lowest-energy excitonic transition (E_{11}) [24].

This phenomenon can be illustrated in Fig. 2b [46] which shows absorption and photoluminescence spectra of (6, 5)-SWCNT-$C_6H_4NO_2$, functionalized with small amounts of 4-nitrobenzenediazonium tetrafluoroborate. As the aryl defects are introduced, the original exciton PL (E_{11}, 981 nm) decreases and the defect PL (E_{11}^-) develops at a red-shifted position (1141 nm), whereas the E_{11} absorption peak only slightly weakens. An exciton that is trapped at this defect site can radiatively decay to produce a red-shifted emission (E_{11}^-) or escape the trap thermally. Such exciton trapping and de-trapping dynamics can be monitored by PL spectroscopy. Kim et al. have studied the influence of the aryl substituents and the concentration ratio ([Diazonium]/[Carbone]). Both the optical gap and the de-trapping energy reveal a linear correlation with the electron-withdrawing/donating capability of the aryl substituents [46].

Powell et al. [42] have demonstrated that it is possible to use visible light, which is tuned in resonance with SWCNT, to drive the nanotubes' functionalization by aryl diazonium. This reaction is then significantly accelerated until 154-fold. The intensity and the position of the defect PL can be tunable via the chemical nature

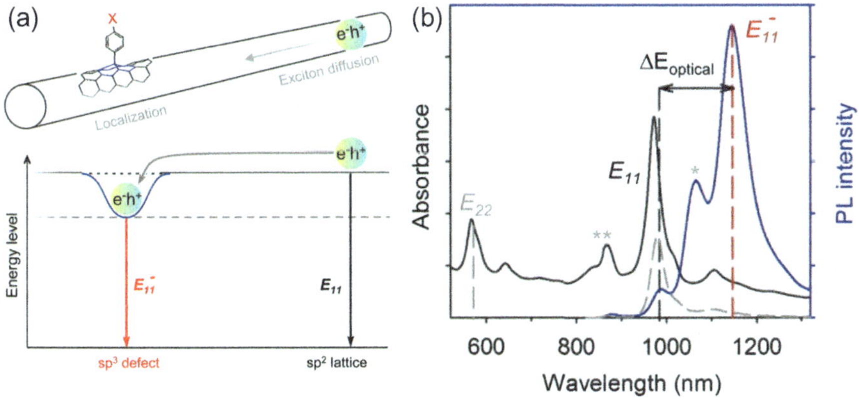

Fig. 2 **a** Schematic of defect-modulated exciton diffusion and recombination. **b** UV−vis−NIR absorption and vis−NIR photoluminescence spectra of (6, 5)-SWCNT-$C_6H_4NO_2$ at room temperature. The dashed line is the PL spectrum of the unfunctionalized starting (6,5)-SWCNTs. The asterisked peaks are due to E_{11}^- PL and E_{11} absorption of (6, 4)-SWCNTs. (6, 5)-SWCNT-$C_6H_4NO_2$ (synthesized at [Diazonium]:[Carbone] = 1:500 [46]: Reused with permission from Reference [46] Copyright © 2016, American Chemical Society

and the concentration of the defect introduced and then leads the defect PL to be a quantitative indicator of the functionalization progression. The addition of a para-nitrobenzenediazonium tetrafluoroborate solution to the suspension of SWCNTs (1% m/v SDS in D_2O) and the irradiation of this solution by visible light initiated the reaction. *In-situ* monitoring of the reaction consists of following the evolution of the defects' PL at E_{11}^- (1140 nm, 1.09 eV in the case of nitroaryl defects) over time, it is important to minimize carefully the exposure to light for the control of the functionalization. The comparison of the defect PL obtained for the bulk thermal grafting activation (protected from light) and the localized thermal grafting activation (irradiated) shows a contribution both of the local heating and the optical excitation in the second case [42].

3.4 Electrical and Thermoelectric Properties

Due to the one-dimensional nature of CNTs, electron conduction occurs in the nanotubes without diffusion. This absence of scattering is named ballistic transport and leads to the electronic conduction in the nanotube forgo dissipated energy such as thermal energy [21]. The charge and heat transports of SWCNTs networks depend on many parameters such as network morphology, defects of the CNTs, and chemical doping [50]. Furthermore, the uncontrolled production of m-SWCNTs and sc-SWCNTs with different electronic structures systematically prevents their specific use. Untowardly, covalent functionalization of CNTs has a negative effect on their structural, electrical, and mechanical properties. This negative impact becomes worse as the degree of functionalization of the nanotubes increases [51].

Piao et al. studied the influence of the chemical functionalization via 4-bromobenzene diazonium salt of monodisperse population of m-SWCNT and sc-SWCNT on the thermoelectronic properties. Due to the higher selectivity of the diazonium reaction for m-SWCNT over sc-SWCNT and consequently their higher degree of functionalization, the deterioration of the electrical conductivity for m-SWCNT networks was much fiercer than that of sc-SWCNT species. On the contrary, it demonstrated a positive impact on the thermoelectric properties (with an increase of their Seebeck coefficient) after grafting the functional groups on the sidewall of sc-SWCNT owing to the energy filtering effect [50].

4 Characterization of the Functionalized CNT

The techniques classically used for the characterization of functionalized materials can usually be applied to the characterization of grafted CNT [20]. In the case of dispersed grafted CNTs, it is also possible to determine the functionalization degree by thermogravimetric analysis, as is also applied for the characterization of grafted pulverulent materials such as functionalized charcoal for example. The use of TGA coupling to the mass spectroscopy MS improved the accuracy of the functionalized degree determination [52] and highlights the covalent nature between the CNT and the grafted groups even in presence of adsorbed molecules (diazonium or surfactant for example). Shirowski et al. [52] studied the functionalization of SWCNT by 4-iodophenyl diazonium in presence of SDS and characterized the functionalized SWCNT via TGA-MS, TGA-GC, and TGA-GC-MS apparatus (TGA: thermogravimetric analysis, GC: gas Chromatography, MS: Mass Spectroscopy). They found that the covalently bonded 4-iodophenyl groups get cleaved off the SWCNTs in the main mass loss region between 150 and 400 °C. The simultaneous presence of physisorbed iodophenyl species has been demonstrated and quantified in the functionalized SWCNT, which detached from the nanotubes around its boiling point of 188 °C. In contrast, covalent C –C bonds of the aryl-SWCNT junctions are broken at 286 °C. A logarithmic correlation between the concentration of diazonium and the functionalization degree was demonstrated using different concentrations of diazonium salt during the functionalization of the same amount of SWCNTs. The maximum possible functionalization degree obtained for this reaction is close to 1%. Overwhelmingly enhancing the concentration of the diazonium reactant (n (diazonium): n (carbon) > 8:1) has a very negative impact: the concentration of 4-iodophenyl moieties grafted on the SWCNT sidewall decreases and the side reactions are promoted. In addition, Raman spectroscopy showed a hyperbolic behavior for the I_D/I_G ratios with maximum I_D/I_G values obtained at a degree of functionalization of 0.9%. This example shows that Raman spectroscopy is a semi-quantitative method applicable for low levels of functionalization (e.g., < 0.9%), but unnecessary if used alone and if precise quantification is required.

Sc-SWCNTs present characteristic near-infrared photoluminescence. This property increases the potential applications of sc-SWCNTs in areas such as sensing, imaging, optoelectronics, and quantum cryptography. Nevertheless, the quantum yield of PL is unexpectedly weak in nanotubes. As for all semiconductors, the optical properties of carbon nanotubes are very sensitive to the presence of defects in the structure. Initially considered as negative to the quantum yield of nanotubes defects are now emerging as a tool to control photoluminescence. Covalent functionalization of semiconducting SWCNTs with a controlled number of aryl functional groups produces a new photoluminescence peak substantially brighter than the original E11 emission [24]. On the other hand, PL can be used to characterize the optical and electronic properties of the functionalized SWCNT.

Raman Spectroscopy

The complexity of the Raman spectra of carbon-based materials is closely correlated with the complexity of their structure [53]. Graphite has very simple Raman spectra due to its highly uniform structure. Graphite structure consists of planar and perfectly aligned layers stacking. Its Raman spectrum shows two characteristic bands; in the first-order region is the G band (at about 1580 cm^{-1}) and in the second-order region is the 2D (or G_0) band (between 2600 and 2700 cm^{-1}). The existence of some defects in the graphite structure leads to the appearance of a weak D band around 1350 cm^{-1}. Consequently, the D band is considered as the disorder or the defects band and its intensity is usually directly proportional to the level of defects in the carbon samples. For example, the D band is present in the case of covalently bound function to the graphite layer or for oxidized graphite. The 2D band corresponds to the second-order of the D band and corresponds to its overtone. A two-phonon lattice vibrational process is at the origin of this 2D band but it is not activated by the vicinity of defects in contrast to the D band.

Significantly, SWCNTs and MWCNTs have different Raman spectra with each having specific characteristics. SWCNTs Raman spectra usually are similar to those of graphite. They show a high G and 2D bands and a weak D band with a slight broadening effect. In contrast, MWCNTs have Raman spectra similar to those of carbon blacks and chars with a very high D band showing a noticeable broadening effect, even for MWCNTs with very few defects. In the case of SWCNTs, the degree of covalent functionalization can be determined by the ratio of the two bands' intensity I_D/I_G [54]. While in the case of covalent functionalized MWCNTs, no notable difference is usually observed in the I_D/I_G ratio [55]. To evaluate the functionalization degree, the identification of the disorder due to the covalent functionalization and consequently to the transformation of sp^2 carbon into sp^3 carbon is essential and must be distinguished from the other structural defects resulting in difference with the ideal graphitic lattice [53]. The Radial Breathing Mode (RBM), in which all carbon atoms of the SWCNT vibrate radially in phase, is observed at low Raman shifts between 100 and 350 cm^{-1}. The energy of the RBM depends inversely on the SWCNT diameter. The RBM intensity of SWCNTs is sensitive to the doping level [56].

Rebelo et al. have shown that diazonium functionalization of MWCNTs led to aromaticity disruption within the external layer. The resulting Raman spectra show the increase of the relative area of satellite bands and a concomitant decrease of the relative areas of graphitic bands, in particular the G band.

Recently, the effect of n or p- doping on the electronic properties in functionalized SWCNT via diazonium chemistry, has been studied using an *in-situ* Raman spectro-electrochemical method [56]. The Raman spectro-electrochemical analyses showed an acceptor effect of 4-benzenesulfonic acid (p-doping), and a donor effect (n-doping) in the case of 4-hydroxy benzyl groups. These results have been corroborated via DFT calculation.

5 Applications of the Functionalized CNT

The chemical, mechanical, electrical, and optical properties of carbon nanotubes in general, and SWCNTs in particular, have made it possible to highlight new phenomena that open the field to numerous applications in the new technology including field-effect transistors (FET), electrodes for electrochemical devices such as supercapacitors, batteries and fuel cells, transparent electrodes for solar cells and organic light-emitting diodes (OLEDs), chemical and biochemical sensors, composite reinforcements, catalysis platforms… [34].

5.1 Catalytic Applications

Mahouche-Chergui et al.[57] have developed a platform that efficiently catalyzes the Suzuki C–C coupling reaction. This platform was constructed by functionalizing CNTs with a 4-ethynylbenzenediazonium salt. The ethynyl groups grafted onto the surface of the nanotubes then led to the formation of CNT@polymer hybrids ("hairy" CNTs) through the click chemistry from polymers bearing azide groups. The formation of a heterostructure by loading Pd nanoparticles on the CNT@polymer hybrids completes the construction of the catalytic platform.

The functionalization via diazonium chemistry can enhance the durability of the CNTs for catalytic applications. For example, Pt/functionalized MWCNTs show higher catalytic activity and electrochemical durability than the Pt/ unmodified or oxidized CNTs in the case of PEMFC (Proton Exchange Membrane Fuel Cells) application. The enhanced electrochemical performance of the Pt/thiophenyl functionalized CNT is attributed to the uniform distribution of Pt nanoparticles with smaller size and the strong interaction between Pt and the functionalized CNT support through the formation of Pt-sulfur bonding [58].

5.2 Nanocomposites

Mekki et al. [59] proposed a facile approach to synthesizing core-shell nanocomposites PANI/MWCNT (PANI: polyaniline) highlights the efficiency of the interfacial chemistry of aryl diazonium salts in generating conductive polymer/MWCNT nanocomposites with enhanced conductivity and high surface area which could be used for applications such as supercapacitors or sensors.

5.3 Energy Conversion

In the field of the conversion of chemical energy to electricity, a new concept has recently appeared: the thermopower wave. A nanowire, as a carbon nanotube, presents a large axial thermal diffusivity and is able to accelerate a self-propagating reaction wave consuming a fuel coated or grafted along its length. In experiments using carbon nanotubes, the wave of the reaction leads electrical carriers in a thermopower wave and thus creates a high-power pulse of up to 7 kW/kg [60]. The scheme of the principle is shown in Fig. 3a. Concerning new classes of nano-energetic materials with low dimensionality, the Strano team [61] approach consists of covalently linked energetic molecules, such as nitrobenzene groups, to a CNT that can use as a one-dimension heat conductor (Fig. 3b). The chemical reaction at the one end of the CNT gives energy that can propagate through the conductor CNT and thus initiates reactions with the attached energetic chemical groups at the propagating wave-front. DSC analysis of energetically decorated SWCNTs via mono-, di- and tribenzene diazonium led to the release of heat when they are thermally excited, whereas non-energetic SWCNT (bromobenzene-SWCNT) showed no energy output. Mono-, di-, and trinitrobenzene-functionalized SWCNT had energy densities of 780, 600,

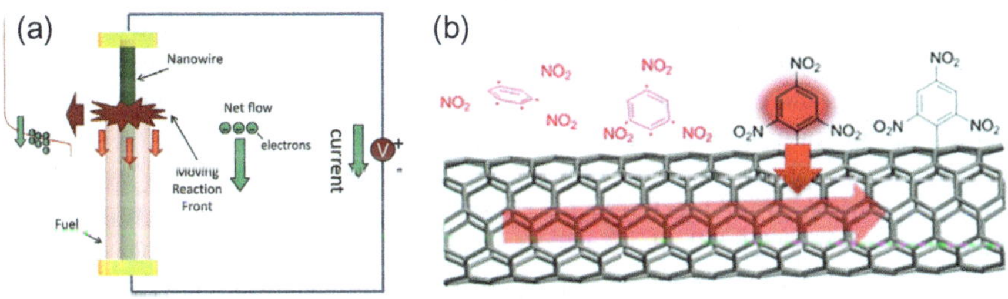

Fig. 3 **a** Schematic of measurement of core/shell nanostructure launching a thermopower wave [60], Copyright © 2010 Elsevier Ltd. **b** Schematic representation of reaction propagation on a SWCNT decorated with energetic molecules (tri-nitrobenzene) from [61]: Reused with permission from Reference [61]. Copyright © 2011 American Chemical Society

and 440 J/g, respectively. These highly promising nano-energetic materials demonstrate 1-D reaction propagation and directional energy release, and they can serve as on-demand nanoscale energy sources that can be remotely activated.

5.4 Sensors and Biosensors

Carbon nanomaterials (CNs) especially nanotubes offer some of the most valuable properties for electrochemical chemical and biochemical sensing applications, such as good electrical conductivity, wide electrochemical stability, high specific surface area, and biocompatibility. Benefits of CNT-based sensors include the speed and reliability obtained from performing multiple assays in parallel [62]. In particular, the modification of CNT by aryl diazonium chemistry is established as a powerful tool that allows tailoring the chemical and electronic properties of the sensing platform [27].

By outlining the results disclosed in the last years, Table 1 provides some examples of the contribution of aryldiazonium grafting to developing CNT-based biochemical sensors and to a lesser extent chemical sensors.

Regardless of the envisioned sensing application, endowing CNT with specific functions through controlled chemical functionalization is fundamental for promoting the specific binding of the analyte. As a versatile and straightforward method of surface functionalization, aryldiazonium chemistry has been successfully used to accommodate in a stable and reproducible way different functionalities. A wide variety of functionalized CNT-sensors from diazonium chemistry were studied based directly on the recognition properties of the grafted function often applied for the chemical sensing such as NH_3 detection for example [63] or used the grafted function to build a more complex recognition system via a coupling with a polymer [64] or a biomolecule such as in enzymatic or antigenic biosensors for example [27]. In addition, the diazonium-generated layer can also act as a barrier and antifouling layer or control the orientation and distribution of biomolecules coupled to the electrode such as enzymes for example. The use of grafted CNTs as charge carriers in the sensors devices increases the loading redox mediator or the enzyme tags and leads to the amplification of the detection signal [27].

Biosensor applications using diazonium salts are very numerous even if the bibliography is restricted to the last 10 years. They may differ in the transduction mode such as electrochemical (cyclic voltammetry (CV), differential pulsed voltammetry (DPV), square wave voltammetry (SWV), chronoamperometry (CA), complex impedance spectroscopy (EIS)), field-effect transistor (FET). Their differences also relate to the variable complexity of the sensor platform architecture from an aligned CNT (ACNT) electrode with only two stages [65] to a multi-stage sandwich platform [66]. The following table gives an overview of the variety of biosensors using diazoniums.

Table 1 Examples of chemical and biochemical sensors based on functionalized CNT obtained from diazonium salt

Diazonium grafting	Support	Detection method	Detection and performances	References
4-carboxybenzene diazonium tetrafluoroborate—Activation via EDC/sulfo-NHS and coupling with antiboby	Antibody-functionalized SWCNT	FET	Detection of lime disease antigen—concentration range 1–3 mg/mL	[68]
In-situ reduction of the diazonium from p-aminobenzoic acid and p-phenylenediamine—Adsorption of cellobiose dehydrogenase	Enzyme-SWCNT adsorbed on glassy carbon	Cyclic voltammetry	Oxidation of lactose—high electron transfer	[69]
In-situ electroreduction of 4-aminobenzene diazonium on BDD electrode—MWCNT coupled to BDD via EDC/NHS then coupled with tyrosinase	Enzyme/MWCNT/BDD	Cyclic voltammetry	Detection of the bisphenol A—Linear range: 0.01–100 nmol/L—LOD 10 pmol/L—tested in spike water river	[70]
In-situ electroreduction of 4-aminobenzene diazonium	Polypyrrole-wrapped MWCNT composite coatings on aminophenyl-modified flexible ITO sheets	DPV (differential pulsed voltammetry	Pb (II) in water Linear range: 10^{-8} to 3 10^{-7} mol $L^{/L}$—LOD = 2, 4 nmol/L	[65]
In-situ electroreduction of the Azure A—coupling with flavin adenine dinucleotide (FAD) dependent glucose dehydrogenase	MWCNT on GC	Cyclic voltammetry	High stability of the Azure A grafted electrode—electrooxidation of the glucose	[71]
In-situ electroreduction of hydrazide-phenyldiazonium on SWCNTox—Then Site-oriented immobilization of antibodies was carried out through hydrazone bond formation	Oxidized antibody/SWCNT/ AuSPE	DPV	Virus (antiMS2 bacteriophage)—9.3 and 9.8 pfu/mL in buffer and in river water, respectively	[72]

(continued)

Table 1 (continued)

Diazonium grafting	Support	Detection method	Detection and performances	References
In-situ chemical grafting of the SWCNT via the 4-amino benzoic acid then reaction of carboxyphenyl-SWCNT with viologen	Sandwich assay HRP-Ab2-V-CP-SWCNT-antiTGF	Chronoamperometric immunosensor	determination of the transforming growth factor β1 (TGF-β1) cytokine (biomarker) Linear range: 2,5 to 1000 pg/mL—LOD: 0,95 pg/mL in saliva	[73]
In-situ electrochemical grafting on nitrogen-doped ACNT via 4-amino benzoic acid	Transducer material, biotinylated lysozyme aptamer covalently immobilized onto the N-ACNTs via EDC/NHS	DPV/detection by the decrease on the $Fe(CN)_6$ $3^-/4^-$ redox signal	Lysozyme—linear range 0.1–7 pmol/L—LOD = 100 fmol/L in serum	[65]
In-situ electroreduction of the 4-amino benzoic acid on MWCNT-SPE	Aptamer-antibody sandwich assay on carboxy phenyl-SWCNT	Electrochemical impedance (EIS) in ferri/ferrocyanide PBS solution	Cytochrome c—Linear range: 25–100 pmol/L—LOD: 12 pmol/L in spike human serum	[66]
Electroreduction of the 4-nitrobenzene diazonium on SWCNT/GC following by a electroreduction of nitro group in acidic medium NH-SWCNT then by the electropolymerization of polyaniline (PANI) in presence of glucose oxidase (GOx)	PANI-GOx-NH-SWCNT-PB-GCE—nb: as catalyst of the H_2O_2 reduction, prussian blue is deposited on GCE	CA	Glucose via H_2O_2 oxidation producted during the enzymatic reaction—Range 0,01 to 5 mmol/L—LOD 0,02 mmol/L—functionalization improved the stability of the electrode platform	[74]

We can conclude this part with an example of promising transduction devices that should lead to detection by FETs consisting of an isolated CNT. The Point-functionalized carbon nanotube (CNT) devices have emerged as all-electronic devices [67] for single-molecule detection platforms. The conductance of the single-point-functionalized CNT field-effect transistor (CNT-FET) is sensitive to the charges localized within few debye lengths on a point defect that is created on the CNT sidewall. A single-point-functionalization method for CNT-FET arrays has been developed. This method is based on the electrochemical control of the functionalization via diazonium salt to generate sp^3 defects on the CNT surface coupled to a scalable spin-casting method for the fabrication of large arrays of devices. In this example, a probe DNA is attached to the functionalized nanotube allowing single-molecule sensing of DNA hybridization with complementary targeted molecule, and thus highlighted the single-point functionalization.

6 Conclusion

During the last decade, the publication of the carbon nanotube functionalization via the diazonium chemistry continued to grow steadily but the field of the concerned applications seems to be dominated by the biosensors. The interesting properties of the carbon nanotubes are currently competing with a more popular carbon nanomaterial: graphene in spite of a polemic about its toxicity. Nevertheless, its intrinsic properties are particularly exciting for fundamental research and for the development of nanodevices based on a single nanotube. Certainly, diazonium chemistry will be the best CNT partner in this challenging competition offering great versatility in methods and grafted functions and now to localize the functionalization.

References

1. Delamar M, Hitmi R, Pinson J, Saveant JM (1992) J Am Chem Soc 114(14):5883–5884. https://doi.org/10.1021/ja00040a074
2. Jenkins J, Kawamura K (1971) Nature 231:176–177. https://doi.org/10.1038/231175a0
3. Pesin L (2002) Review: structure and properties of glass-like carbon. J Mater Sci 37:1–28
4. Shiell TB, Wong S, Yang W, Tanner CA, Haberl B, Elliman RG, McKenzie DR, McCulloch DG, Bradby JE (2019) J Non-Cryst Solids 522:119561. https://doi.org/10.1016/j.jnoncrysol.2019.119561
5. Delamar M, Désarmot G, Fagebaume O, Hitmi R, Pinson J, Savéant J-M (1997) Carbon 35(6):801–807. https://doi.org/10.1016/S0008-6223(97)00010-9
6. Ray K, McCreery RL (1997) Anal Chem 69(22):4680–4687. https://doi.org/10.1021/ac9705531
7. Belmont JA, Amici RM, Galloway P (1998) United States Patent, 5851280 December 22, 1998. (Cabot Corporation)
8. Ranganathan S, McCreery RL (2001) Anal Chem 73(5):893–900. https://doi.org/10.1021/ac0007534

9. Bahr JL, Yang J, Kosynkin DV, Bronikowski MJ, Smalley RE, Tour JM (2001) J Am Chem Soc 123(27):6536–6542. https://doi.org/10.1021/ja010462s
10. Dyke CA, Tour JM (2003) J Am Chem Soc 125(5):1156–1157. https://doi.org/10.1021/ja0289806
11. Barroso-Bujans F, Fierro JLG, Rojas S, Sánchez-Cortes S, Arroyo M, López-Manchado MA (2007) Carbon 45(8):1669–1678. https://doi.org/10.1016/j.carbon.2007.03.039
12. Pagona G, Karousis N, Tagmatarchis N (2008) Carbon 46(4):604–610. https://doi.org/10.1016/j.carbon.2008.01.007
13. Li Z, Yan W, Dai S (2005) Langmuir 21:11999–12006. https://doi.org/10.1021/ja0289806
14. Bekyarova E, Itkis ME, Ramesh P, Berger C, Sprinkle M, de Heer WA, Haddon RC (2009) J Am Chem Soc 131(4):1336–1337. https://doi.org/10.1021/ja8057327
15. Pognon G, Brousse T, Bélanger D (2011) Carbon 49(4):1340–1348. https://doi.org/10.1016/j.carbon.2010.11.055
16. Flavin K, Chaur MN, Echegoyen L, Giordani S (2010) Org Lett 12(4):840–843. https://doi.org/10.1021/ol902939f
17. Mirzaei P, Bastide S, Aghajani A, Bourgon J, Leroy É, Zhang J, Snoussi Y, Bensghaier A, Hamouma O, Chehimi MM, Cachet-Vivier C (2019) Langmuir 35(45):14428–14436. https://doi.org/10.1021/acs.langmuir.9b01911
18. Hwang E, Hwang HM, Shin Y, Yoon Y, Lee H, Yang J, Bak S, Lee H (2016) Sci Rep 6(1):39448. https://doi.org/10.1038/srep39448
19. Gutiérrez-Sánchez C, Mediavilla M, Guerrero-Esteban T, Revenga-Parra M, Pariente F, Lorenzo E (2020) Carbon 159:303–310. https://doi.org/10.1016/j.carbon.2019.12.053
20. Bensghaïer A, Mousli F, Lamouri A, Postnikov PS, Chehimi MM (2020) Chemistry Africa 3(3):535–569. https://doi.org/10.1007/s42250-020-00144-5
21. Karousis N, Tagmatarchis N, Tasis D (2010) Chem Rev 110(9):5366–5397. https://doi.org/10.1021/cr100018g
22. Graff RA, Swanson TM, Strano MS (2008) Chem Mater 20(5):1824–1829. https://doi.org/10.1021/cm702577h
23. Goldsmith BR, Mitala JJ, Josue J, Castro A, Lerner MB, Bayburt TH, Khamis SM, Jones RA, Brand JG, Sligar SG, Luetje CW, Gelperin A, Rhodes PA, Discher BM, Johnson ATC (2011) ACS Nano 5(7):5408–5416. https://doi.org/10.1021/nn200489j
24. Piao Y, Meany B, Powell LR, Valley N, Kwon H, Schatz GC, Wang Y (2013) Nature Chem 5(10):840–845. https://doi.org/10.1038/nchem.1711
25. Hartmann NF, Yalcin SE, Adamska L, Hároz EH, Ma X, Tretiak S, Htoon H, Doorn SK (2015) Nanoscale 7(48):20521–20530. https://doi.org/10.1039/C5NR06343D
26. Mohamed AA, Salmi Z, Dahoumane SA, Mekki A, Carbonnier B, Chehimi MM (2015) Adv Coll Interface Sci 225:16–36. https://doi.org/10.1016/j.cis.2015.07.011
27. Pilan L (2021) Bioelectrochemistry 138:107697. https://doi.org/10.1016/j.bioelechem.2020.107697
28. Yang J, Xu Y, Zhang R, Wang Y, He P, Fang Y (2009) Electroanalysis 21(15):1672–1677. https://doi.org/10.1002/elan.200904591
29. Joyeux X, Mangiagalli P, Pinson J (2009) Adv Mater 21(43):4404–4408. https://doi.org/10.1002/adma.200900121
30. Arias de Fuentes O, Ferri T, Frasconi M, Paolini V, Santucci R (2011) Angew Chem Int Ed 50(15):3457–3461. https://doi.org/10.1002/anie.201006743
31. Liu G, Guo W, Song D (2014) Biosens Bioelectron 52:360–366. https://doi.org/10.1016/j.bios.2013.09.009
32. Lipińska ME, Rebelo SLH, Pereira MFR, Gomes JANF, Freire C, Figueiredo JL (2012) Carbon 50(9):3280–3294. https://doi.org/10.1016/j.carbon.2011.12.018
33. Schmidt G, Gallon S, Esnouf S, Bourgoin J-P, Chenevier P (2009) Chem Eur J 15(9):2101–2110. https://doi.org/10.1002/chem.200801801
34. Hodge SA, Bayazit MK, Coleman KS, Shaffer MSP (2012) Chem Soc Rev 41(12):4409. https://doi.org/10.1039/c2cs15334c

35. Blanch AJ, Lenehan CE, Quinton JS (2012) J Phys Chem C 116(2):1709–1723. https://doi.org/10.1021/jp208191c
36. Doyle CD, Tour JM (2009) Carbon 47(14):3215–3218. https://doi.org/10.1016/j.carbon.2009.07.035
37. Mévellec V, Roussel S, Tessier L, Chancolon J, Mayne-L'Hermite M, Deniau G, Viel P, Palacin S (2007) Chem Mater 19(25):6323–6330. https://doi.org/10.1021/cm071371i
38. Pandurangappa M, Lawrence NS, Compton RG (2002) Analyst 127(12):1568–1571. https://doi.org/10.1039/b209711g
39. Price BK, Hudson JL, Tour JM (2005) J Am Chem Soc 127(42):14867–14870. https://doi.org/10.1021/ja053998c
40. Liu J, Zubiri MR, Vigolo B, Dossot M, Humbert B, Fort Y, McRae E (2007) J Nanosci Nanotech 7(10):3519–3523. https://doi.org/10.1166/jnn.2007.819
41. Pan Y, Tong B, Shi J, Zhao W, Shen J, Zhi J, Dong Y (2010) J Phys Chem C 114(17):8040–8047. https://doi.org/10.1021/jp909904t
42. Powell LR, Piao Y, Wang Y (2016) J Phys Chem Lett 7(18):3690–3694. https://doi.org/10.1021/acs.jpclett.6b01771
43. Strano MS (2003) Science 301(5639):1519–1522. https://doi.org/10.1126/science.1087691
44. Do Y-J, Lee J-H, Choi H, Han J-H, Chung C-H, Jeong M-G, Strano MS, Kim W-J (2012) Chem Mater 24(21):4146–4151. https://doi.org/10.1021/cm302227t
45. Lin S, Hilmer AJ, Mendenhall JD, Strano MS, Blankschtein D (2012) J Am Chem Soc 134(19):8194–8204. https://doi.org/10.1021/ja301635e
46. Kim M, Adamska L, Hartmann NF, Kwon H, Liu J, Velizhanin KA, Piao Y, Powell LR, Meany B, Doorn SK, Tretiak S, Wang Y (2016) J Phys Chem C 120(20):11268–11276. https://doi.org/10.1021/acs.jpcc.6b02538
47. Lohmann S-H, Trerayapiwat KJ, Niklas J, Poluektov OG, Sharifzadeh S, Ma X (2020) ACS Nano 14(12):17675–17682. https://doi.org/10.1021/acsnano.0c08782
48. Ghosh S, Bachilo SM, Simonette RA, Beckingham KM, Weisman RB (2010) Science 330(6011):1656–1659. https://doi.org/10.1126/science.1196382
49. Zhang Y, Valley N, Brozena AH, Piao Y, Song X, Schatz GC, Wang Y (2013) J Phys Chem Lett 4(5):826–830. https://doi.org/10.1021/jz400167d
50. Piao M, Li C, Chu J, Wang X, Zhang H, Chi Y (2018) J Mater Sci 53(10):7648–7656. https://doi.org/10.1007/s10853-018-2063-4
51. Milowska KZ, Burda M, Wolanicka L, Bristowe PD, Koziol KKK (2019) Nanoscale 11(1):145–157. https://doi.org/10.1039/C8NR07521B
52. Schirowski M, Abellán G, Nuin E, Pampel J, Dolle C, Wedler V, Fellinger T-P, Spiecker E, Hauke F, Hirsch A (2018) J Am Chem Soc 140(9):3352–3360. https://doi.org/10.1021/jacs.7b12910
53. Rebelo SLH, Guedes A, Szefczyk ME, Pereira AM, Araújo JP, Freire C (2016) Phys Chem Chem Phys 18(18):12784–12796. https://doi.org/10.1039/C5CP06519D
54. Hines D, Rümmeli MH, Adebimpe D, Akins DL (2014) Chem Commun 50(78):11568–11571. https://doi.org/10.1039/C4CC03702B
55. Botti S, Laurenzi S, Mezi L, Rufoloni A, Santonicola MG (2015) Phys Chem Chem Phys 17(33):21373–21380. https://doi.org/10.1039/C4CP05075D
56. Santidrián A, González-Domínguez JM, Diez-Cabanes V, Hernández-Ferrer J, Maser WK, Benito AM, Anśon-Casaos A, Cornil J, Da Ros T, Kalbáč M (2019) Phys Chem Chem Phys 21(7):4063–4071. https://doi.org/10.1039/C8CP06961A
57. Mahouche Chergui S, Ledebt A, Mammeri F, Herbst F, Carbonnier B, Ben Romdhane H, Delamar M, Chehimi MM (2010) Langmuir 26(20):16115–16121. https://doi.org/10.1021/la102801d
58. Kim JY, Kim YS, Lee S, Pak C, Kim H-T (2015) Electrochim Acta 154:63–69. https://doi.org/10.1016/j.electacta.2014.12.033
59. Mekki A, Samanta S, Singh A, Salmi Z, Mahmoud R, Chehimi MM, Aswal DK (2014) J Colloid Interface Sci 418:185–192. https://doi.org/10.1016/j.jcis.2013.11.077

60. Choi W, Abrahamson JT, Strano JM, Strano MS (2010) Mater Today 13(10):22–33. https://doi.org/10.1016/S1369-7021(10)70183-8
61. Abrahamson JT, Song C, Hu JH, Forman JM, Mahajan SG, Nair N, Choi W, Lee E-J, Strano MS (2011) Chem Mater 23(20):4557–4562. https://doi.org/10.1021/cm201947y
62. Chikkaveeraiah BV, Bhirde A, Malhotra R, Patel V, Gutkind JS, Rusling JF (2009) Anal Chem 81(21):9129–9134. https://doi.org/10.1021/ac9018022
63. Huang J, Ng AL, Piao Y, Chen C-F, Green AA, Sun C-F, Hersam MC, Lee CS, Wang Y (2013) J Am Chem Soc 135(6):2306–2312. https://doi.org/10.1021/ja310844u
64. Lo M, Seydou M, Bensghaïer A, Pires R, Gningue-Sall D, Aaron J-J, Mekhalif Z, Delhalle J, Chehimi MM (2020) Sensors 20(3):580. https://doi.org/10.3390/s20030580
65. Wang Q, Subramanian P, Schechter A, Teblum E, Yemini R, Nessim GD, Vasilescu A, Li M, Boukherroub R, Szunerits S (2016) ACS Appl Mater Interfaces 8(15):9600–9609. https://doi.org/10.1021/acsami.6b00663
66. Ocaña C, Lukic S, del Valle M (2015) Microchim Acta 182(11–12):2045–2053. https://doi.org/10.1007/s00604-015-1540-6
67. Lee Y, Trocchia SM, Warren SB, Young EF, Vernick S, Shepard KL (2018) ACS Nano 12(10):9922–9930. https://doi.org/10.1021/acsnano.8b03073
68. Lerner MB, Dailey J, Goldsmith BR, Brisson D, Charlie Johnson AT (2013) Biosens Bioelectron 45:163–167. https://doi.org/10.1016/j.bios.2013.01.035
69. Tasca F, Harreither W, Ludwig R, Gooding JJ, Gorton L (2011) Anal Chem 83(8):3042–3049. https://doi.org/10.1021/ac103250b
70. Zehani N, Fortgang P, Saddek Lachgar M, Baraket A, Arab M, Dzyadevych SV, Kherrat R, Jaffrezic-Renault N (2015) Biosens Bioelectron 74:830–835. https://doi.org/10.1016/j.bios.2015.07.051
71. Tsujimura S, Tanaka S, Gross A, Holzinger M (2021) J Phys Energy 3(2):024004. https://doi.org/10.1088/2515-7655/abd298
72. Prieto-Simón B, Bandaru NM, Saint C, Voelcker NH (2015) Biosens Bioelectron 67:642–648. https://doi.org/10.1016/j.bios.2014.09.089
73. Sánchez-Tirado E, Arellano LM, González-Cortés A, Yáñez-Sedeño P, Langa F, Pingarrón JM (2017) Biosens Bioelectron 98:240–247. https://doi.org/10.1016/j.bios.2017.06.063
74. Pilan L, Raicopol M (2014) UPB Sci Bull Ser B 76:155–166. https://www.researchgate.net/publication/286232992

Covalent Modification of Graphite and Graphene Using Diazonium Chemistry

Miriam C. Rodríguez González, Kunal S. Mali, and Steven De Feyter

Abstract The production of graphene with controlled properties and structure is one of the most challenging aspects for a chemist. Covalent functionalization is one of the common approaches to obtain well-defined and robust modification of carbon materials. Different protocols have been proposed for carrying out this functionalization step. However, aryl diazonium salts chemistry should be highlighted due to its efficiency and simplicity. In this book chapter we focus on the modification of carbon materials with sp^2 hybridization (graphite and graphene) by using aryl diazonium salts. The on-surface chemistry of diazonium salts on model substrates is explored with a focus on the attempts that have been done to improve the fundamental knowledge about the aryl-carbon interface. Recent developments include control of the structure and the spatial distribution of the aryl moieties on the surface. Finally, the expansion of the protocols to bulk dispersions of graphene and the advantages for the mass production and development of applications based on this material are highlighted.

1 Functionalization of Graphite—Graphene

Graphene is a 2D material which was isolated in 2004 by Novoselov and Geim [1]. It is a single layer of sp^2-hybridized carbon atoms forming a honeycomb hexagonal lattice. This model 2D material exhibits a unique set of properties, including high carrier mobilities, high thermal conductivity, and high mechanical strength. The exceptional mechanical, thermal, and electronic properties of graphene make it a suitable material for a wide range of applications such as sensing, optoelectronic devices, catalysts, or energy storage. Generally, graphene or graphitic materials are quite inert and exhibit good stability in acidic and basic environments. There are different methods to obtain graphene which can be divided into bottom-up

M. C. Rodríguez González · K. S. Mali · S. De Feyter (✉)
Department of Chemistry, Division of Molecular Imaging and Photonics, KU Leuven, Celestijnenlaan 200F, B-3001 Leuven, Belgium
e-mail: steven.defeyter@kuleuven.be

M. M. Chehimi et al. (eds.), *Aryl Diazonium Salts and Related Compounds*, Physical Chemistry in Action, https://doi.org/10.1007/978-3-031-04398-7_8

or top-down approaches. When using a bottom-up approach, graphene is synthetically produced from a variety of small molecule precursors, while in a top-down approach, graphitic substrates are exfoliated to yield single layer graphene. The method selected for the synthesis often depends on the requirements of the application in terms of crystallite size, defect density or scalability. For the exfoliation of graphite mechanical (ball milling, sonication), chemical (liquid phase exfoliation), and electrochemical protocols have been developed. The bottom-up approach mainly centers around the chemical vapor deposition (CVD) and to a lesser extent synthesis from molecular precursors. CVD grown graphene can be transferred to arbitrary substrates with interesting applications in electronic devices. Finally, the bottom-up fabrication of graphene from molecular precursors provides access to narrow strips of graphene, the so-called graphene nanoribbons, with atomically precise size, shape, and more importantly the edge structure [2].

In terms of reactivity, the bulk properties of graphite drastically change if we consider materials formed by 10 layers or less. The different reactivity of single layer (SLG), bilayer (BLG), and trilayer (TLG) graphene has been previously reported [3]. Additionally, the edges and intrinsic defects show a different reactivity as well. Highly oriented pyrolytic graphite (HOPG), a highly ordered and pure synthetic form of graphite, can be seen as a 3D stack of individual graphene sheets stabilized through π-π interactions. The chemical reactivity of HOPG and graphene are comparable, with graphene tending to be even more reactive.

Chemical functionalization of graphite/graphene can be used to tune the chemical and physical properties of the material. There are two different approaches for this functionalization: non-covalent and covalent chemistries [4]. For the non-covalent modification of graphene, a wide variety of molecules, polymers, metals or nanoparticles have been used. One of the advantages of the non-covalent interactions between the material and the molecules is that their weak nature preserves the unique electronic band structure of graphene. Both fundamentals and applied aspects have been explored for the functionalization of graphene via molecular physisorption. In contrast, the covalent attachment of organic molecules onto the basal plane of graphene has been proposed as the more straightforward approach for the surface functionalization. Although the disruption of the sp^2 hybridized structure often implies a decrease in the charge carrier mobility, for applications where this is not a requirement, covalent functionalization has been widely explored. Different properties such as reactivity, solubility, doping, and even the band structure of graphene can be tuned using the covalent functionalization approach. Following this approach, some drawbacks of graphene such as the zero bandgap (small on/off ratios in field effect transistors), the poor solubility of graphene in most solvents or the restacking of the graphene sheets in liquid dispersions can be overcome with the chemical functionalization of graphene.

Among the strategies proposed for the covalent functionalization, we can find mainly nucleophilic addition, cycloaddition, free radical addition, substitution, and rearrangement reactions [5, 6]. Free radicals, carbenes, nitrenes, and arynes have been employed (Fig. 1) [7]. These reactive species can form a covalent bond with the surface. Additionally, graphene can also be used as a Diels–Alder substrate where it

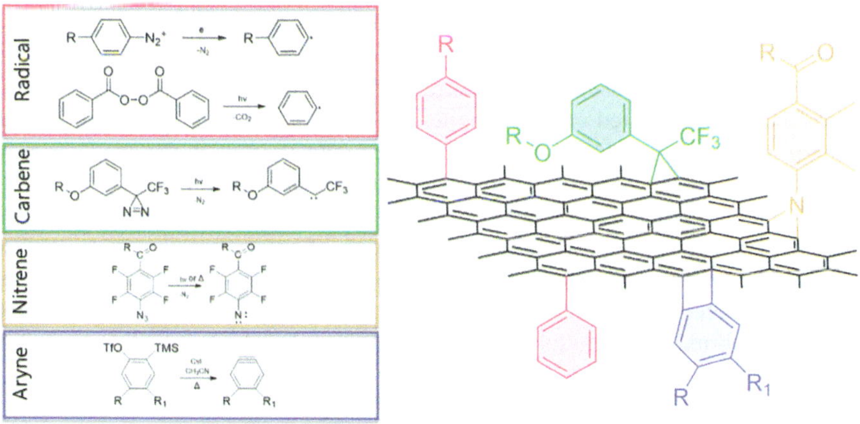

Fig. 1 Methods for the functionalization of graphene. Adapted with permission from Ref. [7] Copyright (2013) American Chemical Society

can act as diene or dienophile. Other covalent approaches which involve the incorporation of (hetero)atoms into the graphene lattice are also used, but this type of substitutional doping is not discussed here. Figure 1 shows a summary of the different strategies proposed for the covalent functionalization of graphite or graphene.

2 Aryl Diazonium Salts as Modifiers of sp^2 Materials

Diazonium chemistry has become popular for the attachment of aryl groups to the poorly reactive basal plane of graphite/graphene. The reaction of diazonium cations proceeds via electron transfer from graphite/graphene to the diazonium salt, forming an aryl radical which can be attached to the carbon network. Although this reaction can proceed spontaneously, it can be promoted as well by using electrochemical, photochemical, or chemical methods.

Electrochemical reduction of aryl diazonium salts on a carbon surface was reported for the first time by Savéant and coworkers [8, 9]. The passivation of the electron transfer on the electrode shows the formation of a thick film on the surface. The mechanism of formation of these thick films has been widely discussed in literature. The generally accepted mechanism consists of the formation of polyphenylene films by continuous attack of radicals to the already grafted moieties. The presence of azo bonds in the structure was confirmed by time of flight secondary ion mass spectroscopy (TOF–SIMS) experiments [10]. Other mechanisms have been proposed for the formation of multilayers on metallic surfaces based on the reaction of aryl cations with the surface [11]. These films have been characterized in-depth by different surface characterization techniques such as atomic force microscopy (AFM), scanning tunneling microscopy (STM), X-ray photoelectron spectroscopy

(XPS), infrared spectroscopy, Raman spectroscopy, surface-enhanced Raman scattering (SERS), TOF–SIMS or contact angle measurements. For sp^2-carbon materials, Raman spectroscopy is usually the preferred technique to detect the presence of covalently attached or grafted units. Graphite and graphene have characteristic Raman spectra, where the generation of sp^3 hybridized defects leads to the appearance of a D-band at around 1350 cm^{-1} with concomitant changes in the G and 2D bands [12]. Raman spectroscopy has been undoubtedly the most often used technique for the characterization of covalent functionalization of sp^2 carbon materials, though some care has to be taken in case of overlapping bands [13]. Among the many types of sp^2-hybridized materials that have been modified, we will focus on HOPG and surface-supported graphene substrates.

3 Functionalization of Graphite Surfaces

HOPG has been widely used in surface science studies, among other reasons because it is a convenient substrate for a variety of analytical and imaging tools, also under ambient conditions. Diazonium cations lead to a higher functionalization efficiency of HOPG surfaces when compared to amine or azide units as was probed using electrochemical and electrochemical scanning tunneling microscopy (EC-STM) [14]. The study of different inner-sphere redox probes on the modified electrodes confirmed the presence of a dense modified surface with a blocking effect against redox reactions. In line with these results, EC-STM showed a densely packed film of randomly grafted units which were multilayers with the thickness of a few nanometers. The thickness of these films can be easily determined by scratching the surface using contact mode AFM, revealing for instance a film thickness up to 6.3 nm for films formed by grafting of tetrafluoroborate diazonium salts of 4-nitroazobenzene.

Although AFM is a suitable technique to establish the thickness and overall morphology of the films, it doesn't distinguish between chemisorbed and physisorbed species. To clarify the real nature of the interaction of aryl diazonium units with the surface of HOPG, electrografted films from 4-NBD ions were characterized not only by AFM but also by ultrahigh vacuum (UHV)-STM [15]. STM images revealed the adsorption of molecular units on the step edges of the surface and some mobile features on the basal plane. Based on the apparent presence of mobile material on the basal plane and its removal during STM scanning, the authors concluded that there was no evidence of the covalent attachment of the aryl groups to the surface.

Greenwood et al. helped to shine light on the covalent modification with aryl diazonium chemistry on sp^2-carbon-based materials [16]. A combination of local microscopy techniques (AFM and STM) and Raman spectroscopy in ambient conditions was used for studying well-defined films. With this purpose, sterically hindered units such as 3,5-di-*tert*-butyl-benzene diazonium (3,5-TBD) chloride were used and compared to the popular NBD (Fig. 2a). The covalent attachment on the surface was supported by Raman spectroscopy indicating a higher grafting density compared to NBD (Fig. 2b–c). This study revealed the power of STM under ambient conditions

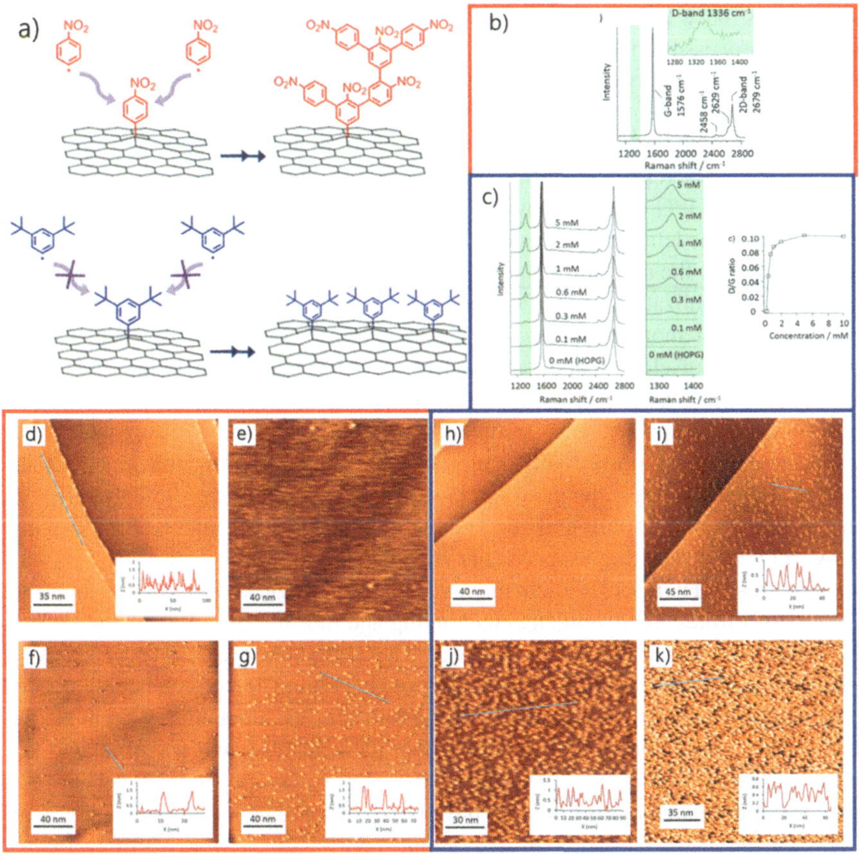

Fig. 2 **a** Scheme for the surface modification of graphite/graphene with 4-NBD (red) and 3,5-TBD (blue). **b** Raman spectra for HOPG substrates modified with NBD from 1 mM solution. **c** Raman spectra for HOPG substrates modified with 3,5-TBD from 0.1, 0.3, 0.6, 1, 2, and 5 mM solutions. **d–g** STM images after grafting of 4-NBD on HOPG from **d** and **e** 0.01 mM, **f** 0.1 mM, and **g** 1 mM solutions. **h–k** STM images after grafting of 3,5-TBD on HOPG from **h** 0.01 mM, **i** 0.1 mM, **j** 1 mM, and **k** 2 mM solutions. Reused with permission from Ref. [16] Copyright (2015) American Chemical Society

as a tool for the visualization (and manipulation) of the grafted units with molecular resolution. STM showed the covalently attached moieties as bright spots (Fig. 2d–k). The surface coverage is in line with the defect density estimated using Raman spectroscopy. By changing the electrodeposition conditions and the concentration of the precursor it was possible to tune the degree of functionalization.

Scanning the covalently functionalized surface with STM tip at high tunneling currents causes the detachment of the covalently bonded units. This process is called nanoshaving and it leads to the regeneration of pristine sp^2 hybridized graphene and graphite, of well-defined size and shape. These regenerated pristine areas, called nanocorrals, have in subsequent studies been used for the study of molecular self-assembly processes under nanoconfinement conditions [17].

The thermal stability of the aryl-based films has been evaluated as well using AFM, STM, and Raman spectroscopy[18]. Interestingly, it is possible to follow the kinetics of desorption with temperature for different substituted aryl groups by using Raman spectroscopy. The results show that under ambient conditions the bond between aryl moieties and HOPG breaks below 200 °C, after which the breakdown products are physisorbed. The temperature dependence for each functional group is different and an estimation of the activation energy was achieved.

The functionalization of HOPG with diazonium salts has been employed for the incorporation of more complex functional units to a surface. This is the case of the attachment of electrochemically active perchlorinated triphenylmethyl (PTM) moieties [19]. EC-STM showed a preferential adsorption of the organic units at the HOPG edges. The presence of a redox reversible couple corresponding to the PTM radical and PTM anion redox process confirms the successful attachment on the surface. Regarding their magnetic properties, DFT calculations show that they are preserved as well. Formation of such robust films could be of interest for the design of hybrid materials acting as molecular switches.

In contrast to HOPG, electrochemical reactions on single layer graphene supported by solid substrates are technically challenging and thus not straightforward, limiting their potential for on-surface reactions. Therefore, chemical approaches using reducing agents to promote the reduction of diazonium ions have been explored. In homogeneous reactions, where the diazonium species and reducing agent are part of the same phase, iodide [20], hypophosphorus acid [21], and ascorbic acid [22] have been employed as reducing agents. Two-phase processes using heterogeneous reducing agents such as iron powder have been explored as well [23].

Iodide mediated reduction of diazonium ions has been shown to be effective for the functionalization of graphite and graphene[20]. This strategy consists of a straightforward, mild, and water-based protocol. The influence of the reaction time, concentration, and stoichiometry of the reagents has been studied showing an impact on the thickness of the film and the grafting density. Interestingly, AFM images revealed the formation of a thin film with empty areas. Their surface coverage and size could be controlled. Nanobubbles or the physisorption of side products from the reaction are potentially responsible for the formation of these empty areas. This process has been shown to be suitable as well for the modification of single layer graphene (SLG) substrates on SiO_2.

Other chemical reducing agents such as ascorbic acid were demonstrated to be efficient as well for the grafting of aryl units on carbon surfaces (Fig. 3a) [22]. In this case, the reaction mechanism is more complex, as it implies the formation of an adduct (diazoether) between the diazonium cation and the ascorbic acid. The characterization by Raman spectroscopy showed different grafting densities dependent on the functional group on the aryl moiety (Fig. 3b). These different limiting grafting density values seem to be related to the position of the SOMO level with respect to the graphite Fermi energy level [24]. This trend could be observed in the STM images as well (Fig. 3c–f). The use of ascorbic acid to promote the formation of aryl-based

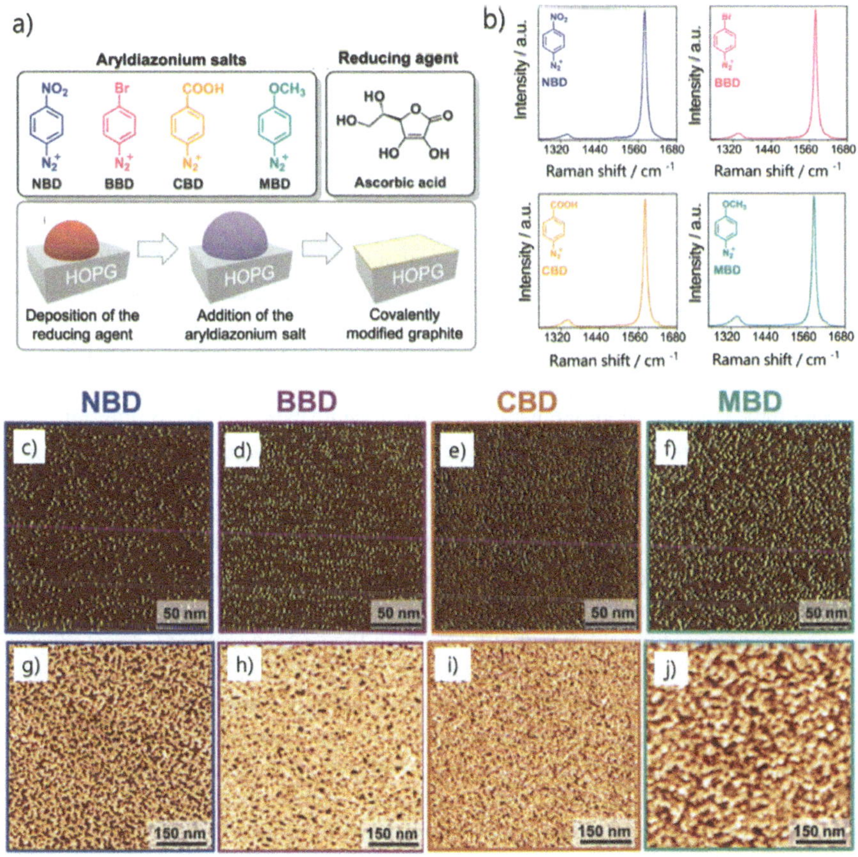

Fig. 3 **a** Scheme of the protocol for the modification using chemical activation of diazonium ions with ascorbic acid on HOPG surfaces. **b** Raman spectra for the grafting of different precursors 4-nitrobenzene diazonium (NBD, blue), 4-bromobenzene diazonium (BBD, pink), 4-carboxybenzene diazonium (CBD, orange), and 4-methoxybenzene diazonium (MBD, green). **c**–**f** STM images of 200 nm × 200 nm for **c** NBD, **d** BBD, **e** CBD, and **f** MBD. **g**–**j** AFM images for **g** NBD, **h** BBD, **i** CBD, and **j** MBD. Reused with permission from Ref. [22] Copyright (2020) Royal Society of Chemistry

films has an important advantage. It is a self-limiting process where films of 1 nm thickness are obtained (Fig. 3g–j). Therefore, the characteristic polyaryl growth for these grafting processes is avoided and films with low thickness and high density of grafting are obtained.

4 On-Surface Functionalization of Graphene

While HOPG can be used to a certain extent as a model substrate for graphene, the reactivity of both substrates is different. This is clear from the difference in the structural and electronic changes upon covalent attachment. For instance, periodic wave density functional theory calculations have revealed that when an aryl unit is attached to the basal plane, the carbon atom of the basal plane is displaced out of the surface by 0.7 Å [25] while this displacement is only 0.46 Å for HOPG [26]. This highlights the influence of the underlying stacked graphene sheets of HOPG on the C–C bond formation on the basal plane. Theoretically it has been shown that the strongest bonding occurs at the graphene edges when compared to the basal plane [25].

The first example of spontaneously grafted 4-nitrobenzene diazonium tetrafluoroborate (4-NBD) on epitaxially grown few-layers graphene on SiC was reported by Bekyarova et al. [27]. The functionalization was confirmed by XPS. Interestingly, the functionalization has an impact in the electronic structure and transport properties of the material. Covalent modification reduced the charge carrier mobility resulting in a change of the resistance at room temperature from near metallic values to semiconducting ones.

The reaction of the aryl diazonium unit with the surface implies a direct electron transfer from the substrate to the diazonium salt. This electron transfer is influenced by several factors including the number of layers, the number of defects or contribution of edges, and the environment of the graphene which can be influenced by the substrate graphene is deposited on. Wang et al. studied the influence of the substrate on the reactivity of monolayer graphene with 4-NBD [28]. CVD graphene was transferred to SiO_2, Al_2O_3, h-BN, and alkyl terminated SiO_2 substrates. The functionalization occurs to a larger extent on graphene on SiO_2 and Al_2O_3 compared to h-BN or alkyl-terminated SiO_2 (Fig. 4a–c). As the rate limiting step of the functionalization process is the electron transfer between the substrate and the diazonium unit, the density of states and the Fermi level of the substrates play a crucial role. By Raman analysis of the substrates, it could be confirmed that the charge impurity puddles are stronger in substrates where the functionalization is more successful. The doping of graphene by these electron–hole puddles results in a shifting of the Fermi level which influences the functionalization process.

The number of layers of graphene strongly influences the reactivity toward aryl diazonium salts. A detailed study on the reactivity of single layer and bilayer graphene flakes was carried out by Koehler et al. [3]. The efficiency of the covalent attachment on the SLG seems to be higher than on bilayer graphene flakes, the reaction being highly favored at the edges (Fig. 4d–f). The process could be followed by using time-dependent confocal Raman microscopy. From the time-dependent D/G ratio, rate constants were extrapolated to quantify the differences in reactivity (Fig. 4d–f). The rate constant changed from 0.241 min^{-1} for the edges of the graphene to 0.044 min^{-1} for the basal plane of the bilayer, which was explained by the fact that it is easier to generate a sp^3 defect on SLG than in bilayer graphene. The number

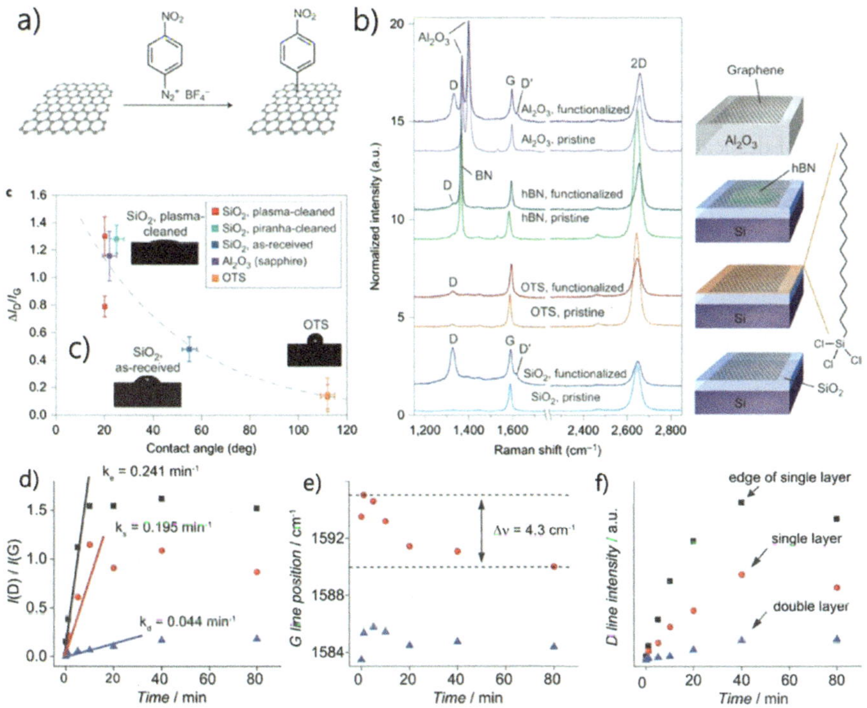

Fig. 4 **a** Reaction scheme of the functionalization of Graphene with 4-NBD. **b** Representative Raman spectra of CVD-grown graphene deposited on different substrate materials before and after diazonium functionalization, normalized to the G peak height. **c** Change in intensity ratio of Raman D and G peaks (ID/IG) after diazonium functionalization (difference between functionalized and unfunctionalized ratios) plotted as a function of water contact-angle of the substrate before graphene deposition. Reused with permission from Ref. [28] Copyright (2012) Nature Publishing Group. **d** Ratio of D-line intensity over G-line intensity for bilayer (triangle), single layer (circles), and the edge of single layer (rectangles) graphene. **e** Position of G line is a measure of the doping level of graphene. **f** D-line intensity at the specified times. Reused with permission from Ref. [3] Copyright (2010) Wiley–VCH

of layers also influences the value of the work function and hence the Fermi level which determines the electron transfer for the functionalization protocol [29]. These thickness-dependence in the reactivity of graphene has been observed as well in other types of reactions such as the photodecomposition of benzoyl peroxide [30].

The curvature of the substrate can have a strong influence on the grafting process as well. A graphene sheet was deformed by depositing it on top of SiO_2 nanoparticles on a silicon surface [31]. SEM images showed a different curvature of the graphene sheet which was treated by NBD. The Raman mapping showed a higher increase in the D/G ratio on top of the deformed areas when compared with the non-deformed ones, revealing a higher efficiency of the reaction on the curved regions. Deformation of the graphene sheet can be obtained through mechanical strain as well [32]. A graphene layer deposited on polydimethylsiloxane (PDMS) was subjected to mechanical strain through stretching of the PDMS and then functionalized with nitro-, bromo-, and methoxy phenyl diazonium (Fig. 5a). The graphene samples under mechanical strain showed a higher reactivity than unstrained ones as shown for the 4-NBD in Fig. 5b.

The formation of sp^3 defects on the graphene lattice implies changes in the conjugation of delocalized electrons and hence in the band structure and electronic properties of the graphene surface. Theoretically it has been proposed that a band gap can

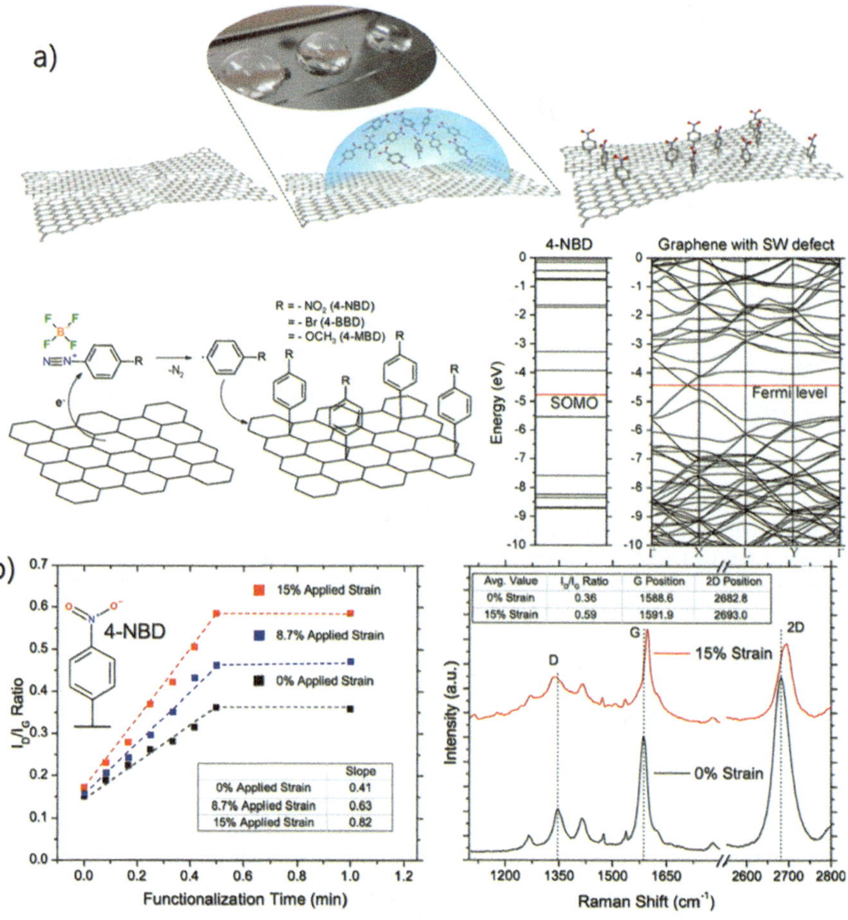

Fig. 5 Schematic of strained graphene functionalization. **a** After being transferred to PDMS, the polycrystalline graphene is strained by elongation. After strain is applied, the aqueous solution of aryl diazonium salt is pipetted onto the graphene surface. The inset shows a photograph of several droplets on a graphene/PDMS substrate. After functionalization, the solution is removed, and the substrate rinsed and dried. **b** Plots of ID/IG ratio for unstrained and strained graphene with 20 mM solution of 4-NBD. Plots comparing the Raman spectra after 30 s of functionalization for both strained and unstrained graphene with 4-NBD. Reused with permission from Ref. [32] Copyright (2013) American Chemical Society

be opened when the surface is covalently functionalized due to mainly two mechanisms: (a) the formation of a large band gap (1–2 eV) around the disordered region and (b) a relatively small band gap (100 mV) in the areas near the disordered region due to the quantum interference phenomena [33].

In an experimental approach, Niyogi et al. combined the information obtained from Raman spectroscopy and ARPES for the characterization of EG/SiC substrates modified with 4-NBD [34], showing a band gap of around 0.4 eV. Some transport measurements were done on exfoliated single layer graphene sheets supported on SiO_2 after functionalization with 4-NBD [35]. These functionalized sheets behaved like a granular metal with a band gap of around 100 meV at low temperatures (4 K), leading to a decreased field effect mobility and the device conductance. The electrochemical functionalization of graphene deposited on SiC and copper with 3,4,5-trimethoxybenzenediazonium (TMeOD) and the impact on the electronic properties have been studied as well [36, 37]. ARPES experiments show that the electronic structure of the Dirac cone is preserved in this case. However, *n*-type doping of the substrate has been shown to trigger a downshift of the Dirac cone which confirms a charge transfer from the molecules to graphene. These results show that the chemical functionalization is a promising route to band gap engineering and tuning of graphene electronics.

5 Control in Thickness. From Multilayers to Monolayers

The inherent high reactivity of the radical units implies a lack of control on the structure at the nanometer scale. Furthermore, polymerization leads to the formation of disordered layers few nanometers thick. Some applications in the area of sensing and electrocatalysis require monolayer thin films though, which has activated research in this area [38]. Approaches to control the film thickness have focused on strategies to avoid or limit radical polymerization process on the surface. Daasberj and Pedersen [39] used a diazonium salt carrying a disulfide unit. After grafting, the S–S bond was cleaved. Using a protocol with protecting groups such as trimethylsilyl (TMS), triethylsilyl (TES), and tri(isopropyl)silyl (TIPS), Hapiot et al. were able to restrict the surface modification to monolayers too [40]. An additional interesting aspect of this approach is the possibility to control the density of the layer (distance between molecules) by changing the size of the protecting group.

Pinson and Podvorica proposed the use of sterically hindering groups such as *tert*-butyl groups in 3,5-di-*tert*-butyl benzene diazonium (TBD) tetrafluoroborate, which avoid the attachment of radicals to the already grafted units [41]. The validity of this concept was confirmed for the functionalization of graphite and graphene using scanning probe techniques such as STM and AFM, in combination with Raman spectroscopy [16].

The use of ionic liquids as alternative to the traditional solvents to carry out the electrografting has been proposed as well. Viscous liquids are expected to limit the diffusion rate of the radicals and hence their concentration at the interface. A decrease in the concentration of radicals should lead to the production of thinner layers. The first reports for the grafting of 4-NBD using ionic liquids showed indeed that the surface coverage decreases when compared to conventional solvents, though the exact mechanism should be investigated in more depth [42].

The use of radical scavengers such as 2,2-diphenyl-1-picrylhydrazyl (DPPH) to control the process of (electro)grafting has been proposed by Breton and coworkers [43]. The film thickness can be controlled by the DPPH concentration and DPPH to diazonium salt ratio. The efficiency of this approach to control the thickness of NBD-based films on HOPG has been reported [44]. Films with a thickness around 0.8 nm and sub-monolayer coverage resulted after 8 electrografting cycles. This coverage depends on the number of cycles. However, this method is quite sensitive to the nature of the substituent on the aryl group [45]. An activating group in the para position lowers the efficiency of the radical and multilayers can still be obtained. Thus, the use of DPPH as radical scavenger is limited to aryl diazonium ions with a deactivating group.

As it can be seen, most of the efforts to control the structure of the films at the nanoscale have been related to grafting processes using electrochemistry as activation tool. As commented before, an approach based on the activation of diazonium ions with ascorbic acid has been reported. The AFM study showed self-limiting growth with the formation of monolayers for a number of functional groups, but only if the ascorbic acid and diazonium salt are subsequently added to the surface [22]. If the components are premixed before dropcasting on the substrate, ill-defined films with exceptionally low density of grafting are found. These findings suggest that the surface plays a key role in the mechanism. When diazonium ions are mixed with ascorbic acid, there are two possible pathways for obtaining the aryl radicals: inner sphere, through the decomposition of a diazoether, and other sphere, via direct reduction. The dominance of the inner sphere mechanism on a reductive surface such as HOPG could explain the results obtained. Finally, this approach has been shown to be efficient for the functionalization of other surfaces such as graphene, but also gold or molybdenum sulfide [22].

6 Control in Spatial Distribution—Patterning

The control of the electronic properties of graphene has been a hot topic in materials research as discussed earlier. The attachment of grafted units to the surface creates a change in the conjugation length accessible to delocalized electrons. The patterning of these grafted units at the molecular level could be a way of patterning the conjugation network and hence inducing changes in the band structure of graphene. The change in the band structure of graphene has been obtained previously through random grafting on the graphene lattice but the values obtained are quite far from the ones predicted through theoretical calculations. DFT results predict a band gap as high as 2 eV for covalently functionalized graphene. However, the band gap found for SLG on SiO_2 has a value below 0.1 meV. This difference is related to some experimental limitations such as the spatially inhomogeneous attachment of aryl groups or the creation of mid-gap states in the substrate during covalent functionalization [33]. Periodic incorporation of grafted units has been suggested as a practical way of producing this band gap opening in a controllable way. Other applications might also benefit from non-random functionalization. For instance, patterned covalent grafting at the micro or nanoscale of functionalities may provide the surface with interesting wetting, catalytic or optical properties [46].

There has been a trend to combine conventional top-down lithography techniques with covalent chemistry based on aryl diazonium ions for the chemical patterning of graphite and graphene sheets deposited on substrates. One such important contribution is the localized patterning, at the micrometer level, of HOPG with 4-carboxybenzenediazonium tetrafluoroborate (4-CBD) [47]. To this end, they used scanning electrochemical cell microscopy (SECCM) to carry out the functionalization under electrochemical control. SECCM uses a dual channel borosilicate glass pipette with the solution of interest and a quasi-reference counter electrode in each channel (Fig. 6a). When the pipette is approached to the surface in a controlled way through xyz piezoelectric positioners, a confined-meniscus electrochemical cell can be created, and the modification of the substrate can be done in specific positions of the substrate. A controlled film density, thickness, and grafting density could be achieved by varying the applied potential and the meniscus contact time (Fig. 6b).

Other approaches are based on providing access of the diazonium salts to the substrate in a spatially controlled way. The use of microspheres to block the surface during the functionalization step and create a patterned surface at the micrometer scale has been such popular approach [48–50]. Pioneering work by Stevenson and co-workers successfully demonstrated this approach for the grafting of diazonium cations onto indium-tin oxide (ITO) electrodes [48]. The surface consisted of

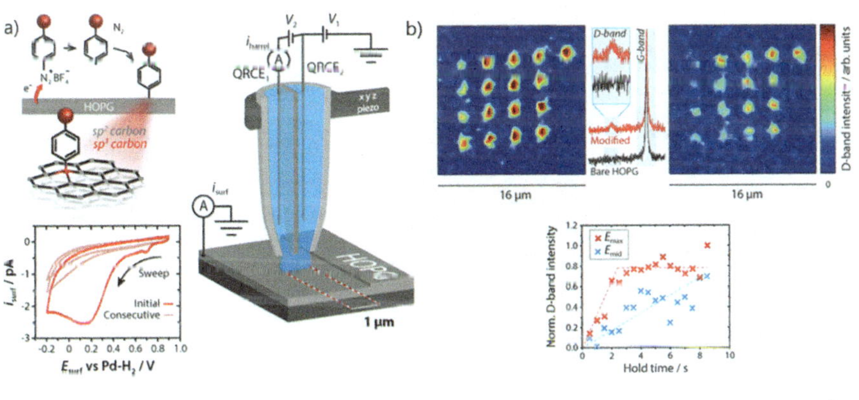

Fig. 6 **a** Schematic of the diazonium modification using the SECCM setup and corresponding cyclic voltammetry showing where the reduction of the 4-carboxybenzenediazonium ions occurs. **b** Raman maps showing the D-band intensity at different potentials and times. Plot of the dependence of the D-band intensity with time of deposition at these two different potentials. Reused with permission from Ref. [47] Copyright (2014) American Chemical Society

regularly spaced holes of 700 nm in diameter surrounded by functionalized areas. Removal of the beads, thereby exposing ungrafted surface areas, followed by electrografting of another diazonium salt on a glassy carbon surface [49] or self-assembly of molecular species on HOPG [50], is a strategy to form hybrid nanostructured surfaces. This strategy for the formation of hybrid nanostructured surfaces was shown to be interesting in the production of bifunctional electrodes [51]. Using diazonium salts with complementary redox properties, such modified electrode can filter the electron transfer with a low potential gap and the electrochemical oxidation/reduction of redox probes on these nanostructured surfaces occurs in completely different areas.

Photolithography protocols have been used as well to control the distribution of grafted units. Using a photoresist mask, Khoeler et al. created a patterned HOPG surface which they exposed to diazonium reagents [52]. Square-shaped functionalized areas with a size of 20 μm × 20 μm were achieved for different functional groups. Kelvin probe force microscopy (KPFM) and SEM experiments showed a different surface potential value for electron withdrawing ($-NO_2$) and donating ($-OCH_3$) groups. Hirsch et al. used an electron beam lithography (ELB)-based protocol for the formation of patterned surfaces at the micrometer level (Fig. 7a) [53]. Shortly, a graphene monolayer was deposited on SiO_2/Si and a layer of PMMA was deposited on top of the graphene. Through EBL, some regions of the graphene were exposed and activated by using Na/K alloy. This treatment creates negative charges that promote the reaction with electrophiles such as nitro or bromo benzene diazonium tetrafluoroborate. After the reaction, the PMMA is removed and then the covalent pattern can be visualized using optical images (Fig. 7b) and Raman spectroscopy mapping (Fig. 7c). As the covalent bonds are relatively weak, annealing

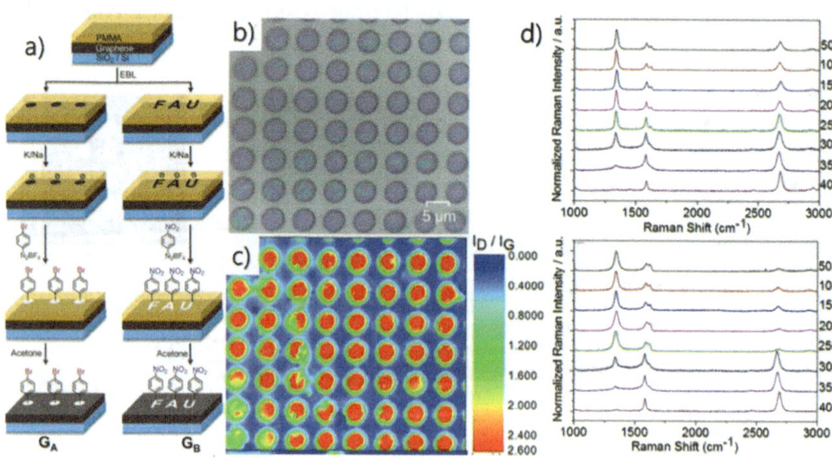

Fig. 7 **a** Schematic illustration of the reaction sequence for the patterned functionalization of monolayer graphene. **b** Optical images of PMMA-patterned graphene where blue regions represent exposed graphene. **c** Corresponding Raman ID/IG mapping images after addend binding and removal of the PMMA layer. **d** Temperature-dependent Raman spectra for G_A (top) and G_B (bottom). Reused with permission from Ref. [53] Copyright (2020) Wiley–VCH

induces the removal of the covalently grafted molecules, a feature which is of interest for writing/erasing cycles (Fig. 7d). Other examples imply the functionalization of the graphene at both sides [54]. In this case AgF was used as fluorination source and introduced in an array of channels produced by classical lithography. The graphene sheet was placed on top and by annealing a short time the fluorination takes place on one side of the sheet. The other side of the graphene undergoes antaratopic addition by using a diazonium salt. Recently, a chemical gradient on defined areas of the surface was created by a complex combination of classical lithography and chemical functionalization [55]. Periodic concentric circles containing bromobenzene-, deutero-, and chloro-addends were formed on the graphene surface. The covalent patterning was confirmed by statistical Raman spectroscopy (SRS) and scanning electron microscopy/energy dispersive X-ray spectroscopy (SEM–EDS), showing areas of different composition. Interestingly, temperature-dependent Raman measurements revealed that the covalent bond formation is reversible, being restored to the sp^2 network after degrafting.

All discussed strategies are useful when the patterning targeted is in the range of micrometers. Quasi-periodic surfaces at the few tens to few hundred nanometer level are obtained by bottom-up approaches where nanobubbles and/or areas covered by physisorbed species act as templates [56]. Nevertheless, when controlled spatial distribution of molecules at the nanometer level is required, these methods or lithographic techniques cannot be employed. Two approaches have been reported. The first one is based on the on-surface self-assembly of reactive aryldiazonium precursors, followed by the generation of the radicals and their covalent attachment to the surface [57, 58]. Ideally, such preassembly strategy would lead to the formation of rows of grafted units separated by few nanometers only. Samori and coworkers designed and synthesized a diazonium salt specifically with this aim. The 4-docosyloxybenzenediazonium tetrafluoroborate (DBT) includes a long aliphatic chain which promotes the physisorption on graphene, forming an ordered template on graphite or CVD graphene. After the self-assembly of the diazonium salts, the sample is transferred into an aqueous electrolyte for covalent attachment under electrochemical control. The results obtained by STM, AFM, XPS and conductivity measurements suggested that the grafting of diazonium ions may occur in a patterned way. A similar "preassembly strategy" was used in case of the covalent attachment of a heteroleptic polypyridyle Ru(II) complex, resulting in parallel linear stripes of grafted molecules separated by 3.8 nm as revealed by STM [58].

A potential disadvantage of the preassembly approach is the fact that any change in grafting periodicity implies the synthesis of a new aryl diazonium compound. Therefore, an alternative strategy was proposed by Tobe, Tahara, De Feyter et al. where the tasks of assembly and covalent grafting are not combined in one species [59, 60]. In their approach, a self-assembled molecular network is formed by a non-reactive species. That network acts as template for the electrochemically activated spatially controlled covalent attachment of the aryl diazonium salt of choice (Fig. 8a) Linear alkanes (e.g. $C_{30}H_{62}$, $C_{40}H_{82}$, $C_{50}H_{102}$) assemble in parallel rows, leading to templated linear grafting of the aryl diazonium salt on graphite and graphene as revealed by STM (Fig. 8b–c). Linearly aligned aryl groups with a lateral periodicity of

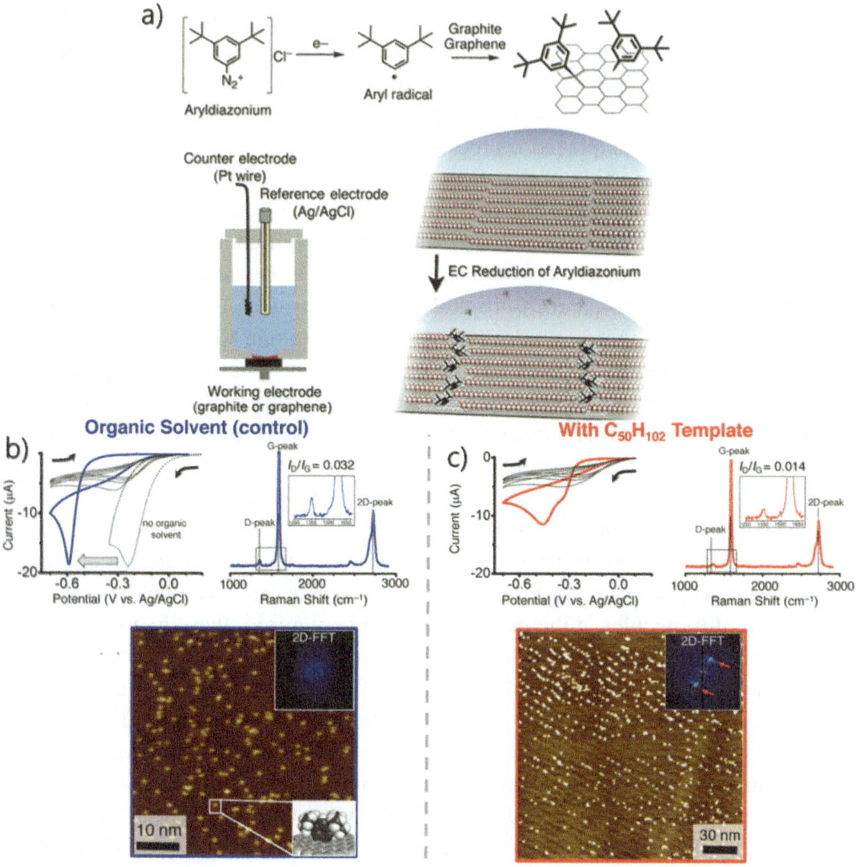

Fig. 8 Illustration of template-directed linear nanopatterning process. **a** Schematic representation for chemical covalent functionalization of graphitic surfaces by an aryl radical generated from an aryldiazonium ion (3,5-di-*tert*-butylbenzene diazonium chloride), illustration of the electrochemistry setup, and schematic representation for linear grafting through the self-assembled molecular network formed by the alkane template. **b** Grafting of 3,5-TBD with a layer of organic solvent on the surface (random grafting). **c** Grafting of 3,5-TBD with a layer of organic solvent on the surface containing the template of self-assembled *n*-alkanes (templated grafting). Reused with permission from Ref. [59] Copyright (2018) American Chemical Society

5–7 nm could be achieved. The periodicity is determined by the length of the alkanes. An attractive feature of this strategy is the large variety of surface patterns that can be formed. For instance, using template molecules that self-assemble into porous hexagonal networks, patterns of hexagonally aligned grafted units were obtained, the periodicity of which depends on the size of the template molecules [60]. Interestingly, the pattern transfer fidelity depends on the pore size, which implies that the quality of the process can be improved by designing pores with optimized shape and size.

7 From Surfaces to Bulk

Functionalization of graphene dispersed in liquid media presents a major challenge as the van der Waals interactions between graphene flakes promote restacking of the sheets through pi-pi interactions. Typically, compared to solid-supported graphene, the structure and properties of liquid-processed materials are less well defined.

Tour and coworkers reported the production of dispersed graphene sheets functionalized with aryl diazonium salts for the first time [61]. For this, they used different aryl diazonium ions such as chlorobenzene, nitrobenzene, methoxybenzene, and bromobenzene on chemically reduced graphene. The degree of functionalization was estimated by TGA analysis which revealed covalent attachment of around 1 functional group per 55 carbon atoms. Interestingly, the dispersion of the material in dimethylformamide (DMF), dimethylacetamide (DMA), and *N*-methyl-2-pyrrolidone is better after functionalization compared to the unfunctionalized material. The same protocol was applied for the functionalization of expanded graphite with bromobenzene, which yielded similar results [62]. Pumera and coworkers reported the influence of halogen atoms and their position in the ring on the electronic and electrochemical properties of graphene [63]. Thermally reduced graphene was modified by the spontaneous reaction of diazonium ions containing halogen atoms in different positions of the ring. The results show that the position in the benzene ring and the type of halogen atom has a strong influence on graphene properties due to inductive and mesomeric effects. These changes in the electronic structure were evaluated by theoretical calculations which revealed differences in the electron density depending on the diazonium ion. Electrochemical measurements showed that the resistivity depends on the electronegativity of the halogen atom as well as its position on the benzene ring, as expected by the theoretical calculations.

An interesting approach is the covalent functionalization of graphene flakes in a dispersion produced by typical intercalation methods (Fig. 9a) [64, 65]. This strategy consists of the use of alkaline metals to produce graphite intercalation compounds. In the next step the dispersion of graphenides is quenched by electrophiles, in this case aryl diazonium salts. The dispersion in a suitable solvent results in complete exfoliation of the material in graphene monolayers. As a result, using this protocol the production of well-defined dispersions of functionalized graphene monolayer flakes is achieved, as shown by AFM and Raman spectroscopy results (Fig. 9b). The reaction conditions require strict control of the presence of oxygen or water. Through changes in the concentration of diazonium salt in the solution materials with different density of grafting were obtained. The dispersion of the graphene flakes in organic solvents such as chloroform was improved after functionalization, reaching concentrations of 27 mg/mL.

Among the methods to produce bulk dispersions of graphene, the electrochemical exfoliation of carbon substrates such as graphite is a popular one. Hapiot and coworkers proposed the cathodic exfoliation of graphite in DMF containing alkylammonium ions [66]. The intercalation of these bulky ions results in an efficient

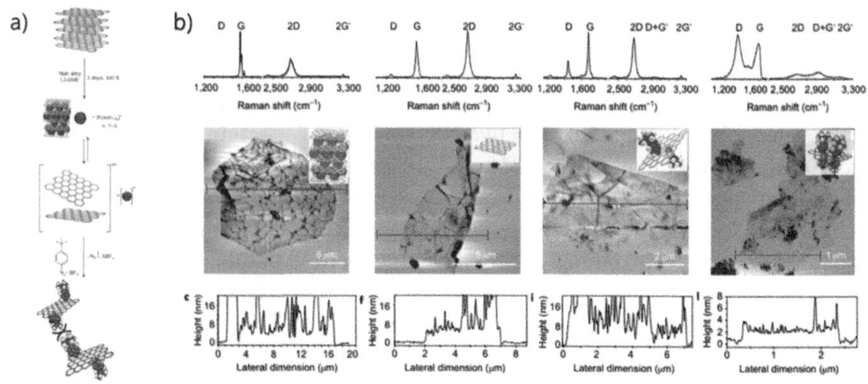

Fig. 9 **a** Representation of the intercalation and exfoliation of graphite with subsequent functionalization. **b** Raman and AFM experiments conducted at different stages of the reduction/exfoliation/functionalization sequence. Reused with permission from Ref. [64] Copyright (2011) Nature Publishing Group

and fast exfoliation of the substrate to produce graphene sheets. When DMF solution contains aryl diazonium ions the simultaneous functionalization of the graphene sheet can occur. Importantly, better results were obtained when HOPG was used probably due to a more efficient intercalation. Bélanger and coworkers reported the one-pot synthesis of graphene sheets modified with anthraquinone by using electrochemical exfoliation [67]. Functionalization led to an increased surface area which was attributed to reduced restacking and aggregation of the graphene sheets.

8 Aryl Linkers to Build Complex Architectures. Toward the Application of Grafted Surfaces

Due to the versatility of aryl diazonium chemistry, diazonium salts are important building stones to construct complex interfaces covalently anchored to carbon surfaces [68]. When carrying an appropriate functional group, the grafted aryl groups act as platform to bind other species via a wide range of reactions including electrostatic interactions, coordination bonding, click chemistry, amide coupling or nucleophilic addition/substitution among others [68]. The formation of complex interfaces is of great importance for example in the design of biosensors [69], where HOPG and graphene are often used as substrates. Using such strategy, the incorporation of biomolecules on the surface of graphite electrodes has been reported [70, 71]. For this, diazonium ions bearing amino or azide groups as terminal group were grafted onto graphite substrates. The reaction of these reactive groups with a counterpart in the DNA moieties [70] or Fe–Fe H2ase model [71] leads to the formation of biosensors or catalytic surfaces respectively. Others nanoobjects such as nanoparticles can be anchored to the previously grafted graphite surfaces as well [72]. For instance, the attachment of silicon nanoparticles to the grafted graphite produces a nanocomposite with interesting properties as negative electrode in lithium-ion batteries. Silicon particles can be grafted by diazotization of an amino group exposed to the interface which can form a covalent bond with the silicon surface.

HOPG has been used as model substrate for these post-functionalization reactions. The functionalization of HOPG with amino groups allows the immobilization of a hydrogenase with catalytic purposes [73]. The targeted orientation of the enzyme is reached by electrostatic interactions directing the orientation for direct electron transfer (DET). A high stability of the hydrogenase-modified carbon electrode was obtained still reaching 90% of the catalytic performance after 1 week. Functionalization of HOPG was used as substrate for the study and design of artificial solid electrolyte interfaces for batteries [74]. In this case, the electrografting of a protected alkyne moiety was carried out and, after the deprotection, a reaction based on thiol-yne click chemistry was done with a thioether on the surface. The effect of the coating on the formation of the solid electrolyte layer was investigated electrochemically. The most important result is the reduction of irreversible capacity loss of the battery.

Graphene nanocomposites have been reported, the grafting of diazonium ions being the first step in the functionalization process. Graphene flakes have been covalently modified with a conjugated polymer, poly[(9,9 dihexylfluorene)-co-alt-(9,9-bis-(6-azidohexyl)fluorene)] (PFA) (Fig. 10a) [75]. The coupling of azo groups with alkynes, a key step in this process, has been used as well for the coupling of azlactones on graphene surfaces [76]. Azalactones are heterocycles usually employed for the synthesis of heterocyclic or natural products. In this case, the reactive azlactone was subjected to ring-opening reaction and the coupling of amine derivatives allowed the incorporation of biological fragments. Additionally, this reaction has been of interest for the fabrication of self-organized three-dimensional graphene electrode [77].

The covalent attachment of macrocycles such as phthalocyanines with relevant catalytic and optical properties to graphene surfaces has been explored as well [78]. In this case, a phenyl carboxylic modified graphene dispersion reacts with a symmetrically substituted zinc phthalocyanine through the formation of an amide bond. CVD graphene modified with carboxyphenyl moieties was used for the immobilization of *N*,*N*-bis(carboxymethyl)-L-lysine hydrate through an amide bond which results in the formation nitrilotriacetic acid unis. These groups can chelate metals such as Ni(II) as proved after the addition of $NiCl_2$ (Fig. 10b) [28]. In this case, a post-functionalization with a fluorescent molecule serves as a way to see the pattern on the surface. Graphene modified with amino or carboxylic groups has been used as platform for the design and fabrication of electrochemical immunosensors based on graphene[79]. The aptamers and antibody covalent attachment could be achieved using the modified graphene.

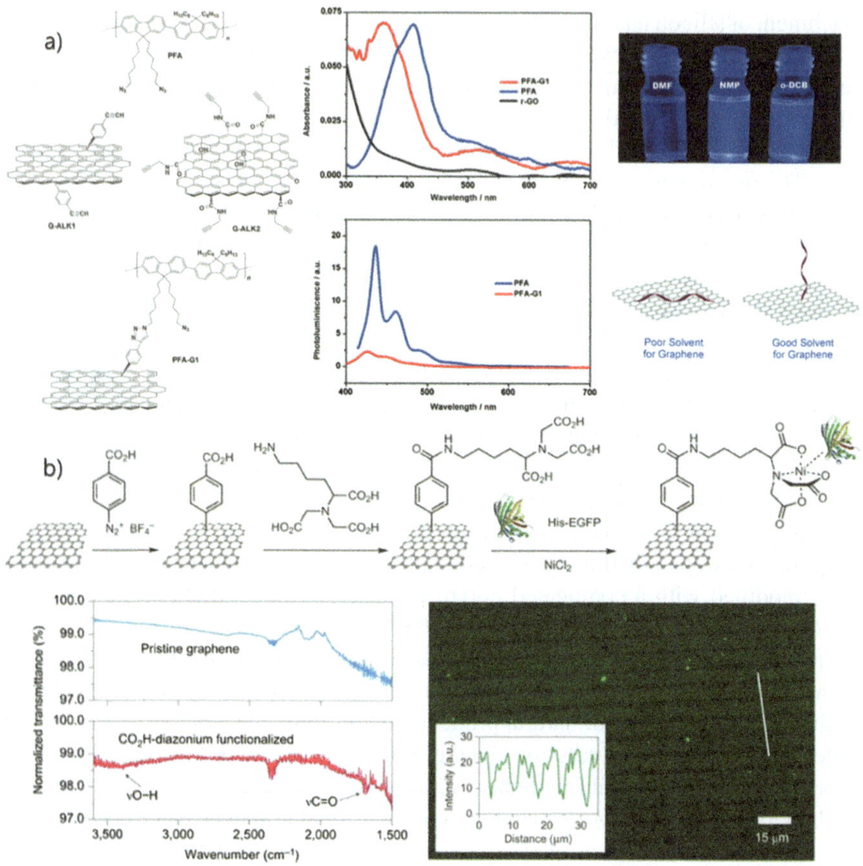

Fig. 10 **a** Schematic of the reaction of PFA with functionalized graphene. Absorption (top) and emission spectra (bottom) of PFA and PFA-G1 in DMF. Solutions of PFA-G1 in various solvents photographed under exposure to UV light and schematic representation of the changes in the structure of PFA-G1 depending on the environment. Reused with permission from Ref. [75] Copyright (2012) Wiley–VCH. **b** Schematic illustration of protein-attachment chemistry. ATR–IR spectra of pristine CVD graphene (blue curve) and CO_2H-diazonium functionalized CVD graphene (red curve), confocal fluorescence microscope image of EGFP attached to graphene. The bright green stripes indicate a higher concentration of EGFP attachment. Reused with permission from Ref. [28] Copyright (2011) Nature Publishing Group

9 Conclusions

Chemical functionalization of graphite/graphene is a hot topic in nanoscience. Aryl diazonium has become one of the preferred precursors to carry out the modification of this type of materials. From the first reports focusing on the functionalization and characterization of the materials, we have seen in the last few years an interest to develop strategies to control the growth of the organic films, both in relation to the thickness and spatial distribution. Controlling film structure offers the possibility to finetune the properties of a functionalized material, for instance the band gap of graphene. Additionally, surface modification of these materials is of great relevance for building more complex structures by post functionalization protocols. The knowledge acquired about chemical functionalization of graphite and graphene using diazonium chemistry will be key for understanding and developing these protocols to new materials.

References

1. Novoselov KS, Geim AK, Morozov SV et al (2004) Electric field in atomically thin carbon films. Science 306:666–669. https://doi.org/10.1126/science.1102896
2. Narita A, Wang XY, Feng X, Müllen K (2015) New advances in nanographene chemistry. Chem Soc Rev 44:6616–6643
3. Koehler FM, Jacobsen A, Ensslin K et al (2010) Selective chemical modification of graphene surfaces: distinction between single- and bilayer graphene. Small 6:1125–1130. https://doi.org/10.1002/smll.200902370
4. Mali KS, Greenwood J, Adisoejoso J et al (2015) Nanostructuring graphene for controlled and reproducible functionalization. Nanoscale 7:1566–1585
5. Kiang Chua C, Pumera M (2013) Covalent chemistry on graphene. Chem Soc Rev 42:3222–3233. https://doi.org/10.1039/c2cs35474h
6. Criado A, Melchionna M, Marchesan S, Prato M (2015) The covalent functionalization of graphene on substrates. Angew Chem Int Ed 54:10734–10750. https://doi.org/10.1002/anie.201501473
7. Park J, Yan M (2013) Covalent functionalization of graphene with reactive intermediates. Acc Chem Res 46:181–189. https://doi.org/10.1021/ar300172h
8. Allongue P, Delamar M, Desbat B et al (1997) Covalent modification of carbon surfaces by aryl radicals generated from the electrochemical reduction of diazonium salts. J Am Chem Soc 119:201–207. https://doi.org/10.1021/ja963354s
9. Delamar M, Hitmi R, Pinson J, Savéant J (1992) Covalent modification of carbon surfaces by grafting of functionalized aryl radicals produced from electrochemical reduction of diazonium salts. J Am Chem Soc 114:5883–5884. https://doi.org/10.1021/ja00040a074
10. Doppelt P, Hallais G, Pinson J et al (2007) Surface modification of conducting substrates. Existence of azo bonds in the structure of organic layers obtained from diazonium salts.https://doi.org/10.1021/cm0700551
11. Mesnage A, Lefèvre X, Jégou P et al (2012) Spontaneous grafting of diazonium salts: chemical mechanism on metallic surfaces. Langmuir 28:11767–11778. https://doi.org/10.1021/la3011103
12. Ferrari AC, Basko DM (2013) Raman spectroscopy as a versatile tool for studying the properties of graphene. Nat Nanotechnol 8:235–246
13. Sampathkumar K, Diez-Cabanes V, Kovaricek P et al (2019) On the suitability of Raman spectroscopy to monitor the degree of graphene functionalization by diazonium salts. J Phys Chem C 123:22397–22402. https://doi.org/10.1021/acs.jpcc.9b06516
14. Tanaka M, Sawaguchi T, Sato Y et al (2011) Surface modification of GC and HOPG with diazonium, amine, azide, and olefin derivatives. Langmuir 27:170–178. https://doi.org/10.1021/la1035757
15. Ma H, Lee L, Brooksby PA et al (2014) Scanning tunneling and atomic force microscopy evidence for covalent and noncovalent interactions between aryl films and highly ordered pyrolytic graphite. J Phys Chem C 118:5820–5826. https://doi.org/10.1021/jp411826s

16. Greenwood J, Phan TH, Fujita Y et al (2015) Covalent modification of graphene and graphite using diazonium chemistry: tunable grafting and nanomanipulation. ACS Nano 9:5520–5535. https://doi.org/10.1021/acsnano.5b01580
17. Verstraete L, De Feyter S (2021) 2D Self-assembled molecular networks and on-surface reactivity under nanoscale lateral confinement. Chem Soc Rev. https://doi.org/10.1039/D0CS01338B
18. van Gorp H, Walke P, Teyssandier J et al (2020) On the thermal stability of aryl groups chemisorbed on graphite. J Phys Chem C 124:1980–1990. https://doi.org/10.1021/acs.jpcc.9b09808
19. Seber G, Rudnev AV, Droghetti A et al (2017) Covalent modification of highly ordered pyrolytic graphite with a stable organic free radical by using diazonium chemistry. Chem Eur J 23:1415–1421. https://doi.org/10.1002/chem.201604700
20. Xia Y, Martin C, Seibel J et al (2020) Iodide mediated reductive decomposition of diazonium salts: towards mild and efficient covalent functionalization of surface-supported graphene. Nanoscale 12:11916–11926. https://doi.org/10.1039/d0nr03309j
21. Pandurangappa M, Ramakrishnappa T (2008) Derivatization and characterization of functionalized carbon powder via diazonium salt reduction. J Solid State Electrochem 12:1411–1419. https://doi.org/10.1007/s10008-007-0470-6
22. Rodríguez González MC, Brown A, Eyley S et al (2020) Self-limiting covalent modification of carbon surfaces: diazonium chemistry with a twist. Nanoscale 12:18782–18789. https://doi.org/10.1039/d0nr05244b
23. Mévellec V, Roussel S, Tessier L et al (2007) Grafting polymers on surfaces: a new powerful and versatile diazonium salt-based one-step process in aqueous media. Chem Mater 19:6323–6330. https://doi.org/10.1021/cm071371i
24. Tahara K, Kubo Y, Lindner B et al (2019) Steric and electronic effects of electrochemically generated aryl radicals on grafting of the graphite surface. Langmuir 35:2089–2098. https://doi.org/10.1021/acs.langmuir.8b03339
25. Jiang D, Sumpter BG, Dai S (2006) How do aryl groups attach to a graphene sheet? https://doi.org/10.1021/JP065980
26. González MCR, Carro P, Vázquez L, Creus AH (2016) Mapping nanometric electronic property changes induced by an aryl diazonium sub-monolayer on HOPG. Phys Chem Chem Phys 18:29218–29225. https://doi.org/10.1039/c6cp05910d
27. Bekyarova E, Itkis ME, Ramesh P et al (2009) Chemical modification of epitaxial graphene: spontaneous grafting of aryl groups. J Am Chem Soc 131:1336–1337. https://doi.org/10.1021/ja8057327
28. Wang QH, Jin Z, Kim KK et al (2012) Understanding and controlling the substrate effect on graphene electron-transfer chemistry via reactivity imprint lithography. Nat Chem 4:724–732. https://doi.org/10.1038/nchem.1421
29. Mathieu C, Barrett N, Rault J et al (2011) Microscopic correlation between chemical and electronic states in epitaxial graphene on SiC(000-1). Phys Rev B—Condens Matter Mater Phys 83.https://doi.org/10.1103/PhysRevB.83.235436
30. Liu H, Ryu S, Chen Z et al (2009) Photochemical reactivity of graphene. J Am Chem Soc 131:17099–17101. https://doi.org/10.1021/ja9043906
31. Wu Q, Wu Y, Hao Y et al (2013) Selective surface functionalization at regions of high local curvature in graphene. Chem Commun 49:677–679. https://doi.org/10.1039/c2cc36747e
32. Bissett MA, Konabe S, Okada S et al (2013) Enhanced chemical reactivity of graphene induced by mechanical strain. ACS Nano 7:10335–10343. https://doi.org/10.1021/nn404746h
33. Shih CJ, Wang QH, Jin Z et al (2013) Disorder imposed limits of mono- and bilayer graphene electronic modification using covalent chemistry. Nano Lett 13:809–817. https://doi.org/10.1021/nl304632e
34. Niyogi S, Bekyarova E, Itkis ME et al (2010) Spectroscopy of covalently functionalized graphene. Nano Lett 10:4061–4066. https://doi.org/10.1021/nl1021128
35. Zhang H, Bekyarova E, Huang JW et al (2011) Aryl functionalization as a route to band gap engineering in single layer graphene devices. Nano Lett 11:4047–4051. https://doi.org/10.1021/nl200803q

36. Ambrosio G, Brown A, Daukiya L et al (2020) Impact of covalent functionalization by diazonium chemistry on the electronic properties of graphene on SiC. Nanoscale 12:9032–9037. https://doi.org/10.1039/d0nr01186j
37. Ambrosio G, Drera G, di Santo G et al (2020) Interface chemistry of graphene/Cu grafted by 3,4,5-tri-methoxyphenyl. Sci Rep 10:1–9. https://doi.org/10.1038/s41598-020-60831-8
38. Breton T, Downard AJ (2017) Controlling grafting from aryldiazonium salts: a review of methods for the preparation of monolayers. Aust J Chem 70:960–972
39. Nielsen LT, Vase KH, Dong M et al (2007) Electrochemical approach for constructing a monolayer of thiophenolates from grafted multilayers of diaryl disulfides. J Am Chem Soc 129:1888–1889. https://doi.org/10.1021/ja0682430
40. Leroux YR, Hapiot P (2013) Nanostructured monolayers on carbon substrates prepared by electrografting of protected aryldiazonium salts. Chem Mater 25:489–495. https://doi.org/10.1021/cm303844v
41. Combellas C, Kanoufi F, Pinson J, Podvorica FI (2008) Sterically hindered diazonium salts for the grafting of a monolayer on metals. J Am Chem Soc 130:8576–8577. https://doi.org/10.1021/ja8018912
42. Actis P, Caulliez G, Shul G et al (2008) Functionalization of glassy carbon with diazonium salts in ionic liquids. Langmuir 24:6327–6333. https://doi.org/10.1021/la703714a
43. Menanteau T, Levillain E, Downard AJ, Breton T (2015) Evidence of monolayer formation via diazonium grafting with a radical scavenger: electrochemical, AFM and XPS monitoring. Phys Chem Chem Phys 17:13137–13142. https://doi.org/10.1039/c5cp01401h
44. González MCR, Orive AG, Salvarezza RC, Creus AH (2016) Electrodeposition of gold nanoparticles on aryl diazonium monolayer functionalized HOPG surfaces. Phys Chem Chem Phys 18:1953–1960. https://doi.org/10.1039/c5cp06415e
45. Menanteau T, Dias M, Levillain E et al (2016) Electrografting via diazonium chemistry: the key role of the aryl substituent in the layer growth mechanism. J Phys Chem C 120:4423–4429. https://doi.org/10.1021/acs.jpcc.5b12565
46. Wei T, Bao L, Hauke F, Hirsch A (2020) Recent advances in graphene patterning. ChemPlusChem 85:1655–1668. https://doi.org/10.1002/cplu.202000419
47. Kirkman PM, Güell AG, Cuharuc AS, Unwin PR (2014) Spatial and temporal control of the diazonium modification of sp^2 carbon surfaces. J Am Chem Soc 136:36–39. https://doi.org/10.1021/ja410467e
48. Maldonado S, Smith TJ, Williams RD et al (2006) Surface modification of indium tin oxide via electrochemical reduction of aryldiazonium cations. Langmuir 22:2884–2891. https://doi.org/10.1021/la052696l
49. Corgier BP, Bélanger D (2010) Electrochemical surface nanopatterning using microspheres and aryldiazonium. Langmuir 26:5991–5997. https://doi.org/10.1021/la904521w
50. van Gorp H, Walke P, Bragança AM et al (2018) Self-assembled polystyrene beads for templated covalent functionalization of graphitic substrates using diazonium chemistry. ACS Appl Mater Interfaces 10:12005–12012. https://doi.org/10.1021/acsami.7b18969
51. Nguyen VQ, Schaming D, Martin P, Lacroix JC (2019) Nanostructured mixed layers of organic materials obtained by nanosphere lithography and electrochemical reduction of aryldiazonium salts. Langmuir 35:15071–15077. https://doi.org/10.1021/acs.langmuir.9b02811
52. Koehler FM, Luechinger NA, Ziegler D et al (2009) Permanent pattern-resolved adjustment of the surface potential of graphene-like carbon through chemical functionalization. Angew Chem Int Ed 48:224–227. https://doi.org/10.1002/anie.200804485
53. Wei T, Kohring M, Chen M et al (2020) Highly efficient and reversible covalent patterning of graphene: 2D-management of chemical information. Angew Chem Int Ed 59:5602–5606. https://doi.org/10.1002/anie.201914088
54. Bao L, Zhao B, Lloret V et al (2020) Spatially resolved bottom-side fluorination of graphene by two-dimensional substrate patterning. Angew Chem Int Ed 59:6700–6705. https://doi.org/10.1002/anie.202002508
55. Wei T, Kohring M, Weber HB et al (2021) Molecular embroidering of graphene. Nat Commun 12:1–8. https://doi.org/10.1038/s41467-020-20651-w

56. Phan TH, van Gorp H, Li Z et al (2019) Graphite and graphene fairy circles: a bottom-up approach for the formation of nanocorrals. ACS Nano 13:5559–5571. https://doi.org/10.1021/acsnano.9b00439
57. Xia Z, Leonardi F, Gobbi M et al (2016) Electrochemical functionalization of graphene at the nanoscale with self-assembling diazonium salts. ACS Nano 10:7125–7134. https://doi.org/10.1021/acsnano.6b03278
58. Nguyen VQ, Sun X, Lafolet F et al (2016) Unprecedented self-organized monolayer of a Ru(II) complex by diazonium electroreduction. J Am Chem Soc 138:9381–9384. https://doi.org/10.1021/jacs.6b04827
59. Tahara K, Ishikawa T, Hirsch BE et al (2018) Self-assembled monolayers as templates for linearly nanopatterned covalent chemical functionalization of graphite and graphene surfaces. ACS Nano 12:11520–11528. https://doi.org/10.1021/acsnano.8b06681
60. Tahara K, Kubo Y, Hashimoto S et al (2020) Porous self-assembled molecular networks as templates for chiral-position-controlled chemical functionalization of graphitic surfaces. J Am Chem Soc 142:7699–7708. https://doi.org/10.1021/jacs.0c02979
61. Lomeda JR, Doyle CD, Kosynkin DV et al (2008) Diazonium functionalization of surfactant-wrapped chemically converted graphene sheets. J Am Chem Soc 130:16201–16206. https://doi.org/10.1021/ja806499w
62. Sun Z, Kohama S, Zhang Z et al (2010) Soluble graphene through edge-selective functionalization. Nano Res 3:117–125. https://doi.org/10.1007/s12274-010-1016-2
63. Bouša D, Pumera M, Sedmidubský D et al (2016) Fine tuning of graphene properties by modification with aryl halogens. Nanoscale 8:1493–1502. https://doi.org/10.1039/c5nr06295k
64. Englert JM, Dotzer C, Yang G et al (2011) Covalent bulk functionalization of graphene. Nat Chem 3:279–286. https://doi.org/10.1038/nchem.1010
65. Abellán G, Schirowski M, Edelthalhammer KF et al (2017) Unifying principles of the reductive covalent graphene functionalization. J Am Chem Soc 139:5175–5182. https://doi.org/10.1021/jacs.7b00704
66. Leroux YR, Bergamini JF, Ababou S et al (2015) Synthesis of functionalized few-layer graphene through fast electrochemical expansion of graphite. J Electroanal Chem 753:42–46. https://doi.org/10.1016/j.jelechem.2015.06.013
67. Ossonon BD, Bélanger D (2017) Functionalization of graphene sheets by the diazonium chemistry during electrochemical exfoliation of graphite. Carbon 111:83–93. https://doi.org/10.1016/j.carbon.2016.09.063
68. Gautier C, López I, Breton T (2021) A post-functionalization toolbox for diazonium (electro)-grafted surfaces: review of the coupling methods. Mater Adv. https://doi.org/10.1039/D1MA00077B
69. Hetemi D, Noël V, Pinson J (2020) Grafting of diazonium salts on surfaces: application to biosensors. Biosensors 10
70. Hajdukiewicz J, Boland S, Kavanagh P, Leech D (2010) An enzyme-amplified amperometric DNA hybridisation assay using DNA immobilised in a carboxymethylated dextran film anchored to a graphite surface. Biosens Bioelectron 25:1037–1042. https://doi.org/10.1016/j.bios.2009.09.020
71. Ahmed ME, Dey S, Mondal B, Dey A (2017) H_2 evolution catalyzed by a FeFe-hydrogenase synthetic model covalently attached to graphite surfaces. Chem Commun 53:8188–8191. https://doi.org/10.1039/c7cc04281g
72. Martin C, Alias M, Christien F et al (2009) Graphite-grafted silicon nanocomposite as a negative electrode for lithium-ion batteries. Adv Mater 21:4735–4741. https://doi.org/10.1002/adma.200900235
73. Rüdiger O, Abad JM, Hatchikian EC et al (2005) Oriented immobilization of Desulfovibrio gigas hydrogenase onto carbon electrodes by covalent bonds for nonmediated oxidation of H_2. J Am Chem Soc 127:16008–16009. https://doi.org/10.1021/ja0554312
74. Moock DS, Steinmüller SO, Wessely ID et al (2018) Surface functionalization of silicon, HOPG, and graphite electrodes: toward an artificial solid electrolyte interface. ACS Appl Mater Interfaces 10:24172–24180. https://doi.org/10.1021/acsami.8b04877

75. Castelaín M, Martínez G, Merino P et al (2012) Graphene functionalisation with a conjugated poly(fluorene) by click coupling: striking electronic properties in solution. Chem Eur J 18:4965–4973. https://doi.org/10.1002/chem.201102008
76. Neri G, Scala A, Barreca F et al (2015) Engineering of carbon based nanomaterials by ring-opening reactions of a reactive azlactone graphene platform. Chem Commun 51:4846–4849. https://doi.org/10.1039/c5cc00518c
77. Fortgang P, Tite T, Barnier V et al (2016) Robust electrografting on self-organized 3D graphene electrodes. ACS Appl Mater Interfaces 8:1424–1433. https://doi.org/10.1021/acsami.5b10647
78. Peng C, Xiong Y, Liu Z et al (2013) Bulk functionalization of graphene using diazonium compounds and amide reaction. Appl Surf Sci 280:914–919. https://doi.org/10.1016/j.apsusc.2013.05.094
79. Wang R, Xue C (2013) A sensitive electrochemical immunosensor for alpha-fetoprotein based on covalently incorporating a bio-recognition element onto a graphene modified electrode via diazonium chemistry. Anal Methods 5:5195–5200. https://doi.org/10.1039/c3ay40739j

Aryldiazonium Tetrachloroaurate(III) Salts: Synthesis, Structure, and Fundamental Applications

Ahmed A. Mohamed

Abstract The synthesis of aryldiazonium tetrachloroaurate(III) salts [X–4–$C_6H_4N{\equiv}N$]$AuCl_4$ (X=F, Cl, Br, I, CN, NO_2, COOH, bisaniline, triazine-based dendrimers, C_8F_{17}, C_6H_{13}) is reported by the protonation of anilines with chloroauric acid in acetonitrile followed by one-electron oxidation using nitrosonium salt [NO]X (X=tetrafluoroborate, hexafluorophosphate). X-ray crystal structure of X=CN, NO_2, COOH, C_8F_{17}, C_6H_{13} from acetonitrile or water (X=COOH) displayed $[-N{\equiv}N]^+$ bond distance typical of a triple bond. Electrochemical reduction of the salts showed low or even positive potential values versus silver/silver chloride reference electrode. The robust gold-aryl nanoparticles, termed organometallic nanoparticles, were constructed using green and mild chemical reduction routes of the diazonium gold(III) salts for forensic, environmental, and nanomedicine engineering applications. Specifically, the salts were used in, for example, the development of latent fingerprints on copper and nickel coins, formation of gold-silver alloys, gold-aryl core-tine oxide shell structures, formation of protein and amino acid bioconjugates, and electrochemical synthesis of gold-aryl nanoparticles stabilized with polyaniline.

1 Introduction

Aryldiazonium salts which are stable under ambient conditions and easily soluble in different solvents are interesting because of their need to modify nanoparticles; however, long-term stability has been a major concern [1–6]. To tackle this critical hurdle, many attempts were investigated to synthesize and isolate diazonium salts with considerable stability and solubility in organic solvents and water [5, 6]. However, instead of using isolated diazonium salts, one can achieve high grafting densities of aryl layers using in situ generated salts. Recently, the emerged stabilizing effect for the diazonium salts is the metal-based anion. The efforts to utilize tetrachloroaurate(III) anion significantly enhanced the stability of diazonium salts [4, 7].

A. A. Mohamed (✉)
Department of Chemistry, University of Sharjah, Sharjah 27272, UAE
e-mail: ah.mohamed@sharjah.ac.ae

M. M. Chehimi et al. (eds.), *Aryl Diazonium Salts and Related Compounds*,
Physical Chemistry in Action, https://doi.org/10.1007/978-3-031-04398-7_9

In this chapter, our contribution will be addressed in the synthesis of stable diazonium salts stabilized using tetrachloroaurate(III) anion at ambient conditions [7]. X-ray structures of some of the salts are presented. Some of the interesting applications of the salts are discussed particularly those for the fabrication of gold-aryl nanoparticles. The electrochemistry of the diazonium salts is discussed to better understand the spontaneity of their grafting on flat surfaces.

2 Synthesis and X-Ray Structure of Diazonium Gold(III) Salts

Stable diazonium tetrachloroaurate(III) salts $[X–4–C_6H_4N{\equiv}N]AuCl_4$ (X=CN, C_8F_{17}, NO_2, F, Cl, Br, I, COOH, bisaniline, C_6H_{13}, triazine dendrons) were synthesized in a high yield [1, 2, 7–10], Table 1. Initially, the synthesis started with the protonation of $CN–4–C_6H_4NH_2$ with the weak acid chloroauric acid in water which resulted in the coordination compound $CN–4–C_6H_4NH_2{\cdot}AuCl_3$, Fig. 1 [1]. However, when acetonitrile was used as a solvent the protonation by chloroauric acid formed the anilinium tetrachloroaurate(III) salt $[CN–4–C_6H_4NH_3]AuCl_4$, which was then oxidized by $[NO]PF_6$ to form the diazonium salt $[CN–4–C_6H_4N{\equiv}N]AuCl_4$. The literature diazotization using $NaNO_2$ was also tried followed by the exchange with $H[AuCl_4]$. Using $[NO]^+$ was more convenient in acetonitrile. Using the common textbook sodium nitrite as an oxidizing agent is quite cheaper and easier to handle than the unstable and expensive nitrosonium ion. The gold precursor tetrachloroaurate plays several important roles in this chemistry, (1) provides the needed acidic medium, (2) stabilizes the diazonium cation, (3) gold surface for aryl grafting, (4) dissolves the anilines.

Diazonium salts stabilized with tetrachloroaurate(III) were tried for surface modification and nanoparticles fabrication. The sodium borohydride reduction of $[C_8F_{17}–4–C_6H_4N{\equiv}N]AuCl_4$, for example, in acetonitrile produced ruby red gold-carbon nanoparticles [7]. In this approach, it is not required to use a phase-transfer catalyst such as tetraoctylammonium bromide. The nanoparticles did not gradually lose their colloidal stability in acetonitrile or water when exposed to air for even two years. The nanoparticles can withstand prolonged sonication in ethanol and acetonitrile and indefinite exposure to ambient conditions in the solid-state under laboratory conditions. Single crystals of X=CN, NO_2, C_8F_{17}, C_6H_{13} were grown from acetonitrile by slow evaporation at room temperature, however, in the case of X=COOH [8], the crystals were grown from water. X-ray structure of diazonium salt (X=CN) is shown in Fig. 2 [1] and space-filling of X=NO_2 is shown in Fig. 3 [7]. The N≡N distance of 1.098(7) Å [1, 7].

Attempts to synthesize diazonium salts stabilized with tetrachloroaurate(III) were extended to macromolecules such as triazine dendrimers [9]. Using NMR spectroscopy investigation, it was noticed a deprotection of the BOC groups which is most pronounced for G0 and decreases to G2. FTIR spectroscopy showed the

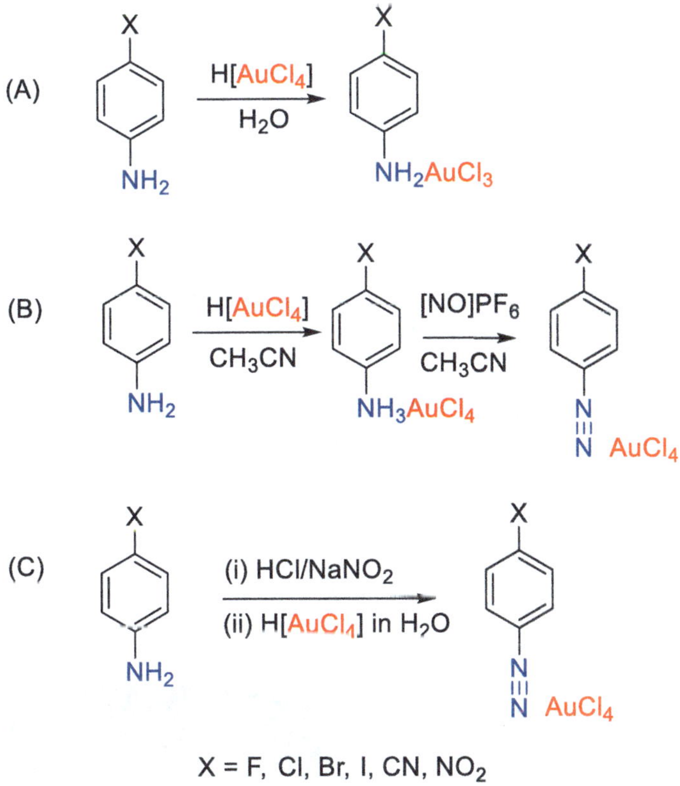

Fig. 1 Synthesis of aryldiazonium tetrachloroaurate(III) salts. **a** Synthesis in water, **b** synthesis using $[NO]PF_6$ in acetonitrile, **c** synthesis using $NaNO_2$ in HCl. Reproduced with permission from Reference [7]. Copyright (2018) Elsevier

appearance of an N≡N stretch at 2279 cm^{-1}. The fabrication of gold nanoparticles occurred using sodium borohydride in acetonitrile solvent. The acetonitrile solution of G0 turned deep purple while those derived from G1 and G2 turned ruby red with absorption maxima at 533 nm, 517 nm, and 515 nm, respectively, Fig. 4. The fabricated G0–G2 nanoparticles showed different solubility behavior. Based on the NMR analysis, a loss in the protecting groups was noticed which is reflected in the solubility of the nanoparticles in water and toluene. When added to a biphasic mixture of toluene and water, G0@AuNPs migrated to the aqueous phase while G1@AuNPs and G2@AuNPs migrated to the organic phase. These results can be explained by the almost complete deprotection of the two BOC groups during the synthesis in the presence of chloroauric acid. Reduced levels of deprotection are seen with G1@AuNPs and G2@AuNPs [9].

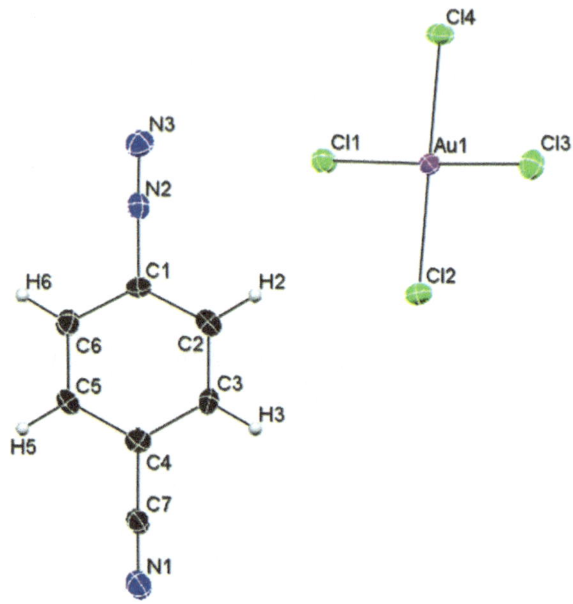

Fig. 2 X-ray crystal structure of [NC–4–$C_6H_4N{\equiv}N$]$AuCl_4$. Reproduced with permission from Reference [1]. Copyright (2012) American Chemical Society

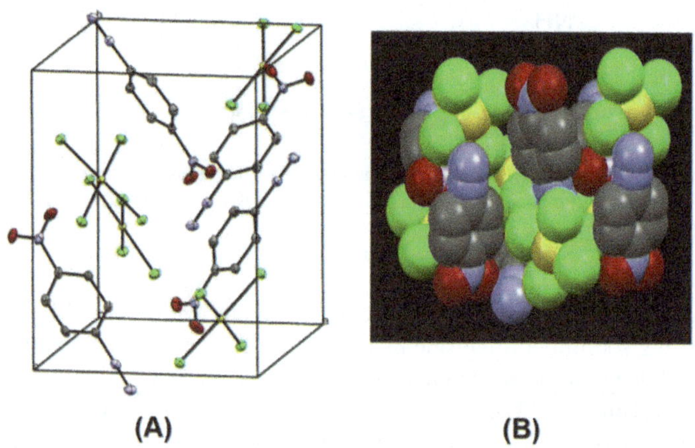

Fig. 3 **a** Crystal packing and **b** space-filling of [O_2N–4–$C_6H_4N{\equiv}N$]$AuCl_4$. Reproduced with permission from Reference [7]. Copyright (2018) Elsevier

Two types of diazonium salt isomers such as cis and trans were synthesized (Fig. 5). The syn (*cis*) and anti (*trans*) isomeric forms of the organic ligand, 4,4′–(1,3–phenylenediisopropylidene) bisaniline and 4,4′–(1,4–phenylenediisopropylidene) bisaniline were used to form the organic shell modifier [10]. The organometallic gold-carbon nanoparticles were formed and characterized using TEM and XPS. This is the first time to report the use of ligands with two different syn (*cis*) and anti

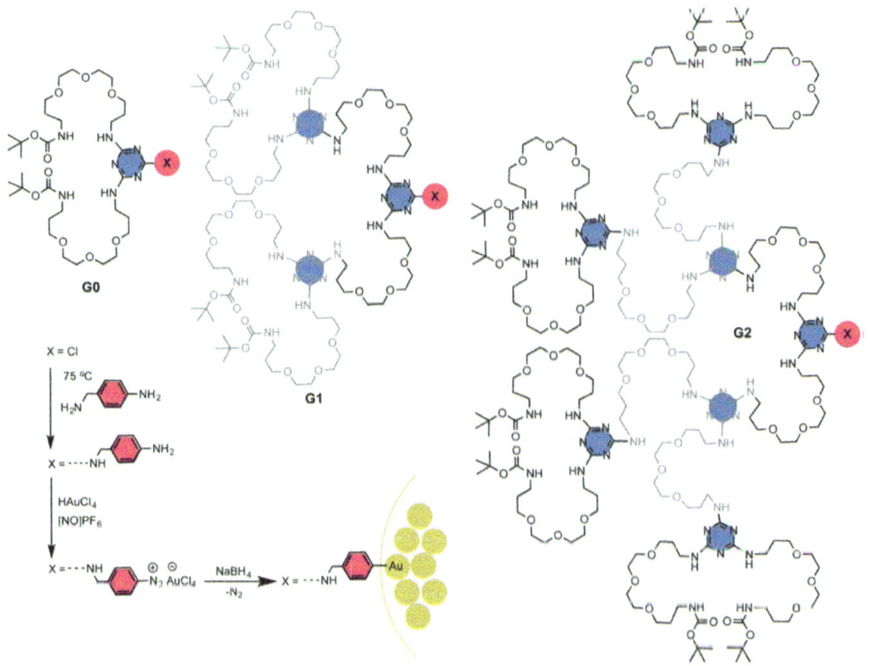

Fig. 4 Dendrimers used in this study (G0–G2) and the synthetic route adopted. Reproduced with permission from Reference [9]. Copyright (2017) Royal Society of Chemistry

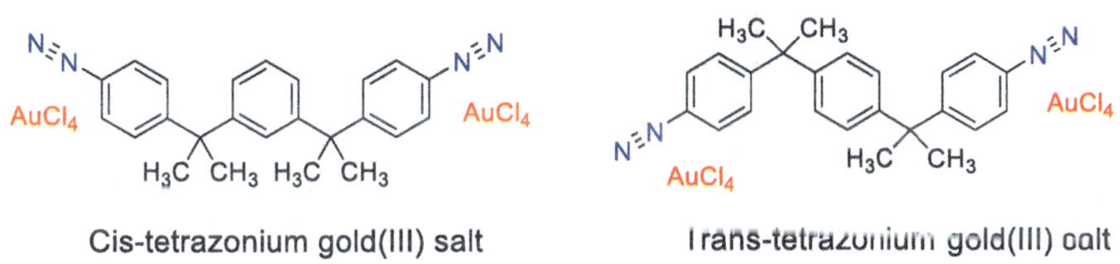

Fig. 5 Tetrazonium tetrachloroaurate(III) structures. Reproduced with permission from Reference [10]. Copyright (2021) Elsevier

(*trans*) isomeric forms. The isolated nanoparticles were studied for photothermal properties. Interestingly, TEM studies showed a larger size of trans than cis nanoparticles. Probing the electrochemical reduction and blocking properties of the two salts displayed a strong blocking effect of the film. The as-formed film displayed a thermal diffusivity studied using the photothermal mirror method [10].

Overall, the diazonium salts synthesized so far using the schemes above are indefinitely stable in the solid-state at ambient conditions and are soluble in dimethyl sulfoxide, acetone, chloroform, and dichloromethane however insoluble in water except when X=COOH [8]. Electron-donating substituted diazonium salts such as X=C_6H_{13}, CH_3, showed low stability under ambient conditions. Moreover, these

Table 1 ATR-FTIR and Raman stretching frequencies in cm^{-1} for the diazonium function in aryldiazonium gold salts $[X–4–C_6H_4N\equiv N]AuCl_4$ (X=F, Cl, Br, I, CN, NO_2)

X	ATR-FTIR	Raman
F	2261	2270
Cl	2253	2260
Br	2252	2262
I	2247	2254
CN	2274	2278
NO_2	2282	2282

diazonium salts exhibit long-term stability in the solid state when kept in the fridge, however, some white precipitate is formed when dissolved in water after an extended period in X=COOH. Thermal gravimetric analysis of diazonium tetrachloroaurate(III) salts displayed a sudden breakage above 120 °C. Also, the thermal analysis of an aryldiazonium gold salt X=C_8F_{17} using residual gas analysis technique showed the major released gases of the decomposition is N_2 in addition to chloroaromatic compounds [2].

3 Electrochemistry of Aryldiazonium Tetrachloroaurate(III) Salts

Cyclic voltammetry studies of 1.0 mM diazonium gold(III) salts $[X–4–C_6H_4N\equiv N]AuCl_4$ (X=F, Cl, Br, I, CN, NO_2) showed a highly irreversible reduction peak at 100 mV/s scan rate in the range of 0.20 to −0.40 V versus Ag/AgCl, KCl (satd.) in 0.1 M TBAHFP-CH_3CN on platinum and glassy carbon working electrodes, Fig. 6 [1]. For all salts, multicyclic voltammetric studies using 10 consecutive cycles at both the glassy carbon (GC) and Pt working electrodes exhibited a decrease in the reduction current with an increase in the cycle number, and for the tenth cycle, there is almost no reduction current illustrating that the grafting and construction of the gold-aryl film are complete. To determine the blocking properties of the grafted gold-aryl film, cyclic voltammetry studies of 0.2 mM $[Fe(CN)_6]^{3-/4-}$ reversible redox couple in 1.0 M KNO_3 were conducted using GC and Pt working electrodes before and after deposition using 10 consecutive cycles in the potential range from 0.80 to −0.80 V versus Ag/AgCl, KCl (satd.), Fig. 7a [11]. For a typical case of X=Br salt, the redox peaks of $[Fe(CN)_6]^{3-}$ were completely attenuated due to the complete inhibition of its redox electron transfer by the grafted film on the GC electrode confirming the grafted aryl film was efficient. This demonstrates the presence of a nearly compact and dense grafted layer on the GC electrode surface. For the grafted film of the Pt electrode, a typical result of the X=Br salt exhibits that the reduction peak of $[Fe(CN)_6]^{3-/4-}$ is not completely disappearing illustrating that the blocking and the efficiency of the grafted film of the Pt electrode are not as good as the GC electrode.

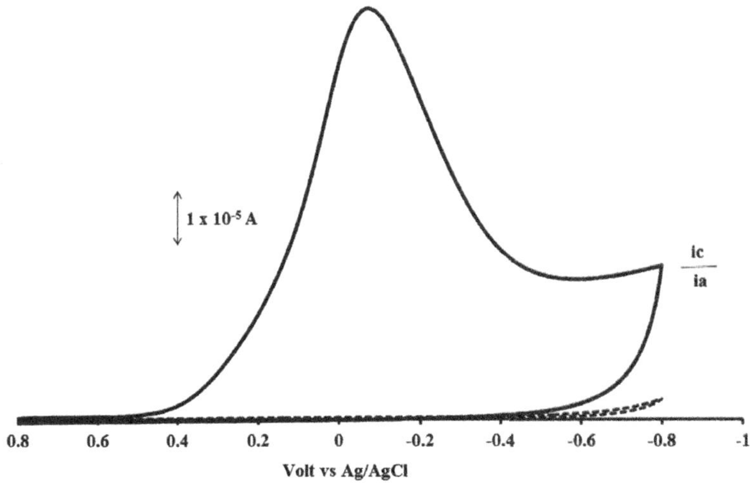

Fig. 6 Cyclic voltammogram of 0.1 mM [CN–4–$C_6H_4N{\equiv}N$]$AuCl_4$ at a GC electrode in CH_3CN/0.1 M [Bu_4N]PF_6 at 100 mV/s versus Ag/AgCl. Top: Reduction at the first cycle. Bottom: Reduction at the second cycle. Reproduced with permission from Reference [1]. Copyright (2012) American Chemical Society

The reduction potential of 1.0 mM [C_8F_{17}–4–$C_6H_4N{\equiv}N$]$AuCl_4$ occurs at −0.08 V versus Ag/AgCl in 0.1 M [Bu_4N]PF_6/CH_3CN at a GC working electrode [11]. The gold-organic film stability was studied using time-dependent open-circuit potential in pH buffer values 4, 5, 7, 8, and 10 (Fig. 7b). In acidic media, a gradual decrease in the potential occurred because of the removal of some of the film. The measured potential is relatively constant after about 600 s following the sudden increase in acid media. The behavior in alkaline media, pH values 8 and 10, showed the minimum change. There is no strictly linear relationship between the potential and pH change.

Electrochemical synthesis of gold-aryl nanoparticles stabilized by polyaniline (PANI) was achieved [12]. In this study, two graphite electrodes cell under direct current was used. To synthesize the gold-aryl nanoparticles, the electrochemical reduction of [HOOC–4–$C_6H_4N{\equiv}N$]$AuCl_4$ salt was tried in water at 0.2, 0.6, and 1.0 V. Overall, the electrodeposition at all tried voltage values in the presence of 42, 100, and 200 mg PANI [12]. All AuNPs/PANI nanocomposites displayed blue colors in DMSO. The PANI-coated gold nanoparticles were further soaked to nano sand and used for methylene blue dye removal from textile wastewater. The adsorption kinetics showed a typical pseudo-second order.

4 Forensic Science Applications

Diazonium salts are promising in forensic science to develop latent fingerprints [13–15]. They react preferentially on the reducing metal surfaces to generate fingerprint

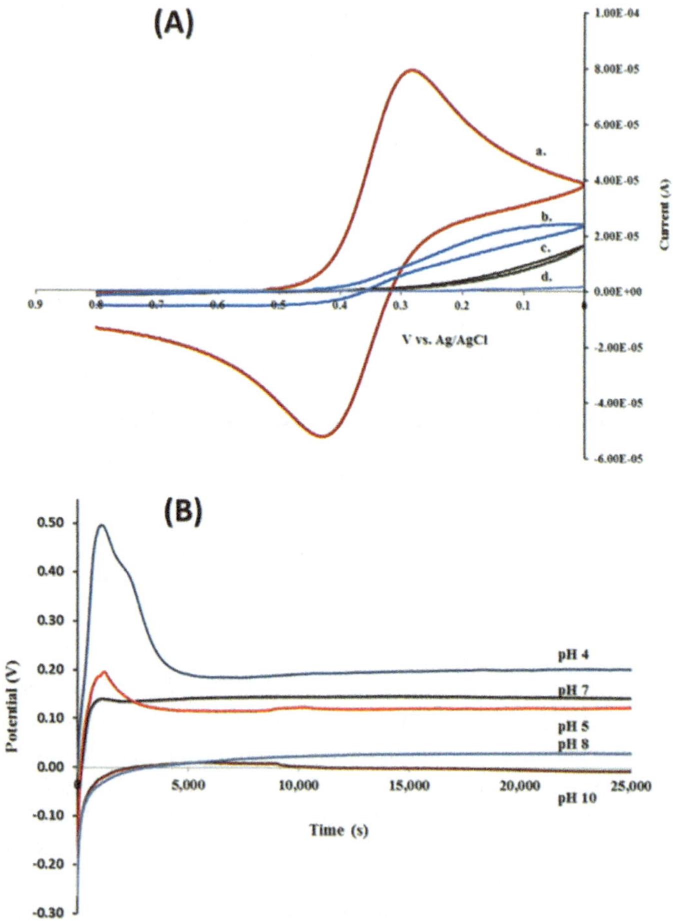

Fig. 7 **a** Voltammetric scans of 6.0 mM $[Fe(CN)_6]^{3-/4-}$ in 1.0 M KNO_3 aqueous solution at 100 mV/s, at **a** bare glassy carbon electrode surface, **b** glassy carbon electrode surface modified with gold-organic film, **c** after sonication for 1 h in acetonitrile, and **d** after sonication for 1 h in milli-Q water. **b** OCP was measured as a function of time on a GC electrode modified with gold-organic film from $[C_8F_{17}–4–C_6H_4N{\equiv}N]AuCl_4$ in different buffer pH solutions. Reproduced with permission from Reference [11]. Copyright (2015) Elsevier

pictures. Aryldiazonium gold(III) salts were evaluated in forensic chemistry for latent fingerprint development on metal surfaces such as copper, aluminum, zinc, and lead. $[4–NO_2–C_6H_4N_2]AuCl_4$ salt was applied on the metal surfaces using the components of human sweat as the reducing agent to construct gold deposits on the eccrine fingerprints [13]. SEM studies of the developed fingerprints showed well-separated ridges. One advantage of the diazonium salts in forensic science is the facile deposition of gold metal on the latent fingerprints that can be traced by XRF and EDS.

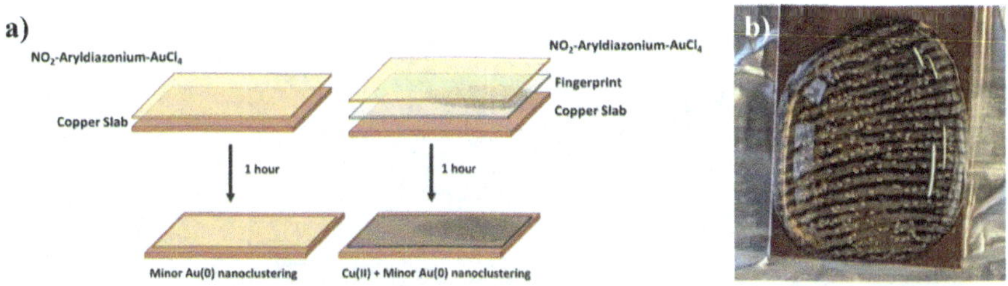

Fig. 8 **a** Reaction of clean and fingerprint-coated copper samples with $[4\text{–}NO_2\text{–}C_6H_4N{\equiv}N]AuCl_4$ aqueous solution, **b** Latent fingerprint demonstration using diazonium solution on a copper sheet. Bubbles correspond to the release of N_2 due to the spontaneous reduction of diazonium cation. Reproduced with permission from Reference [14]. Copyright (2020) American Chemical Scoiety

X-ray photoelectron spectrometry (XPS) was also used to evaluate diazonium gold salts as surface modifiers of latent fingerprints on copper surfaces [14]. The reaction of the aqueous solution with the copper surface formed positive dark lines and bubbles were seen between the fingerprint lines, Fig. 8. The formation of gold(0) was supported by the presence of Au $4d_{5/2}$. Forensic studies on the copper surfaces were extended to nickel coins. XPS was used to study the progress of the spontaneous reduction of the diazonium gold salt on the nickel coins. SEM and EDS were used to follow the morphology of the surfaces [15].

5 Medical Applications

The few literature examples of the biocompatible gold nanoparticles AuNPs–C_6H_4–4–COOH coated with proteins have benefited the nanomedicine engineering applications [16]. Gold-aryl nanoparticles were coated with bovine serum albumin (BSA), collagen, zein, and lysozyme proteins. In all the synthesis routes, protein-coated gold nanoparticles have been fabricated using green and chemical reduction routes. The coated nanoparticles were efficiently uptaken by MG-63 cells and can be used as diagnostic and therapeutic agents for osteosarcoma as shown in Fig. 9 [16]. Cellular uptake of the bioconjugates was studied by MG-63 osteosarcoma cells using laser confocal fluorescence microscopy (LCFM) and flow cytometry. Results showed that zein and lysozyme coated AuNPs prepared by the green method were not efficiently uptaken. We believe that the major factors such as the difference in hydrophobicity and charge that influence the internalization [16].

Gold bioconjugates of amino acids were fabricated by the incubation of the water-soluble and easily reducible $[HOOC\text{–}4\text{–}C_6H_4N{=}N]AuCl_4$ salt with tyrosine, tryptophan, and cysteine [17]. The reaction occurred at room temperature and formed the distinct purple color of gold colloidal solutions in a few minutes with tryptophan and tyrosine but in 2 h in the case of cysteine. TEM and agarose gel electrophoresis were used to confirm the fabrication of the bioconjugates. The constructed bioconjugate

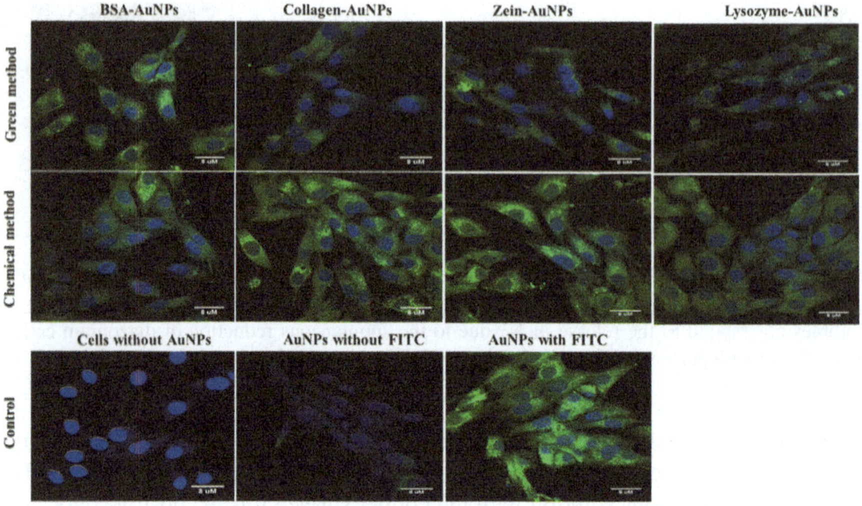

Fig. 9 Confocal microscopy images of the uptake of protein-coated gold-carbon nanoparticles synthesized with green and chemical reduction methods. The more intense the fluorescence the higher uptake of the AuNPs by MG-63 cells. The FITC-conjugated nanoparticles give green fluorescence while the nucleus stained with DAPI gives blue fluorescence. Reproduced with permission from Reference [16]. Copyright (2020) American Chemical Society

particle sizes were 27.2 ± 5.4 nm, 14.6 ± 7.7 nm, and 8.6 ± 2.6 nm for tyrosine, tryptophan, and cysteine, respectively. The biomimetic route for the synthesis of the amino acids bioconjugates is a great addition to the applications in the biomedical field and cellular imaging as they showed negligible cytotoxicity to human dermal normal fibroblast cell lines. The isolated and characterized gold nanoparticles bioconjugates of amino acids have shown outstanding FRET-based fluorescent ability in drug assay [17]. The bioconjugates were used specifically for the quantitative estimation of ranitidine (RNH), a peptic ulcer and gastroesophageal reflux drug, with a limit of detection of 0.174, 0.56, and 0.332 μM for Tyr@AuNPs, Trp@AuNPs, and Cys@AuNPs bioconjugates, respectively.

6 Fabrication of Alloy and Core–Shell Structures

The rich chemistry of diazonium salts and their use in the fabrication of gold-aryl nanoparticles for many applications has been described. Past, current, and future applications are diverse. We reported some in the titles before in some details, moreover, some are summarized in this section briefly. The applications under investigation include alloys [18] and core–shell [19] structures, for example. The gold-silver alloys were synthesized by a few routes such as immobilization, co-reduction, and seeding. All routes resulted in the alloy nanostructures which exhibited antibacterial properties [18]. The core–shell structures were synthesized by simple mixing of $SnCl_2$ with the diazonium gold salt, X=COOH, in water [19]. The characterized structures were formed immediately because of the exceptionally redox-active properties of both salts. This research is expected to be expanded to several metal salts rather than tin chloride.

7 Miscellaneous Applications

The applications of aryldiazonium gold salts in various fields have been expanded recently to include energy aspects such as hydrogen evolution reaction upon their reduction in the presence of clay. Moreover, the gold-aryl nanoparticles have shown outstanding catalytic properties in the reduction of the environmental pollutant nitrophenol. The full strategy has been summarized in Langmuir invited feature review [20].

8 Conclusions

Aryldiazonium gold salts were synthesized and showed high stability in the solid-state at ambient conditions. The salts displayed spontaneous reduction on metal surfaces and by proteins and amino acids in an aqueous solution. Organometallic nanoparticles, gold-aryl, were synthesized from the diazonium gold salts and have shown outstanding robustness in physiological media and under harsh conditions such as sonication in organic solvents. This chapter provided a narrative of the progress in metal–carbon nanoparticles chemistry from our laboratory, identified potential applications we have achieved so far, and highlighted the future chemistries based on the diazonium gold precursors and the fabricated gold nanoparticles. From our point of view, there are opportunities in basic chemistry, organic shell modifiers to fine-tune the colloidal properties, photophysics, and computation to make valuable contributions in current and future applications.

Acknowledgments A.A.M. acknowledges the University of Sharjah support of SEED grant VC-GRC-SR-83-2015, competitive grants 160-2142-029-P and 150-2142-017-P, Organometallic Research Group grant RISE-046-2016, and Functionalized Nanomaterials Synthesis Lab grant 151-0039.

References

1. Overton AT, Mohamed AA (2012) Inorg Chem 51:5500–5502
2. Neal SN, Orefuwa SA, Overton AT, Staples RJ, Mohamed AA (2013) Inorganics 1:70–84
3. Mohamed AA, Salmi Z, Dahoumane SA, Mekki A, Carbonnier B, Chehimi MM (2015) Adv Colloid Interf Sci 225:16–36
4. Ahmad AAL, Workie B, Mohamed AA (2020) Surfaces 3:182–196
5. Filimonov VD, Krasnokutskaya EA, Kassanova AZ, Fedorova VA, Stankevich KS, Naumov NG, Bondarev AA, Kataeva VA (2019) Eur J Org Chem 2019:665–674
6. Filimonov VD, Trusova M, Postnikov P, Krasnokutskaya EA, Lee YM, Hwang HY, Kim H, Chi KW (2008) Org Lett 10:3961–3964
7. Mohamed AA, Neal SN, Atallah B, AlBab ND, Alawadhi HA, Pajouhafsar Y, Abdou HE, Workie B, Sahle-Demessie E, Han C, Monge M, Lopez-de-Luzuriaga JM, Reibenspies JH, Chehimi MM (2018) J Organomet Chem 877:1–11
8. Ahmad AAL, Panicker S, Chehimi MM, Monge M, Lopez-De-Luzuriaga JM, Mohamed AA, Bruce AE, Bruce MRM (2019) Catal Sci Technol 9:6059–6071
9. Enciso AE, Doni G, Nifosì R, Palazzesi F, Gonzalez R, Ellsworth AA, Coffer JL, Walker AV, Pavan GM, Mohamed AA, Simanek EE (2017) Nanoscale 9:3128–3132
10. Panicker S, Marcano A, Isah S, Kenney B, Workie B, Han C, Lee H, Chehimi MM, Mohamed AA (2021) J Organomet Chem 935:121681
11. Neal S, Workie B, McCandless B, Mohamed AA (2015) J Electroanal Chem 757:73–79
12. AlMashrea BA, Abla F, Chehimi MM, Workie B, Han C, Mohamed AA (2020) Synth Met 269:116528
13. Ahmad AAL, Alawadhi AH, Park J, Abdou HE, Mohamed AA (2019) Forensic Chem 13:100144
14. Almheiri S, Ahmad AAL, Droumaguet BL, Pires R, Mohamed AA, Chehimi MM (2020) Langmuir 36:74–83
15. Almheiri S, Ahmad AAL, Droumaguet B, Pires R, Chehimi MM, Mohamed AA (2021). Surf Interface Anal. https://doi.org/10.1002/sia.6941
16. Hameed M, Panicker S, Abdallah SH, Khan AA, Han C, Chehimi MM, Mohamed AA (2020) Langmuir 36:11765–11775
17. Hameed MK, Parambath JBM, Kanan SM, Mohamed AA (2021) Anal Bioanal Chem 413:1117–1125
18. Abla F, Kanan SM, Park Y, Han C, Omastova M, Chehimi MM, Mohamed AA (2021) Colloids Surf A Physicochem Eng Asp 616:126266
19. Panicker S, Ahmady IM, Han C, Chehimi M, Mohamed AA (2020) Mater Today Chem 16:100237
20. Ahmad A, Parambath J, Postnikov P, Guselinkova O, Chehimi M, Bruce M, Bruce A, Mohamed A (2021) Langmuir 37:8897–8907

Modification and Uses of Synthetic and Biobased Polymeric Materials

Julien Vieillard, Franck Le Derf, Charlène Gadroy, and Brahim Samir

Abstract The surface engineering of polymers is still in development in order to modulate their properties to meet the requirements of the targeted applications. Lignocellulosic and agro-waste materials are increasingly studied as alternative to conventional ones employed for pollution remediation but also to develop original composites. In both cases, the surface of the material could be modified by coating or by covalent grafting. The diazonium chemistry has been successfully employed to modify the surface of carbon and metallic materials but its application to organic surfaces is still incipient. This chapter provides an overview of the recent developments in diazonium chemistry to modify polymer and biomass surfaces and its application in biosensor design, catalysis, and pollution removal.

1 Introduction

The surface properties of a material (wettability, adhesion, biocompatibility, chemical resistance) depend to a large extent on the chemical composition of its surface. This composition can be tuned by adding molecular groups at the surface to modify the solid-liquid or solid gas interface. Surface chemical modification can be achieved by chemical treatment (acid or basic treatment), impregnation, and covalent coating. Covalent grafting has the advantage of being resistant to chemicals, and mechanical stress so that the coating remains stable over the long term. In 1992, Pinson et al. demonstrated that aryldiazonium salt could be used to modify a glassy carbon surface covalently [1]. The synthesized diazonium is mainly selected with a functional group in para-position of the diazonium function. Diazonium salts can be synthesized and used later using the isolated form or directly in situ of the reaction as described by Bélanger et al. [2]. Diazonium chemistry presents numerous advantages such as ease of preparation from an aniline precursor, a good stability of the modified surface, a limited cost, and robustness to chemical treatment. In the literature, diazonium

J. Vieillard (✉) · F. Le Derf · C. Gadroy · B. Samir
Normandie Université, UNIROUEN, INSA Rouen, CNRS, COBRA, UMR CNRS 6014, Normandie University, 55 rue Saint Germain, 27000 Evreux, France
e-mail: julien.vieillard@univ-rouen.fr

M. M. Chehimi et al. (eds.), *Aryl Diazonium Salts and Related Compounds*, Physical Chemistry in Action, https://doi.org/10.1007/978-3-031-04398-7_10

chemistry is mainly applied for modifying carbon and semi-conductor surfaces by electroreduction, spontaneous grafting, or chemical reduction [3]. Surface modification by diazonium salt can be reached by applying various treatments (microwave, mechanical grafting, heating, ultrasonication) and by dediazonation under different conditions (with a reducing agent, with a reducing material, spontaneously, by electrografting, or by photoinitiation). The chemical reduction of diazonium salt in the absence of electro-induction has much to offer because it broadens the application of diazonium salt to polymeric and lignocellulosic surfaces.

This chapter reviews the application of diazonium salt to the grafting of polymeric and lignocellulosic surfaces. The chapter is organized as follows:

- (1) An overview of the application of diazonium salt to the grafting of polymeric surfaces.
- (2) An overview of the application of diazonium salt to the grafting of natural lignocellulosic materials.

1.1 Polymers

Compared to other materials, polymers are special because they have a macromolecular structure mainly composed of C–C and C–O bonds. Moreover, polymers have a viscoelastic behavior that is highly dependent on the presence of additives in the matrix and on environmental parameters such as temperature, solvent, and gas [4]. Therefore, their surfaces have to be adapted and controlled for various processes and applications. Polymer surfaces can be modified by physical (gamma or ionic irradiation, vapor phase infiltration, atomic layer deposition, mechanical), chemical (hydrolysis, reduction, oxidation, silanization, phosphonatation, radical-induced surface modification), electrochemical (only for conductive polymers like polyaniline) and photochemical (photoinduction) treatment [5]. Such surfaces can also be modified with diazonium salt to add organic groups to the polymeric surface. Table 1 presents different diazonium salts employed so far to graft aryl groups to polymeric surfaces. Even if the number of commercially available diazonium salts is limited, various chemical groups have been successfully grafted on polymers according to the requested application.

The wettability of a polymer is crucial for a large range of applications such as filtration, and it is fully dependent on surface properties. Chehimi et al., Picot et al., and Brisset et al. tried to modulate the hydrophilicity of PMMA [8], polyethersulfone [14], and COC [16], respectively. Fluoro-, nitro- and bromobenzene groups improved hydrophobicity while carboxyl-, amine-, phosphonium- and hydroxyl-groups promoted hydrophilicity [8, 16] (Fig. 1). The wettability of electrically conductive PEDOT:PSS polymer surface has also been tuned by grafting aryl groups bearing CF_3, NO_2, NH_2, or COOH groups from diazonium tosylate salts [18].

The hydrophilicity of the polymer can be tuned to form a membrane dedicated to filtering aqueous-based or solvent-based solutions. However, experimental conditions have to be optimized to avoid clogging the porous membrane and keep a

Table 1 Modification of polymeric surfaces by diazonium salts

Diazonium	support	Potential application	Reference
Calix[4]tetra-*O*-$(CH_2)_3C_4F_9$ tetradiazonium salt	Polypropylene (PP)	Develop surfaces with controlled wettability for anti-biofouling, anti-frosting, anti-fogging, self-cleaning, anti-corrosion or water-proofing	[6]
Calix[4]tetra-*O*-$(CH_2)_3CF_3$ tetradiazonium salt Calix[4]tetra-*O*-CH_2COO tetradiazonium salt 4-carboxymethoxybenzene diazonium salt	Polypropylene (PP) Polyethylene terephthalate (PET) Polystyrene (PS)	Control the modification of polymeric surfaces, grafting without drastic modifications of their bulk properties and introducing a large range of chemicals (biomolecules, nanoparticles, chromophores)	[7]
4-hydroxybenzene diazonium salt 4-hydroxymethylbenzene diazonium salt 4-carboxybenzene diazonium salt 4-Sulfonamidobenzene diazonium salt 4-Nitrobenzene diazonium salt 3,5-dimethoxybenzene diazonium salt 3,5-dimethylbenzene diazonium salt 3,5-Bis-trifluoromethylbenzene diazonium salt 4-perfluorohexylbenzene diazonium salt 4-perfluorodecylbenzene diazonium salt	Polymethylmethacrylate (PMMA)	Control the hydrophilicity/hydrophobicity of the surface for microfluidic application	[8]
4-Nitrobenzene diazonium salt 3,5-bis-trifluoromethylbenzene diazonium salt	Polyvinylchloride (PVC)	Protect the surface of materials and improve their aesthetics	[9]
4-carboxybenzene diazonium salt	Polylactic acid (PLA)	Compare surface plasma modifications or diazonium treatment of PLA for post-functionalization	[10]
4-carboxybenzene diazonium salt	polytetrafluoroethylene (PTFE)	Graft silane derivates on PTFE using a diazonium linker	[11]

(continued)

Table 1 (continued)

Diazonium	support	Potential application	Reference
4-aminobenzene diazonium salt	Acrylonitrile–butadiene–styrene (ABS)	Copolymerization with acrylic acid to metalize ABS plastic without using a reducer	[12]
4-sulphobenzene diazonium salt 4-cyanobenzene diazonium salt 4-aminobenzene diazonium salt 4-nitrobenzene diazonium salt 1-naphtylbenzene diazonium salt 2- methoxybenzene diazonium salt	Polyaniline (PANI)	Functionalization of conductive polymers	[13]
4-benzyltriphenylphosphonium diazonium salt 4-nitrophenyl diazonium salt 4-benzonitrile diazonium salt 4-phenylacetic acid diazonium salt	Polyethersulfone (PES)	Modification of PES membranes for filtering	[14]
Ethynylbenzene diazonium salt	Cyclic olefin copolymer (COC)	Isatin immobilization on COC surface	[15]
4-nitrobenzene diazonium salt 4-bromobenzene diazonium salt 4-(carboxymethyl)benzene diazonium salt 4-(2-bromoethyl)benzene diazonium salt 4-hydroxybenzene diazonium salt 4-mercaptobenzene diazonium salt	Cyclic olefin copolymer (COC)	Robust method for COC surface functionalization	[16]
benzene diazonium salt	Polyethylene (PE)	Improve the mechanical properties of raw rice-polyethylene composites	[17]
4-nitrobenzene diazonium tosylate 4-aminobenzene diazonium tosylate 4-carboxybenzene diazonium tosylate 3,5-ditrifluoromethylbenzene diazonium tosylate	Poly(3,4-ethylenedioxythiophene): polystyrene sulfonate (PEDOT:PSS)	Tuning electrical, optical and wetting properties of PEDOT:PSS	[18]

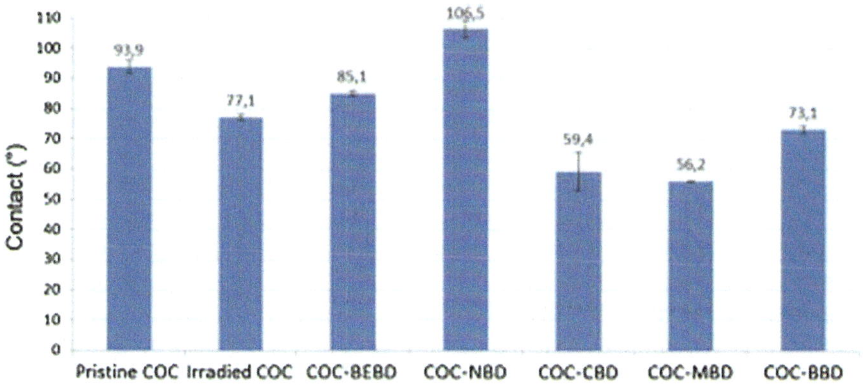

Fig. 1 Effects of various aryl layers on COC wettability. Wettability was analyzed by water droplet contact angle measurements. Reprinted with permission from [16] Copyright (2015), Elsevier (license: 5,084,780,847,331)

streaming current and a good yield of rejection of the targeted molecule (Fig. 2) [14].

COC, PMMA, PET, and PVC are transparent in the visible range, so they are frequently used in optical applications (optical fibers, catheters for instance). If such transparency is needed after diazonium grafting, a few chemical groups such as methoxybenzene or bromobenzene have to be avoided because they can generate autofluorescence of the polymer. This phenomenon can be associated with the nature of the chemical group but also with the thickness of the aryl layer that can be controlled during the chemical reduction. Thus, the concentration of diazonium salt, the diazonium/polymer ratio, and the reaction time can influence the thickness of the

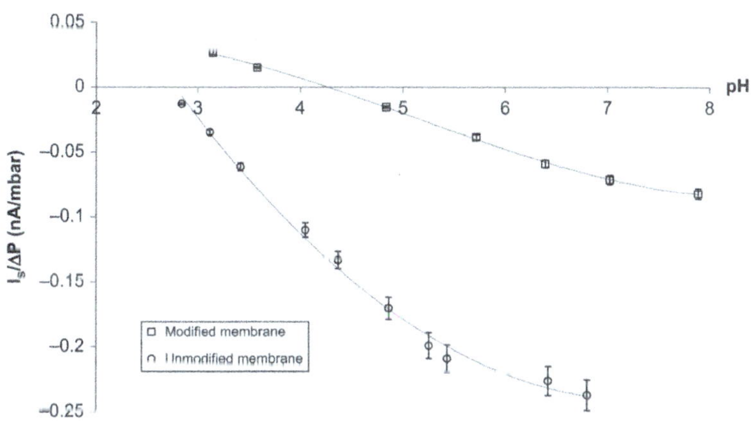

Fig. 2 pH dependence of the streaming current coefficient (Is/ΔP) measured with different membranes; circles: unmodified PES membrane; squares: PES membrane modified by N_2^+. Reprinted with permission from [14] Copyright (2012) Elsevier (license: 5,030,810,219,921)

deposit. For conductive polymers, thickness is estimated by cyclic voltammetry. For non-conductive polymers, cyclic voltammetry cannot be used except if the polymer is coated on a metallic electrode [19]. Thickness can be also estimated by XPS, atomic force microscopy, UV–visible and IR spectroscopy [8, 20]. On polymers, the thickness of the aryl deposit is estimated from 1–5 layers (few nm) to ten layers (28 nm) as a function of the reaction time. Unfortunately, the molecular structure of the grafted film is difficult to identify. Some mechanisms have been emphasized but they are difficult to prove unless powerful bulk or surface specific techniques are employed such as solid state-NMR [21] or time-of-flight secondary ion mass spectroscopy (ToF–SIMS) [22]. These techniques brought strong supporting evidence for surface-aryl bonds and aryl oligomerization. Moreover, Atmospheric solid analysis probe-ion mobility mass spectrometry (ASAP-IM-MS) was previously used to analyze the molecular composition of polyolefin or poly(ether ketone) [23, 24]. In a recent work, we demonstrated that atmospheric solids analysis probe (ASAP) can be used also to characterize the surface of modified COC [20]. The material is melted and then introduced into the ionization source of an ionic mobility chamber hyphenated with a mass spectrometer (IM-MS). As the grafted film is really thin compared to COC thickness, a combination of these orthogonal techniques is required. To facilitate the detection of the aryl film, we grafted an aryl layer bearing heteroatoms like bromobenzene and mercaptobenzene. We detected dibromo-biphenyl, tribromo-triphenyl, and tetrabromo-tetraphenyl confirming that up to four layers were grafted on COC. ASAP-IM-MS analysis also described a mixed phenyl layer with phenyl molecules substituted with Br or not resulting in a rough COC surface. Similar results were obtained with mercaptobenzene: up to five layers could be detected. ASAP-IM-MS analysis was so sensitive that we are able to detect the difference between COC surfaces grafted by bromobenzene diazonium or by bromoethylbenzene diazonium. The presence of the ethyl group in the diazonium salt modified the organization of the grafted film and limited the thickness to two layers. This is related to the flexibility of the ethyl group that hinders the access of the aryl radical to the grafted film Only the m/z ion attributed to dibromoethyl-diphenyl was detected on the MS spectrum (Fig. 3). However, the molecular composition of the carboxybenzene layer was so close to the molecular signal of COC that we did not clearly detect it.

For post-functionalization with biomolecules such as isatin, for instance, two strategies are available with diazonium chemistry; if the biomolecule has an aniline group, it can be transformed to a diazonium group, dediazotized, and directly grafted on the polymeric surface. If not, the surface has to be grafted with amine, carboxylic acid, or alkyne and then post-functionalized by amide, Sonogashira, or click chemistry [15]. To optimize the immobilization of the biomolecules, the surface has to be smooth and an even distribution of the chemical group over the surface is required to guarantee a good surface coverage by the biomolecule. If the control of the experimental parameters is not sufficient to control thickness, a radical trap can be added or a calixarene platform can be used. Due to its macrocycle structure, calixarene permits dense grafting with a low thickness (Fig. 4).

The tailoring of diazonium modification of polymeric surfaces is currently limited. The proof of concept has been validated with chromophores or fluorinated groups

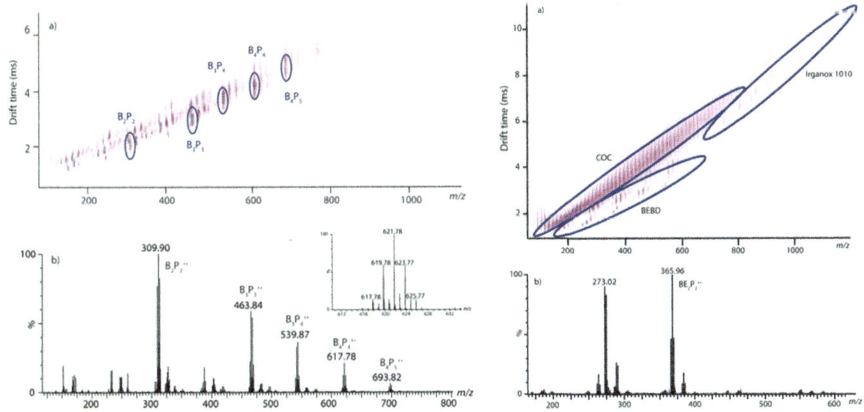

Fig. 3 ASAP IM-MS analysis of COC surface modified by bromobenzene diazonium (left) and bromoethylbenzene diazonium (right). Comparison of the drift time versus MS plots (a) and mass spectra (b). Reprinted with the permission from [20] Copyright 2015 American Chemical Society

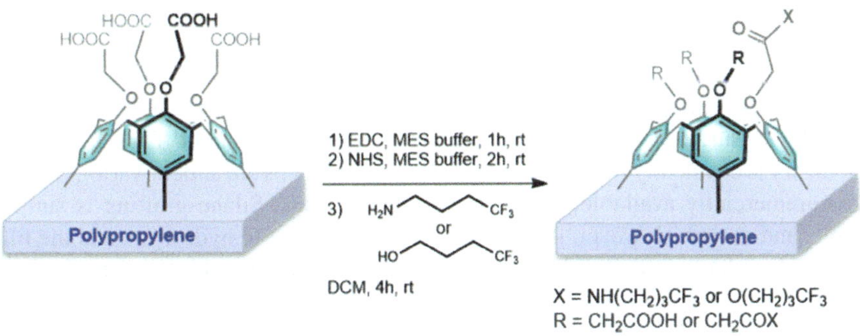

Fig. 4 Post-functionalization of polypropylene surface modified with calix [4] arene-tetradiazonium. Reprinted with permission from [7] Copyright (2016) American Chemical Society

but more investigations are required for immobilized nanoparticles and biomolecules among other species.

For post-functionalization of the surface with biomolecules (proteins, DNA), surface roughness has to be limited and the presence of carboxylic and amine groups on the surface is required. These chemical groups can be grafted from diazonium salt; however, polymers are not all equivalent as concerns their resistance to diazonium treatment. Thus, polypropylene and COC present a smooth surface after carboxybenzene coating while PLA surfaces are considerably damaged (Fig. 4).

1.2 Biomass

The development of high-performance material from marine or terrestrial renewable resources has recently gained interest, contrary to fossil resources. However, the expanding world population requires more and more agricultural resources to feed people and animals so that the agricultural resources used to develop materials are limited to biogenic residue and waste. These residues (grass, coir, cocoa shell, corn stover, straw, olive pit, rice husk, etc.) have some advantages such as low cost, bioavailability, biocompatibility, biodegradability, and they are not required as a food counterpart. The valorization of these wastes products is currently limited to garden mulch, soil conditioners, or food additives for animals. These materials can be converted into activated carbon and then modified with diazonium salt for further applications [25–28]. Therefore, there is still room for valorization. The main difficulty with these materials is their heterogeneity. For instance, lignocellulosic materials are composed of carbohydrate polymers (cellulose, hemicellulose), aromatic polymers (lignin), and proteins, fatty acids, or phenol. As a result, the chemical groups at the solid/liquid or solid/gas interface are difficult to control. Lignocellulosic material can be treated by physical (polyelectrolyte) or wet (acidic, basic) impregnation, but it is reversible and may require harsh conditions to modify porosity and surface roughness. Lignocellulosic materials can also be treated by chemical grafting using silane agents bearing two reactive groups: one group reacts with the hydroxyl layer at the polymer surface and the other one reacts with the environment. Thus, the surface chemistry is covalently modified and new chemical groups are added. Various silanes are commercially available, with different substituents. Silane grafting is simple, stable, and varied. However, this coating is also sensitive to hydrolysis and the film can break under specific conditions. Silane purity, the water yield concentration, and the reaction conditions have to be carefully controlled to avoid a rough and inhomogeneous film. As an alternative to silane grafting, diazonium grafting can modify different biomasses and improve their chemical, mechanical or thermal properties. As described in Table 2, different diazonium compounds have been synthesized to modify the surface chemistry and the morphology of biomass for different applications. Table 2 clearly shows that there exists a wide combination of biobased materials and diazonium salts for particular applications, and the flexible interface chemistry of diazonium salts has better days ahead.

The surface modification of cellulose by diazonium salt can be achieved by a Graftfast® process [29, 37–39]. In this case, diazonium salt is chemically reduced by ascorbic acid and the aryl radical is coated on cellulose by dipping in diazonium solution or by inkjet printing of the reactive solution. This second process is eco-friendlier and more cost-effective because waste is limited thanks to the patterning application.

Based on this approach, some researchers tried to modify agricultural wastes with diazonium salt. Thus, Kabil et al. coated jute fiber with diazonium salt. Jute fibers were simply impregnated with diazonium salt, but this was sufficient to increase fiber

Table 2 Applications of diazonium salts for modifying lignocellulosic material

Diazonium	Support	Potential application	Reference
4-nitrobenzene diazonium salt 4-azidobenzene diazonium salt 4(2-ammonioethyl)benzene diazonium salt 4-carboxybenzene diazonium salt 4-mercaptobenzene diazonium salt	Cellulose paper	Cellulose modification for lateral flow immunoassay	[29]
4-methoxyphenyl diazonium salt	Coumarin-function-nalized cellulose sheet	Direct modification of cellulose	[30]
4-nitrobenzene diazonium salt 4-bromobenzene diazonium salt	Cocoa shell	Improvement of surface properties	[31]
benzene diazonium chloride	Jute fiber	Improvement of the physico-mechanical properties of jute fiber during its treatment	[32]
4-mercaptobenzene diazonium salt 4-carboxybenzene diazonium salt 4-aminobenzene diazonium salt	Olive pit particle	Immobilization of nanocatalysts	[33]
benzene diazonium salt	Abaca	Improvement of the mechanical properties of cellulose-polypropylene composites	[34]
benzene diazonium salt	Sawdust	Reinforcement of the mechanical properties of chemically treated sawdust-polypropylene composites	[35]
O-hydroxybenzene diazonium salt	Coir fiber	Reinforcement of the mechanical properties of coir-PP composites	[36]
anthraquinone diazonium salt	Activated carbon	Influence of diazonium grafting on electrochemical double layer capacitors	[26, 27]
Catechol diazonium salt	Activated carbon	Active composite electrode material in an aqueous electrochemical capacitor	[28]

elongation. However, the tensile strength was lower than in the control after coating [32].

Fioresi et al. investigated the modification of cocoa shell surface with nitrobenzene and bromobenzene diazonium salts [31], using hypophosphorous acid as a chemical reducer. The surface modification induced a modification of the morphology and hydrophilicity of cocoa shell. During the process, cocoa shell became more hydrophobic and moved at the surface of the reactor. Its surface modification was confirmed by the detection of the vibrations associated with nitro and aromatic group in IR spectroscopy and the appearance of a thin film on scanning electron microscopy (SEM) images (Fig. 5a). Diazonium grafting was really robust as it resisted chemical and ultrasonic treatments. The thickness of the film was adjusted by controlling the reaction time. The energy-dispersive spectroscopy mapping presented in Fig. 6 illustrates the homogeneity of the bromobenzene film on the surface.

As cocoa shell is lignocellulosic, the authors tried to identify if the coating was deposited mainly on lignin or on the cellulosic counterpart. They eliminated lignin by alkaline treatment and tried to graft diazonium. The grafting was unsuccessful, demonstrating that it was essentially restricted to lignin. Starting from this work, Belbekhouche et al. developed original biobased catalysts using diazonium salt and metallic (Ag, Au) nanoparticles [33]. They synthesized different diazonium salts with amine, thiol, and carboxylic acid to modify olive pits. The thermogravimetric analysis of the modified material demonstrated that diazonium grafting modifies the thermal stability of the material. The material displayed better thermal resistance than the control at high temperatures whereas the control was more stable at low temperatures. They also demonstrated that the grafting of amine and thiol was successful

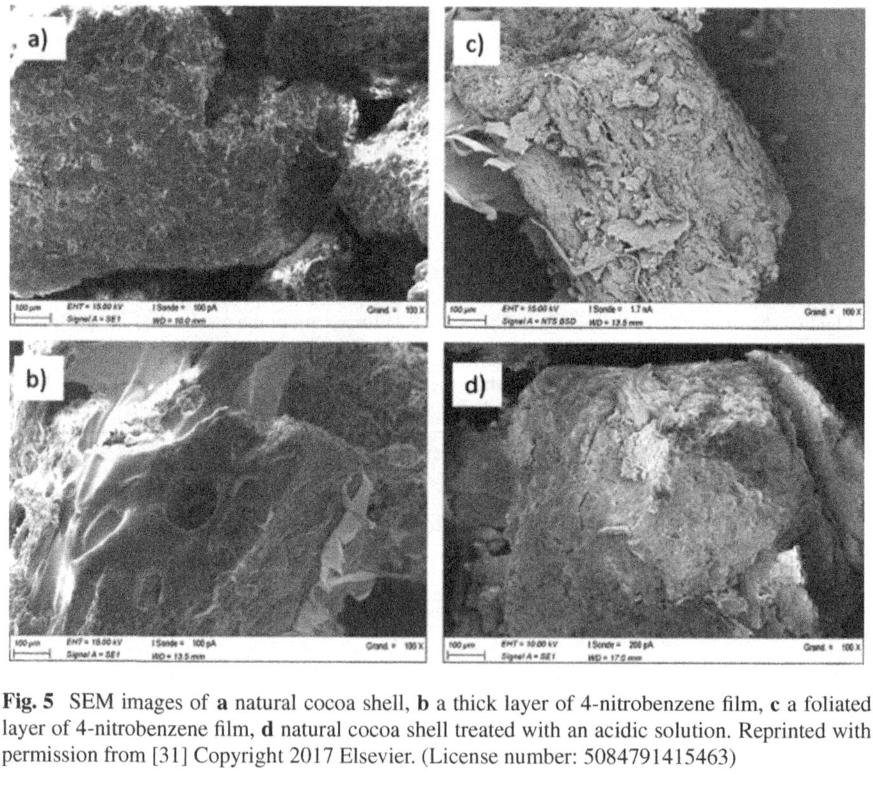

Fig. 5 SEM images of **a** natural cocoa shell, **b** a thick layer of 4-nitrobenzene film, **c** a foliated layer of 4-nitrobenzene film, **d** natural cocoa shell treated with an acidic solution. Reprinted with permission from [31] Copyright 2017 Elsevier. (License number: 5084791415463)

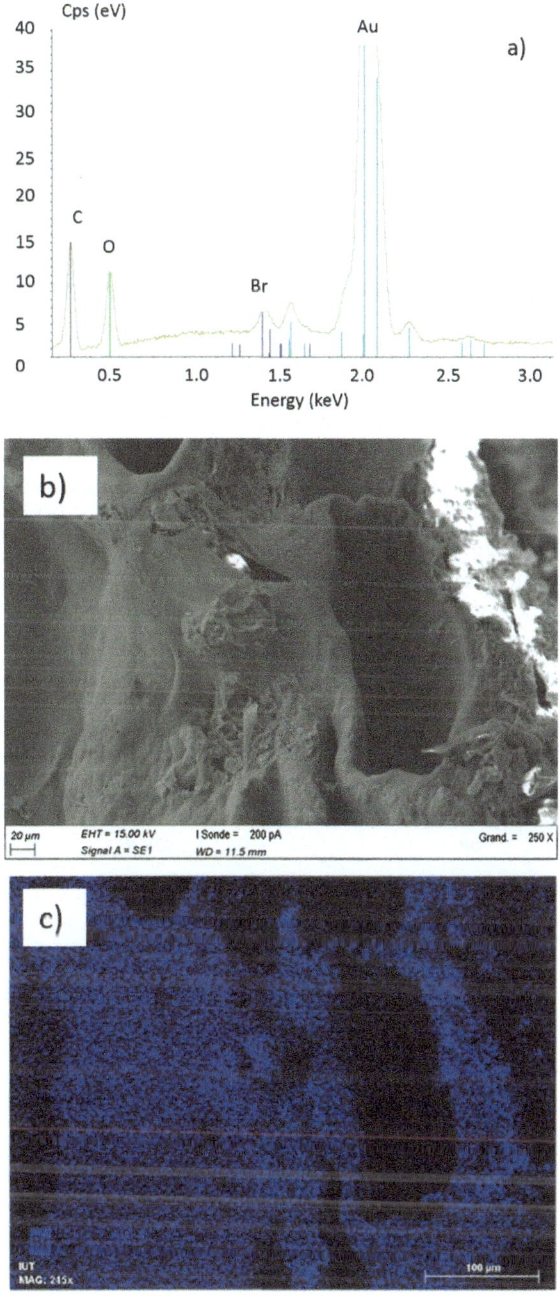

Fig. 6 EDX analysis (a), SEM image (b), and elemental mapping (c) of bromobenzene layer grafted on the cocoa shell surface. Reprinted with permission from [31] Copyright 2017 Elsevier (License number: 5084791415463)

in converting simple olive pit into a performing heterogeneous catalyst. Thus, they developed a low-cost catalyst combining gold and silver nanoparticles and biobased materials (Fig. 7).

According to the previous section, we demonstrated that diazonium salt could be used to modify biobased materials and polymeric surfaces. In polymer chemistry, the mechanical properties of a material are highly dependent on its solid/liquid interface. This interface can be modified using different fillers, and a linker has to be carefully selected to control filler-matrix adhesion. The role of the bifunctional diazonium salt in the control of this interface is substantially documented. As diazonium chemistry is versatile, different fillers can be used, e.g., silica fillers can be grafted on phenol-formaldehyde composite through hydroxymethyl diazonium salt [40]. In general, silanes are preferred as a coupling agent but the structure of silica is quite different from the structure of resin, and this decreases the stability of the interface. Adding an aromatic ring improves stability and increases flexural strength by about 35%. To increase mechanical performance, lignocellulosic material (sawdust, cellulose, coir, abaca) could be a promising alternative to silica as a filler and diazonium salt could be used to bridge the filler and polypropylene (PP). Through these approaches, the fiber agglomeration and the presence of void at the filler–polymer interface were limited, and the mechanical properties of PP were improved [34, 36, 41].

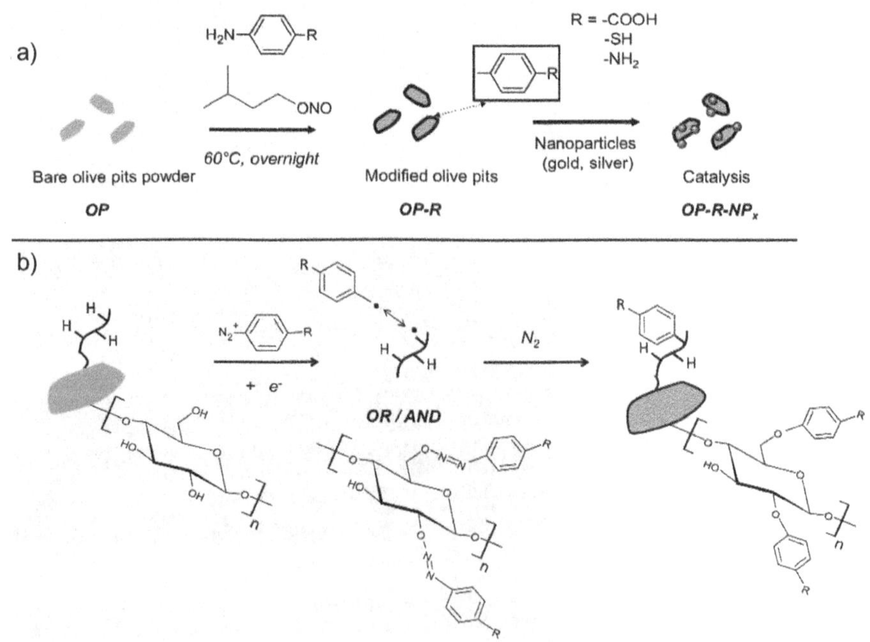

Fig. 7 **a** Synthetic route for the in situ grafting of an aryl layer for a catalytic application. **b** Possible mechanism of aryl grafting onto olive pits. Reprinted with permission from [33] Copyright 2017 Elsevier (license 5,084,800,404,991)

2 Conclusion

The use of aryldiazonium salts to modify surface chemistry has been advancing at an incredible speed because it is quick and simple to apply to a wide range of materials.

This chapter focuses on polymeric and lignocellulosic surfaces modified with diazonium salts and their use in composites. We reviewed scientific publications and presented information about the possible application of these modified materials.

The flexibility of diazonium salt grafting on polymers and biobased materials as well as the variety of possible functional groups can open up new opportunities in several fields such as catalysis, gas capture, inorganic and organic pollutant removal, and the synthesis of heteroatom-doped carbon for electrocatalysis.

This overview shows the huge impact of surface modification using diazonium salts on the thermal stability and mechanical properties of the modified surfaces (interfacial shear strength, compressive strength, flexural strength).

The reactivity of the diazonium function makes it possible to consider it as a future prospect for the development of a new generation of reactive and functional hybrid materials at low cost.

This capacity would allow integrating a diversity of functions, under ambient conditions and over a wide range of surface areas.

We anticipate that diazonium modification of lignocellulosic surfaces and polymers will continue to attract researchers and be renewed on a regular basis.

References

1. Delamar M, Hitmi R, Pinson J, Saveant JM (1992) Covalent modification of carbon surfaces by grafting of functionalized aryl radicals produced from electrochemical reduction of diazonium salts. J Am Chem Soc 114:5883–5884. https://doi.org/10.1021/ja00040a074
2. Baranton S, Bélanger D (2005) Electrochemical derivatization of carbon surface by reduction of in situ generated diazonium cations. J Phys Chem B 109:24401–24410. https://doi.org/10.1021/jp054513+
3. Hetemi D, Noël V, Pinson J (2020) Grafting of diazonium salts on surfaces: application to biosensors. Biosensors 10. https://doi.org/10.3390/bios10010004
4. Nemani SK, Annavarapu RK, Mohammadian B, Raiyan A, Heil J, Haque MA, Abdelaal A, Sojoudi H (2018) Surface modification of polymers: methods and applications. Adv Mater Interfaces 5:1801247. https://doi.org/10.1002/admi.201801247
5. Pinson J, Thiry D (2019) Surface modification of polymers: methods and applications. Wiley
6. Mattiuzzi A, Troian-Gautier L, Mertens J, Reniers F, Bergamini J-F, Lenne Q, Lagrost C, Jabin I (2020) Robust hydrophobic gold, glass and polypropylene surfaces obtained through a nanometric covalently bound organic layer. RSC Adv 10:13553–13561. https://doi.org/10.1039/D0RA01011A
7. Troian-Gautier L, Martínez-Tong DE, Hubert J, Reniers F, Sferrazza M, Mattiuzzi A, Lagrost C, Jabin I (2016) Controlled modification of polymer surfaces through grafting of calix[4]arene-Tetradiazoate salts. The J Phys Chem C 120:22936–22945. https://doi.org/10.1021/acs.jpcc.6b06143
8. Chehimi MM, Lamouri A, Picot M, Pinson J (2014) Surface modification of polymers by reduction of diazonium salts: polymethylmethacrylate as an example. J Mater Chem C 2:356–363. https://doi.org/10.1039/C3TC31492H

9. Bouriga M, Chehimi MM, Combellas C, Decorse P, Kanoufi F, Deronzier A, Pinson J (2013) Sensitized photografting of diazonium salts by visible light. Chem Mater 25:90–97. https://doi.org/10.1021/cm3032994
10. Durán IR, Vanslambrouck S, Chevallier P, Hoesli CA, Laroche G (2020) Atmospheric pressure cold plasma versus wet-chemical surface treatments for carboxyl functionalization of polylactic acid: a first step toward covalent immobilization of bioactive molecules. Colloids Surf, B 189:110847. https://doi.org/10.1016/j.colsurfb.2020.110847
11. Bagheri H, Bayat P, Piri-Moghadam H (2013) Grafting the sol–gel based sorbents by diazonium salts: a novel approach toward unbreakable capillary microextraction. J Chromatogr A 1318:58–64. https://doi.org/10.1016/j.chroma.2013.10.033
12. Garcia A, Berthelot T, Viel P, Mesnage A, Jégou P, Nekelson F, Roussel S, Palacin S (2010) ABS polymer electroless plating through a one-step poly(acrylic acid) covalent grafting. ACS Appl Mater Interfaces 2:1177–1183. https://doi.org/10.1021/am1000163
13. Acevedo DF, Salavagione HJ, Miras MC, Barbero CA (2005) Synthesis, properties and aplications of functionalized polyanilines. J Braz Chem Soc 16:259–269
14. Picot M, Rodulfo R, Nicolas I, Szymczyk A, Barrière F, Rabiller-Baudry M (2012) A versatile route to modify polyethersulfone membranes by chemical reduction of aryldiazonium salts. J Membr Sci 417–418:131–136. https://doi.org/10.1016/j.memsci.2012.06.025
15. Soulignac C, Cornelio B, Brégier F, Le Derf F, Brière JF, Clamens T, Lesouhaitier O, Estour F, Vieillard J (2019) Heterogeneous-phase Sonogashira cross-coupling reaction on COC surface for the grafting of biomolecules—application to isatin. Colloids Surf, B 181:639–647. https://doi.org/10.1016/j.colsurfb.2019.06.001
16. Brisset F, Vieillard J, Berton B, Morin-Grognet S, Duclairoir-Poc C, Le Derf F (2015) Surface functionalization of cyclic olefin copolymer with aryldiazonium salts: a covalent grafting method. Appl Surf Sci 329:337–346. https://doi.org/10.1016/j.apsusc.2014.12.060
17. Rahman MR, Islam MN, Huque MM, Hamdan S, Ahmed AS (2010) Effect of chemical treatment on rice husk reinforced polyethylene composites. Bioressources 5:854–869
18. Guselnikova OA, Postnikov PS, Fitl P, Tomecek D, Sajdl P, Elashnikov R, Kolska Z, Chehimi MM, Švorčík V, Lyutakov O (2017) Tuning of PEDOT:PSS properties through covalent surface modification. J Polym Sci, Part B: Polym Phys 55:378–387. https://doi.org/10.1002/polb.24282
19. Combellas C, Kanoufi F, Mazouzi D, Thiébault A, Bertrand P, Médard N (2003) Surface modification of halogenated polymers. 4. Functionalisation of poly(tetrafluoroethylene) surfaces by diazonium salts. Polymer 44:19–24. https://doi.org/10.1016/S0032-3861(02)00736-X
20. Vieillard J, Hubert-Roux M, Brisset F, Soulignac C, Fioresi F, Mofaddel N, Morin-Grognet S, Afonso C, Le Derf F (2015) Atmospheric solid analysis probe-ion mobility mass spectrometry: an original approach to characterize grafting on cyclic olefin copolymer surfaces. Langmuir 31:13138–13144. https://doi.org/10.1021/acs.langmuir.5b03494
21. Li H, Kopiec G, Müller F, Nyßen F, Shimizu K, Ceccato M, Daasbjerg K, Plumeré N (2021) Spectroscopic evidence for a covalent sigma Au–C bond on Au surfaces using 13C isotope labeling. JACS Au 1:362–368. https://doi.org/10.1021/jacsau.0c00108
22. Combellas C, Kanoufi F, Pinson J, Podvorica FI (2005) Time-of-flight secondary ion mass spectroscopy characterization of the covalent bonding between a carbon surface and aryl groups. Langmuir 21:280–286. https://doi.org/10.1021/la048106l
23. Cossoul E, Hubert-Roux M, Sebban M, Churlaud F, Oulyadi H, Afonso C (2015) Evaluation of atmospheric solid analysis probe ionization coupled to ion mobility mass spectrometry for characterization of poly(ether ether ketone) polymers. Anal Chim Acta 856:46–53. https://doi.org/10.1016/j.aca.2014.12.013
24. Barrère C, Selmi W, Hubert-Roux M, Coupin T, Assumani B, Afonso C, Giusti P (2014) Rapid analysis of polyester and polyethylene blends by ion mobility-mass spectrometry. Polym Chem 5:3576–3582. https://doi.org/10.1039/C4PY00164H
25. Lebègue E, Madec L, Brousse T, Gaubicher J, Levillain E, Cougnon C (2011) Modification of activated carbons based on diazonium ionsin situ produced from aminobenzene organic acid without addition of other acid. J Mater Chem 21:12221–12223. https://doi.org/10.1039/C1JM11538C

26. Pognon G, Brousse T, Bélanger D (2011) Effect of molecular grafting on the pore size distribution and the double layer capacitance of activated carbon for electrochemical double layer capacitors. Carbon 49:1340–1348. https://doi.org/10.1016/j.carbon.2010.11.055
27. Pognon G, Brousse T, Demarconnay L, Bélanger D (2011) Performance and stability of electrochemical capacitor based on anthraquinone modified activated carbon. J Power Sources 196:4117–4122. https://doi.org/10.1016/j.jpowsour.2010.09.097
28. Pognon G, Cougnon C, Mayilukila D, Bélanger D (2012) Catechol-modified activated carbon prepared by the diazonium chemistry for application as active electrode material in electrochemical capacitor. ACS Appl Mater Interfaces 4:3788–3796. https://doi.org/10.1021/am301284n
29. Credou J, Volland H, Dano J, Berthelot T (2013) A one-step and biocompatible cellulose functionalization for covalent antibody immobilization on immunoassay membranes. J Mater Chem B 1:3277–3286. https://doi.org/10.1039/C3TB20380H
30. Schroll P, Fehl C, Dankesreiter S, König B (2013) Photocatalytic surface patterning of cellulose using diazonium salts and visible light. Org Biomol Chem 11:6510–6514. https://doi.org/10.1039/C3OB40990B
31. Fioresi F, Vieillard J, Bargougui R, Bouazizi N, Fotsing PN, Woumfo ED, Brun N, Mofaddel N, Le Derf F (2017) Chemical modification of the cocoa shell surface using diazonium salts. J Colloid Interface Sci 494:92–97. https://doi.org/10.1016/j.jcis.2017.01.069
32. Kabir MA, Huque MM, Islam MR (2004) Modification of jute fibre by bifunctional diazonium salts in the presence of various mordants. Biosci, Biotech. Res Asia 2:79–84
33. Belbekhouche S, Kebe SI, Mahouche-Chergui S, Guerrouache M, Carbonnier B, Jaziri M, Chehimi MM (2017) Aryl diazonium-modified olive waste: A low-cost support for the immobilization of nanocatalysts. Colloids Surf, A 529:541–549. https://doi.org/10.1016/j.colsurfa.2017.06.011
34. Rahman MR, Huque MM, Islam MN, Hasan M (2009) Mechanical properties of polypropylene composites reinforced with chemically treated abaca. Compos A Appl Sci Manuf 40:511–517. https://doi.org/10.1016/j.compositesa.2009.01.013
35. Rezaur Rahman M, Nazrul Islam M, Monimul Huque M (2010) Influence of fiber treatment on the mechanical and morphological properties of sawdust reinforced polypropylene composites. J Polym Environ18:443–450. https://doi.org/10.1007/s10924-010-0230-z
36. Islam MN, Rahman MR, Haque MM, Huque MM (2010) Physico-mechanical properties of chemically treated coir reinforced polypropylene composites. Compos A Appl Sci Manuf 41:192–198. https://doi.org/10.1016/j.compositesa.2009.10.006
37. Credou J, Faddoul R, Berthelot T (2014) One-step and eco-friendly modification of cellulose membranes by polymer grafting. RSC Adv 4:60959–60969. https://doi.org/10.1039/C4RA11219A
38. Berthelot T, Garcia A, Le XT, El Morsli J, Jégou P, Palacin S, Viel P (2011) "Versatile toolset" for DNA or protein immobilization: toward a single-step chemistry. Appl Surf Sci 257:3538–3546. https://doi.org/10.1016/j.apsusc.2010.11.071
39. Mévellec V, Roussel S, Tessier L, Chancolon J, Mayne-L'Hermite M, Deniau G, Viel P, Palacin S (2007) Grafting polymers on surfaces: a new powerful and versatile diazonium salt-based one-step process in aqueous media. Chem Mater 19:6323–6330. https://doi.org/10.1021/cm071371i
40. Sandomierski M, Strzemiecka B, Chehimi MM, Voelkel A (2016) Reactive diazonium-modified silica fillers for high-performance polymers. Langmuir 32:11646–11654. https://doi.org/10.1021/acs.langmuir.6b02891
41. Haque MM, Hasan M, Islam MS, Ali ME (2009) Physico mechanical properties of chemically treated palm and coir fiber reinforced polypropylene composites. Biores Technol 100:4903–4906. https://doi.org/10.1016/j.biortech.2009.04.072

Surface Modification of Plasmonic Nanomaterials with Aryl Diazonium Salts

Da Li, Théo Geronimi, Nordin Felidj, Philippe Nizard, Delphine Onidas, Yun Luo, and Claire Mangeney

Abstract This chapter provides an overview of the various surface modification strategies based on the use of aryl diazonium salts for the functionalization of plasmonic nanoparticles. Aryl diazonium salts appear as a valuable alternative to commonly used thiol self-assembled monolayers due to several advantages, including the formation of robust interfacial metal-surface bonds, simple synthesis, and surface grafting protocols, a large range of functional groups, and the possibility to form multilayers around the nanoparticle surface. Owing to these outstanding features, the combination of aryl diazonium salts and plasmonic nanoparticles is receiving increasing attention for potential applications in various fields such as nanosensors, bioimaging, environment, forensic science, and antimicrobial materials. This chapter describes the chemical processes involved in the surface modification of plasmonic nanoparticles by diazonium salts, the resulting nanohybrids, and the targeted applications.

1 Properties of Plasmonic Nanoparticles

Plasmonic nanoparticles (NPs) with unique chemical, physical and optical properties have attracted a lot of attention over the past decades. These NPs sustain a collective coherent oscillation of the free electrons at their surface upon electromagnetic excitation, called localized surface plasmon (LSP), as illustrated in Fig. 1 [1]. When the frequency of the incident radiation matches the frequency of LSP modes, this results in LSP resonance (LSPR), which can be observed as intense extinction bands in the visible and near-infrared spectral range. The plasmon band wavelength depends on the NPs' composition, size, shape, and surrounding dielectric environment. Therefore, the presence of analytes in the vicinity of plasmonic NPs can be

D. Li · P. Nizard · D. Onidas · Y. Luo (✉) C. Mangeney (✉)
Université Paris Cité, CNRS,LCBPT, 75006 Paris, France
e-mail: claire.mangeney@parisdescartes.fr

T. Geronimi · N. Felidj
Université Paris Cité, CNRS, ITODYS, 75013 Paris, France

M. M. Chehimi et al. (eds.), *Aryl Diazonium Salts and Related Compounds*, Physical Chemistry in Action, https://doi.org/10.1007/978-3-031-04398-7_11

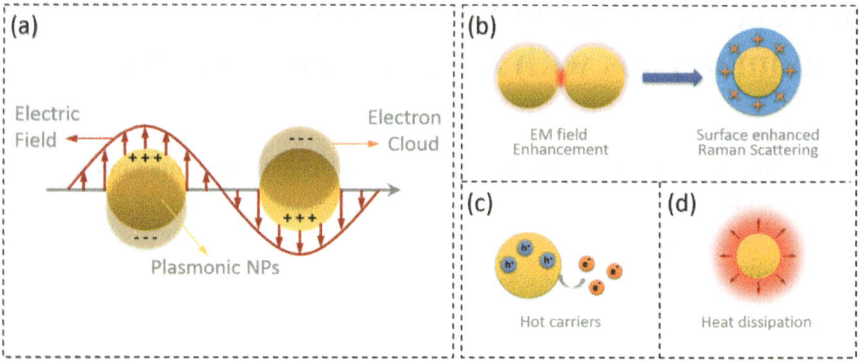

Fig. 1 **a** Representative scheme of the collective oscillation of free electrons at the surface of plasmonic NPs. The excitation of LSP results in a cascade of effects, including, **b** a strong electromagnetic (EM) field enhancement which can be exploited for SERS, **c** the generation of hot carriers and **d** heat dissipation leading to local heating

detected by spectral shifts of the LSP wavelength, induced by changes in the local medium refractive index, allowing real-time chemical and biological sensing. The excitation of LSP is also accompanied by the generation of highly enhanced electromagnetic fields, which have been widely used in surface-enhanced spectroscopies, such as Surface-Enhanced Raman spectroscopy (SERS) to detect trace amount of analytes, down to the single-molecule level [2]. Besides, nonradiative relaxation of LSP results in heat energy release and production of hot carriers. The plasmonic NPs can thus act as nanosources of heat for photothermal applications or charge carriers (hot electrons and holes) to induce chemical reactions locally [3].

Among the different types of plasmonic NPs, gold and silver NPs (Au NPs and Ag NPs) have been the most widely studied. While the size and shape of the NPs are crucial parameters to optimize their optical properties, the control of their surface chemistry is also a critical issue as it determines the interaction between the NPs and their environment. Therefore, considerable efforts have been made in order to develop appropriate surface modification strategies, with the aim to strengthen colloidal stability and to add new functionalities, thereby broadening the application scope of plasmonic NPs.

2 Common Strategies for Surface Functionalization of Plasmonic NPs

The main strategies for surface functionalization of plasmonic NPs are based on either physical adsorption or covalent bonding. Compared to physical adsorption, covalent bonding offers more robust interactions between the NPs and the immobilized molecules, thus improving interfacial stability. The most common strategy consists in using thiol ligands resulting in the formation of thiolate self-assembled monolayers (SAMs), anchored on the NPs surface via gold–sulfur bonds [4]. Thanks to the rich chemistry of gold–sulfur bonds, a wide variety of functional groups have been attached around the plasmonic cores leading to innovative hybrid nanomaterials [5]. However, issues regarding the chemical stability of SAMs have been identified. For example, they have been shown to desorb spontaneously from surfaces after few days in aqueous solutions [5, 6]. Moreover, they degrade upon laser excitation [7] and can be exchanged in biological media, jeopardizing their efficacy and reliability. Although significant improvements have been obtained via the use of longer-chain thiols, multidentate sulfur-based adsorbates, silica coatings, or small amounts of amphiphilic surfactants, the development of alternative approaches to obtain strong covalent anchoring of surface functions onto the plasmonic cores still remains an important challenge.

3 Surface Functionalization of Plasmonic NPs with Aryl Diazonium Salts

A recent approach has emerged for the functionalization of plasmonic nanomaterials based on the use of aryl diazonium salts. General formula of aryl diazonium salts can be expressed as A^-, $^+N_2$-C_6H_4-R, where A^- is the counter-ion and R stands for the chemical function on the aromatic ring. Since Jean Pinson and coworker's pioneering work in 1992 described the mechanism of electrochemical reduction of aryl diazonium salts to modify glassy carbon electrodes [8], these compounds have attracted widespread attention in the field of surface chemistry. Indeed, the cleavage of N_2 from diazonium salts occurs at low potential and on reducing surfaces opening promising prospects as surface functionalization agents in various applications including nanomedicine, forensic science, sensors, catalysis, and energy. These compounds offer several advantages over other common strategies, such as their ease of synthesis, their fast grafting on surfaces, the large choice of available functional groups, and the formation of robust covalent interfacial bonds with the surface. Generally, their grafting on NPs involves a two-stage mechanism. Firstly, the decomposition of the diazonium salts generates an aryl radical able to bind on the surface. In a second step, the formed highly reactive radicals react with the already grafted groups resulting in nanometer to micrometer thick polyaryl layers. To obtain functionalized NPs, diazonium salts can be used either as isolated compounds or can be generated in situ. The use of isolated salts allows to control the initial concentration of the reactants. However, the long-term stability of the diazonium salt solids and solutions is a major concern. The use of in situ generated aryldiazonium salts from the corresponding aromatic amines can overcome this issue and achieve high grafting densities on the NPs while preserving colloidal stability. Several approaches have been proposed to modify plasmonic nanomaterials by diazonium salts [9]. The main

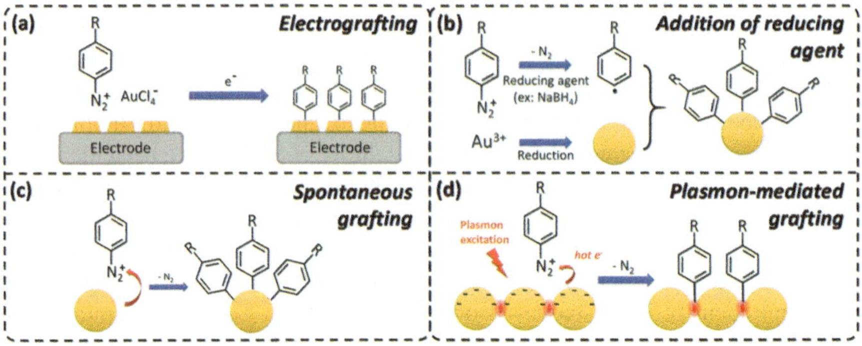

Fig. 2 Representative scheme of the main approaches for the surface modification of plasmonic NPs by aryldiazonium salts, via **a** electrografting; **b** addition of a reducing agent[1]; **c** spontaneous reaction between the diazonium salt and the NPs and **d** plasmon-mediated reduction of diazonium salts

ones based on electrochemistry, addition of reducing agents, spontaneous reaction, or plasmon-mediated chemistry, are described in (Fig. 2).

3.1 *Electrografting*

As aryl radicals can be generated by electrochemical route, electrografting of aryl diazonium salts has become a powerful method to functionalize supported plasmonic NPs. For example, arenediazonium tosylates could be deposited on plasmon-active Au and Ag thin films via either spontaneous reactions or electrochemistry (see Fig. 3) [10]. Interestingly, the spontaneous grafting results in thin monomolecular layers with potential interest for Surface Plasmon Resonance (SPR) technique or SERS sensors while electrochemical activation generates thick polyaryl layers opening promising prospects for applications in tunable plasmonic devices and plasmon-based lasers. In another example, aryl diazonium salts were electrografted via pulse deposition and used as antifouling coatings onto Au-based SPR sensors [11]. The antifouling performances of the coatings obtained from diazonium salts bearing different functional groups or alkanethiolate self-assembled monolayers were compared, revealing an enhanced antifouling behavior in the case of 4-phenylalanine diazonium chloride with a significant reduction of the nonspecific adsorption surface coverage of crude

[1] For the sake of clarity, the scheme (b) depicts only the case of the simultaneous reduction of auric and isolated diazonium salts. The mechanisms based on the phase transfer of diazonium cations to $[AuCl_4]^-$ phase or on the reduction of benzenediazoniumtetrachloroaurate(III) complexes by sodium borohydride are not represented.

For the sake of clarity, the scheme (b) depicts only the case of the simultaneous reduction of auric and isolated diazonium salts. The mechanisms based on the phase transfer of diazonium cations to [AuCl4]- phase or on the reduction of benzenediazoniumtetrachloroaurate(III) complexes by sodium borohydride are not represented.

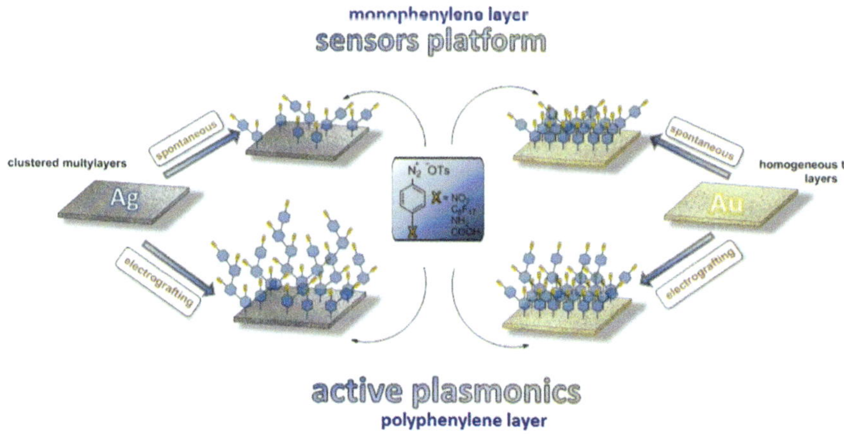

Fig. 3 Representative scheme of the functionalization of plasmon-active Au and Ag thin films by arenediazonium tosylates: spontaneous reactions or electrochemistry. Reproduced with permission from Ref. [10]. Copyright, 2017, Elsevier

bovine serum proteins (down to 42 ± 4 ng/cm^2). Proteins, modified with an aryl-diazonium adduct, could be also electroadressed onto gold SPR imaging (SPRi) chips for the detection of antibodies and ovalbumin [12]. This approach yielded multiple spots of robustly anchored proteins distributed over the SPRi biochip for the detection of antigen/antibody interactions.

3.2 *Spontaneous Grafting or Chemical Reduction*

Contrary to macroscopic surfaces for which electrografting of diazonium salts is mostly used, it becomes tricky in the case of colloidal NPs. Therefore, for colloidal NPs, spontaneous reactions or addition of a reducing agent in the medium appear as simple alternatives to generate radicals from aryldiazonium salts. The spontaneous grafting is a straightforward approach consisting to mix the NPs with the diazonium salts and to let them react at room temperature, in the presence of air, as described by McDermott et al. [13]. This simple process was proved to preserve the colloidal stability of the nanohybrids and to yield strongly anchored organic shells with the desired functionality on the NP surface. The possibility to obtain, with this approach, multifunctional layers with various chemical groups around the plasmonic cores, as illustrated in Fig. 4, was assessed using SERS, with the presence of different tags around the NPs [14]. Au NPs functionalized by aryl groups were also obtained via the phase transfer of diazonium cations to $[AuCl_4]^-$ phase [15], the simultaneous reduction of auric and isolated diazonium salts [16], or through the reduction of the benzenediazoniumtetrachloroaurate (III) complexes by sodium borohydride ($NaBH_4$) [17]. The resulting NPs were shown to withstand long periods

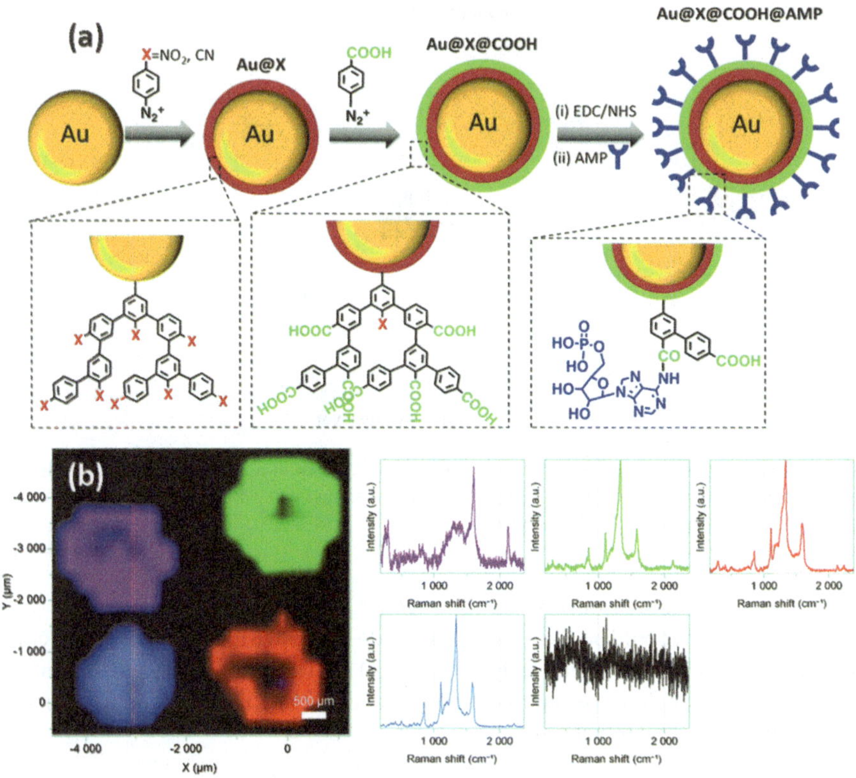

Fig. 4 **a** Representative scheme of the spontaneous grafting of aryl diazonium salts on Au NPs and formation of multifunctional layers; **b** Raman images of dried drops of Au@CN (magenta), Au@NO_2 (blue), Au@CN@NO_2 (green) and Au@NO_2@CN (red), and their corresponding Raman spectra. Adapted from Ref. [14]. Copyright, 2020, Royal Society of Chemistry

of sonication in various solvents and remained perfectly stable in the solid-state, in ambient conditions.

3.3 *Plasmon-Mediated Grafting*

Another approach to reduce aryldiazonium salts at the surface of plasmonic NPs consists of taking advantage of the "hot electrons" produced by the excitation of surface plasmons. Hot electrons are generated in nano-localized areas around the NP surface, sustaining a strong near field enhancement, resulting in site-selective hot-electron mediated reduction of diazonium salts. This plasmon-mediated surface functionalization approach thus offers an unprecedented "bottom-up" approach to place molecules or nanomaterials into reactive areas or high electromagnetic field regions (hot spots). It was used for the regioselective functionalization of plasmonic nanostructures, including nanotriangles [18], nanorods [19, 20], and nanodisks [21]. For example, aryl layers derived from bisthienylbenzene diazonium salts were regioselectively grafted on gold nanotriangles under visible-light illumination (Fig. 5a), in the absence of reducing agent [18].

Spatially selective grafting could be also achieved on colloidal gold nanorods using both iodonium and diazonium salts. Plasmon-assisted homolysis of the C–I bond of iodonium salts under excitation of longitudinal plasmon resonances resulted in the regioselective grafting of organic layers on the tips of the rods while their sides were functionalized using spontaneous diazonium salt grafting [20]. In another example, lithographic gold nanorods were irradiated by a polarized laser beam along

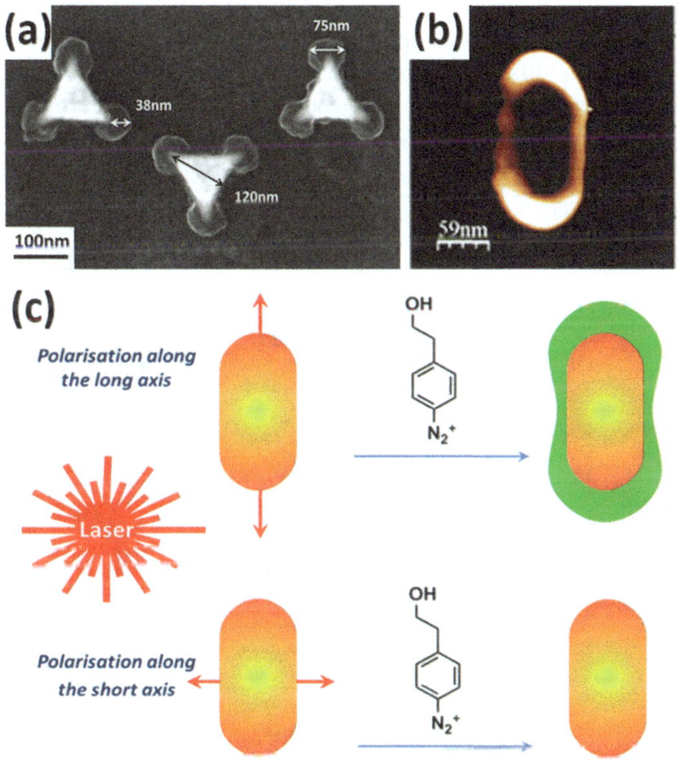

Fig. 5 **a** SEM image of Au triangles after plasmon-induced grafting during 4 min under visible-light irradiation, in the presence of bisthienylbenzene diazonium salt aqueous solution; **b** Result of the difference between AFM images of Au nanorod recorded before and after plasmon-mediated grafting under illumination polarized along the long axis; **c** Representative picture of gold nanorod modified via regioselective surface functionalization under plasmon excitation with polarized light. **a** Reproduced and readapted with permission from Ref. [18] Copyright, 2017, American Chemical Society. **b** and **c** Reproduced and readapted with permission from Ref. [19]. Copyright, 2017, Royal Society of Chemistry

their long axis, resulting in the preferential grafting of polyaryl layers on the nanorod tips (Fig. 5b, c) [19]. Interestingly, a multi-step process using successive illuminations under perpendicular polarization directions, in the presence of two different diazonium salts, enabled the formation of multifunctional patterned layers grafted at the surface of gold nanodisks [21].

4 Surface Immobilization of Small Molecules, Polymer Films, and Biomolecules at the Surface of Plasmonic Nanoparticles

Triggering the surface immobilization of functional species on plasmonic NPs is of great importance for advanced applications in (bio-)sensing, environment, catalysis, and optics. The different strategies based on diazonium salts to attach small molecules, polymer coatings, and biomolecules on the surface of metallic NPs are summarized (Fig. 6).

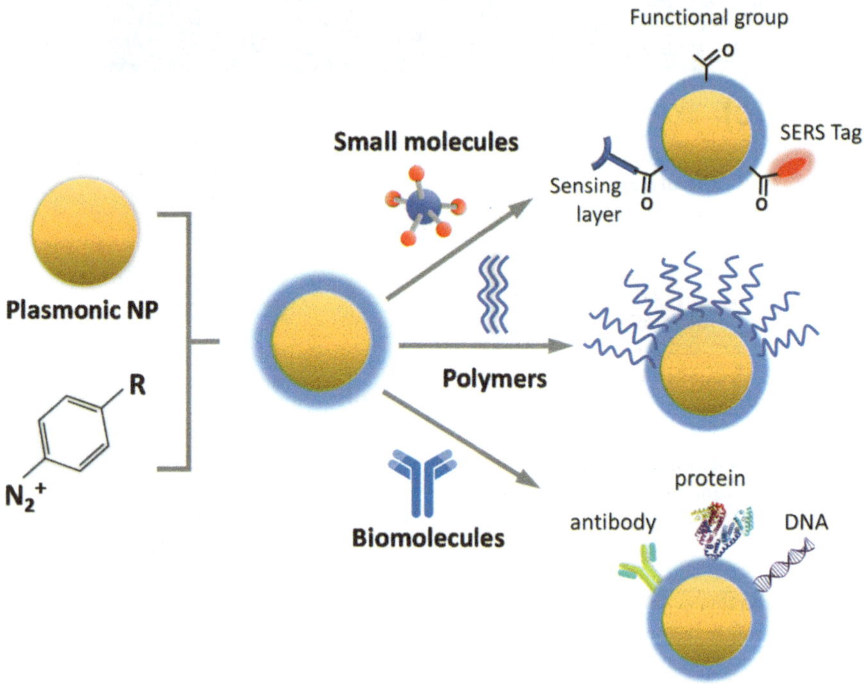

Fig. 6 Representative scheme of the surface immobilization of small molecules, polymer layers, and biomolecules at the surface of plasmonic NPs, via diazonium salt chemistry

4.1 Surface Immobilization of Small Molecules

Small molecules (with weight less than 900 daltons) based on aryl diazonium salts [22], can be grafted onto metallic NPs via reduction of diazonium salts. The resulting organic mono- or multilayers attached to the NPs have been considered as promising interfaces for various applications, such as chemical and bio-sensors or bioimaging tags. For example, macrocyclic compounds, such as calix[4]arenes bearing diazonium groups on their large rim were grafted on Au NPs [23]. Remarkably, these molecules can be decorated with various functions, such as oligo (ethylene glycol) chains providing an excellent colloidal stability to the nanohybrids. Mixtures of calixarenes could be also grafted on the NPs, with a fine control over the final surface chemical composition, opening the way to the design of Au NPs coated with well-defined functional and post-functionalizable groups [24]. Thanks to this strategy, carboxyl groups from PEG layers were conjugated to amine-containing biomolecules such as peptide aptamers, for the detection of cancer biomarkers [25]. Regarding bioimaging applications, the grafting of Raman reporter groups at the surface of plasmonic NPs is an effective means to design SERS tags. Among the possible Raman reporter groups, nitro- and cyanophenyl moieties are particularly attractive as they display intense and characteristic Raman peaks, which can be easily detected by Raman imaging. These Raman labels were spontaneously grafted onto Au NPs for Raman imaging of tumor cells after NPs internalization [14].

4.2 Surface Immobilization of Polymers

The development of strategies to obtain plasmonic NPs with surface-grafted polymers has stimulated a wide interest for potential application in (bio-)sensors [26], drug release [27], or environmental remediation. They can be classified into two main categories: (i) polymers can be "grafted to" the metallic NPs by adding a diazonium end group on the backbone of pre-formed polymer chains. For example, calix[4]arene-tetra-diazonium bearing four oligo (ethylene glycol) chains on their small rim [24] were used to functionalize Au NPs and provide them water solubility, steric stabilization, and increased circulation lifetime in the blood; (ii) the polymerization can be driven "from" the surface where active species initiate the covalent polymerization. "Grafting from" polymerization techniques offer an accurate control of polymer coating thickness, homogeneity, and grafting density. Different polymerization mechanisms have been initiated from the surface of plasmonic NPs, including atom transfer radical polymerization (ATRP) [28], reversible addition/fragmentation chain transfer polymerization (RAFT), and polymerization controlled by an iniferter. Interestingly, the surface-initiated growing of stimuli-responsive (thermo- or pH-sensitive) polymer brushes, such as poly (N-isopropylacrilamide) (PNIPAM) by ATRP on plasmonic nanostructures was shown to yield active optical hybrid devices, which properties can be modified via an external stimulus [29]. Another strategy was based on plasmon-activated "from surface" RAFT-controlled radical polymerization [30]. The RAFT agent was grafted on gold gratings via aryl diazonium salts and plasmon excitation was used to trigger surface-assisted growth of PNIPAM. The process appeared to be self-limited offering new prospects for the formation of ultrathin smart polymer coatings on gold nanostructures. The iniferter method from surface-grafted diazonium salts-derived initiators was exploited to immobilize molecularly imprinted polymers on gold nanorods to yield water-soluble plasmonic nanosensors[31] or on nanostructured copper surfaces to obtain anti-icing coatings [32].

4.3 Surface Immobilization of Biomolecules

The flexible chemistry provided by the substituent on the aryl group offers many opportunities to immobilize on plasmonic NPs a wide range of biomolecules modified by diazonium end groups or aryl layers for post-functionalization. For example, Au NPs bioconjugates with insulin [33] or bovine serum albumin (BSA) [34] were elaborated using aryl diazonium salt chemistry, demonstrating a high hemocompatibility with human red blood cells. Au NPs modified by aryl diazonium salts bearing carboxylic acid groups were coated with polydiallyldimethylammonium chloride [35], a biodegradable cationic polyelectrolyte to form conjugates with deoxyribonucleic acid (DNA) and obtain non-viral gene systems. Capture antibodies, such as anti-B-type natriuretic peptide (anti-BNP), could also be immobilized on Au NPs grafted through aryl diazonium salt chemistry to obtain sandwich-type immunoassays on screen-printed carbon electrodes for the accurate detection of clinically relevant levels of BNP in human serum samples [36]. In another strategy, the amino groups of Au NPs modified by 4-aminothiophenol were transformed to diazonium salts in order to bind a tumor marker, the human p53 protein. The nanoconjugates were then captured on a sensor surface modified by another protein, azurin, which enables the formation of stable complexes with p53 protein. The transduction process was based on the SERS activity of Au NPs, resulting in intense SERS signals in the presence of p53 proteins in human serum, even at low concentrations (down to 500 fM) [37].

5 Targeted Applications

Thanks to the outstanding properties of plasmonic NPs and their easy surface functionalization by aryl diazonium salts, a considerable scope for application research is open up, as illustrated in Fig. 7. It will take advantage of the enhanced colloidal stability of plasmonic NPs provided by functionalization with diazonium salts, robust interfacial links between the inorganic cores and the organic coatings, and the wide variety of available functional groups. The targeted applications cover a broad range

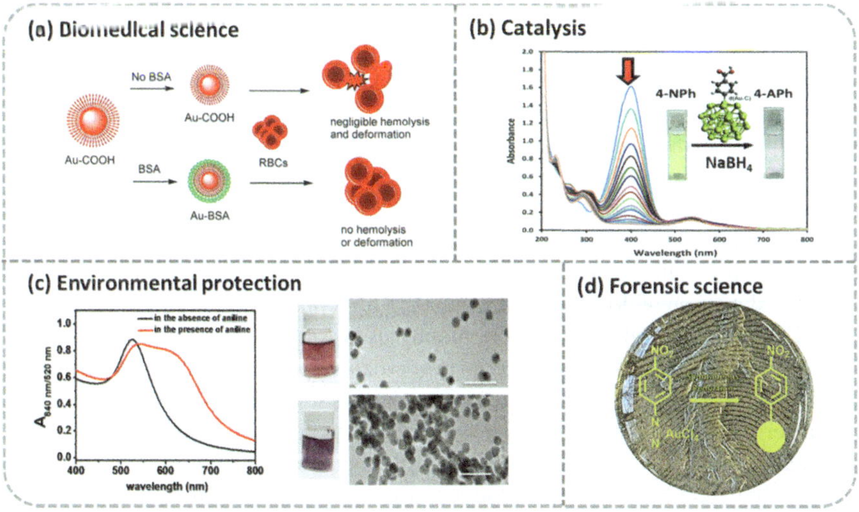

Fig. 7 **a** Representative scheme of the effect of Au-BSA bioconjugates upon hemocompatibility and comparison with uncoated Au-COOH NPs. **b** UV–vis spectra recorded during the reduction of 4-NPh with $NaBH_4$ in the presence of AuNPs–COOH. The inset shows photos of the solution during the reduction reaction from 0 to 5 min. **c** UV–vis spectra (left) and photos (center) of Au NPs functionalized by diazonium salts, in the absence (top) and presence (bottom) of aniline (100 μM). TEM images of the corresponding Au NPs solutions in the absence (up) and the presence (down) of aniline (100 μM). Scale bars: 50 nm. **d** Latent fingerprint image obtained using diazonium solution on a nickel coin. **a** Reproduced and readapted with permission from Ref. [34]. Copyright, 2018, Elsevier. **b** Reproduced and readapted with permission from Ref. [38]. Copyright, 2019, Royal Society of Chemistry. **c** Reproduced and readapted with permission from Ref. [39]. Copyright, 2017, Elsevier. **d** Reproduced and readapted with permission from Ref. [40]. Copyright, 2021, Wiley

from (i) biomedical and antibacterial science, (ii) environmental protection, (iii) forensic science, and (iv) catalysis.

5.1 *Biomedical Science*

The use of plasmonic NPs in biomedical science has been the subject of intense research. In this field, the NPs must be modified with a variety of functional (bio-) molecules, such as drugs, targeting groups, chemical tags, or antibacterial agents. For example, Au NPs were modified via aryl diazonium salts by BSA conjugates for drug delivery applications [34]. A remarquable hemocompatibility was observed after surface modification with BSA offering promising opportunities for intravenous drug delivery, as illustrated in Fig. 7a. Nanosensors based on molecularly imprinted polymers (MIPs) could be elaborated by combining aryl diazonium salts, iniferter polymerization, and gold nanorods for the detection of analytes by SERS. The gold nanorods coated by MIPs layers enabled the SERS label-free detection of folic acid down to 0.1 μM concentration [41].

Optical transduction based on SERS was also reported for the detection of p53 proteins in human serum at low concentrations [37]. In another approach, Au NPs were trapped within nanocomposite films of MIPs deposited on gold electrodes [42]. For this, immobilized Au NPs were functionalized with benzophenone, via aryl diazonium salts, to initiate the photopolymerization of methacrylic acid and ethylene glycol dimethacrylate, in the presence of dopamine, used as the template molecule. The resulting nanocomposite layers were used for the electrochemical label-free detection of dopamine, down to 0.35 nM concentration. Regarding antimicrobial applications, silver NPs were mainly used for their capacity to release Ag^+ ions, demonstrating strong antibacterial properties. Nevertheless, Ag NPs suffer from colloidal instability and tend to aggregate in aqueous solutions, thereby affecting their antibacterial effect. Aryl diazonium salt chemistry was exploited to improve their colloidal stability by introducing surface-grafted carboxylic groups acting as water-soluble stabilizers, resulting in efficient antibacterial nanohybrids toward *S. Aureus* [43].

5.2 Environmental Protection

Nanocomposites were elaborated from the electrochemical reduction of [HOOC-4-$C_6H_4N{\equiv}N$] $AuCl_4$ salts in the presence of polyaniline (PANI) emeraldine salt for the removal of synthetic dyes in textile wastewater treatment. The composites were shown to exhibit a high removal rate of methylene blue dye onto nanosand paving the way toward environmental remediation applications [44]. Diazonium salts and Au NPs were also used as building blocks for the preparation of hybrid thin films on indium tin oxide (ITO) electrodes to obtain heavy metal sensors. Square wave voltammetry (SWV) measurements showed a high sensitivity for the electrochemical detection of Cu^{2+} ions, with a linear behavior between mM to pM concentrations and a detection limit of about 1 pM [45]. In another example, diazonium salt-decorated Au NPs were used to detect aniline, an important raw material in many industries, which presence in waters raises pollution concerns. The presence of aniline resulted in a fast and visible red-to-blue color change of the Au NPs solutions (Fig. 7c) allowing naked-eyes detection of arylamine-polluted water and quantitative measurements by absorbance changes [39].

5.3 Forensic Science

Diazonium salts combined with metallic NPs were evaluated as innovative developers for latent fingerprints on crime tools [46, 40], This strategy, based on the in situ reductions of aryldiazonium $[4\text{-}NO_2\text{-}C_6H_4N_2]^+AuCl_4$ by the components of human sweat, allowed the formation of gold deposits on eccrine secretions present at the surface of copper, lead, nickel and aluminum surfaces. The gold deposits appeared efficient to visualize the primary, secondary, and tertiary details of the developed fingerprints. This approach was combined with the XPS technique [47], a surface-sensitive technique for chemical analysis. The formation of robust gold–aryl films from simultaneous reduction of diazonium moieties and gold (III) in $[AuCl_4]^-$ resulted in an increased gold(0) quantity on the surface of copper coins and model flat sheets. Diazonium gold (III) salts thus provide a promising alternative to common methods for latent fingerprint detection and development.

5.4 *Catalysis*

Water-soluble polyaryl-coated Au NPs were used as catalysts for the reduction of an environmental pollutant, 4-nitrophenol (Fig. 7b). It is noteworthy that nitrophenols are toxic organic pollutants widely spread in industrial and agricultural wastewaters. Nanohybrids elaborated from the mild reduction of $[HOOC\text{-}4\text{-}C_6H_4N{\equiv}N]AuCl_4$ displayed high catalytic activity for the reduction of 4-nitrophenol to 4-aminophenol by sodium borohydride [38]. A series of para-substituted phenyl groups ($-CH_3$, $-F$, $-Cl$, $-OCF_3$, and $-CF_3$) derived from diazonium salts were successfully grafted on the surface of Pt NPs, by co-reduction of diazonium salts and H_2PtCl_4 [48]. $-CF_3$ functionalized Pt NPs were shown to display the highest catalytic activity for oxygen reduction reactions (ORR), with ~ 3 times higher ORR specific activity than the value found for commercial Pt/C catalysts. In another example, palladium NP-cored G1 dendrimers obtained from the reduction of Pd^{II} in the presence of diazodendrons were used for the chemoselective room temperature hydrogenation catalysis of carbon-carbon multiple bonds, with good recyclability and short reaction times [49]. The Pd-G1 nanohybrids appeared to be also efficient catalysts for Suzuki, Stille, and Hiyama coupling reactions [50].

Table 1 summarizes the main characteristics of the various plasmonic nanomaterials functionalized by aryl diazonium salts described in this review and their targeted applications.

6 Conclusion

This book chapter provides a general overview of the various surface modification strategies based on aryl diazonium salts for the functionalization of plasmonic nanomaterials. Aryl diazonium salts appear as a valuable alternative to commonly used thiol self-assembled monolayers for the modification of plasmonic nanoparticles. Indeed, they offer added-value characteristics such as (i) friendly synthesis and surface grafting protocols, (ii) broad choice of available terminal groups, (iii) robust covalent surface-C interfacial bonds, and (iv) formation of multifunctional

Table 1 Summary of plasmonic NPs functionalized by aryl diazonium salts and their targeted applications

Type of NPs	Shape of NPs	Functional groups on aryl rings	Immobilized objects	Targeted application	References
Au	Sphere	$-NO_2$, $-CN$, $-COOH$	Aryl mono- or multilayers	SERS tags for cell labeling	[14]
Au	Sphere	$-C_8F_{17}$	Polyaryls	–	[17]
Au	Sphere	$-Br$	PNIPAM brushes	Optical sensors	[28]
Au	Nanorod	$-O-CH_2-CH_2-$DEDTC	Molecularly imprinted polymers	Capture and detection of FA by SERS	[31]
Au	Nanorod	$-C_2H_4OH$	Polyaryls	Regioselective grafting for nano-optics	[19]
Au	Nanodisk	$-COOH$, $-C_2H_4OH$	Polyaryls		[21]
Au	Nanotriangle	−BTB	Polyaryls	Plasmonic electrochemistry	[18]
Ag	Spheroid	$-COOH$	Polyaryls	Antimicrobial applications	[43]
Pt	Sphere	$-CH_3$, $-F$, $-Cl$, $-OCF_3$, $-CF_3$,	Polyaryls	Electrocatalysis for oxygen reduction reactions	[48]
Pd	Sphere	Dendron	G_1-dendrimer	Chemoselective hydrogenation catalyst	[49]

layers. Owing to these outstanding properties, surface agents based on aryl diazonium salts were applied to the functionalization of plasmonic NPs for a wide range of applications, including nanosensors, nano-optics, bioimaging, latent fingerprints, environment, and antimicrobial materials.

However, this research area is still in an early stage with respect to real-life applications and several challenges need to be addressed to promote industrial development. From a methodological point of view, the mechanisms of covalent binding on the plasmonic NP surface and formation of polyaryl multilayers should be fully elucidated to achieve well-controlled nanohybrids. In practical terms, the long-term stability of the modified nanomaterials in various environments and upon external stimulations, such as light, temperature, or pH should be examined in depth. From a toxicological standpoint, the biosafety of plasmonic NPs functionalized by diazonium salts should be assessed. To tackle these issues and boost the technology transfer of diazonium salt chemistry for plasmonic NPs functionalization, it will be crucial to establish coordinated and collaborative investigations covering chemical, toxicological and industrial aspects.

References

1. Xia Y, Halas NJ (2005) MRS Bull 30:338–348. https://doi.org/10.1557/mrs2005.96
2. Schlücker S (2014) Angew Chem 53:4756–4795. https://doi.org/10.1002/anie.201205748
3. Baffou G, Quidant R (2014) Chem Soc Rev 43:3898–3907. https://doi.org/10.1039/C3CS60364D
4. Kang H, Buchman JT, Rodriguez RS, Ring HL, He J, Bantz KC et al (2018) Chem Rev 119:664–699. https://doi.org/10.1021/acs.chemrev.8b00341
5. Vericat C, Vela ME, Benitez G, Carro P, Salvarezza RC (2010) Chem Soc Rev 39:1805–1834. https://doi.org/10.1039/B907301A
6. Yang G, Amro NA, Starkewolfe ZB, Liu G-Y (2004) Langmuir 20:3995–4003. https://doi.org/10.1021/la0499160
7. Walia J, Guay J-M, Krupin O, Variola F, Berini P, Weck A (2018) Phys Chem Chem Phys 20:238–246. https://doi.org/10.1039/C7CP07024A
8. Delamar M, Hitmi R, Pinson J, Saveant JM (1992) J Am Chem Soc 114:5883–5884. https://doi.org/10.1021/ja00040a074
9. Mohamed AA, Salmi Z, Dahoumane SA, Mekki A, Carbonnier B, Chehimi MM (2015) Adv Colloid Interface Sci 225:16–36. https://doi.org/10.1016/j.cis.2015.07.011
10. Guselnikova O, Postnikov P, Elashnikov R, Trusova M, Kalachyova Y, Libansky M et al (2017) Colloids Surf A Physicochem Eng Asp 516:274–285. https://doi.org/10.1016/j.colsurfa.2016.12.040
11. Zou Q, Kegel LL, Booksh KS (2015) Anal Chem 87:2488–2494. https://doi.org/10.1021/ac504513a
12. Corgier BP, Bellon S, Anger-Leroy M, Blum LJ, Marquette CA (2009) Langmuir 25:9619–9623. https://doi.org/10.1021/la900762s
13. Laurentius L, Stoyanov SR, Gusarov S, Kovalenko A, Du R, Lopinski GP et al (2011) ACS Nano 5:4219–4227. https://doi.org/10.1021/nn201110r
14. Luo Y, Xiao Y, Onidas D, Iannazzo L, Ethève-Quelquejeu M, Lamouri A et al (2020) Chem Commun 56:6822–6825. https://doi.org/10.1039/D0CC02842H
15. Mirkhalaf F, Paprotny J, Schiffrin DJ (2006) J Am Chem Soc 128:7400–7401. https://doi.org/10.1021/ja058687g
16. Xu M, Qi W, Zhang L, Lai J, Majeed S, Xu G (2014) Microchim Acta 181:737–742. https://doi.org/10.1007/s00604-013-1111-7
17. Orefuwa SA, Ravanbakhsh M, Neal SN, King JB, Mohamed AA (2014) Organometallics 33:439–442. https://doi.org/10.1021/om400927g
18. Nguyen V-Q, Ai Y, Martin P, Lacroix J-C (2017) ACS Omega 2:1947–1955. https://doi.org/10.1021/acsomega.7b00394
19. Nguyen M, Kherbouche I, Gam-Derouich S, Ragheb I, Lau-Truong S, Lamouri A et al (2017) Chem Commun 53:11364–11367. https://doi.org/10.1039/C7CC05974D
20. Olshtrem A, Guselnikova O, Postnikov P, Trelin A, Yusubov M, Kalachyova Y et al (2020) Nanoscale 12:14581–14588. https://doi.org/10.1039/D0NR02934C
21. Tijunelyte I, Kherbouche I, Gam-Derouich S, Nguyen M, Lidgi-Guigui N, de la Chapelle ML et al (2018) Nanoscale Horiz 3:53–57. https://doi.org/10.1039/C7NH00113D
22. Macielag MJ (2012) Antibiotic discovery and development. Springer pp 793–820. http://doi.org/https://doi.org/10.1007/978-1-4614-1400-1_24
23. Troian-Gautier L, Valkenier H, Mattiuzzi A, Jabin I, Van den Brande N, Van Mele B et al (2016) Chem Commun 52:10493–10496. https://doi.org/10.1039/C6CC04534K
24. Valkenier H, Malytskyi V, Blond P, Retout M, Mattiuzzi A, Goole J et al (2017) Langmuir 33:8253–8259. https://doi.org/10.1021/acs.langmuir.7b02140
25. Retout M, Blond P, Jabin I, Bruylants G (2021) Bioconjug Chem 32:290–300. https://doi.org/10.1021/acs.bioconjchem.0c00669
26. Mejía-Salazar JR, Oliveira ON Jr (2018) Chem Rev 118:10617–10625. https://doi.org/10.1021/acs.chemrev.8b00359

27. Kim M, Lee JH, Nam JM (2019) Adv Sci 6:1900471. https://doi.org/10.1002/advs.201900471
28. Gehan H, Fillaud L, Chehimi MM, Aubard J, Hohenau A, Felidj N et al (2010) ACS Nano 4:6491–6500. https://doi.org/10.1021/nn101451q
29. Gehan H, Mangeney C, Aubard J, Lévi G, Hohenau A, Krenn JR et al (2011) J Phys Chem Lett 2:926–931. https://doi.org/10.1021/jz200272r
30. Erzina M, Guselnikova O, Postnikov P, Elashnikov R, Kolska Z, Miliutina E et al (2018) Adv Mater Interfaces 5:1801042. https://doi.org/10.1002/admi.201801042
31. Ahmad R, Félidj N, Boubekeur-Lecaque L, Lau-Truong S, Gam-Derouich S, Decorse P et al (2015) Chem Commun 51:9678–9681. https://doi.org/10.1039/C5CC01489A
32. Gam-Derouich S, Pinson J, Lamouri A, Decorse P, Bellynck S, Herbaut R et al (2018) J Mater Chem A 6:19353–19357. https://doi.org/10.1039/C8TA06944A
33. AlBab ND, Hameed MK, Maresova A, Ahmady IM, Arooj M, Han C et al (2020) Colloids Surf, A 586:124279. https://doi.org/10.1016/j.colsurfa.2019.124279
34. Hameed MK, Ahmady IM, Alawadhi H, Workie B, Sahle-Demessie E, Han C et al (2018) Colloids Surf A Physicochem Eng Asp 558:351–358. https://doi.org/10.1016/j.colsurfa.2018.09.004
35. Panicker S, Ahmady IM, Almehdi AM, Workie B, Sahle-Demessie E, Han C et al (2019) Appl Organomet Chem 33:e4803. https://doi.org/10.1002/aoc.4803
36. Serafín V, Torrente-Rodríguez R, González-Cortés A, De Frutos PG, Sabaté M, Campuzano S et al (2018) Talanta 179:131–138. https://doi.org/10.1016/j.talanta.2017.10.063
37. Domenici F, Bizzarri AR, Cannistraro S (2012) Anal Biochem 421:9–15. https://doi.org/10.1016/j.ab.2011.10.010
38. Ahmad AA, Panicker S, Chehimi MM, Monge M, Lopez-de-Luzuriaga JM, Mohamed AA et al (2019) Catal Sci Technol 9:6059–6071. https://doi.org/10.1039/C9CY01402K
39. Du J, Du H, Li X, Fan J, Peng X (2017) Sens Actuators B Chem 248:318–323. https://doi.org/10.1016/j.snb.2017.03.067
40. Almheiri S, Ahmad AAL, Le Droumaguet B, Pires R, Chehimi MM, Mohamed AA (2021) Surf Interface Anal 53:543–549. https://doi.org/10.1002/sia.6941
41. Ahmad R (2015) Griffete Nbw, Lamouri A, Felidj N, Chehimi MM, Mangeney C. Chem Mater 27:5464–5478. https://doi.org/10.1021/acs.chemmater.5b00138
42. Gam-Derouich S, Mahouche-Chergui S, Truong S, Hassen-Chehimi DB, Chehimi MM (2011) Polymer 52:4463–4470. https://doi.org/10.1016/j.polymer.2011.08.007
43. Kawai K, Narushima T, Kaneko K, Kawakami H, Matsumoto M, Hyono A et al (2012) Appl Surf Sci 262:76–80. https://doi.org/10.1016/j.apsusc.2012.02.018
44. AlMashrea BA, Abla F, Chehimi MM, Workie B, Han C, Mohamed AA (2020) Synth Met 269:116528. https://doi.org/10.1016/j.synthmet.2020.116528
45. Ait-Touchente Z, Falah S, Scavetta E, Chehimi MM, Touzani R, Tonelli D et al (2020) Molecules 25:3903. https://doi.org/10.3390/molecules25173903
46. Ahmad AA, Alawadhi AH, Park J, Abdou HE, Mohamed AA (2019) Forensic Chem 13:100144. https://doi.org/10.1016/j.forc.2019.100144
47. Almheiri S, Ahmad AA, Le Droumaguet B, Pires R, Mohamed AA, Chehimi MM (2019) Langmuir 36:74–83. https://doi.org/10.1021/acs.langmuir.9b03390
48. Zhou Z-Y, Kang X, Song Y, Chen S (2012) J Phys Chem C 116:10592–10598. https://doi.org/10.1021/jp300199x
49. Kumar VKR, Gopidas KR (2011) Tetrahedron Lett 52:3102–3105. https://doi.org/10.1016/j.tetlet.2011.04.011
50. Kumar VR, Krishnakumar S, Gopidas KR (2012) Eur Org Chem 2012:3447–3458. https://doi.org/10.1002/ejoc.201101749

Diazonium Electroreduction and Molecular Electronics

Jean Christophe Lacroix

Abstract This chapter describes the advantages of diazonium electroreduction for generating molecular junctions and the main electronic functions obtained using such systems. It is an updated version of a review article entitled "Electrochemistry Does the Impossible: Robust and Reliable Large Area Molecular Junction" published in "Current Opinion in Electrochemistry" in 2018 [1]. Part of this chapter is reproduced with permission from Ref. (Lacroix in Curr Opin Electrochem 7:153–160, 2018) Copyright 2018 Elsevier.

1 Introduction

Molecular junction (Mj) consists of an assembly of many molecules or a single molecule between two conducting electrodes and is the basic component of molecular electronics [1–5]. Current versus potential curves characterize Mj transport properties and depend mainly on the distance between the two electrodes and on the coupling of the molecules to the contacts which can be weak when little interactions exist, or strong when covalent bonds are created between the electrodes and the molecules. Initial proposals [6, 7] were mainly theoretical and did not pay attention to the binding of the molecules to the electrodes. This chapter describes the advances made in the past ten years when the electroreduction of diazonium compounds is used to generate molecular junctions. It describes some of the many electronic functions that can be obtained and gives some perspectives.

J. C. Lacroix (✉)
Université Paris Cité, CNRS, ITODYS, 75013 Paris, France
e-mail: lacroix@u-paris.fr

M. M. Chehimi et al. (eds.), *Aryl Diazonium Salts and Related Compounds*,
Physical Chemistry in Action, https://doi.org/10.1007/978-3-031-04398-7_12

2 Advantages of Diazonium Electroreduction for Generating Molecular Junctions

The first step in the fabrication of molecular junctions is the deposition of monolayers or ultrathin molecular layers on a bottom electrode, generally made of carbon, gold, and silver [2, 3]. The flatness of the bottom electrode has a strong impact on the MJ performance, and as a consequence, its roughness needs to be below 1 nm [8, 9]. In a second step, a top electrode is deposited on the organic layer.

Monolayers based on thiols adsorption on gold or silver have been widely used in MJs. They are easily generated by dipping a gold or silver electrode into a solution during few hours. Such self-assembled monolayers (SAM) are densely packed and highly ordered with total thickness between 1 and 4 nm [10]. One of the most widely used top electrodes on these layers is an eutectic indium–gallium alloy (EGaIn), which is liquid at room temperature and behaves as a metal [11]. It can be reproducibly deposited on SAM, and the contact area between the SAM and the liquid top electrode can be controlled. Using such top electrodes has given numerous results of interest for academic work and allowed the first statistical studies of transport properties in large-area molecular junctions[12, 13]. As an example, MJs using an EGaIn top electrode and showing very large rectification ratios were recently reported [14]. Electric field-driven dual-functional molecular switches in tunnel junctions were also recently demonstrated [15].

EGaIn top electrodes are not compatible with the development of commercial electronic devices which needs direct metal evaporation onto the molecular layer to generate the top electrode of the MJ for production in massively parallel processes. Unfortunately, the direct gold deposition on SAM leads to a very low fabrication yield (1.2%) of operating Mjs [16]. In fact, metal deposition from the vacuum on an organic monolayer is doomed to failure, the metal goes through the organic monolayer like “a hot knife in butter” and leads to the formation of metallic filaments between the top and the bottom electrodes, which generates short circuits. Therefore, SAM based on thiol adsorption on gold is not compatible with current microelectronic fabrication requirements.

Different protective strategies have been developed to avoid the formation of metallic filaments. The use of a conductive polymer (PEDOT: PSS) as a protective layer [17, 18] improved the fabrication yields of the devices above 50%. Soft contact deposition [19] and transfer printing were also used to deposit electrodes of different metals deposited on a monolayer, and 90% of non-short-devices were obtained by these techniques which are unfortunately difficult to use at the submicron scale. The use of new carbon paint layers was also recently reported [20]. Many academic results were also obtained using the tip of a conductive AFM [21–24] or of a scanning tunneling microscope but such systems are also not compatible with massively parallel fabrication processes and single-molecule devices generated by STM break junction method are unstable [25–27]. Overall top electrode deposition on monolayers and SAM-based layers has been the main fabrication bottleneck that has prevented the commercialization of molecular electronic devices.

Diazonium electrochemical reduction is an important way to solve these problems. It leads to aryl radicals which bind covalently to the electrode [28, 29]. Disorganized multilayers with film thickness ranging from ca. 1–20 nm are usually obtained but strategies based on the use of bulky groups [30] associated with protection/deprotection steps [31, 32] organometallic complexes, [33] ionic liquids, [34] radical scavenger in the solution [35, 36] can restrict the growth to monolayers. Compact, pinhole free, strongly adherent, and stable under ambient conditions, layers are easily electrogenerated. When grown from a flat surface their roughness remains low and close to that of the initial surface. Because C–C is one of the strongest known bonds, its use yields extremely robust C–C-bonded molecular film [37–40]. Compared to organothiolate assemblies on Au or Ag, the films grafted onto carbon from aryldiazonium salts are strongly coupled with the substrate, thermally and chemically stable. Other methods to generate robust large-area MJs and to circumvent the widely documented limitations of using thiol-based layers in the fabrication of MJs [16, 41] have also recently been developed [42]. They will not be described in this chapter but developing such methods has become a critical issue for practical applications in molecular electronics [43].

Many molecular layers based on various conjugated structures have been immobilized on gold or on carbon-based surfaces through the diazonium electroreduction. They have been mostly generated in situ from aromatic amines bearing various moieties. Layers-based oligophenylenes have been widely used in early studies [44, 45] but as polyparaphenylene is a wide bandgap conjugated material such layers are not the most useful in molecular electronics. Other layers based on azobenzenes, [46] p dopable oligomers incorporating thiophene, [47, 48], 3,4-ethylenedioxythiophene (EDOT) [49] or aniline moieties [50, 51] n dopable oligomers incorporating phenylene, [52] thienopyrazines [53] or naphthalene diimides, electroactive organometallic compounds [54, 55] and more generally layers incorporating redox active building blocks have been deposited on surface, incorporated in diazonium-based Mjs and have yielded to devices with interesting electronic properties. These layers are electroactive and can act as electron transport relay or can be switched between an insulating and a conductive state. As a consequence, they can act as organic electrodes, [47, 52, 56] for a second electrodeposition process and several bottom-up electrochemical processes for fabricating ultrathin layers have been developed [50, 57, 58]. As an example, ultrathin layers of covalently grafted oligo(bisthienylbenzene) (oligo(BTB)) have been used as switchable organic electrodes, and 3,4-ethylenedioxythiophene (EDOT) was oxidized and grafted on this layer. Adding only a few EDOT moieties (5–6 units, less than 3 nm) changed the electrochemical properties of the layer. New materials with tunable switching properties were obtained. They consist of oligo(BTB)-oligo(EDOT) diblock oligomers of various relative lengths covalently grafted onto the underlying electrode. These films, with overall thickness below 11 nm, can still switch reversibly between insulating and conductive states, and their switching potential can be finely tuned between + 0.6 and −0.3 V/SCE [57]. Successive electroreduction of two different diazonium salts and click chemistry has also been used to generate various bilayers that have been engaged in MJs [59–61].

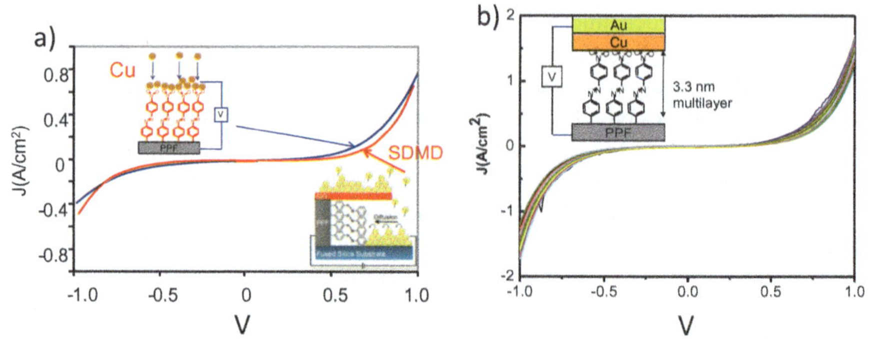

Fig. 1 **a** Characterization (I(V) curves) of MJs obtained with direct copper evaporation on the molecular layer (blue curve) and surface diffusion-mediated deposition (SDMD) method (red curve). **b** Overlay of 32 curves of the 32 devices fabricated in a CMOS compatible massively parallel process using direct metal evaporation. Adapted with permission from references [66] Copyright (2008) IOP Science and [69] Copyright (2010) American Chemical Society

Diazonium-based molecular layers were initially incorporated in MJs by McCreery et al. [45, 62–65]. This group was the first to show that Cu or Si can be e-beam evaporated directly on top of a 5 nm-thick-diazonium-based layers [66–69]. To do so they compared the I(V) curve of a device in which the top electrode is deposited by surface diffusion-mediated deposition, SDMD (a mild deposition method) compared to that of one where copper was directly evaporated onto the layer (Fig. 1). The transport properties of the two devices are similar and the stability and the reproducibility of the device are good (Fig. 1b). The resistance to high temperature and to scan rates was excellent, and the fabrication yield was reported to be close to 90% despite direct evaporation of the metal on the diazonium-based organic layer. These results clearly opened the way toward commercial molecular junction fabricated using massively parallel CMOS compatible processes with conventional equipment. First products were recently released on the market [70].

Titanium/gold top electrodes have also been directly evaporated onto the diazonium-based layer and used as top electrode (the titanium layer impedes gold penetration and avoids short circuits) [63, 71]. Ti evaporation must be performed with caution. Indeed, high pressure convert titanium, in titanium dioxide, a large bandgap semiconductor which modifies drastically the electronic behavior of the MJ [71, 72]. When low pressure is used, titanium is partially converted into oxide layers of low oxygen content, known to be metallic [73, 74]. The formation of titanium carbide at molecule/titanium interface has also been evidenced [75, 76]. As a consequence, the organic layer is strongly coupled to both electrodes which play a key role in the transport properties of the MJs as strong internal build-in electric fields can be obtained when the two electrodes have different work functions [77, 78].

Overall, reproducible and robust MJs have been obtained using Au bottom electrodes and Ti/Au top contacts directly deposited on diazonium-based organic layers

[77–79] and both the Au/molecule/Ti/Au and "all-carbon" devices have been thoroughly characterized, are directly amenable to the variety of molecular structures, and are compatible with current microelectronic manufacturing with minor changes in processing and packaging.

3 Various Electronic Functions Demonstrated in Large-Area Diazonium-Based Mjs

As already stated, the thickness of diazonium-based layers can be controlled from few nanometers to tens of nanometers. The effect of layer thickness on transport properties in Mjs was therefore first reported, and several important results were obtained. Fabrication yield is high and reached 95% when the thickness of the used layer is above 5 nm. It drops to 30–40% when using thinner organic layers but remains high when bulky inorganic complexes are deposited as monolayers [80].

The 2 and 5 nm thickness range was first investigated. In this range, direct tunneling between the two electrodes is the main transport mechanism and the current is given by $J = J_0 \exp(-\beta d)$ where β is the attenuation factor. Many studies measured attenuation factor between 2 and 3 nm^{-1} by changing the thickness of the layer for many conjugated layers [46]. These results are in agreement with those obtained using a C-AFM top contact [22, 81, 82] or in single molecule junction generated by STM break-junction [83, 84]. When the molecular structure of the layer was changed with HOMOs and LUMOS varying in gas phase by as much as 3 eV little variation in behavior for these different molecular structures was observed. Transport was shown to be mainly controlled by the thickness of the layer in between the two electrodes [85]. These surprising and important results were explained by the strong electronic coupling between the grafted layer and the bottom electrode.

In parallel, we used bisthienylbenzene-based (BTB) molecular layers with thicknesses above 5 nm and gold and titanium/gold electrodes to develop rectifiers. No rectification was observed with thicknesses below 5 nm while devices using 7 nm BTB film showed rectification ratios over 50 [77, 86]. The effect was attributed to a mechanism similar to that proposed by Nijhius et al. for SAM-based devices incorporating ferrocene units; [12] i.e., a decrease in the effective tunneling barrier distance by hole injection into the BTB layer and to the use of two different contacts (gold and titanium/gold). Next, the transport properties of BTB in all-carbon devices were investigated. Symmetric *I*(*V*) curves were obtained for film thicknesses between 2 and 22 nm (Fig. 2a, b), and a change in the attenuation factor above 7 nm was observed. This indicates a transition in the dominant transport mechanism (Fig. 2c, d) [87]. Frisbee et al. already observed such a mechanistic transition, when thicknesses increase [22, 88] but with BTB-based layers, the electron transport was surprisingly independent of temperature, and activation-less transport across 20 nm, was demonstrated (Fig. 2b, d). Such behavior was attributed to the ionization of the molecules by the applied electric field associated with activation-less intra-chain hopping [87].

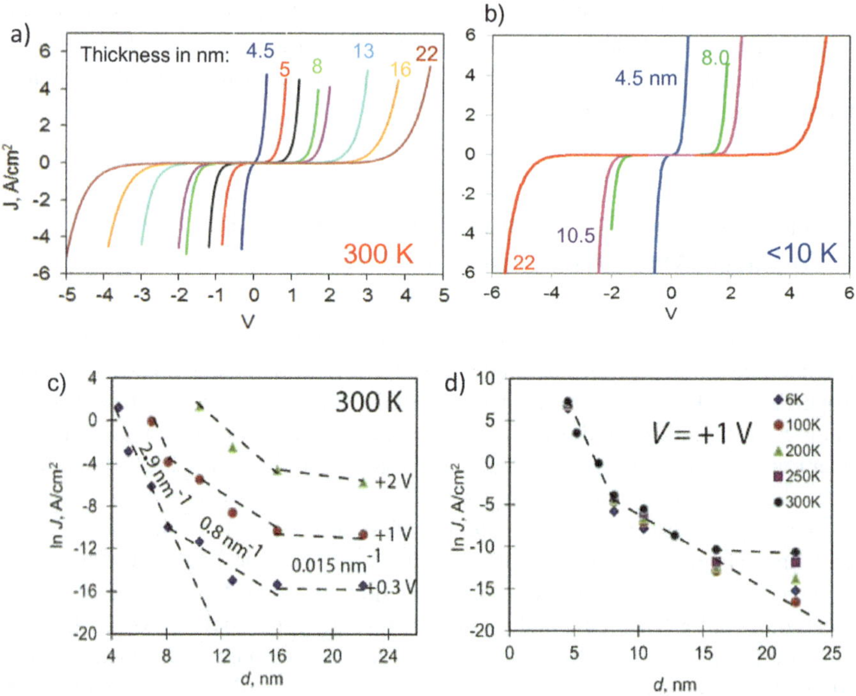

Fig. 2 JV Overlay of PPF/BTB/eC MJs for various thicknesses of the BTB layer **a** at 300 K, **b** below 10 K. **c** Attenuation plots taken at various potential V of PPF/BTB/eC Mjs at 300 K showing transition in dominant transport mechanisms, **d** attenuation plots taken at 1 V of PPF/BTB/eC Mjs at various temperature (from 6 to 300 K) showing the existence of an activation-less transport mechanism up to 22 nm. Adapted with permission from reference [87] Copyright (2013) National Academy of Sciences

Moreover, activated hopping when $d > 16$ nm for high temperatures and low bias attributed to interchain hopping was also evidenced (Fig. 2d). Overall this study bridges the gap between short-range tunneling in molecular junctions and activated hopping in bulk organic films.

Transport properties of other molecular systems were also investigated. MJs using 3 and 14 nm-thick viologen-based layers in Au/molecule/Ti-Au structures were studied [89]. They show highly efficient long-range transport, with an attenuation factor as small as 0.25 nm^{-1} and no rectification (symmetric J–V curves). This was attributed both to the fact that the viologen LUMO energy lies between the energies of the Fermi levels of the two contacts and to strong electronic coupling between molecules and contacts. Consequently, resonant tunneling is likely to be the dominant transport mechanism within these MJs, but the temperature dependence of the transport properties suggests that activated redox hopping plays a role at high temperature. In other studies, MJs incorporating inorganic complexes deposited by diazonium electroreduction were studied. As an example, molecular junctions

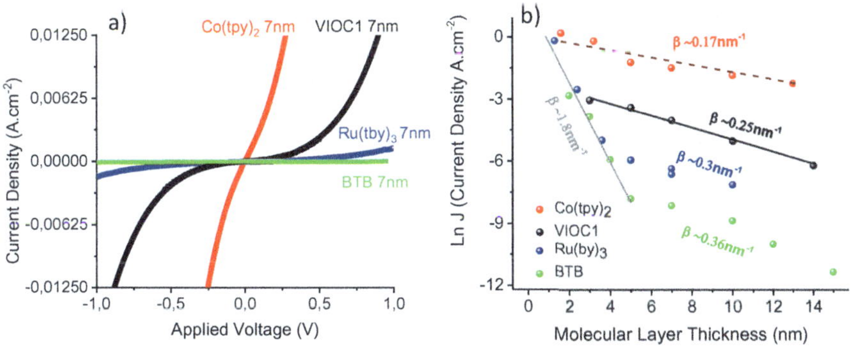

Fig. 3 **a** JV overlay of [Co(tpyr)$_2$]$_{7\,nm}$ (red), VIOC1$_{7nm}$ (black), [Ru(bpy)$_3$]$_{7\,nm}$ (blue) and BTB$_{7nm}$ green MJs. **b** Attenuation plots taken at 1 V of MJs for different molecular units. β value is the slope of the curve and is indicated for each molecule. Adapted with permission from reference [91] Copyright (2020) Wiley

based on a cyclometalated ruthenium complex (Ru(bpy)$_2$(ppy)) were studied and compared to MJs based on Ru(bpy)$_3$ [90]. A strong influence of the redox potential of the metallic center on the transport properties of the Mjs was evidenced. Cobalt terpyridine oligomers, which have low redox potentials, were also compared with π-conjugated and ruthenium-centered layers in MJs with identical contacts (Fig. 3) [91, 92]. Strong molecular signature on charge transport was found with a variation of four orders of magnitude of current density (J) for different molecules and d = 7 nm (Fig. 3a). Attenuation plots for different molecular layers were compared. For a Ru(bpy)$_3$ complex and BTB MJs, the attenuation plot showed a transition between two different dominant transport mechanisms. On the contrary, Co(tpy)$_2$-based MJs showed no transition in the attenuation plot (Fig. 3b) with film thickness, and very low attenuation factors (β of 0.17 nm^{-1} from 2 to 14 nm). This β value indicates highly efficient long-range transport and was again attributed to the fact that the energy levels of the Co(tpy)$_2$ frontier orbital involved in transport are between, and thus almost in resonance with, the Fermi levels of the electrodes [89, 91].

Bilayers of donor–acceptor components whose total thickness is less than 20 nm were used to generate robust molecular diodes with rectification ratio reaching 500. The rectification was clearly assigned to the used molecules as the two electrodes were made of carbon. Control of the direction of rectification of these devices by reversing the order of the deposited layers was demonstrated [59]. An electron donor–acceptor dyad, consisting of a Ruthenium(terpyridine)$_2$ complex with one terpyridine linked to a naphthalene diimide unit, was also designed [80]. Diazonium electrografting was used to form well-controlled monolayers. The robustness of this monolayer enabled the creation of MJs (Fig. 4a) using direct top-coat evaporation with a high yield of operating devices. Moreover, the low thickness of the device enabled very high current densities and the molecular structure associated with the use of two different electrodes (Au and Ti/Au) induced large rectification ratio (Fig. 4b) [80].

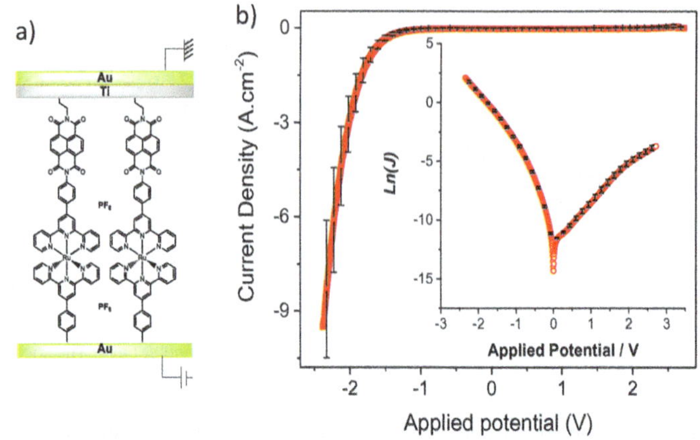

Fig. 4 **a** Schematic illustration of the junction based using electron donor–acceptor dyad, consisting of a Ruthenium(terpyridine)$_2$ complex with one terpyridine linked to a naphthalene diimide unit. **b** JV and semilogarithmic JV curve for Au/dyad monolayer/Ti/Au MJ. Adapted with permission from reference [80] Copyright (2017) Royal Society of Chemistry

Next rectifying diodes using Au/molecules/Ti/Au structure with rectification factors varying over 5 orders of magnitude depending on the type of molecules were obtained (Fig. 5) [78]. MJs with 9 nm BTB layers showed averaged rectification ratio above 1000 at 2.7 V. The current is higher when the bottom gold electrode is biased positively, and titanium/gold is negative (Fig. 5a). When the molecular layer is based on tetrafluororobenzene, a molecule with high HOMO–LUMO gap, no rectification is observed (Fig. 5b) while MJs with naphthalene diimides, (with terminal alkyl moieties), i.e., molecules with a low LUMO energy level and a terminal decoupling

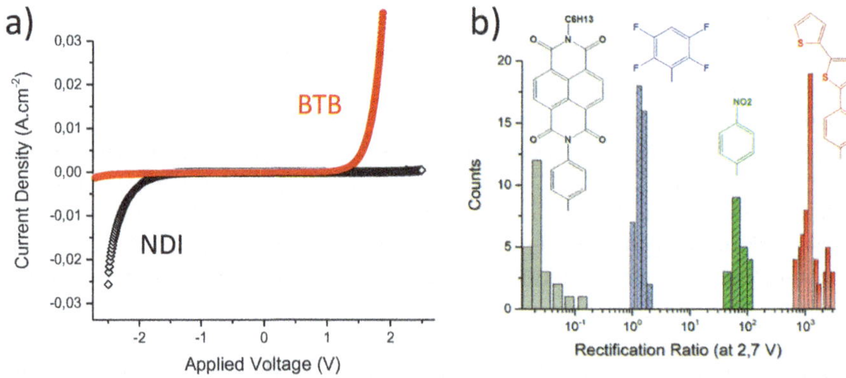

Fig. 5 **a** *JV* curves for Au/BTB$_9$/Ti$_2$/Au (red curve) and Au/NDI$_7$/Ti$_2$/Au (gray curve) MJ with subscripts indicating layer thicknesses in nm, **b** rectification ratio ($J(+2.7\text{ V})/J(-2.7\text{ V})$) histograms for 63 BTB (red), 21 NB (green) and 42 FB (blue) devices and 26 NDI (gray) junctions. Adapted with permission from reference [78]. Copyright (2017) American Chemical Society

unit, the direction of rectification is reversed (Fig. 5a) These results show that rectification is directly affected by orbital energies of the molecular layer even though it is also induced by the asymmetric contacts [78].

Destructive quantum interferences were also used to control transport in diazonium-based MJs [79]. Cross-conjugated anthraquinone layer was grafted electrochemically. A pronounced dip in the differential conductance close to zero voltage bias is the signature of these quantum interferences (minimum > 2 orders of magnitude lower than the conductance at higher bias at 4 K). Enhanced signatures of electron–phonon interaction at low temperature and vibrational modes excited by inelastic events are observed in such devices and superposed on the conductance suppression induced by destructive quantum interference [93].

Resistive memory behavior was also demonstrated using diazonium-based MJs [94, 95]. These memristors were generated by electrografting specific redox polymer based on inorganic complexes onto the surface. Long retention time multi-level conductivity, high R_{max}/R_{min} ratio, and potential of these devices to store analog synaptic weights in neural network circuit strategies were demonstrated [95].

Ultrathin light-emitting diode incorporating $Ru(bipyridine)_3{}^{2+}$ films and using symmetrical carbon contacts were obtained (Fig. 6). The emission was attributed to bipolar transport, (Fig. 6c, e) in 7 nm-thick layer and above [96, 97]. The emission observed is that of the excited state of $Ru(bipyridine)_3{}^{2+}$ and demonstrates the existence of redox phenomena within the layer.

In the past few years, the response to light illumination of various photoactive diazonium-based Mjs was studied.

Mjs, consisting of nitroazobenzene oligomers in "all-carbon" devices, were illuminated with UV–vis light through a partially transparent carbon top electrode [98]. The effects of incident wavelength, light intensity, molecular layer thickness, and temperature on photocurrents and junction conductance with a DC bias imposed were studied. The photocurrent spectrum was found to track the in-situ absorption spectrum of nitroazobenzene, to increase linearly with light intensity, and to depend exponentially on applied bias. The electronic characteristics of the photocurrent were compared to those of the same device in the dark. Very weak attenuation with molecular layer thickness ($\beta = 0.14$ nm^{-1} for thickness above 5 nm) and orders of magnitude higher conductance and were observed. The photocurrent was similar to the dark current for thin molecular layers but greatly exceeded the dark current for low bias and thick molecular layers. It was concluded that photoexcited carriers are transported without thermal activation for a thickness range of 5–10 nm. These results showed that Mjs with partially transparent top contacts could act as photosensors. Photocurrent, photovoltage, and rectification in Mjs containing oligomers of two different aromatic molecules in bilayer were also studied [99] and previously reported Mjs containing nitroazobenzene oligomers, and bilayers were then evaluated for sensitivity, dark signal, responsivity, and limits of detection [100]. Discussions on how to improve specific detectivity by 3–4 orders of magnitude in order to reach the range for commonly used photodetectors were presented along with the advantages of such systems (operation outside the 300–1000 nm range of silicon detectors and very low dark currents).

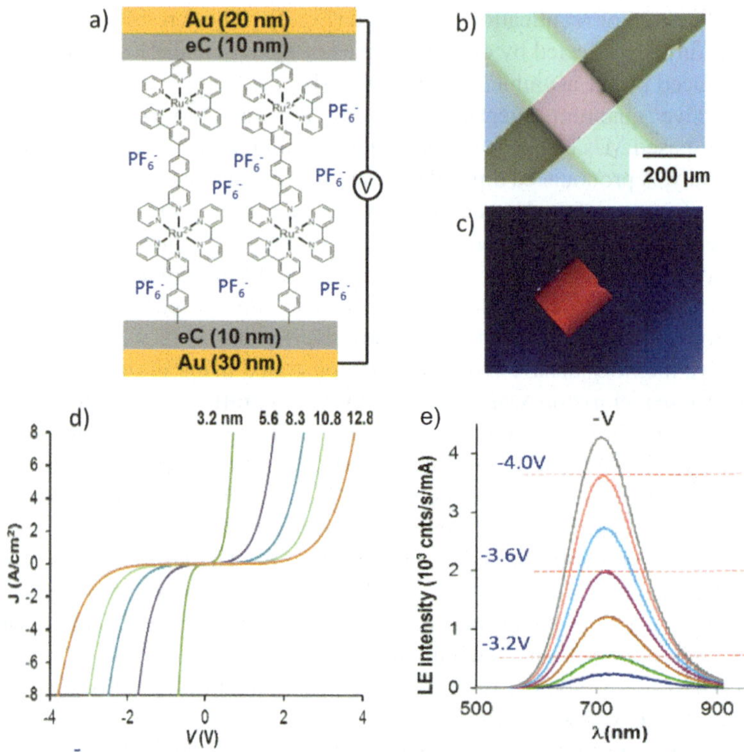

Fig. 6 **a** Schematic illustration of $Au_{30}/eC_{10}/Ru(bpy)_3/eC_{10}/Au_{20}$ molecular junction. with subscripts indicating layer thicknesses in nm, **b** optical image of 12.8 nm $Ru(bpy)_3$ molecular junctions: no bias in room light and **c** under 3.4 V bias in darkness. **d** *JV* curves for $Au_{30}/eC_{10}/Ru(bpy)_3P/eC_{10}/Au_{20}$ junctions (average of 8 MJs, in vacuum) with molecular layer thickness indicated for each curve. **e** Light emission spectra from $Ru(bpy)_3$ junctions for progressively applied negative, with V_{app} from −3.0 V, to −4.2 V in order from low to high intensity. Adapted from with permission reference [94] and [95] Copyright (2017 and 2019) American Chemical Society

Photochromic units have also been recently incorporated in diazonium-based Mjs. In a first study, thin layers of diarylethene oligomers (oligo(DAE)), with total layer thicknesses fixed at 2–3 nm and 8–9 nm, i.e., below and above the direct tunneling limit, were deposited by electrochemical reduction of a diazonium salt on gold electrodes [101]. C-AFM tip was used to complete molecular junctions. DAE was then photoswitched between its open and closed forms. Oligo(DAE) MJs can be switched between highly resistive (using the open form of DAE) and more conductive states (with DAE in the closed form). ON/OFF ratios of 2–3 and 200–400 were obtained for 3 and 9 nm-thick DAE MJs, respectively. Next 4 nm-thick diarylethene (DAE) and 5 nm-thick bisthienylbenzene (BTB) layers were used to improve the properties of the photoactive molecular junctions [102]. Total thickness was above the direct tunneling limit and in the hopping regime. The DAE/BTB bilayer structure exhibited new electronic functions combining photoswitching and photorectification. The

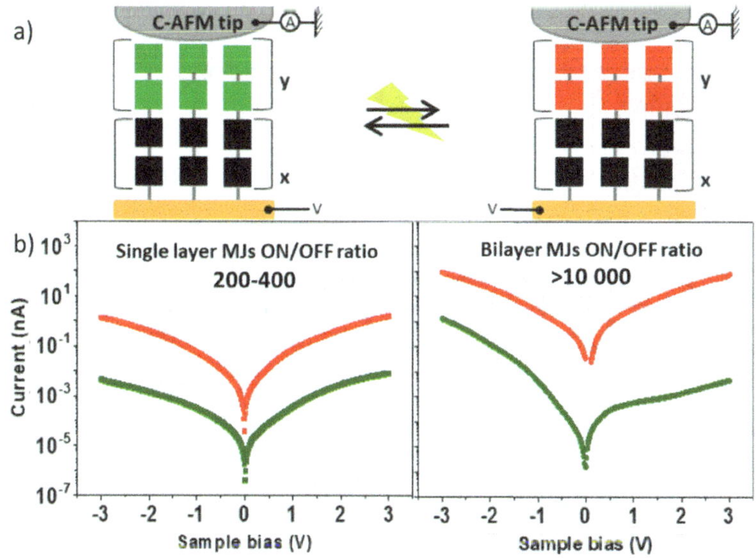

Fig. 7 **a** Schematic illustration of Au/BTB/DAE/C-AFM tip (Pt/Ir) molecular junction and the photoswitch of the oligo(DAE) units. The switch is turned ON by UV and OFF by visible light. **b** log(I) versus V characteristics before (green) and after (red) UV irradiation of: DAE_{9nm} single layer and DAE_{4nm}/BTB_{5nm} bilayer junctions measured by C-AFM up to +3 V. Reproduced with permission from reference [102] Copyright (2021) American Chemical Society

open form of DAE/BTB showed low conductance and asymmetric I–V curves while the closed form showed symmetric I–V curves and high conductance. More importantly, unprecedented ON/OFF current ratios of over 10 000 at 1 V were reproducibly measured [102] (Fig. 7).

Finally, we, and others, [103–105] have reported recently the use of diazonium-based layers in the fabrication of single molecule junction (SMJ). Thin layers (thicknesses between 2 and 8 nm) of cobalt and ruthenium polypyridyl-oligomers were deposited on gold. Such layers were previously studied in large-area molecular junction with a Ti/Au top electrode. Their conductance has been studied by a scanning tunneling microscopy break junction (STM-bj) technique. The charge transport of the Au-$[Co(tpy)_2]_n$-Au (n = 1–4) SMJs did not depend markedly on the oligomer length, and an extremely low attenuation factor ($\beta \sim 0.19\ nm^{-1}$) was measured at the single molecule level. Moreover, no thickness-dependent transition between two mechanisms was observed. Resonant charge transport was proposed as the main transport mechanism. Changing the metal from Co to Ru decreases the SMJ conductance by 1 order of magnitude. Moreover, a charge transport transition from direct tunneling to hopping was evidenced by a break in the length-dependent β-plot in Au-$[Ru(tpy)_2]_n$-Au and Au-$[Ru(bpy)_3]_n$-Au SMJs. Most importantly, these results were in good agreement with those obtained on large-area molecular junctions [91, 103]. Next rigid oligomers of variable length, based on porphyrin derivatives (free base or cobalt complex forms), were grafted on the surface and their conductance has been

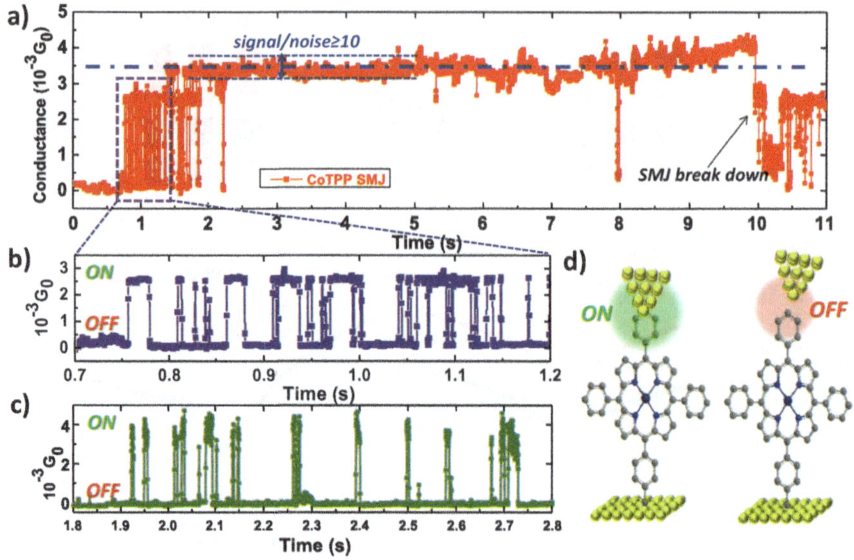

Fig. 8 Conductance versus time traces from Au-[CoTPP]-Au junction **a** G(t) trace showing initial blinking and subsequent stabilization of the SMJ. **b** Zoom in the first second of SMJ formation showing random telegraph signals. **c** Au-[TPP]-Au junction telegraph signals. **d** Scheme showing the SMJ ON (tip is in contact with the molecule) and OFF (tip is not in contact with the molecule) states. Reproduced with permission from reference [104] Copyright (2021) American Chemical Society

studied by STM-bj and G(t) measurements. The lifetime of the SMJs was also investigated [104]. Charge transport in the porphyrins SMJs was shown to be relatively efficient and enhanced by the presence of the cobalt center. In G(t) measurements random telegraph G(t) signals showing SMJ on/off states were first recorded with both systems. The SMJs then stabilize and exhibit surprisingly long lifetimes around 10 s, and attenuation plots, obtained by both G(t) and STM-bj measurements, give identical values. These last works show that highly stable SMJs can be prepared using a diazonium grafting approach which has been attributed to the advantages of the diazonium electroreduction grafting; that is, the oligomers are covalently bonded to the bottom electrode in a highly compact and robust film which impedes molecular motion (Fig. 8).

4 Conclusion

The main bottleneck of molecular electronics in terms of practical issues was to deposit top electrode using CMOS compatible processes. Layers generated by diazonium electroreduction make this possible and are compatible with massively parallel fabrication techniques. This achievement has paved the route for molecular junction commercialization.

Important academic results were also obtained with diazonium-based layers. The thickness of the layer was shown to have a major impact on transport properties, and strong "molecular signature" for the electronic response of the devices is clearly observed when using layers with thicknesses between 5 and 20 nm. Many electronic devices including rectifiers, memories, switches, and dual-functional devices have been demonstrated. Single molecule junction using diazonium-based layers has shown unprecedented stability and robust photoactive diazonium-based MJs with potential applications in sensors or in photoswitches have started to be developed. Further variations of molecular layer structure, order, and ionic content may enable a variety of electronic functions which are difficult or impossible with conventional semiconductor materials.

Acknowledgements Professor Richard McCreery is highly thanked for numerous discussions and fruitful collaborations.

References

1. Lacroix JC (2018) Electrochemistry does the impossible: robust and reliable large area molecular junctions. Curr Opin Electrochem 7:153–160. https://doi.org/10.1016/j.coelec.2017.11.017
2. Xiang D, Wang X, Jia C, Lee T, Guo X (2016) Molecular-scale electronics: from concept to function. Chem Rev 116:4318–4440. https://doi.org/10.1021/acs.chemrev.5b00680
3. Metzger RM (2015) Unimolecular electronics. Chem Rev 115:5056–5115. https://doi.org/10.1021/cr500459d
4. McCreery RL, Yan H, Bergren AJ (2013) A critical perspective on molecular electronic junctions: there is plenty of room in the middle. Phys Chem Chem Phys 15:1065–1081. https://doi.org/10.1039/c2cp43516k
5. Vilan A, Aswal D, Cahen D (2017) Large-area, ensemble molecular electronics: motivation and challenges. Chem Rev 117:4248–4286. https://doi.org/10.1021/acs.chemrev.6b00595
6. Aviram A, Ratner MA (1974) Molecular rectifiers. Chem Phys Lett 29:277–283. https://doi.org/10.1016/0009-2614(74)85031-1
7. Aviram A (1988) Molecules for memory, logic, and amplification. J Am Chem Soc 110:5687–5692. https://doi.org/10.1021/ja00225a017
8. Weiss EA, Chiechi RC, Kaufman GK, Kriebel JK, Li Z, Duati M, Rampi MA, Whitesides GM (2007) Influence of defects on the electrical characteristics of mercury-drop junctions: self-assembled monolayers of n-alkanethiolates on rough and smooth silver. J Am Chem Soc 129:4336–4349. https://doi.org/10.1021/ja0677261
9. Yuan L, Jiang L, Thompson D, Nijhuis CA (2014) On the remarkable role of surface topography of the bottom electrodes in blocking leakage currents in molecular diodes. J Am Chem Soc 136:6554–6557. https://doi.org/10.1021/ja5007417
10. Love JC, Estroff LA, Kriebel JK, Nuzzo RG, Whitesides GM (2005) Self-assembled monolayers of thiolates on metals as a form of nanotechnology. Chem Rev 105:1103–1169. https://doi.org/10.1021/cr0300789
11. Chiechi RC, Weiss EA, Dickey MD, Whitesides GM (2008) Eutectic gallium-indium (EGaIn): a moldable liquid metal for electrical characterization of self-assembled monolayers. Angew Chemie Int Ed 47:142–144. https://doi.org/10.1002/anie.200703642

12. Nijhuis CA, Reus WF, Whitesides GM (2010) Mechanism of rectification in tunneling junctions based on molecules with asymmetric potential drops. J Am Chem Soc 132:18386–18401. https://doi.org/10.1021/ja108311j
13. Yuan L, Nerngchamnong N, Cao L, Hamoudi H, del Barco E, Roemer M, Sriramula RK, Thompson D, Nijhuis CA (2015) Controlling the direction of rectification in a molecular diode. Nat Commun 6:6324. https://doi.org/10.1038/ncomms7324
14. Chen X, Roemer M, Yuan L, Du W, Thompson D, del Barco E, Nijhuis CA (2017) Molecular diodes with rectification ratios exceeding 105 driven by electrostatic interactions. Nat Nanotechnol 12:797–803. https://doi.org/10.1038/nnano.2017.110
15. Han Y, Nickle C, Zhang Z, Astier HPAG, Duffin TJ, Qi D, Wang Z, del Barco E, Thompson D, Nijhuis CA (2020) Electric-field-driven dual-functional molecular switches in tunnel junctions. Nat Mater. https://doi.org/10.1038/s41563-020-0697-5
16. Kim T-W, Wang G, Lee H, Lee T (2007) Statistical analysis of electronic properties of alkanethiols in metal–molecule–metal junctions. Nanotechnology 18:315204. https://doi.org/10.1088/0957-4484/18/31/315204
17. Akkerman HB, Blom PWM, de Leeuw DM, de Boer B (2006) Towards molecular electronics with large-area molecular junctions. Nature 441:69
18. Wang G, Yoo H, Na S-I, Kim T-W, Cho B, Kim D-Y, Lee T (2009) Electrical conduction through self-assembled monolayers in molecular junctions: Au/molecules/Au versus Au/molecule/PEDOT:PSS/Au. Thin Solid Films 518:824–828. https://doi.org/10.1016/j.tsf.2009.07.094
19. Shimizu KT, Fabbri JD, Jelincic JJ, Melosh NA (2006) Soft deposition of large-area metal contacts for molecular electronics. Adv Mater 18:1499–1504. https://doi.org/10.1002/adma.200600195
20. Karuppannan SK, Neoh EHL, Vilan A, Nijhuis CA (2020) Protective layers based on carbon paint to yield high-quality large-area molecular junctions with low contact resistance. J Am Chem Soc 142:3513–3524. https://doi.org/10.1021/jacs.9b12424
21. Engelkes VB, Beebe JM, Frisbie CD (2004) Length-dependent transport in molecular junctions based on SAMs of alkanethiols and alkanedithiols: effect of metal work function and applied bias on tunneling efficiency and contact resistance. J Am Chem Soc 126:14287–14296. https://doi.org/10.1021/ja046274u
22. Choi SH, Kim BS, Frisbie C. (2008) Electrical resistance of long conjugated molecular wires. Science (80-)320:1482–1486
23. Dalla Francesca K, Lenfant S, Laurans M, Volatron F, Izzet G, Humblot V, Methivier C, Guerin D, Proust A, Vuillaume D (2019) Charge transport through redox active [H7P8W48O184]33−polyoxometalates self-assembled onto gold surfaces and gold nanodots. Nanoscale 11:1863–1878. https://doi.org/10.1039/C8NR09377F
24. Bonnet R, Lenfant S, Mazérat S, Mallah T, Vuillaume D (2020) Long-range electron transport in Prussian blue analog nanocrystals. Nanoscale 12:20374–20385. https://doi.org/10.1039/d0nr06971j
25. Chen F, Hihath J, Huang Z, Li X, Tao NJ (2007) Measurement of single-molecule conductance. Annu Rev Phys Chem 58:535–564. https://doi.org/10.1146/annurev.physchem.58.032806.104523
26. Venkataraman L, Klare JE, Nuckolls C, Hybertsen MS, Steigerwald ML (2006) Dependence of single-molecule junction conductance on molecular conformation. Nature 442:904–907. https://doi.org/10.1038/nature05037
27. Kamenetska M, Quek SY, Whalley AC, Steigerwald ML, Choi HJ, Louie SG, Nuckolls C, Hybertsen MS, Neaton JB, Venkataraman L (2010) Conductance and geometry of pyridine-linked single-molecule junctions. J Am Chem Soc 132:6817–6821. https://doi.org/10.1021/ja1015348
28. Pinson J, Podvorica F (2005) Attachment of organic layers to conductive or semiconductive surfaces by reduction of diazonium salts. Chem Soc Rev 34:429–439
29. Bélanger D, Pinson J, Belanger D, Pinson J, Bélanger D, Pinson J (2011) Electrografting: a powerful method for surface modification. Chem Soc Rev 40:3995. https://doi.org/10.1039/c0cs00149j

30. Combellas C, Kanoufi F, Pinson J, Podvorica FI (2008) Sterically hindered diazonium salts for the grafting of a monolayer on metals. J Am Chem Soc 130:8576–8577. https://doi.org/10.1021/ja8018912
31. Leroux YR, Fei H, Noël J-M, Roux C, Hapiot P (2010) Efficient covalent modification of a carbon surface: use of a silyl protecting group to form an active monolayer. J Am Chem Soc 132:14039–14041. https://doi.org/10.1021/ja106971x
32. Malmos K, Dong M, Pillai S, Kingshott P, Besenbacher F, Pedersen SU, Daasbjerg K (2009) Using a hydrazone-protected benzenediazonium salt to introduce a near-monolayer of benzaldehyde on glassy carbon surfaces. J Am Chem Soc 131:4928–4936. https://doi.org/10.1021/ja809816x
33. Nguyen VQ, Sun X, Lafolet F, Audibert JF, Miomandre F, Lemercier G, Loiseau F, Lacroix JC (2016) Unprecedented self-organized monolayer of a Ru(II) complex by diazonium electroreduction. J Am Chem Soc 138:9381–9384. https://doi.org/10.1021/jacs.6b04827
34. Fontaine O, Ghilane J, Martin P, Lacroix J-C, Randriamahazaka H Ionic liquid viscosity effects on the functionalization of electrode material through the electroreduction of diazonium. Langmuir 26:18542–18549. https://doi.org/10.1021/la103000u
35. Menanteau T, Levillain E, Breton T (2013) Electrografting via diazonium chemistry: from multilayer to monolayer using radical scavenger. Chem Mater 25:2905–2909
36. Breton T, Downard AJ (2017) Controlling grafting from aryldiazonium salts: a review of methods for the preparation of monolayers. Aust J Chem 70:960–972
37. Ranganathan S, McCreery RL (2001) Electroanalytical performance of carbon films with near-atomic flatness. Anal Chem 73:893–900. https://doi.org/10.1021/ac0007534
38. Yan H, Bergren AJ, McCreery RL (2011) All-carbon molecular tunnel junctions. J Am Chem Soc 133:19168–19177. https://doi.org/10.1021/ja206619a
39. Morteza Najarian A, Szeto B, Tefashe UM, McCreery RL (2016) Robust all-carbon molecular junctions on flexible or semi-transparent substrates using "process-friendly" fabrication. ACS Nano 10:8918–8928. https://doi.org/10.1021/acsnano.6b04900
40. Gupta R, Jash P, Sachan P, Bayat A, Singh V, Mondal PC (2021) Electrochemical potential-driven high-throughput molecular electronic and spintronic devices: from molecules to applications. Angew Chemie Int Ed n/a:. https://doi.org/10.1002/anie.202104724
41. Kong GD, Yoon HJ (2016) Influence of air-oxidation on rectification in thiol-based molecular monolayers. J Electrochem Soc 163:G115–G121. https://doi.org/10.1149/2.0091609jes
42. Kang S, Park S, Kang H, Cho SJ, Song H, Yoon HJ (2019) Tunneling and thermoelectric characteristics of N-heterocyclic carbene-based large-area molecular junctions. Chem Commun 55:8780–8783. https://doi.org/10.1039/C9CC01585J
43. Jeong H, Kim D, Xiang D, Lee T (2017) High-yield functional molecular electronic devices. ACS Nano 11:6511–6548. https://doi.org/10.1021/acsnano.7b02967
44. Anariba F, McCreery RL (2002) Electronic conductance behavior of carbon-based molecular junctions with conjugated structures. J Phys Chem B 106:10355–10362. https://doi.org/10.1021/jp026285e
45. Solak AO, Ranganathan S, Itoh T, McCreery RL (2002) A mechanism for conductance switching in carbon-based molecular electronic junctions. Electrochem Solid-State Lett 5:E43. https://doi.org/10.1149/1.1490716
46. Bergren AJ, McCreery RL, Stoyanov SR, Gusarov S, Kovalenko A (2010) Electronic characteristics and charge transport mechanisms for large area aromatic molecular junctions. J Phys Chem C 114:15806–15815. https://doi.org/10.1021/jp106362q
47. Fave C, Leroux Y, Trippé G, Randriamahazaka H, Noel V, Lacroix J-C, Trippé G, Randriamahazaka H, Noel V, Lacroix J-C (2007) Tunable electrochemical switches based on ultrathin organic films. J Am Chem Soc 129:1890–1891. https://doi.org/10.1021/ja068143u
48. Stockhausen V, Ghilane J, Martin P, Trippe-Allard G, Randriamahazaka H, Lacroix JC (2009) Grafting oligothiophenes on surfaces by diazonium electroreduction: a step toward ultrathin junction with well-defined metal/oligomer interface. J Am Chem Soc 131:14920–14927. https://doi.org/10.1021/ja963354s

49. Stockhausen V, Trippe-Allard G, Nguyen VQ, Ghilane J, Lacroix J-C (2015) Grafting π-conjugated oligomers incorporating 3,4-ethylenedioxythiophene (EDOT) and thiophene units on surfaces by diazonium electroreduction. J Phys Chem C 119:19218–19227. https://doi.org/10.1021/acs.jpcc.5b05456
50. Santos L, Ghilane J, Fave C, Lacaze P-C, Randriamahazaka H, Abrantes L, Lacroix J-C Electrografting polyaniline on carbon through the electroreduction of diazonium salts and the electrochemical polymerization of aniline. J Phys Chem C 112:16103–16109. https://doi.org/10.1021/jp8042818
51. Nguyen L-L, Le Q-H, Pham V-N, Bastide M, Gam-Derouich S, Nguyen V-Q, Lacroix J-C (2021) Confinement effect of plasmon for the fabrication of interconnected AuNPs through the reduction of diazonium salts. Nanomaterials 11.https://doi.org/10.3390/nano11081957
52. Solak AO, Eichorst LR, Clark WJ, McCreery RL (2003) Modified carbon surfaces as "organic electrodes" that exhibit conductance switching. Anal Chem 75:296–305
53. Bastide M, Frath D, Gam-Derouich S, Lacroix J-C (2021) Electrochemical and plasmon-induced grafting of n-dopable π-conjugated oligomers. ChemElectroChem 8:2512–2518. https://doi.org/10.1002/celc.202100563
54. Jousselme B, Bidan G, Billon M, Goyer C, Kervella Y, Guillerez S, Hamad EA, Goze-Bac C, Mevellec J-YY, Lefrant S (2008) One-step electrochemical modification of carbon nanotubes by ruthenium complexes via new diazonium salts. J Electroanal Chem 621:277–285. https://doi.org/10.1016/j.jelechem.2008.01.026
55. Agnès C, Arnault J-C, Omnès F, Jousselme B, Billon M, Bidan G, Mailley P (2009) XPS study of ruthenium tris-bipyridine electrografted from diazonium salt derivative on microcrystalline boron doped diamond. Phys Chem Chem Phys 11:11647–11654. https://doi.org/10.1039/B912468C
56. Fave C, Noel V, Ghilane J, Trippé-Allard G, Randriamahazaka H, Lacroix JC (2008) Electrochemical switches based on ultrathin organic films: from diode-like behavior to charge transfer transparency. J Phys Chem C 112:18638–18643. https://doi.org/10.1021/jp806827a
57. Stockhausen V, Nguyen VQ, Martin P, Lacroix JC (2017) Bottom-up electrochemical fabrication of conjugated ultrathin layers with tailored switchable properties. ACS Appl Mater Interfaces 9:610–617. https://doi.org/10.1021/acsami.6b08754
58. Villemin E, Lemarque B, Vũ TT, Nguyen VQ, Trippé-Allard G, Martin P, Lacaze P-C, Lacroix J-C (2019) Improved adhesion of poly (3, 4-ethylenedioxythiophene)(PEDOT) thin film to solid substrates using electrografted promoters and application to efficient nanoplasmonic devices. Synth Met 248:45–52
59. Bayat A, Lacroix JC, McCreery RL (2016) Control of electronic symmetry and rectification through energy level variations in bilayer molecular junctions. J Am Chem Soc 138:12287–12296. https://doi.org/10.1021/jacs.6b07499
60. James DD, Bayat A, Smith SR, Lacroix J-C, McCreery RL (2018) Nanometric building blocks for robust multifunctional molecular junctions. Nanoscale Horizons 3:45–52
61. Sayed SY, Bayat A, Kondratenko M, Leroux Y, Hapiot P, McCreery RL (2013) Bilayer molecular electronics: all-carbon electronic junctions containing molecular bilayers made with "click" chemistry. J Am Chem Soc 135:12972–12975. https://doi.org/10.1021/ja4065443
62. Ranganathan S, Steidel I, Anariba F, McCreery RL (2001) Covalently bonded organic monolayers on a carbon substrate: a new paradigm for molecular electronics. Nano Lett 1:491–494. https://doi.org/10.1021/nl015566f
63. McCreery R, Dieringer J, Solak AO, Snyder B, Nowak AM, McGovern WR, DuVall S (2003) Molecular rectification and conductance switching in carbon-based molecular junctions by structural rearrangement accompanying electron injection. J Am Chem Soc 125:10748–10758. https://doi.org/10.1021/ja0362196
64. Nowak AM, McCreery RL (2004) In situ raman spectroscopy of bias-induced structural changes in nitroazobenzene molecular electronic junctions. J Am Chem Soc 126:16621–16631. https://doi.org/10.1021/ja045763r
65. McCreery RL, Viswanathan U, Prasad Kalakodimi R, Nowak AM (2006) Carbon/molecule/metal molecular electronic junctions: the importance of "contacts." Faraday Discuss 131:33–43. https://doi.org/10.1039/B505684P

66. Bergren AJ, Harris KD, Deng F, McCreery RL (2008) Molecular electronics using diazonium-derived adlayers on carbon with Cu top contacts: critical analysis of metal oxides and filaments. J Phys Condens Matter 20:374117. https://doi.org/10.1088/0953-8984/20/37/374117
67. Kumar R, Yan H, McCreery RL, Bergren AJ (2011) Electron-beam evaporated silicon as a top contact for molecular electronic device fabrication. Phys Chem Chem Phys 13:14318–14324. https://doi.org/10.1039/C1CP20755E
68. Bonifas AP, McCreery RL (2010) Soft Au, Pt and Cu contacts for molecular junctions through surface-diffusion-mediated deposition. Nat Nanotechnol 5:612–617. https://doi.org/10.1038/nnano.2010.115
69. Ru J, Szeto B, Bonifas A, McCreery RL (2010) Microfabrication and integration of diazonium-based aromatic molecular junctions. ACS Appl Mater Interfaces 2:3693–3701. https://doi.org/10.1021/am100833e
70. Bergren AJ, Zeer-Wanklyn L, Semple M, Pekas N, Szeto B, McCreery RL (2016) Musical molecules: the molecular junction as an active component in audio distortion circuits. J Phys Condens Matter 28:094011. https://doi.org/10.1088/0953-8984/28/9/094011
71. McCreery R, Dieringer J, Solak AO, Snyder B, Nowak AM, McGovern WR, DuVall S (2004) Molecular rectification and conductance switching in carbon-based molecular junctions by structural rearrangement accompanying electron injection. J Am Chem Soc 2003, 125:10748–10758. J Am Chem Soc 126:6200. https://doi.org/10.1021/ja048667z
72. Wu J, Mobley K, McCreery RL (2007) Electronic characteristics of fluorene/TiO_2 molecular heterojunctions. J Chem Phys 126:24704. https://doi.org/10.1063/1.2423011
73. Strachan JP, Pickett MD, Yang JJ, Aloni S, David Kilcoyne AL, Medeiros-Ribeiro G, Stanley Williams R (2010) Direct identification of the conducting channels in a functioning memristive device. Adv Mater 22:3573–3577. https://doi.org/10.1002/adma.201000186
74. Bartholomew RF, Frankl DR (1969) Electrical properties of some titanium oxides. Phys Rev 187:828–833. https://doi.org/10.1103/PhysRev.187.828
75. Walker AV, Tighe TB, Stapleton J, Haynie BC, Upilli S, Allara DL, Winograd N (2004) Interaction of vapor-deposited Ti and Au with molecular wires. Appl Phys Lett 84:4008–4010. https://doi.org/10.1063/1.1748844
76. Blackstock JJ, Donley CL, Stickle WF, Ohlberg DAA, Yang JJ, Stewart DR, Williams RS (2008) Oxide and carbide formation at titanium/organic monolayer interfaces. J Am Chem Soc 130:4041–4047. https://doi.org/10.1021/ja710448e
77. Martin P, Della Rocca ML, Anthore A, Lafarge P, Lacroix JC (2012) Organic electrodes based on grafted oligothiophene units in ultrathin, large-area molecular junctions. J Am Chem Soc 134:154–157. https://doi.org/10.1021/ja209914d
78. Nguyen QV, Martin P, Frath D, Della Rocca ML, Lafolet F, Barraud C, Lafarge P, Mukundan V, James D, McCreery R, Lacroix J-C Control of rectification in molecular junctions: contact effects and molecular signature. J Am Chem Soc 139:11913–11922. https://doi.org/10.1021/jacs.7b05732
79. Rabache V, Chaste J, Petit P, Della Rocca ML, Martin P, Lacroix J-C, McCreery RL, Lafarge P (2013) Direct observation of large quantum interference effect in anthraquinone solid-state junctions. J Am Chem Soc 135:10218–10221. https://doi.org/10.1021/ja403577u
80. Frath D, Nguyen QV, Lafolet F, Martin P, Lacroix J-C (2017) Electrografted monolayer based on a naphthalene diimide–ruthenium terpyridine complex dyad: efficient creation of large-area molecular junctions with high current densities. Chem Commun 53:10997–11000. https://doi.org/10.1039/C7CC04972B
81. Choi SH, Risko C, Carmen Ruiz Delgado M, Kim B, Brédas JL, Daniel Frisbie C (2010) Transition from tunneling to hopping transport in long, conjugated oligo-imine wires connected to metals. J Am Chem Soc 132:4358–4368. https://doi.org/10.1021/ja910547c
82. Sangeeth CSS, Demissie AT, Yuan L, Wang T, Frisbie CD, Nijhuis CA (2016) Comparison of DC and AC transport in 1.5–7.5 nm oligophenylene imine molecular wires across two junction platforms: eutectic Ga–In versus conducting probe atomic force microscope junctions. J Am Chem Soc 138:7305–7314. https://doi.org/10.1021/jacs.6b02039

83. He J, Chen F, Li J, Sankey OF, Terazono Y, Herrero C, Gust D, Moore TA, Moore AL, Lindsay SM (2005) Electronic decay constant of carotenoid polyenes from single-molecule measurements. J Am Chem Soc 127:1384–1385. https://doi.org/10.1021/ja043279i
84. Capozzi B, Dell EJ, Berkelbach TC, Reichman DR, Venkataraman L, Campos LM (2014) Length-dependent conductance of oligothiophenes. J Am Chem Soc 136:10486–10492. https://doi.org/10.1021/ja505277z
85. Sayed SY, Fereiro JA, Yan H, McCreery RL, Bergren AJ, Johan A, Bergren AJ, Johan A (2012) Charge transport in molecular electronic junctions: compression of the molecular tunnel barrier in the strong coupling regime. Proc Natl Acad Sci 109:11498–11503. https://doi.org/10.1073/pnas.1201557109
86. Fluteau T, Bessis C, Barraud C, Della Rocca ML, Martin P, Lacroix J-C, Lafarge P (2014) Tuning the thickness of electrochemically grafted layers in large area molecular junctions. J Appl Phys 116:114509–114514. https://doi.org/10.1063/1.4896106
87. Yan H, Bergren AJ, McCreery R, Della Rocca ML, Martin P, Lafarge P, Lacroix JC (2013) Activationless charge transport across 4.5 to 22 nm in molecular electronic junctions. Proc Natl Acad Sci USA 110:5326–5330. https://doi.org/10.1073/pnas.1221643110
88. Luo L, Choi SH, Frisbie CD (2011) Probing hopping conduction in conjugated molecular wires connected to metal electrodes. Chem Mater 23:631–645. https://doi.org/10.1021/cm102402t
89. Nguyen QV, Martin P, Frath D, Della Rocca ML, Lafolet F, Bellinck S, Lafarge P, Lacroix JC (2018) Highly efficient long-range electron transport in a viologen-based molecular junction. J Am Chem Soc 140:10131–10134. https://doi.org/10.1021/jacs.8b05589
90. Nguyen QV, Lafolet F, Martin P, Lacroix JC (2018) Ultrathin molecular layer junctions based on cyclometalated ruthenium complexes. J Phys Chem C 122:29069–29074. https://doi.org/10.1021/acs.jpcc.8b10766
91. Nguyen QV, Tefashe U, Martin P, Della Rocca ML, Lafolet F, Lafarge P, McCreery RL, Lacroix J-C (2020) Molecular signature and activationless transport in cobalt-terpyridine-based molecular junctions. Adv Electron Mater 6:1901416. https://doi.org/10.1002/aelm.201901416
92. Tefashe UM, Nguyen QV, Najarian AM, Lafolet F, Lacroix J-C, McCreery RL (2018) Orbital control of long-range transport in conjugated and metal-centered molecular electronic junctions. J Phys Chem C 122:29028–29038.https://doi.org/10.1021/acs.jpcc.8b09978
93. Salhani C, Della Rocca ML, Bessis C, Bonnet R, Barraud C, Lafarge P, Chevillot A, Martin P, Lacroix J-C (2017) Inelastic electron tunneling spectroscopy in molecular junctions showing quantum interference. Phys Rev B 95:165431. https://doi.org/10.1103/PhysRevB.95.165431
94. Cabaret T, Fillaud L, Jousselme B, Klein J, Derycke V (2014) Electro-grafted organic memristors: properties and prospects for artificial neural networks based on STDP. In: 14th IEEE international conference on nanotechnology, pp 499–504
95. Lin Y-P, Bennett CH, Cabaret T, Vodenicarevic D, Chabi D, Querlioz D, Jousselme B, Derycke V, Klein J-O (2016) Physical realization of a supervised learning system built with organic memristive synapses. Sci Rep 6:31932
96. Tefashe UM, Nguyen QV, Lafolet F, Lacroix J-C, McCreery RL (2017) Robust bipolar light emission and charge transport in symmetric molecular junctions. J Am Chem Soc 139:7436–7439. https://doi.org/10.1021/jacs.7b02563
97. Tefashe UM, Van Dyck C, Saxena SK, Lacroix J-C, McCreery RL (2019) Unipolar injection and bipolar transport in electroluminescent Ru-centered molecular electronic junctions. J Phys Chem C 123:29162–29172. https://doi.org/10.1021/acs.jpcc.9b10076
98. Najarian AM, McCreery RL (2019) Long-Range activationless photostimulated charge transport in symmetric molecular junctions. ACS Nano 13:867–877. https://doi.org/10.1021/acsnano.8b08662
99. Saxena SK, Smith SR, Supur M, McCreery RL (2019) Light-stimulated charge transport in bilayer molecular junctions for photodetection. Adv Opt Mater 7:1–11. https://doi.org/10.1002/adom.201901053

100. Saxena SK, Tefashe UM, Supur M, McCreery RL (2021) Evaluation of carbon based molecular junctions as practical photosensors. ACS Sensors 6:513–522. https://doi.org/10.1021/acssensors.0c02183
101. Hnid I, Frath D, Lafolet F, Sun X, Lacroix J-C (2020) Highly efficient photoswitch in diarylethene-based molecular junctions. J Am Chem Soc 142:7732–7736. https://doi.org/10.1021/jacs.0c01213
102. Hnid I, Liu M, Frath D, Bellynck S, Lafolet F, Sun X, Lacroix J-C (2021) Unprecedented ON/off ratios in photoactive diarylethene-bisthienylbenzene molecular junctions. Nano Lett 21:7555–7560. https://doi.org/10.1021/acs.nanolett.1c01983
103. Yao X, Sun X, Lafolet F, Lacroix JC (2020) Long-range charge transport in diazonium-based single-molecule junctions´. Nano Lett 20:6899–6907. https://doi.org/10.1021/acs.nanolett.0c03000
104. Yao X, Vonesch M, Combes M, Weiss J, Sun X, Lacroix J-C (2021) Single-molecule junctions with highly improved stability. Nano Lett 21:6540–6548. https://doi.org/10.1021/acs.nanolett.1c01747
105. Peiris CR, Vogel YB, Le Brun AP, Aragonès AC, Coote ML, Díez-Pérez I, Ciampi S, Darwish N (2019) Metal–single-molecule–semiconductor junctions formed by a radical reaction bridging gold and silicon electrodes. J Am Chem Soc 141:14788–14797. https://doi.org/10.1021/jacs.9b07125

Modification of Surfaces with Calix[4]arene Diazonium Salts

Ludovic Troian-Gautier, Alice Mattiuzzi, Pascale Blond, Maurice Retout, Gilles Bruylants, Olivia Reinaud, Corinne Lagrost, and Ivan Jabin

Abstract Since their first report in 2012, calix[4]arene tetradiazonium derivatives have experienced a growing interest. They now represent a favored method to obtain robust post-functionalizable monolayers with controlled composition on a wealth of surfaces (conductive, semi-conductive, or insulating, as well on large surfaces as on nanomaterials). These compounds are easily synthesized and handled and, so far, have been used to functionalize surfaces for applications in (bio)sensing, catalysis, as well as for the development of hydrophobic or antifouling coatings. This chapter describes the current synthetic methods, applications, and limitations of these polydiazonium salts and discusses the potential of the field.

1 Introduction

Calix[n]arenes are oligomeric macrocycles composed of "n" *para*-substituted phenolic units (with $n = 4$, 6, or 8 being the most common) linked in the *ortho* position through methylene bridges [1]. Among the so-called remarkable conformations in solution, [2] the "cone" conformation, with its well-defined small and large rims, represents an attractive candidate for surface functionalization using diazonium

L. Troian-Gautier · P. Blond · I. Jabin (✉)
Laboratoire de Chimie Organique, Service de Chimie et PhysicoChimie Organiques, Université libre de Bruxelles (ULB), Avenue F.D. Roosevelt 50, CP160/06, 1050 Brussels, Belgium
e-mail: Ivan.Jabin@ulb.be

A. Mattiuzzi
X4C, 128 Rue du chêne Bonnet, 6110 Montigny-le-Tilleul, Belgium

M. Retout · G. Bruylants
Engineering of Molecular NanoSystems, Ecole Polytechnique de Bruxelles, Université libre de Bruxelles (ULB), Avenue F. D. Roosevelt 50, CP165/64, B-1050 Brussels, Belgium

O. Reinaud
Université Paris Cité, CNRS,LCBPT, Paris, France

C. Lagrost
Université de Rennes, CNRS, ISCR-UMR 6226, F-35000 Rennes, France

M. M. Chehimi et al. (eds.), *Aryl Diazonium Salts and Related Compounds*, Physical Chemistry in Action, https://doi.org/10.1007/978-3-031-04398-7_13

chemistry. Especially, calix[4]arenes offer a rigid cone-constrained structure with a small rim composed of phenolic units which are particularly suited for the easy introduction of functional groups whereas the large rim can be selectively functionalized to introduce diazonium groups. Surface functionalization with molecular macrocycles including calixarenes, cyclodextrins, and cucurbiturils has already been reported notably from self-assembly procedures [3, 4] in order to develop sensors where the molecular cavities defined by the structure of the macrocycles were employed for trapping analytes. Yet, the preparation of functional surfaces with well-ordered and robust structures, an aspect that has hardly been realized in the field, is of great importance in many applications including (bio-)sensors, molecular electronics, and precision engineering. Patterning surfaces with organic thin films through reductive diazonium grafting constitute an attractive approach in this context because of the high robustness of the resulting interface and the large choice of materials that can be modified, both in terms of nature and forms. While many approaches are limited to a few types of materials, diazonium can be grafted on a very wide range of materials: metals (even oxidizable ones), carbon under various allotropic forms, polymers, semi-conductors, oxides, glasses, etc. [5, 6].

However, in some applications, nanostructured ultrathin films are required, and the formation of dense and organized monolayers or ultrathin films are particularly difficult to control through diazonium grafting. Indeed, the grafting process involves highly reactive radicals that are able to bind various substrates, but that are also able to react with already grafted organic moieties, leading to loosely packed aryl multilayers of 10–15 nm thickness. Several strategies have been proposed to produce well-ordered monolayers as recently reviewed [7, 8]. The use of calix[4]arenes represents an original approach to reach a structuring of the interface at the molecular scale [9–11]. The methylene bridges at the ortho position of the phenoxy substituents efficiently prevent the radical reaction responsible for the formation of multilayers. The possible functionalization of the small rim allows the introduction of various chemical function with a spatial control imposed by the geometry of the rim. The spatial arrangement of functional groups at the interface has been less considered than the control of monolayers formation and robustness of the interface but may be crucial in some applications. The rigid structure of calix[4]arene allows very good lateral and spatial controls of interfacial functionality [12]. Another key point is the cone-constrained structure that orients all the diazonium functions at the large rim in the same direction. This specific design is expected to enhance the stability of the monolayers thanks to multiple anchoring points, as evidenced with the self-assembly of multipodal ligands [13]. In this chapter, we describe the synthesis of calix[4]arenes-tetradiazonium salts and their use as highly versatile platforms for surface modification able to bring robust and precise control of functionality of many materials including nanomaterials. The interest of the strategy is finally illustrated by a few examples of applications involving the calix[4]arene-based coatings.

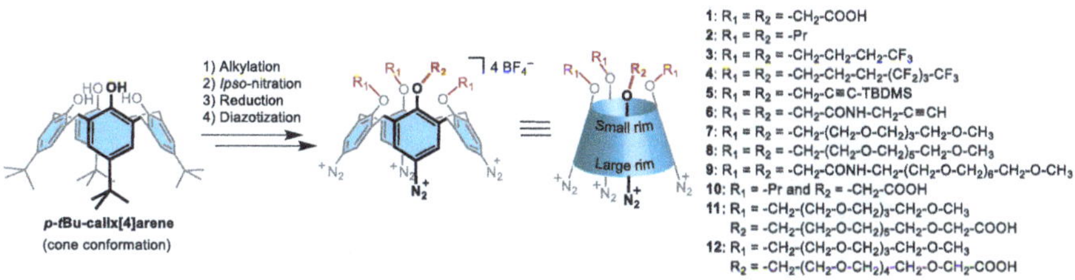

Fig. 1 Structures of calix[4]arene tetradiazonium derivatives **1–12** [9, 10, 14, 16–21]

2 Synthesis of Calix[4]arene Tetradiazonium Derivatives

Most of the calix[4]arene tetradiazonium derivatives have been synthesized through a common route that involves (i) an alkylation of the phenol units of *p*-*t*Bu-calix[4]arene, (ii) an *ipso*-nitration reaction, (iii) a reduction of the resulting nitro groups into the corresponding amines, and finally, (iv) a diazotization reaction. The synthesis of a calix[4]arene bearing a single diazonium group was also reported according to a slightly different strategy [14]. Selective introduction of different functional groups on the small rim can be readily achieved through strategies that are well-known in the field of calix[4]arene chemistry [15]. It is important to mention that substituents larger than an ethyl group have to be introduced on the phenol positions in order to prevent "*through the anulus*" rotation of the aromatic units and thus lock the calix[4]arene in the required cone conformation. The tetradiazonium salts that have been reported so far (*i.e.*, **1–12**) are represented in Fig. 1. All these compounds were easily characterized by NMR and IR spectroscopy and stored in the fridge for months without any noticeable degradation.

3 Surface Modification and Characterization

Surface modification with calix[4]arene derivatives was achieved on conductive (glassy carbon, pyrolyzed photoresist films (PPF), gold), semi-conductive (germanium), and insulating (glass, polypropylene, polyethylene terephthalate, polystyrene) surfaces, as well as on metallic nanoparticles (Fig. 2) [9, 10, 14, 16–21]. The modified surfaces were characterized by several techniques including electrochemistry, atomic force microscopy (AFM), X-ray photoelectron spectroscopy (XPS), ellipsometry, or infrared spectroscopy [9, 10, 16, 18–21]. In all cases, a dense, robust, and homogeneous monolayer of calix[4]arenes was observed.

In the case of conductive surfaces, electrochemical reduction of the calixarene tetradiazonium derivatives was achieved either through cyclic voltammetry, where the potential is cycled through a range of potentials that include the reduction potential of the diazonium moiety, or through chronoamperometry, where a potential more

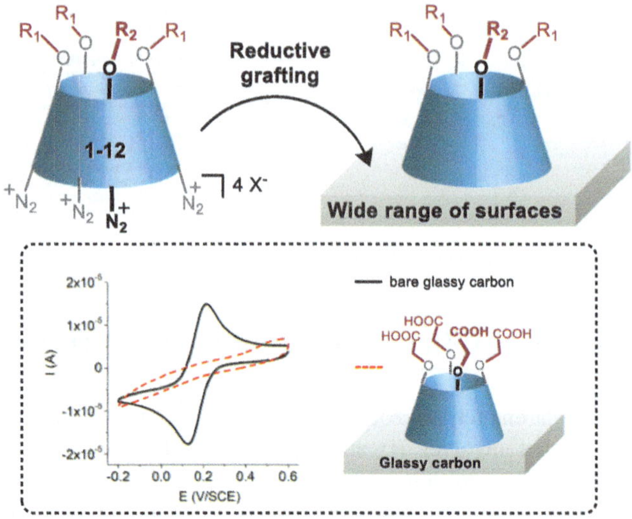

Fig. 2 Grafting of calixarene tetradiazonium derivatives **1–12** on surfaces. The inset shows the cyclic voltammetry of $[Fe(CN)_6]^{3-}$ (1 mM) on bare (black) and modified (red) glassy carbon in aqueous 0.5 M KPF_6 recorded at 100 mV s^{-1}

negative than the reduction potential of the diazonium moiety is applied. The electrochemical reduction leads to the formation of aryl radicals at the vicinity of the electrode, resulting in an efficient grafting at the surface of the conductive electrode. Surface modification can be assessed through several methods, but most commonly, electrochemistry with redox probes such as dopamine or potassium ferricyanide is used [22]. Indeed, the thin layer of calix[4]arenes, if homogeneous, creates a barrier that significantly slows down the electron transfer kinetics at the interface. Experimentally, the efficient grafting is evidenced by a drastic decrease of current intensity along with an increased peak-to-peak separation between redox peaks (Fig. 2).

Reducing agents such as $NaBH_4$ or sodium ascorbate were used for the grafting of calixarene tetradiazonium derivatives on nanoparticles. For example, $NaBH_4$ was used to obtain calix[4]arene-coated gold nanoparticles (AuNPs-calix), either formed in situ from (H/K)$AuCl_4$ or through ligand exchange of citrate-capped AuNPs [20, 21, 23]. Another chemical approach relies on the formation of intermediate diazoates through the reaction between diazonium derivatives and sodium hydroxide. This was used to covalently immobilize calix[4]arenes **1** and **3** onto gold, polypropylene, polyethylene terephthalate, and polystyrene materials [19]. Similar procedures were used to functionalize glass, gold, and polypropylene surfaces with calix[4]arene **4** [24] or germanium surfaces with calix[4]arenes **7**, **8**, and **11** [16]. In the specific case of calix[4]arene **4**, a mixture of CH_3CN/aq. NaOH (0.1 M) was used as it maximized solubility of the polyfluorinated calix[4]arene **4**, while still allowing the generation of diazoates by reaction with sodium hydroxide.

3.1 Formation of True Monolayers

The formation of true monolayers may be crucial for some applications such as (bio)sensing or electrocatalysis. Both of these applications require fast (electronic) communication between the sensitive layer and the analyte or compound to activate. Typical aryl diazonium derivatives, unless specific approaches are used, [7, 8] form multilayers by reaction of the corresponding reactive radicals in solution with already grafted aryl groups. In the case of calix[4]arene tetradiazonium salts, the methylene bridges linking the different phenol units efficiently prevent further side reactions with the macrocyclic scaffold, once these are grafted on a surface. The formation of true monolayers is thus obtained. Ellipsometry and AFM scratching experiments were used to confirm the formation of monolayers of calix[4]arenes **1**, **3**, **4**, **7**, and **8** on gold, germanium, PPF, and polypropylene surfaces [10, 16, 19]. As a representative example, a thickness of 1.3 $\pm$ 0.1 nm (AFM) and 1.09 $\pm$ 0.2 nm (ellipsometry) was measured for a layer of calix[4]arene **3** on gold surfaces, which is in close agreement with the theoretical value of 1.1 nm estimated from MM2 energy minimizations [10].

3.2 Formation of Mixed Monolayers of Controlled Composition

Functionalizing surfaces with different organic molecules in controlled ratios represents a promising, yet challenging endeavor. Control over the surface chemical composition can, for example, have a drastic impact on the sensitivity of a biosensor. Indeed, diluting the component of recognition can increase sensitivity by improving accessibility of the analyte [25, 26]. In addition, a combination of the properties of the individual organic molecules is expected to be transferred on the surfaces when mixed layers are formed. Chemisorption of organic molecules in mixed ratios, such as thiols, is a possible approach, but it suffers from several limitations: (i) surface coverage may differ from solution composition due to different adsorption kinetics [27–29] and (ii) inhomogeneity at the surface may result from segregation upon self-assembly [30, 31]. Thoroughly investigated on surfaces, this difficulty to control the layer composition has also been demonstrated on nanomaterials [32]. The formation of mixed layers of aryl diazonium derivatives in controlled ratios is also challenging. Indeed, the most easily reduced diazonium derivative is usually found in a larger ratio on the surface than in solution [33, 34].

The formation of a mixed monolayer with a controlled composition represents another breakthrough, which was made possible through the use of calix[4]arene tetradiazonium derivatives. Indeed, the common macrocyclic scaffold roughly insulates the diazonium groups from the rest of the molecule and, as a result, the reduction potentials of all the calix[4]arene tetradiazonium salts are quite close. As a representative example, mixtures of calix[4]arenes **1** and **3** (in ratios of 100/0, 50/50, 10/90, and 0/100) were grafted using chronoamperometry at a applied potential of –0.5 V versus SCE for 5 min [18]. Control over the composition of the resulting mixed layers was then evidenced through determination of the static contact angle. The angle gradually increased from 68 ± 3° to 89 ± 2°, as the proportion of non-polar calix[4]arene **3** used in the grafting mixture increased. The formation of mixed monolayers was also achieved using $NaBH_4$ on gold nanoparticles with calix[4]arenes **1** and **9** in ratios 0/100, 5/95, 10/90, 25/75, 50/50, and 100/0 [21]. Attenuated total reflectance Fourier transform infrared (ATR-FTIR) spectroscopy confirmed that the solution ratio was effectively transferred onto the surface, through the analysis of characteristic IR absorbance bands (asymmetric COC stretching at 1105 cm^{-1} as well as amide-I at 1669 cm^{-1} and amide-II at 1540 cm^{-1} for calix[4]arene **9** and asymmetric (1604 cm^{-1}) and symmetric (1420 cm^{-1}) COO^- stretching for calix[4]arene **1**).

3.3 Post-functionalization of Calixarene-Based Monolayers

A particular benefit of using calix[4]arene derivatives is that the small rim can be substituted with groups particularly suited for post-functionalization (*i.e.,* functionalization of the calixarenes once grafted), such as carboxyl, azide, and alkyne [9, 10, 17–21]. Therefore, the covalent immobilization of various (bio)molecules can be achieved under classical conjugation reaction conditions. For example, amino-containing compounds were anchored to grafted calixarenes bearing carboxyl groups (*i.e.,* **1**, **11,** and **12**) via classical peptide-type coupling reactions. This strategy was used to conjugate small amines or alcohols, [19, 20] as well as larger peptides or proteins [35, 36]. Besides, click chemistry allowed the conjugation of azido-containing molecules on calixarene-based layers bearing alkynes groups [17].

3.4 Robustness of the Calixarene-Based Coatings

Calix[4]arene tetradiazonium salts display four anchoring points and can thus potentially form up to four links with the surface. This leads to a remarkable stability of the calixarene-based coating that outperform that of other classical organic coatings (SAMs of thiols, aryldiazoniums, etc.). This stability is notably assessed by the aggressive post-grafting workup that includes several washing cycles under sonication in various solvents (e.g., water, EtOH, DCM, ACN, THF, 0.1 M HCl, toluene, etc.). In the specific case of germanium surfaces modified by calix[4]arenes **7** and **8**, a continuous 10 μL/min flow of PBS-D_2O buffer over 16 h was used. The IR spectra recorded at intervals during these 16 h were superimposable, confirming the robustness of the grafted layer. The increased stability is particularly striking on nanomaterials [20, 21]. Indeed, typical gold nanoparticles stabilized by citrate readily degrade or aggregate due to changes in pH, ionic strength, or in the presence of large concentration of fluoride. In contrast, calixarene-coated gold nanoparticles

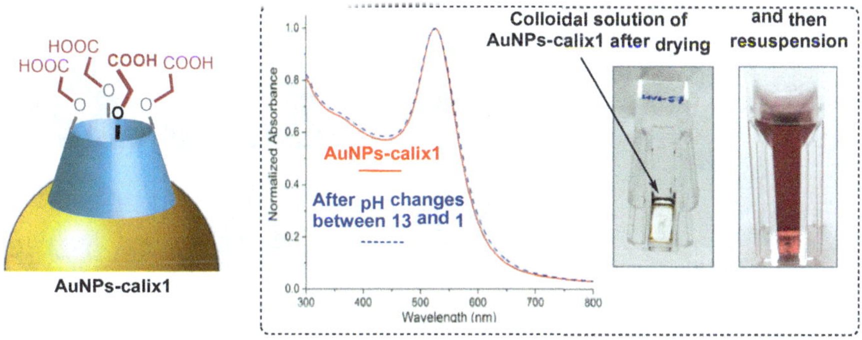

Fig. 3 Schematic representation of AuNPs coated with calix[4]arene **1** (AuNPs-calix1). Inset: normalized UV-visible absorption spectra of AuNPs-calix1 after several pH changes between 13 and 1 (left). Dried AuNPs-calix1 and resuspended in 0.1 M aqueous NaOH (right). Reproduced with permission from reference [20]. Copyright (2016) Royal Society of Chemistry

such as AuNPs-calix**1** remained stable upon extreme pH variations, in the presence of large concentrations of fluoride ions (between 0.15 and 0.75 M) or when the ionic strength was increased (Fig. 3) [20, 21]. Even more remarkably, AuNPs-calix**1** could be completely dried, yielding a gold-colored film, and then resuspended into a 0.1 M aqueous NaOH solution. This unique stability conferred by calixarene-based coatings was exploited for the synthesis of silver and alloyed gold-silver nanoparticles [23]. These nanomaterials are particularly attracting in the biosensing field due to the enhanced optical properties of silver in comparison with gold.

4 Applications of Calix[4]arene Tetradiazonium-Based Coatings

The use of calix[4]arene tetradiazonium derivatives has already found applications in the fields of dewetting, [24] antifouling, [16] (bio)sensing, [16, 17] and electrocatalysis [37].

4.1 Hydrophobicity

Exerting control over wettability [38–40] is a highly relevant endeavor for the development of coatings for anti-frosting and anti-fogging, [41, 42] self-cleaning, [43] anticorrosion, [44, 45] or antibiofouling surfaces [16, 46]. The wettability is assessed by static contact angle measurements. Glass, gold, and polypropylene surfaces were covalently modified with a polyfluorinated calix[4]arene tetradiazonium salt (**4**) (Fig. 4) [24]. A chemical grafting was performed using CH_3CN/0.1 M NaOH aqueous

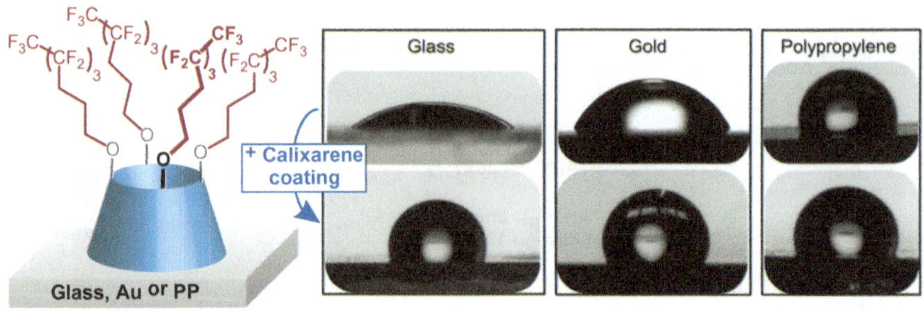

Fig. 4 Left: schematic representation of the surfaces modified with calixarene **4**. Images of 2 μL water droplets in contact with (top) bare glass, gold, and polypropylene surfaces and with (bottom) the same surfaces freshly modified with calix[4]arene **4**. Reproduced with permission from reference [24]. Copyright (2020) Royal Society of Chemistry

solution. For gold surfaces, the grafting of **4** was also performed electrochemically for comparisons purposes. The modified surfaces displayed significant changes in static contact angles that evolved from $24.6 \pm 2.0°$, $64.7 \pm 2.1°$ and $102.9 \pm 3.9°$ to $110.0 \pm 1.8°$, $113.7 \pm 2.2°$ and $112.6 \pm 4.0°$ for glass, gold, and polypropylene surfaces, respectively. These values are not far from the maximum value of 120° that has been estimated for smooth surfaces. Importantly, the coatings remained stable for at least 18 months, as the static contact angles for the aged surfaces were within 5% of the values obtained for freshly modified ones. Such a result is remarkable regarding ultrathin coatings.

4.2 Antifouling

Germanium prisms displaying antifouling properties were developed with the aim of designing ATR-FTIR biosensors (vide supra). For this, calix[4]arene tetradiazonium salts substituted with oligo-(ethylene glycol) (oEGs) chains (*i.e.*, **7** and **8**) were used, as oEGs are known to be nontoxic and non-immunogenic, as well as to prevent the nonspecific adsorption of biomolecules. The efficient grafting was confirmed by AFM and ATR-FTIR measurements, where the typical asymmetric COC stretching (1100 cm^{-1}) from the oEG chains, together with the symmetric COCAr stretching (from 1050 to 1020 cm^{-1}) and the aromatic ring stretching (1460 cm^{-1}) were observed. The nonspecific adsorption of bovine serum albumin (BSA) was then investigated in phosphate buffer media in D_2O (PBS-D_2O) at 22 °C. With an unmodified germanium prism, the nonspecific adsorption of BSA was clearly observed through increased absorbance signals corresponding to the amide-I', C = O(ND) stretching vibrations, and amide-II' in plane ND bending vibrations measured at 1640 cm^{-1} and 1450 cm^{-1}, respectively. Astoundingly, the nonspecific adsorption of BSA was prevented by more than 85% with the calix[4]arene-coated germanium prism.

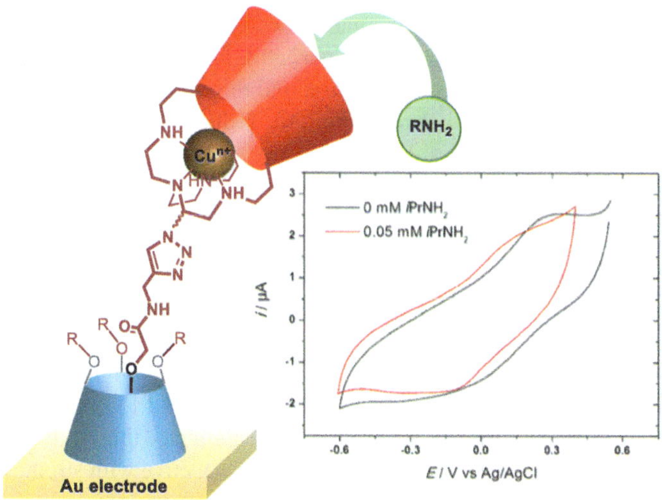

Fig. 5 Selective electrochemical sensing of primary amines on gold electrodes using a calix[6]arene $Cu^{II/I}$ funnel complex. Reproduced with permission from ref [17]. Copyright (2016) American Chemical Society

4.3 *Sensing*

4.3.1 Electrochemical Sensing in Water

Gold electrodes were covalently modified with calix[4]arene **1** and a post-functionalization reaction with propargylamine allowed the introduction of alkyne groups on the surface [17]. A calix[6]arene-based copper *funnel* complex was then conjugated through an electro-click reaction. Funnel complexes are efficient receptors for neutral molecules and mimic the active site of metallo-enzymes by exhibiting a hydrophobic cavity associated to a confined metal center [47]. The resulting sensor was then used for the selective electrochemical detection of primary amines at μM concentrations in aqueous and organic solutions, where the electrochemical signal of the $Cu^{II/I}$ couple was shown to shift by 100 mV upon binding the amine derivative (Fig. 5). Interestingly, the sensor was selective toward primary linear alkylamines that can access to the metal center through the calix[6]arene cavity.

4.3.2 FTIR Detection of Proteins in Biological Media

Detection of proteins in complex media can be performed by Fourier transform infrared (FTIR) spectroscopy [48]. One advantage of this technique is that a multitude of biophysical and chemical information can be obtained from the IR signature of proteins (*i.e.*, identification of their secondary and tertiary structures, post-translational modifications, etc.) [49]. ATR-FTIR-based biosensors make use of an organic layer, directly grafted onto the internal reflection element, to which a biological receptor is bound [50]. Germanium is particularly suitable for these applications. In this context, the grafting of stable, dense, and thin organic layers on germanium surfaces was recently accomplished through the reductive grafting of calix[4]arene tetradiazonium salts [16]. Calix[4]arene **11** was specifically designed to decrease nonspecific adsorption thanks to oEG chains while the introduction of a carboxyl group allows for bioconjugation to a recognition unit [35]. Germanium surfaces modified with calix[4]arene **11** were converted to the corresponding activated esters using EDC/NHS prior to being conjugated to a biotin derivative bearing a terminal amino group (Fig. 6). The resulting calix[4]arene-biotin-based germanium biosensors were then used to detect streptavidin (SA) (100 μg/mL) from a complex medium by ATR-FTIR spectroscopy. Bound SA was identified through the characteristic amide-I and amide-II IR absorption bands at 1637 cm^{-1} and 1535 cm^{-1}, respectively (Fig. 6a). The location of these IR absorption bands indicated that SA was bound in its native β-sheet conformation. Recognition was also effective with streptavidin in the presence of bovine serum albumin (BSA) (100 μg/mL). Very interestingly, following incubation, BSA is not detected in the IR spectra, which highlights the remarkable antifouling properties of the calixarene-coated germanium surfaces. Similar strategies were also developed using fluorescent streptavidin-ATTO655, for which the specific interaction with biotin-spotted germanium surface generated a fluorescent microarray corresponding to the immobilized biotin pattern (Fig. 6b).

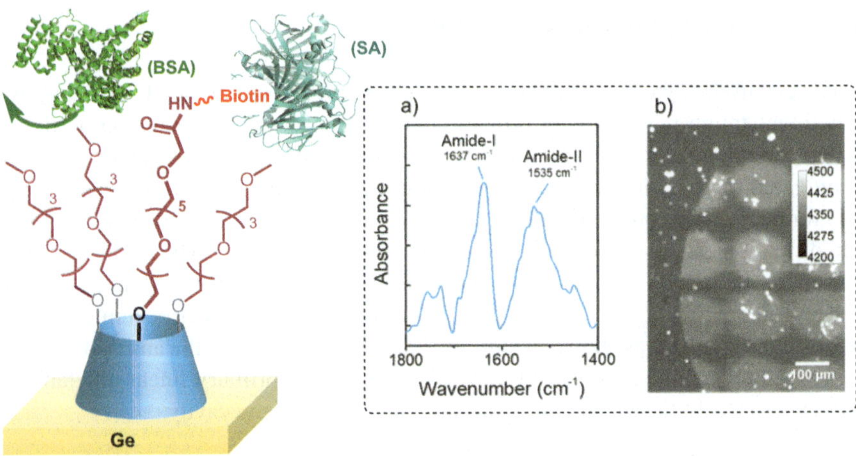

Fig. 6 FTIR-based selective biosensor for streptavidin (SA) composed of a calix[4]arene-biotin-based germanium surface. Inset: **a** ATR-FTIR adsorption spectra obtained after incubation in a solution of SA; **b** fluorescence image acquired on Ge-calix[4]arene-biotin-spotted surface that was incubated with fluorescent streptavidin-ATTO655. The microarray's interstices, where incubation was carried out but biotin is not immobilized, served as a control area to evaluate the SA nonspecific adsorption, which was extremely limited according to the very weak fluorescence intensity in these regions. Reproduced with permission from ref [35]. Copyright (2020) American Chemical Society

Altogether, these results indicated that surface modification with a calix[4]arene-based platform allowed immobilization of receptors that retain their recognition properties and could selectively detect a protein in a medium containing other proteins such as BSA. Additional strong points of this strategy are (i) the possibility to identify the secondary structure, (ii) the small quantity of material necessary for the analysis (1 μL of solution, 100 μg/mL), and (iii) the analysis time (1 min by IR spectra).

4.3.3 Calix[4]arene-Coated Gold Nanoparticles for Sensing of MDM2 in Biological Media

Gold nanoparticles are often employed as colorimetric indicators for in vitro diagnostic systems due to their optical properties. These NP have to be functionalizable, in order to allow coupling of specific recognition ligands, and their colloidal suspension must be stable in the operating medium (*i.e.*, serum). The grafting on AuNPs is mostly relying on the well-known sulfur–gold interaction. However, control over the bioconjugation density represents a major challenge with thiol chemistry [32]. In this context, controlled bioconjugation of peptide aptamers and sensing of a cancer biomarker, *i.e.*, the oncoprotein Mdm2, were recently obtained using calix[4]arene-coated AuNPs [36]. Firstly, AuNPs coated with a dense and stable oEG layer displaying carboxyl groups were readily obtained by using mixtures of calixarenes **7** and **12**. The resulting AuNPs-calix were shown to be more resistant to physical stress, such as temperature increase or drying, than related thiolated AuNPs. The high stability of the AuNPs-calix enabled their dispersion in human serum without any aggregation or degradation. The carboxyl terminal groups of the calixarene-based layer were used to conjugate amino-containing molecules via the formation of amide bonds. Control over the bioconjugation density was demonstrated by the conjugation of a cyanine7.5 dye to different batches of AuNPs prepared with different mixtures of calixarenes **7** and **12**. As shown by UV–Vis spectroscopy, the density of the conjugated dye was proportional to the amount of calix[4]arene **12** used during the modification of the AuNPs, showing that the calixarene-based strategy is suitable for controlling the conjugation density of (bio)molecules on AuNPs. This approach was extended to the conjugation of peptides aptamers. It is remarkable that a positively charged aptamer could be grafted on the particles with a defined density without loss of colloidal stability, which was proven impossible with the classical thiol chemistry on citrate-capped AuNPs [51]. Once modified, the resulting AuNPs were used for the detection of Mdm2 via a dual-trapping strategy (Fig. 7a), leading to an aggregation of the AuNPs which was then quantified by UV–Vis spectroscopy. The aggregation intensity was shown to be proportional to the concentration of Mdm2 present at the AuNP interface (Fig. 7b). This confirmed that biomolecules such as peptide aptamers can be covalently conjugated on the oEG shell of calixarenes-modified AuNPs in a controlled way, without any loss of their recognition abilities.

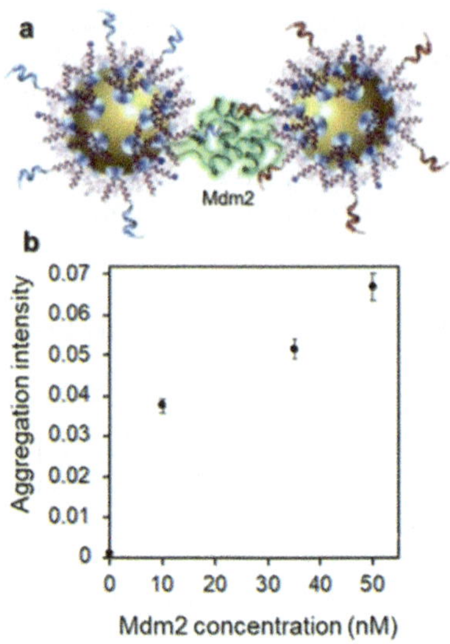

Fig. 7 **a** Illustration of the dual-trapping of Mdm2 by calixarene-coated AuNPs. **b** Aggregation intensity of calixarene-coated AuNPs as a function of Mdm2 concentration. Reproduced with permission from ref [36]. Copyright (2021) American Chemical Society

4.4 Electrocatalysis

Oxygen reduction reaction (ORR) is one of the most important reactions for technologies related to energy conversion. However, it is a complex, multi-electron process, exhibiting sluggish kinetics and, without a suitable catalyst, ORR occurs at high overpotentials. The direct, 4-electron reduction, which produces water, is generally the most desired route, while the 2-electron ORR pathway results in the formation of hydrogen peroxide. Additionally, the catalysts can also strongly influence product selectivity by favoring specific reaction pathways. As efficient catalysts, noble metal nanomaterials have been extensively developed with variations of their size, shape, and topmost layer composition in order to enhance their efficiency, durability, and selectivity. More recently, the deliberate surface modification of nanocatalysts with organic ligands has emerged as a promising strategy to increase the electrocatalytic performance [52, 53]. AuNPs have been shown to act as catalysts for the ORR in alkaline media, mainly through an indirect ($2e^- + 2e^-$) pathway [54]. In this context, gold nanoparticles of 6 nm diameter coated with a monolayer of covalently bound calix[4]arene **1** (AuNPs-**1**) showed enhanced selectivity and stability compared to AuNPs classically stabilized by citrate (AuNPs-citrate) [37]. Rotating-ring disk electrode (RRDE) measurements allowed the monitoring of the number of electrons exchanged (n) together with the proportion of hydrogen peroxide formed during the process (%H_2O_2). While AuNPs-citrate displayed the expected $2e^- + 2e^-$ catalytic mechanism, the calix[4]arene-gold nanohybrids showed a constant value of n (3.9) and very low level of %H_2O_2 (% HO_2- <5%) over a large range of potentials (0.8 to –0.4 V versus RHE) (Fig. 8). AuNPs-**1** clearly exhibit an enhanced selectivity toward the 4-electron pathway, close to what is currently obtained with the archetypal Pt

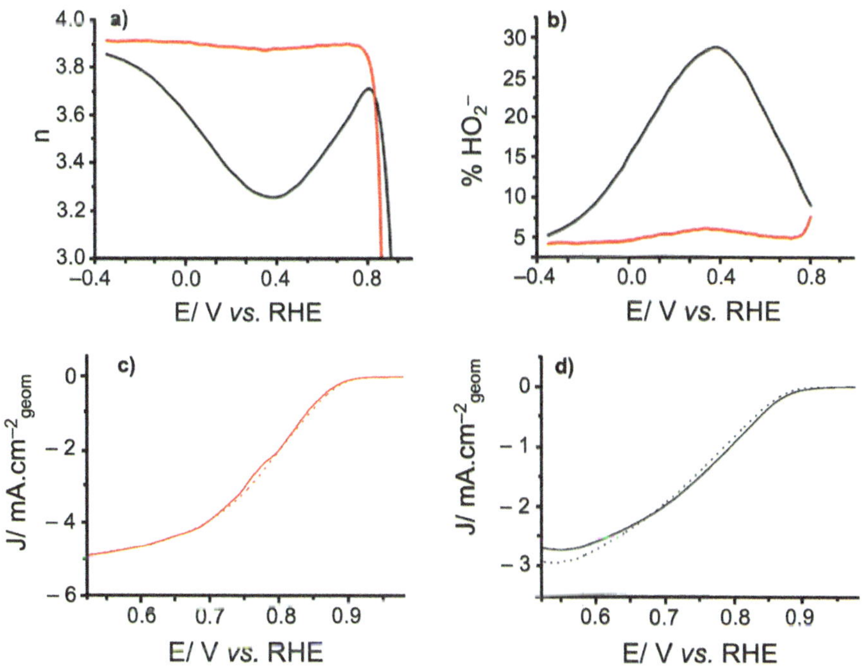

Fig. 8 RRDE measurements showing **a** the variation of n (number of exchanged electrons) and **b** %$HO_2^=$ (proportion of H_2O_2 produced) during ORR process for AuNPs-calix1 (red line) and AuNPs-citrate (black line). LSV curves before (dotted lines) and after (solid lines) the accelerated durability tests for **c** AuNPs-calix1 and **d** AuNPs-citrate. Reproduced with permission for ref [37]. Copyright (2021) John Wiley and Sons

ORR nanocatalysts but in acidic media. The kinetic current density and onset potential (~0.9 V versus RHE) compared well with reported specific activities of noble metal electrocatalysts, both bare and chemically modified [37]. Interestingly, the Tafel plots at low current density (close to the onset potential) were found to have a slope of 42 mV/decade, which is close to the value reported for bare 5 nm AuNPs, but substantially lower than the 60 mV/decade usually observed for metallic catalysts. This value indicates superior activity of the hybrid nanocatalysts [55]. Accelerated durability tests highlighted the robustness of AuNPs-**1** under operation. No variation of the linear sweep voltammograms (LSV) were observed after 1000 cycles in an aqueous 0.1 M KOH O_2-saturated solution. Additionally, X-ray photoelectron spectroscopy analyses further confirmed the robustness of the nanohybrid catalysts as the chemical footprint of the catalytic system remained intact in contrast to AuNPs-citrate that underwent chemical degradation. This strategy is expected to be further developed by modifying the small rim functionalities and the nature of the metallic core.

5 Conclusions and Future Perspectives

This chapter describes the use of calix[4]arene tetradiazonium derivatives constrained in a cone conformation as rigid 4-feeted platforms that present pre-/post-functionalizable arms. The synthesis of the platform is relatively straightforward and scalable. One side of the truncated cone (large rim of the calix[4]arene) is easily functionalized by amino groups that are ideal precursors for diazonium grafting, whereas the other side (small rim) can be pre- or post-functionalized almost at will to introduce specific functions. The large rim is used for the up-to-4-point grafting through the diazonium strategy, which yields the formation of strong chemical bonds with the surface. The molecular structure of the platform prevents the pitfalls of oligomerization during the grafting process that is usually observed with standard aryl diazonium derivatives. This guarantees the formation of clean, compact, robust, dense monolayers on a great variety of materials. The optimal diazonium grafting procedure depends on the material and a variety of procedures have thus been developed for metal, polymers, glass surfaces of different size and shape. One of the most spectacular achievement is the formation of highly stable gold nanoparticles that do not irreversibly coalesce upon drying or with changes in pH or ionic strength. In addition, mixed monolayers of controlled composition are readily obtained when a mixture of two different calix[4]arene platforms with different arm functionalities for the grafting process is used.

Several applications benefiting from these calix[4]arene monolayers have already been reported. These include (i) the development of hydrophobic surfaces with perfluorinated groups grafted at the small rim, (ii) the development of germanium surfaces for protein sensing and antifouling, (iii) the electrochemical sensing or primary amines in water, (iv) the recognition of MDM2 in biological media using a gold nanoparticle-based colorimetric sensor, and (v) the controlled electrocatalysis of oxygen reduction.

Given their huge potential in the field of surface modification, we believe that calix[4]arene tetradiazonium salts will find many applications that could be exploited at the industrial level.

References

1. Gutsche CD (1998) Calixarenes revisited, monographs in supramolecular chemistry. The Royal Society of Chemistry, Cambridge
2. Ikeda A, Shinkai S (1997) Chem Rev 97:1713–1734. https://doi.org/10.1021/cr960385x
3. Li H, Yang Y-W (2013) Chin Chem Lett 24:545–552. https://doi.org/10.1016/j.cclet.2013.04.014
4. Montes-García V, Pérez-Juste J, Pastoriza-Santos I, Liz-Marzán LM (2014) Chem Eur J 20:10874–10883. https://doi.org/10.1002/chem.201403107
5. Bélanger D, Pinson J (2011) Chem Soc Rev 40:3995–4048. https://doi.org/10.1039/C0CS00149J

6. Berisha A, Chehimi MM, Pinson J, Podvorica F (2016) Electrode surface modification using diazonium salts. In: Bard AJ, Zoski CG (eds) Electroanalytical chemistry, a series of advances. CRC Press, Taylor & Francis group, Boca Raton
7. Breton T, Downard AJ (2017) Aust J Chem 70:960–972. https://doi.org/10.1071/CH17262
8. Mattiuzzi A, Lenne Q, Carvalho Padilha J, Troian-Gautier L, Leroux YR, Jabin I, Lagrost C (2020) Frontiers Chem 8. https://doi.org/10.3389/fchem.2020.00559
9. Buttress JP, Day DP, Courtney JM, Lawrence EJ, Hughes DL, Blagg RJ, Crossley A, Matthews SE, Redshaw C, Bulman Page PC, Wildgoose GG (2016) Langmuir 32:7806–7813. https://doi.org/10.1021/acs.langmuir.6b02222
10. Mattiuzzi A, Jabin I, Mangeney C, Roux C, Reinaud O, Santos L, Bergamini J-F, Hapiot P, Lagrost C (2012) Nat Commun 3:1130. https://doi.org/10.1038/ncomms2121
11. Troian-Gautier L, Mattiuzzi A, Reinaud O, Lagrost C, Jabin I (2020) Org Biomol Chem 18:3624–3637. https://doi.org/10.1039/D0OB00070A
12. Valášek M, Mayor M (2017) Chem Eur J 23:13538–13548. https://doi.org/10.1002/chem.201703349
13. Li Z-Q, Tang J-H, Zhong Y-W (2019) Chem Asian J 14:3119–3126. https://doi.org/10.1002/asia.201900989
14. Malytskyi V, Troian-Gautier L, Mattiuzzi A, Lambotte S, Cornelio B, Lagrost C, Jabin I (2018) Eur J Org Chem 2018:6590–6595. https://doi.org/10.1002/ejoc.201801253
15. Lavendomme R, Zahim S, De Leener G, Inthasot A, Mattiuzzi A, Luhmer M, Reinaud O, Jabin I (2015) Asian J Organ Chem 4:710–722. https://doi.org/10.1002/ajoc.201500178
16. Blond P, Mattiuzzi A, Valkenier H, Troian-Gautier L, Bergamini J-F, Doneux T, Goormaghtigh E, Raussens V, Jabin I (2018) Langmuir 34:6021–6027. https://doi.org/10.1021/acs.langmuir.8b00464
17. De Leener G, Evoung-Evoung F, Lascaux A, Mertens J, Porras-Gutierrez AG, Le Poul N, Lagrost C, Over D, Leroux YR, Reniers F, Hapiot P, Le Mest Y, Jabin I, Reinaud O (2016) J Am Chem Soc 138:12841–12853. https://doi.org/10.1021/jacs.6b05317
18. Santos L, Mattiuzzi A, Jabin I, Vandencasteele N, Reniers F, Reinaud O, Hapiot P, Lhenry S, Leroux Y, Lagrost C (2014) J Phys Chem C 118:15919–15928. https://doi.org/10.1021/jp5052003
19. Troian-Gautier L, Martínez-Tong DE, Hubert J, Reniers F, Sferrazza M, Mattiuzzi A, Lagrost C, Jabin I (2016) J Phys Chem C 120:22936–22945. https://doi.org/10.1021/acs.jpcc.6b06143
20. Troian-Gautier L, Valkenier H, Mattiuzzi A, Jabin I, den Brande NV, Mele BV, Hubert J, Reniers F, Bruylants G, Lagrost C, Leroux Y (2016) Chem Commun 52:10493–10496. https://doi.org/10.1039/C6CC04534K
21. Valkenier H, Malytskyi V, Blond P, Retout M, Mattiuzzi A, Goole J, Raussens V, Jabin I, Bruylants G (2017) Langmuir 33:8253–8259. https://doi.org/10.1021/acs.langmuir.7b02140
22. DuVall SH, McCreery RL (1999) Anal Chem 71:4594–4602. https://doi.org/10.1021/ac990399d
23. Retout M, Jabin I, Bruylants G (2021) ACS Omega 6:19675–19684. https://doi.org/10.1021/acsomega.1c02327
24. Mattiuzzi A, Troian-Gautier L, Mertens J, Reniers F, Bergamini J-F, Lenne Q, Lagrost C, Jabin I (2020) RSC Adv 10:13553–13561. https://doi.org/10.1039/D0RA01011A
25. Liu, Gooding JJ (2006) Langmuir 22:7421–30. https://doi.org/10.1021/la0607510
26. Khor SM, Liu G, Fairman C, Iyengar SG, Gooding JJ (2011) Biosens Bioelectron 26:2038–2044. https://doi.org/10.1016/j.bios.2010.08.082
27. Laibinis PE, Fox MA, Folkers JP, Whitesides GM (1991) Langmuir 7:3167–3173. https://doi.org/10.1021/la00060a041
28. Laibinis PE, Nuzzo RG, Whitesides GM (1992) J Phys Chem 96:5097–5105. https://doi.org/10.1021/j100191a065
29. Love JC, Estroff LA, Kriebel JK, Nuzzo RG, Whitesides GM (2005) Chem Rev 105:1103–1170. https://doi.org/10.1021/cr0300789
30. Stranick SJ, Parikh AN, Tao YT, Allara DL, Weiss PS (1994) J Phys Chem 98:7636–7646. https://doi.org/10.1021/j100082a040

31. Imabayashi S-i, Hobara D, Kakiuchi T, Knoll W (1997) Langmuir 13:4502–4.https://doi.org/10.1021/la970447u
32. Retout M, Brunetti E, Valkenier H, Bruylants G (2019) J Colloid Interf Sci 557:807–815. https://doi.org/10.1016/j.jcis.2019.09.047
33. Liu G, Chockalingham M, Khor SM, Gui AL, Gooding JJ (2010) Electroanal 22:918–926. https://doi.org/10.1002/elan.200900539
34. Louault C, D'Amours M, Bélanger D (2008) ChemPhysChem 9:1164–1170. https://doi.org/10.1002/cphc.200800016
35. Blond P, Bevernaegie R, Troian-Gautier L, Lagrost C, Hubert J, Reniers F, Raussens V, Jabin I (2020) Langmuir 36:12068–12076. https://doi.org/10.1021/acs.langmuir.0c02681
36. Retout M, Blond P, Jabin I, Bruylants G (2021) Bioconjug Chem 32:290–300. https://doi.org/10.1021/acs.bioconjchem.0c00669
37. Lenne Q, Mattiuzzi A, Jabin I, Le Poul N, Leroux YR, Lagrost C (2020) Adv Mater Interfaces 7:2001557. https://doi.org/10.1002/admi.202001557
38. Li X-M, Reinhoudt D, Crego-Calama M (2007) Chem Soc Rev 36:1350–1368. https://doi.org/10.1039/B602486F
39. Wang S, Liu K, Yao X, Jiang L (2015) Chem Rev 115:8230–8293. https://doi.org/10.1021/cr400083y
40. Yong J, Chen F, Yang Q, Huo J, Hou X (2017) Chem Soc Rev 46:4168–4217. https://doi.org/10.1039/C6CS00751A
41. Kim A, Lee C, Kim H, Kim J (2015) ACS Appl Mater Interfaces 7:7206–7213. https://doi.org/10.1021/acsami.5b00292
42. Howarter JA, Youngblood JP (2007) Adv Mater 19:3838–3843. https://doi.org/10.1002/adma.200700156
43. Ragesh P, Anand Ganesh V, Nair SV, Nair AS (2014) J Mater Chem A 2:14773–14797. https://doi.org/10.1039/C4TA02542C
44. Xiao F, Yuan S, Liang B, Li G, Pehkonen SO, Zhang T (2015) J Mater Chem A 3:4374–4388. https://doi.org/10.1039/C4TA05730A
45. Zhang B, Zhao X, Li Y, Hou B (2016) RSC Adv 6:35455–35465. https://doi.org/10.1039/C6RA05484F
46. Banerjee I, Pangule RC, Kane RS (2011) Adv Mater 23:690–718. https://doi.org/10.1002/adma.201001215
47. Le Poul N, Le Mest Y, Jabin I, Reinaud O (2015) Acc Chem Res 48:2097–2106. https://doi.org/10.1021/acs.accounts.5b00152
48. Devouge S, Conti J, Goldsztein A, Gosselin E, Brans A, Voué M, De Coninck J, Homblé F, Goormaghtigh E, Marchand-Brynaert J (2009) J Colloid Interf Sci 332:408–415. https://doi.org/10.1016/j.jcis.2008.12.045
49. Nabers A, Ollesch J, Schartner J, Kötting C, Genius J, Haußmann U, Klafki H, Wiltfang J, Gerwert K (2016) J Biophotonics 9:224–234. https://doi.org/10.1002/jbio.201400145
50. Voue M, Goormaghtigh E, Homble F, Marchand-Brynaert J, Conti J, Devouge S, De Coninck J (2007) Langmuir 23:949–955. https://doi.org/10.1021/la061627j
51. Retout M, Valkenier H, Triffaux E, Doneux T, Bartik K, Bruylants G (2016) ACS Sensors 1:929–933. https://doi.org/10.1021/acssensors.6b00229
52. Lenne Q, Leroux YR, Lagrost C (2020) ChemElectroChem 7:2345–2363. https://doi.org/10.1002/celc.202000132
53. Franco F, Rettenmaier C, Jeon HS, Roldan Cuenya B (2020) Chem Soc Rev 49:6884–6946. https://doi.org/10.1039/D0CS00835D
54. Rodriguez P, Koper MTM (2014) Phys Chem Chem Phys 16:13583–13594. https://doi.org/10.1039/C4CP00394B
55. Erikson H, Sarapuu A, Tammeveski K, Solla-Gullón J, Feliu JM (2014) ChemElectroChem 1:1338–1347. https://doi.org/10.1002/celc.201402013

Diazonium Salts and Related Compounds for Biomedical Applications

Ahmed Saad and Marta Cerruti

Abstract Diazonium chemistry has been widely adopted for numerous biomedical applications such as implant materials, tissue engineering scaffolds, and drug delivery. The versatility of substrate materials that can be modified by diazonium chemistry and the wide range of functional groups that can be introduced on the surface are among the advantages of diazonium chemistry over other surface modification techniques. Indeed, it provides a versatile tool to change the surface properties of biomaterials to control their wettability, biocompatibility, protein adsorption, and immune response. Diazonium chemistry is also useful as a linker to attach biomolecules or polymeric layer on biomaterial surfaces. In this chapter, we provide an up-to-date review of the use of diazonium chemistry in biomaterials used in biomedical applications.

1 Introduction

Biomaterials are materials that are meant to interact with biological systems [1]. They can be made of any class of materials including metals, ceramics, polymers or composites of them [2]. The response of the body to such materials is greatly affected by the surface properties of the biomaterials, since the surface is the first region of the biomaterial that the body gets in contact with. Much research is devoted to control biomaterial surface properties mainly to enhance their biocompatibility. For some biomaterials, this means preventing undesired reactions from the body and for other materials, the purpose is to induce specifically desired reactions. For instance, in bone-related implants, it is desired that the biomaterials are bioactive and integrate with the surrounding bones [3]. Bone integration could be inhibited if the body encapsulates the biomaterials with fibrous tissue that isolates the biomaterials from the body. Another class is blood-contacting biomaterials; for these materials, the surface should be hemocompatible, minimize bacterial adhesion, and inhibit biofilm formation [4, 5].

A. Saad · M. Cerruti (✉)
Department of Mining and Materials Engineering, McGill University, Montreal, Canada
e-mail: marta.cerruti@mcgill.ca

M. M. Chehimi et al. (eds.), *Aryl Diazonium Salts and Related Compounds*,
Physical Chemistry in Action, https://doi.org/10.1007/978-3-031-04398-7_14

Different surface modification techniques have been used to modify the surface of a biomaterial. These include physisorptive, chemisorptive, and covalent modification techniques. One or more of these methods can be used to obtain the desired surface properties. We provide in this section a quick summary of each technique, along with the advantages and disadvantages of using it.

Physisorption represents possibly the simplest method for surface modification [6]. In this method, molecules adhere to the biomaterial surface through hydrophobic interactions, hydrogen bonds, and weak electrostatic interactions. Physisorption is versatile and can be used on either metallic, ceramic, or polymeric materials. One of the major drawbacks of physisorption is the instability of physically adsorbed coatings. The adsorbed layer is prone to quick degradation and/or desorption since it only weakly interacts with the surface.

Chemisorption techniques create layers that interact through electrostatic forces with the substrate. An example is self-assembled monolayers (SAMs) using thiol-terminated molecules [7]. To make a SAM, the modifying molecules consist of three parts, a head group, a spacer (mostly alkyl chain), and a terminal functional group [8]. The head group has a specific affinity for substrate, which helps in adsorbing and packing of molecules. The functional group determines the surface properties of the modified surfaces. SAMs are widely exploited for modifying the surface properties of metals. Gold and other metal surfaces are modified by thiols. On the other hand, the modification is not robust and can easily dissociate from the surface [9, 10].

Covalent modification techniques create a covalent bond between the functional layer and the surface of the biomaterial. Three main examples here include SAMs built with silanes; plasma modification; and diazonium reactions.

SAM can form a covalent bond on hydroxylated surfaces such as silica and glass when the modifying molecules are terminated with silanes instead of thiols [7, 10]. Silanes form a Si–O covalent bond on the surface, which is more stable than electrostatic interaction observed with thiols [10]. However, the major limitation of silanes is the substrates that can be functionalized, which are limited to oxides that have hydroxyl functionalities on their surface.

Plasma is a reactive chemical environment that is obtained when gases are excited into highly energetic states and become ionized [11]. These ionized species can react with many different material surfaces and change their properties. Several techniques use plasma for surface modification including etching, plasma sputtering, plasma implantation, plasma deposition, plasma polymerization, and plasma spraying [12]. Plasma-based surface modification techniques have several advantages such as being reproducible, their ability to precisely control the modified area, the ease to scale up to the industrial level with the same efficiency, and applicability to different surface geometries. In addition, plasma can be used to modify the surfaces of different materials including metals, ceramics, and polymers. On the other hand, plasma-based modifications may not be stable for long-term applications, may cause degradation to some polymers due to generated heat, and cannot modify complex three-dimensional materials and porous materials [13, 14].

Diazonium chemistry overcomes several of the limitations discussed for the previous techniques. Diazonium reactions allow a wide range of functional groups to be introduced, including hydroxyl, carboxylic acid, amino, nitro, phosphonate, and other groups [15]. In these reactions, arylamines compounds with the functional group of interest are used as precursors that react with sodium nitrite in the presence of an acid to transform the arylamines into metastable diazonium salts [16]. Diazonium cations are then transformed into radicals by the addition of a reducing agent, such as iron powder or hypophosphorous acid [17]. The radicals can react with almost any surface and create covalent bonds. This method is versatile and can be used to modify the surfaces of carbon, polymers, metals, and metal oxides [16]. Diazonium coupling can be performed using electrochemical, photochemical or chemical methods [18]. Major advantages of diazonium reactions over other surface modifications include the stability of modification, since covalent bonds are formed, and the possibility of modifying complex three-dimensional structures [17]. The technique is also simple and inexpensive which allows its commercial use on a large scale. Finally, the reactions usually occur in aqueous solution at room temperature, thus preventing substrate degradation [19].

An important aspect of diazonium-based surface modification that makes it applicable in biomedical applications is its biocompatibility. Many studies used both in-vitro cell cultures and in-vivo animal studies to confirm the biocompatibility of diazonium functionalization. To mention a few, a study by Oweis et al. confirmed the viability and proliferation of fibroblast cells cultured on dental alloys modified by diazonium chemistry [20]. Another study by Abdallah et al. showed enhanced proliferation of gingival epithelial cells cultured on diazonium-modified polymers compared to control polymers [21]. In-vivo biocompatibility was also shown in a study by Mahjoubi et al. where diazonium-modified polyetheretherketone (PEEK) integrated with bones when implanted in a rat calvarial defect model [22].

This chapter focuses on the use of diazonium reactions in biomedical applications. We have subdivided this chapter in three subsections, namely surface modifications of implantable materials, scaffolds for tissue engineering, and biomaterials for drug delivery. We conclude the chapter with a summary and prospective analysis.

2 Implantable Materials

Diazonium chemistry is used to change the surface properties of implants either as the final coating layer or to create an intermediate adhesive coating onto which another layer is grafted, in the form of a polymer, nanoparticles or biomolecules. When diazonium reactions are used as the final coating layer, the aryl diazonium precursor should have a functional group that will induce the intended surface properties. On the other hand, when diazonium reactions are used to create an adhesive layer, p-phenylenediamine can be used as the precursor. In this case, one of the two amino groups gets reduced and covalently grafted to the surface, while the free amino group can then be reduced in a second step to conjugate the molecules of interest [23]. Figure 1 shows a schematic of the mechanism by which p-phenylenediamine can be used to create a self-adhesive layer. In this section, we present studies that

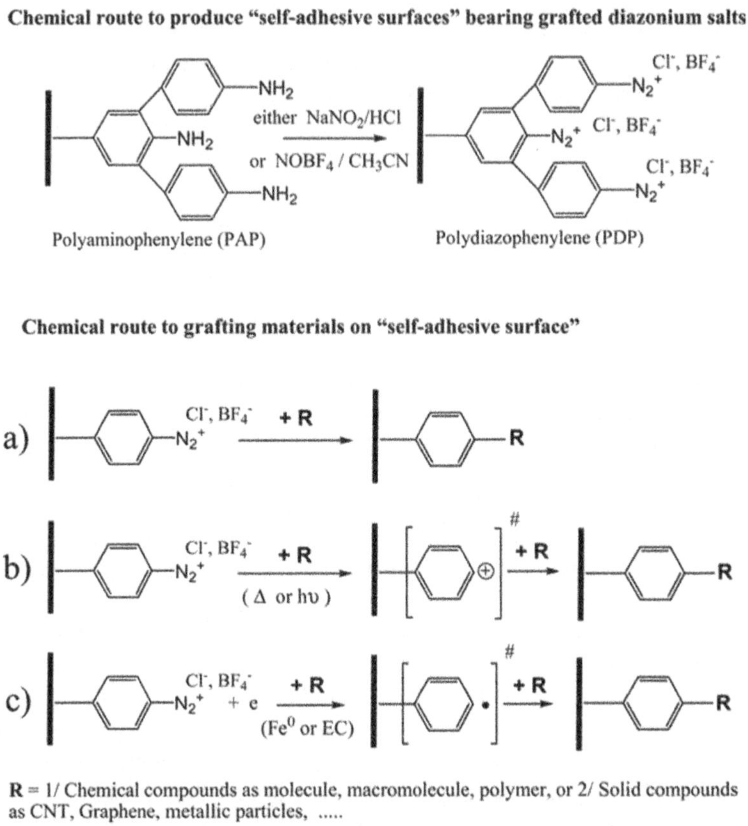

Fig. 1 Synthesis of self-adhesive surfaces with diazonium salts (top) and different routes to graft molecules/materials on diazonium salt bearing surfaces (bottom). **a** Spontaneous reaction between the aryldiazonium grafted salt and nucleophilic compounds. **b** Physical activation by thermal or UV irradiation, creating an intermediate phenyl cation. **c** Chemical activation through electron transfer by a redox reaction or the application of an electrical potential. Reprinted with permission from [23]. Copyright © 2008, the Royal Society of Chemistry [23]

used diazonium chemistry techniques as a linker to attach a secondary coating on implants or as the final layer to modify the implant surface properties. Table 1 summarizes the presented studies including the implant materials that were modified, the attached moieties or coatings, the diazonium precursors that were used, and reason for changing the surface chemistry.

Table 1 A summary of studies that used diazonium chemistry for surface modification of implants

Implant material	Attached moiety/coating	Diazonium precursor	Reason for the modification	Reference
Stainless steel	2-bromoethylbenzene	4-(2-bromoethyl) benzenediazonium tetrafluoroborate	Improve the durability of a drug-in-PMMA poly(methyl methacrylate) coating matrix	[24]
Stainless steel and CrCo	1-dodecyloxy benzene	4-(1-Dodecyloxy)-phenyldiazonium tetrafluoroborate	Improve the adhesion of a drug-in-PMMA coating matrix	[25]
Stainless steel and titanium	Amino groups	P-phenylenediamine	Link PMMA to implants	[26]
CoCr alloy	Amino groups	P-phenylenediamine	Link PMMA to implants	[27]
Stainless steel	4-nitrobenzene	4-nitrobenzene diazonium	Provide durable anti-corrosion polyetheretherketone layer	[28]
Ti6Al4V alloys	4-(Hydroxymethyl) benzene	4-aminobenzyl alcohol	Improve abrasion resistance by coating with polyurethane	[29]
Polyetheretherketone	Phosphonate	2-aminoethylphosphonic acid	Increase biomineralization	[22]
Poly (methyl methacrylate) and poly (D, L-lactic acid)	Amino groups	p-phenylenediamine	Control protein adsorption and enhance soft tissue integration	[21]
Polystyrene	Amines, phosphonic acids, and carboxylic acids	p-phenylenediamine, 4-aminobenzyl phosphonic acid, and 4-aminobenzoic acid	Control protein adsorption and immune response	[30]

2.1 *Using Diazonium Modification to Bind Intermediate Linkers*

(a) **In stents**

Diazonium-based modification was used as an intermediate step to modify the surface of stainless steel stents to improve the durability of a drug-in-polymer matrix coating that was meant to help preventing occlusion of arteries (drug-eluting stents) [24]. In a study by Shaulov et al., 4-(2-bromoethyl) benzenediazonium tetrafluoroborate (BrD) was electrochemically reduced on stainless steel stents to covalently conjugate 2-bromoethylbenzene groups on the stent surface, which served as a radical initiator for surface polymerization of MMA into (PMMA) brushes [24]. The presence of stable covalent layer improved the adhesion of drug-in-polymer matrix as shown in the scanning electron microscope (SEM) images of coated stents with and without the BrD-PMMA layer (Fig. 2(A)). Similarly, Levy et al. modified stainless steel and CoCr stents by the electrochemical grafting of 1-dodecyloxy benzene on the stent surfaces using 4-(1-Dodecyloxy)-phenyldiazonium tetrafluoroborate salt [25]. The stents were further spray-coated with several drug-in-polymer matrices and the coating stability of the different matrices was compared with the unmodified stents. Results showed that the diazonium grafted hydrophobic layer greatly enhanced the adhesion of the drug-in-polymer matrices (Fig. 2(B)). However, the study did not provide a justification for choosing this specific diazonium linker.

Drug-eluting stents can release an antimitotic drug (such as Sirolimus) once implanted in the artery to prevent restenosis (i.e. re-clogging of the arteries) [31]. The French company Alchimedics has patented a process that permits coating of the stent with an electrografted polymer (polybutylmethacrylate) [31]. This coating is achieved by reducing 4-nitro benzenediazonium salt in the presence of butylmethacrylate. A spray coating of polylactic and polyglycolic acid containing the drug is then deposited on the polybutylmethacrylate primer. This process is used by Sinomed, a Chinese company to produce Buma® stents, which have been implanted in more than eight hundred thousand patients.

(b) **In dental and orthopedic devices**

Diazonium modification has shown great potential toward improving the integrity of dental device structure. Partial dentures, prostheses, and dental restorations suffer from mechanical failure due to the weak link between metallic and acrylic components (such as PMMA) in the devices [32, 33]. Changing this link from mere mechanical interaction into chemical covalent bond using diazonium compounds as linkers can greatly improve device lifetime and reduce their failure rate [26]. In a study by Alageel et al., p-phenylenediamine was used to link PMMA to the surface of stainless steel and titanium implants using a redox reaction [26]. The bond strengths between PMMA and stainless steel or titanium implants were 2.5 and 5.2 times higher than untreated groups, respectively showing a significant improvement in

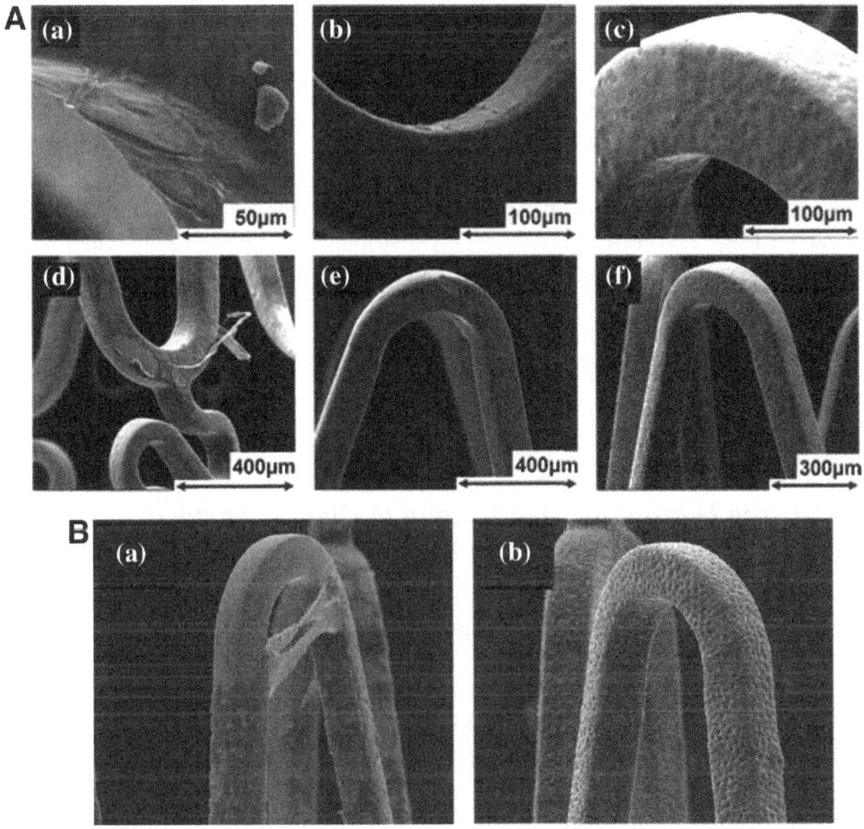

Fig. 2 SEM images of (**A**) (a and d) control stents, (b and e) diazonium-grafted BrD, and (c and f) PMMA-BrD stents coated with a single layer of the drug-in-polymer matrix. Reprinted with permission from [24]. SEM images of (**B**) (a) control stents and (b) stents spray-coated with poly (butyl methacrylate/polyethylene vinyl acetate/paclitaxel) composite [25]. All stents were incubated for 3 days in (phosphate buffer, 0.1 M, pH 7.4, 0.3% SDS at 60 °C, 100 rpm). Reprinted with permission from [25]. Copyright © 2009, American Chemical Society [24]

mechanical strength due to the use of diazonium linkers. Mezour et al. also used p-phenylenediamine precursor to create a poly(aminophenylene) (PAP) layer on the thin oxide layer on the surface of CoCr alloys through electrochemical reduction. PMMA adhered to the coated CoCr surfaces with a strength that was fivefold higher than on uncoated CoCr [27].

Diazonium chemistry was also explored as linking method in another study to bind a coating layer of PEEK on stainless steel to provide durable anti-corrosion protection to stainless steel implants [28]. A number of diazonium linkers were tested but the strength between the PEEK film and stainless steel was only higher when using 4-nitrophenyl-modified stainless steel. Similar results were reported when 4-nitro benzenediazonium linker was used to link PEEK coatings onto titanium implant surfaces [34].

Ti6Al4V alloys are commonly used as orthopedic implants. However, their poor abrasion resistance can lead to the release of toxic vanadium and aluminum ions. A stable and biocompatible organic coating can help overcome this problem. The surface of Ti6Al4V was successfully modified by 4-(hydroxymethyl) benzene and subsequent linking of 10 layers of polyurethane polymer, which has high abrasion resistance [29]. The presence of an evenly distributed coating of the polymer led to a tenfold decrease in the coefficient of friction compared to unmodified alloy.

The studies emphasize that the diazonium chemistry can help in the formation of stable, covalently linked polymer layers and a wide spectrum of biomaterials including metal, metal oxides, and alloys that are used for stents, dental, and orthopedic applications.

2.2 *Using Diazonium Modification to Change Surface Properties Directly*

PEEK is an excellent material for orthopedic implants because its mechanical properties match those of bones and because of its radiolucency; however, its bio-inertness limits integration with surrounding bones [35]. Mahjoubi et al. explored the use of diazonium chemistry to change the surface chemistry of PEEK by grafting phosphonate groups on its surface [22]. They used a two-step method where p-phenylenediamine was used as a precursor to graft a poly(aminophenylene) layer. Diazonium cations were then generated on this layer by reacting the amino groups with sodium nitrite, and reacted with 2-aminoethylphosphonic acid to graft phosphonate groups. Phosphonated surfaces increased hydroxyapatite mineral (HA) deposition and adhesion by 40% compared to a control, unmodified group (Fig. 3a). These results were also supported by in-vivo experiments in a rat calvaria defect showing mineral deposition at the interface between implants and bone was higher in treated PEEK compared to the control.

The surface chemistry of implants greatly influences the adsorption profile of proteins upon implantation, which in turn influences cellular response to the implants. Trying to reproduce the surface chemistry of percutaneous natural organs, Abdallah et al. modified the surface of PMMA and poly(D, L-lactic acid) (PDLLA) by conjugating amino-terminated aryl diazonium molecules [21]. Aminated PMMA and aminated PDLLA showed surface properties adsorbed more key proteins (laminins and nidogen) than their unmodified counterparts. In addition, human gingival epithelial cells showed higher proliferation on modified surfaces compared to unmodified surfaces (Fig. 3b). In another study, Buck et al. investigated the effect of the surface chemistry of biomaterials on protein adsorption and subsequent macrophage response [30]. The authors used diazonium-based redox surface modification reactions to graft common biological functional groups (amines, phosphonic acids, and

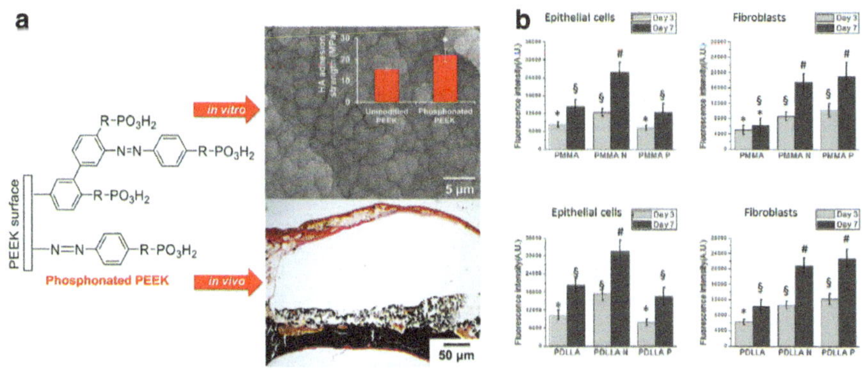

Fig. 3 **a** An illustration of phosphonate functional groups modifying PEEK surface alongside with in-vitro results of HA deposition and adhesion strength and in-vivo mineralization after 3 month implantation in rat cranial defects [22]. Reprinted with permission from [22] **b** Influence of surface amination and phosphonation of PMMA (PMMA-N and PMMA-P, respectively) and PDLLA (PDLLA-N and PDLLA-P, respectively) on cell behavior. Proliferation rates of cells on the biomaterials on days 3 & 7. Differences in the symbols (*, § and #) indicate significant differences between the different biomaterials at $p < 0.05$ [21]. Reprinted with permission from [21]. Copyright © 2017, Acta Materialia Inc.[21]. Copyright © 2016, Acta Materialia Inc.[22]

carboxylic acids) on polystyrene. They evaluated protein adsorption before and after macrophage cell culture, and macrophage adhesion and morphology. Results showed that carboxylic acid-functionalized surfaces showed the highest anti-inflammatory response as well as the highest adsorption of proteins associated with normal wound healing process. Based on this, the authors speculated that carboxylic groups would promote implant tissue integration more than the other groups tested.

These works show to which extent it is possible to control surface properties of implant by simply adding a diazonium-based functional layer on implant surfaces. Diazonium functionalization allows controlling protein adsorption and cell behavior, and ultimately implant integration in-vivo.

3 Scaffolds for Tissue Engineering

Scaffolds are three dimensional platforms that are used in tissue engineering to replace missing tissues or regenerate damaged tissues [36]. They can be loaded with cells, growth factors, and drugs [37]. Upon implantation, proteins and cells interact with the scaffold surfaces, and thus the surface properties greatly influence the final scaffold fate [38]. Since scaffolds are three dimensional and porous structures, diazonium chemistry represents an ideal technique to modify them homogenously and stably [39]. In fact, diazonium modifications have been applied to scaffolds used in bone tissue engineering, extracellular matrix (ECM) mimetics, and modulating immune response. In this section, we discuss the work done so far in using diazonium chemistry in these three categories of scaffolds.

Table 2 summarizes the studies discussed in this section, including scaffold materials, attached moieties or coatings, the diazonium precursors, and aim for changing the surface properties.

3.1 Bone Tissue Engineering

Diazonium chemistry can improve the field of bone tissue engineering by tuning the surface properties to enhance the formation of hydroxyapatite mineral on scaffold surfaces. This was achieved by Mahjoubi et al. through grafting phosphonate groups on the surface of three-dimensional PDLLA scaffolds [17]. As described above, the authors grafted phosphonate groups using a two-step method where p-phenylenediamine precursor was first used to introduce amine functional groups through a redox reaction that subsequently were reacted with aminoethyl phosphonic acid in another redox reaction. Phosphonated PDLLA scaffolds deposited more hydroxyapatite when immersed in a simulated body fluid and showed higher viability and mineralization of osteogenic MC3T3-E1 cells and higher proliferation of chondrogenic ATDC5 cells compared to bare scaffolds.

Murphy et al. exploited diazonium chemistry to graft different functional groups on silk fibroin protein scaffolds to control the protein structure, wettability, and cellular interactions [39]. Aniline precursors were grafted on the tyrosine residue side chains via an electrophilic substitution reaction as shown in Fig. 4a. Five different functional groups were grafted on silk fibroin, i.e. carboxylic acid, amine, ketone, sulfonic acid, and alkyl groups. Targeting tyrosine side chains using diazonium has several advantages including the abundance of tyrosine residues in silk (~10%), the rapid reaction, and the mild conditions of the reaction that allow the recovery of >90% of functionalized protein [39]. Silk scaffolds with different modifications were tested for their effect on proliferation and differentiation of human mesenchymal stem cells (hMSCs). Although the functional groups resulted in range of surface contact angles (wettability), hMSCs osteogenic differentiation was not statistically different among the groups.

In another study, Chernozem et al. modified the surfaces of both polyhydroxybutyrate and polycaprolactone (PCL) electrospun scaffolds by introducing 3,4-dicarboxyaryl moieties to increase hydrophilicity [43]. Diazonium radicals were generated through a photo induction process from 3,4-dicarboxybenzenediazonium tosylate precursor and formed covalent bonds on the scaffold surfaces. The attachment of these hydrophilic functional groups decreased water contact angle from 127 $\pm$ 4° to 82 $\pm$ 1° for PCL and from 126 $\pm$ 4° to 78 $\pm$ 2° for polyhydroxybutyrate and increased osteoblastic cell density and area on both polymers compared to control scaffolds [43].

Chemical modification using diazonium was also used to add carboxyl groups to silk fibroin to be used as bone graft hydrogels. The carboxyl groups served as links

Table 2 A summary of studies that used diazonium chemistry for surface modification of scaffolds

Scaffold material	Attached moiety	Diazonium precursor	Reason for the modification	Reference
PDLLA	Phosphonate	p-phenylenediamine then aminoethyl phosphonic acid	Increase scaffold hydrophilicity and biomineralization	[17]
Silk fibroin	Carboxyl, amine, ketone, sulfonic acid, and heptyloxyl	Aniline precursors of the 5 functional groups	Control wettability and protein structure	[39]
Silk fibroin	Carboxyl	4-aminobenzoic acid	Conjugate β-cyclodextrin on scaffold surface	[40]
Silk fibroin	Carboxyl	4-aminobenzoic acid	Conjugate RGD on scaffold surface	[41]
Silk fibroin	Sulfonic acid	4-sulfanilic acid	Mimic sulfated glycosaminoglycan heparan sulfate	[42]
Polyhydroxy-butyrate and polycaprolactone	3,4-dicaboxyaryl	3,4-dicarboxybenzene-diazonium tosylate	Increase scaffold hydrophilicity	[43]
Silk fibroin coated polycaprolactone	4-*azido phenyl*	4-*azidoaniline hydrochloride salt*	Conjugate *heparin disaccharide to* modulate immune response	[44]
Silk fibroin coated polycaprolactone	4-azido phenyl	4-azidoaniline hydrochloride salt	Conjugate mechano-growth factor to modulate immune response	[45]
Diazoresin-pectin	No attached moiety	4-diazodiphenyl salt of the diazoresin polymer itself	Conjugate insulin on diazoresin	[46]

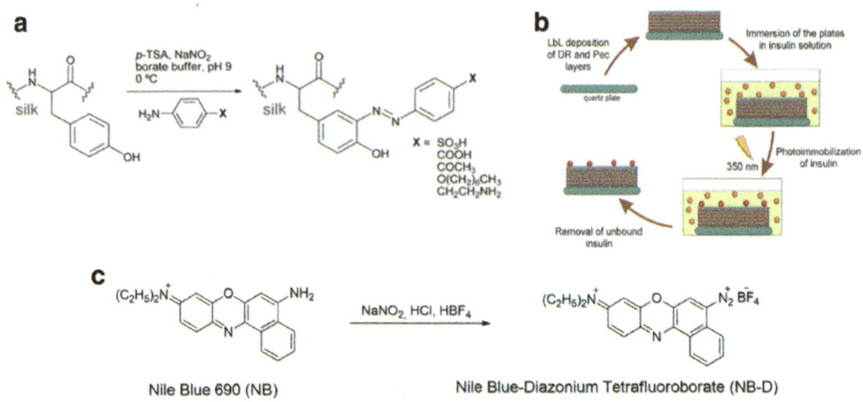

Fig. 4 **a** Schematic of the reaction between aniline derivatives and silk fibroin tyrosine residues side chains [39]. Reprinted with permission from [39] **b** Fabrication of insulin-functionalized diazoresin/pectin-coated surfaces [46]. Copyright © 2008, Elsevier Ltd.[39]

to covalently attach biomolecules through EDC-NHS coupling and formation of an amide bond, exploiting free primary amines of biomolecules (such as β-cyclodextrin) [40]. In another study, carboxyl groups were added to silk using diazonium chemistry to conjugate RGD peptides [41]. The RGD-functionalized silk was used to coat titanium implants to improve fibroblast adhesion and proliferation, hence, improve soft tissue integration.

Mikulska et al. also exploited diazonium chemistry to tether insulin on the surface of three-dimensional scaffolds made of diazoresin, a cationic polymer, and pectin, an anionic polysaccharide [46]. Since the diazoresin polymers inherently possess the diazonium functionality, insulin, in borate buffer, was directly conjugated on the polymer through photo-immobilization as shown in Fig. 4b. The scaffolds were used to culture human mesenchymal stem cells; modified scaffolds increased proliferation and osteogenic differentiation compared to control scaffolds.

3.2 ECM Mimetics

The ECM plays a crucial role in regulating cell behaviors such as migration, proliferation, and differentiation. Biomaterials that mimic the ECM structure have great potentials as tissue engineering scaffolds due to the mechanical and biochemical cues they provide to cells. Adding growth factors further increases scaffold capacity to mimic the natural ECM environment. Diazonium chemistry has been used to achieve these goals in a few instances.

Wenk et al. used silk fibroin to mimic sulfated glycosaminoglycan heparan sulfate, an ECM component, in its ability to bind fibroblast growth factor 2 (FGF-2), a key growth factor in regeneration of several body tissues [42]. They used diazonium chemistry to graft sulfonic acid groups onto silk fibroin surfaces, to mimic glycosaminoglycan heparan sulfate. The diazonium reaction was carried in borate buffer using the same scheme shown in Fig. 4a. A range of molar ratios between diazonium salt and silk fibroin were tested to obtain different densities of sulfonic acid groups per silk fibroin molecule [42]. Silk fibroin decorated with 70 sulfonic acid groups per molecule resulted in a twofold increase in FGF-2 binding compared to unmodified silk fibroin. The authors also found that FGF-2 bound to sulfonated silk protein could be released upon cellular demand, when the films were placed in contact with human mesenchymal stem cells.

3.3 *Immune Response Modulation*

When a foreign material such as a scaffold enters the body, the immune system interacts with the material based on its surface properties. If the immune system recognizes the material as foreign, it can induce a foreign body reaction that can lead to the formation of a fibrous capsule around the scaffold and prevent its integration with surrounding tissues [30]. The hydrophobic nature of some scaffolds leads to non-specific protein adsorption that can in turn cause fibrous tissue encapsulation. To prevent this, Qian et al. investigated the use of silk fibroin to coat PCL hydrophobic scaffolds [44]. Silk fibroin was further decorated with heparin disaccharide (HD) using diazonium chemistry as a link for the covalent conjugation. 4-azidoaniline hydrochloride salt was used a precursor which was reacted with silk fibroin in a borate buffer. Silk fibroin coated PCL was able to non-covalently bind interleukin-4; this molecule was able to modulate macrophage polarization at the early stage of culturing in an in-vitro setting, and inhibited fibrous tissue formation around the scaffold at later stages when implanted subcutaneously in rat model [44]. A similar construct was prepared by Song et al., who decorated silk fibroin with a short peptide called mechano-growth factor (MGF) [45]. Like in the previous study, silk fibroin was functionalized with 4-azidophenyl to tether MGF on its surface. The presence of MGF accelerated macrophage transformation into M2 phenotype in an in-vitro cell culture test and downregulated foreign body reaction in an in-vivo rat model.

These studies show the potential of diazonium chemistry as a method for scaffold modification, with applications including bone tissue engineering, ECM-mimetics, and prevention of fibrous tissue formation. Diazonium chemistry was effective at reducing scaffold hydrophobicity while conjugating functional groups that were bioactive themselves or were used to link specific biomolecules, proving that diazonium chemistry is a simple yet robust method with excellent potential for scaffold surface engineering.

4 Drug Delivery

Several groups have exploited the versatility of diazonium chemistry to generate drug delivery systems based on a variety of substrates. Among them, two classes of materials stand out: carbon-based nanomaterials, which can be easily modified by diazonium chemistry (Chaps. 7 and 8), and protein-based nanocarriers, which can be modified by diazonium chemistry exploiting tyrosine residues (see Sect. 3 and Fig. 4a). In this section we will summarize the work performed in these two areas, and conclude with a section where we discuss other recently published drug-delivery work involving diazonium chemistry.

4.1 Carbon-Based Nanomaterials as Drug Carriers Modified by Diazonium Chemistry

Carbon-based nanomaterials including both graphene and carbon nanotubes have been proposed by many researchers as substrates for drug delivery, for cancer-related applications [47] and beyond [48]. Since diazonium chemistry can covalently bind many functional groups on graphene and other carbon-based nanomaterials (Chaps. 7 and 8), several authors have used this technique to modify these materials and transform them into drug delivery systems.

Wei et al. modified the surface of reduced graphene oxide (rGO) with benzoic acid through diazonium chemistry, using p-aminobenzoic acid as a precursor [49]. This modification allowed increasing the hydrophilicity of rGO. The authors then added polyethyleneimine (PEI) to further enhance water solubility, and folic acid (FA) as a target molecule for cancer cells. They showed that the rGO-PEI-FA nanoparticles were not toxic to cells up to a concentration of 12.5 mg L^{-1}, but when loaded with doxorubicin (DOX), they were able to induce apoptosis in a cell culture model of CBFH7919 cancer cells.

Liu et al. also used diazonium chemistry to introduce carboxylic groups on rGO and increase its water stability; then, they used these groups as anchors to bind RAFT (reversible-addition-fragmentation chain-transfer) polymerization initiators through esterification [50]. After this, the authors were able to polymerize a layer of poly(acrylic acid) (PAA) onto the rGO substrate. The resulting rGO/PAA hydrogel (Fig. 5a) could be loaded with DOX and release it over the course of 10 h in basic solutions, which is interesting for applications targeting the intestine.

Lucherelli et al. used diazonium derivatives to create a trifunctional graphene platform, starting from graphite, intercalating it with potassium ions, and then reacting it with three different diazonium salts in equimolar proportions [51]. In a follow up paper, the same authors used this tri-functional nanographene as a substrate to covalently link a fluorophore, a targeting molecule (FA) and DOX; DOX loading was further increased by sonication in DOX solutions through physisorption [52]. The results showed the graphene-based carrier was highly toxic to HeLa cells and

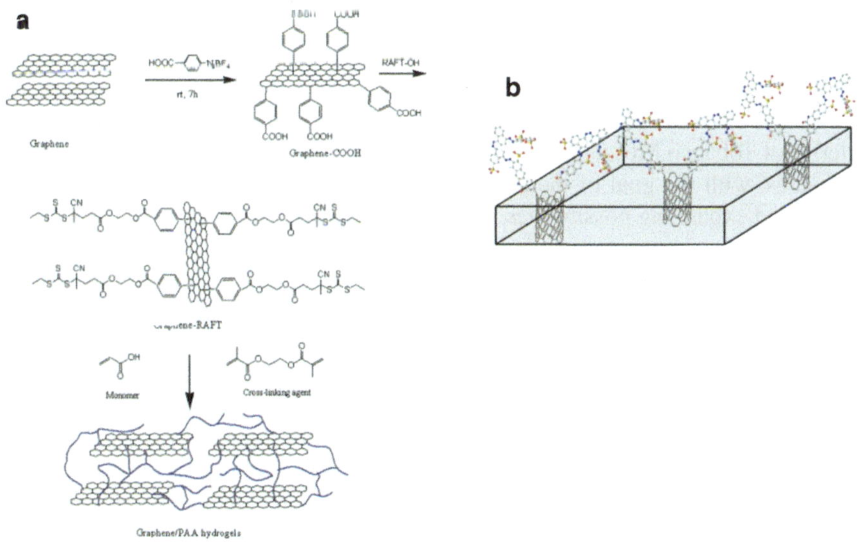

Fig. 5 Carbon-based nanomaterials functionalized by diazonium chemistry to create drug delivery vehicles. **a** Introduction of a RAFT initiator on rGO through diazonium chemistry and synthesis of an rGO/PAA hydrogel composite for drug delivery [50]. Reprinted with permission from [50] **b** Carbon nanotubes functionalized with an anionic dye inserted in a membrane used for electrophoretic nicotine delivery[53]. Copyright © 2013, Elsevier Ltd. [50]

could be monitored in its position through the fluorophore, transforming this system into a theranostic (therapeutic + diagnostic) system.

Lastly, in a very different, non-cancer related application, Wu et al. used diazonium chemistry to introduce sulphonate groups on carbon nanotubes, first through electrochemical diazonium grafting of a monolayer of polybenzoic acid and then coupling to a highly sulphonated dye via carbodiimide coupling reaction [53]. These surface-modified carbon nanotubes were part of a membrane that was used for electrophoretic delivery of nicotine (Fig. 5b). The surface modification was necessary to get efficient electroosmosis since a high charge density increases the number of counter ions adsorbed and decreases energy required for pumping neutral molecules through.

4.2 *Protein-Based Drug Delivery Systems*

Protein-based carriers can be easily modified via diazonium chemistry, exploiting tyrosine residues present on the proteins, as described in Sect. 3 [54]. The aromatic group of tyrosine can be reacted with para-amino benzene compounds to form covalently linked benzenediazonium derivatives as shown in Fig. 4a.

A lot of work in this area relates to the modification of viral capsid nanoparticles as carriers for drug delivery. For example, the Francis group used diazonium tyrosine coupling to modify the interior of MS2 viral capsids [55] with handles that could be used to covalently bind large model drug molecules [56]. They also modified the exterior of Tobacco Mosaic Virus (TMV) nanoparticles with large amounts of PEG molecules, with the goal of increasing blood circulation time of the nanoparticles in-vivo and expanding possibilities for further coupling with drug molecules [57].

Bruckman et al. used a similar strategy to modify external tyrosines of TMV nanoparticles; however, the diazonium salts they used carried alkyne functionalities that allowed for click-chemistry reactions with many different compounds [58], which the group used to modify the behavior of different types of cells in contact with substrates coated with the surface-modified TMVs [59].

The same strategy including diazonium initial functionalization of TMV particles followed by click-chemistry coupling was used by Marin-Caba et al. to bind Cy5 dye on TMV particles, which they then loaded as a coating on mesoporous silica nanospheres [60]. The whole complex was cytocompatible and able to deliver rhodamine B isothiocyanate as a model drug in solutions at pH 5, which the authors explained was used to mimic tumor and lysosomal microenvironment. While rhodamine B was released, Cy5 was retained thanks to its covalent coupling to TMV achieved via diazonium chemistry. This allowed the authors to use the dye to track the position of the TMV/silica complex.

Ma et al. used the same tyrosine-targeted diazonium chemistry to create TMV-based hydrogels, some of them also containing disulfide bonds that could be broken inside cell microenvironments, and proposed their application for in-vivo biocompatible drug delivery systems [61].

While the previous examples are all related to viral substrates, a few researchers used the same strategy of diazonium coupling through tyrosine residues to create silk-based drug delivery vehicles. We already discussed the work by Wenk et al., who showed binding of FGF-2 on silk using this method [42]. Coburn et al. were able to introduce carboxylate and sulfonate groups that altered the secondary structure of silk films, which allowed greatly increasing loading of vincristine and DOX [62]. The effect of the modification on drug release varied depending on the degree of crystallinity of the films, but mostly led to a decreased rate, and hence more sustained drug release from the silk films (Fig. 6).

4.3 Other Substrates Modified by Diazonium Chemistry for Drug Delivery Applications

The concepts described in the previous sections were applied to other substrates in three recent papers: using a procedure first introduced by Chehimi's group [63], Ferrand-Drake del Castillo et al. bound atom transfer radical polymerization (ATRP)

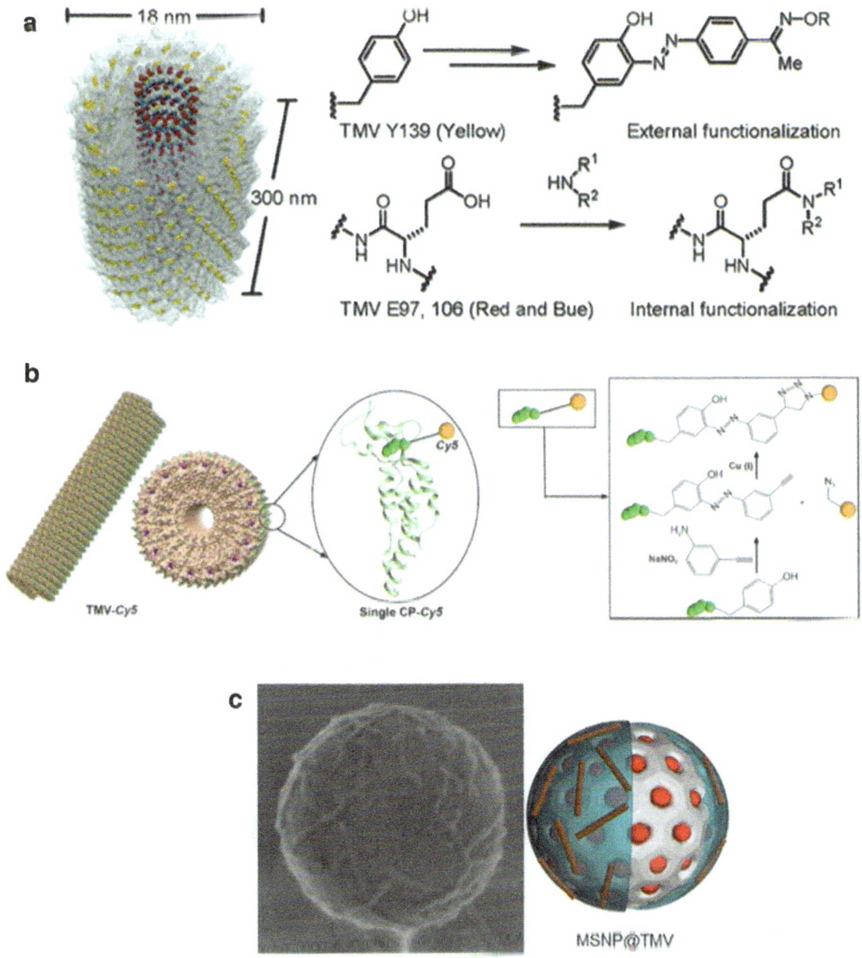

Fig. 6 Tobacco Mosaic Virus (TMV) functionalization via diazonium chemistry to create drug delivery vessels. **a** In the work by Schlick et al., external functionalization of TMV particles exploiting tyrosine amino acids (Y139) happens through diazonium chemistry. Many different residues can be added as functional groups. The functionalization of the interior of TMV particles is done through amidification reactions using glutamic acid residues (E97 and E106). Reprinted with permission from [57]. **b** Marin-Caba et al. adopted a similar strategy to add a click-chemistry initiator on the surface of TMV, used to covalently bind a dye (Cy-5)In this work, the TMV-Cy5 compound becomes part of a theranostic system (panel C), where rhodamine B isothiocyanate molecules (red dots in the schematic) are encapsulated as a model drug in a mesoporous silica sphere, coated with a positively charged polymer (blue layer in the schematic), upon which the TMV-Cy5 particles adhere. Reprinted with permission from [60]. Copyright © 2005, American Chemical Society [57]. Copyright © 2019, American Chemical Society [60]

initiators on gold substrates that were modified with aryl layers prepared by electrochemically reduction of the corresponding diazonium salts [64]. Then through ATRP the authors grafted thin layers of PAA and poly(methacrylic acid) brushes on the gold substrates. Such layers were able to capture large amounts of many proteins with very different charges at pH 5, due to strong hydrogen-bond interactions, and then release them at pH higher than the protein pI.

Another example of an inorganic/organic system is the one devised by Di Martino et al.: the authors functionalized iron oxide nanoparticles with carboxylic groups using diazonium chemistry, and then adsorbed chitosan (CS) or chitosan-poly(lactic acid) (CS-PLA) polymeric chains on them exploiting ionic interactions [65]. DOX was introduced as the polymeric coating produced around the particles. The presence of the polymeric coating slowed down DOX release and made it sensitive to pH.

Finally, we provide an example of functionalization of another bidimensional nanomaterial in addition to what described in Sect. 3.1: Zhao et al. functionalized black phosphorous (BP) nanosheets with Nile Blue (NB) via diazonium chemistry, after having converted NB to NB-diazonium tetrafluoroborate (NB-D) (Fig. 7a); the reaction created a stable C–P bond that enhanced BP stability and gave fluorescent properties to the particles (Fig. 7b). This, coupled with the photothermal properties of BP, made the resulting NB@PB particles excellent platforms for photothermal therapy: the authors showed that particles were able to kill cancer cells and to both image and treat tumors in-vivo in mice animal models [66].

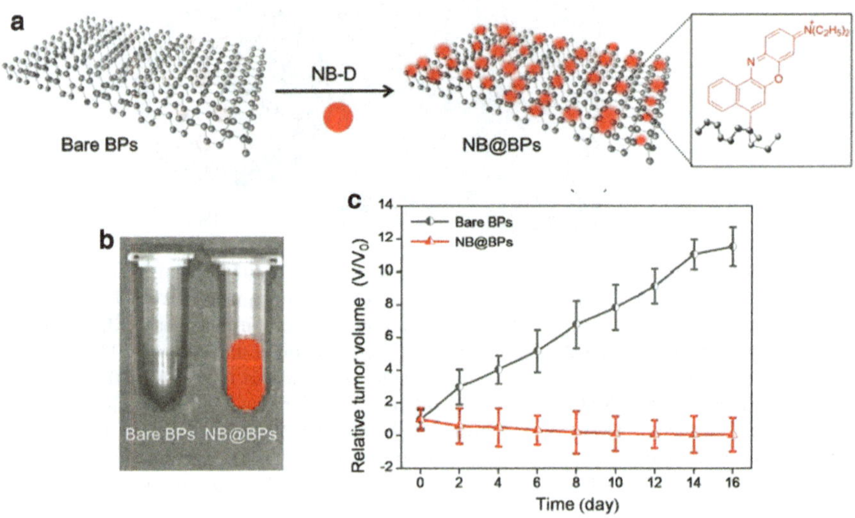

Fig. 7 **a** Functionalization of black phosphorus (BP) nanoparticles with Nile Blue (NB) through diazonium chemistry, leading to particles that can fluoresce upon excitation with 570 nm light (panel B), and that can strongly decrease tumor volume in mice models, as shown in panel C (comparison of the effect observed on nude mice with a tumor after intravenous injection of bare BPs and NB@BPs irradiated with a 808 nm and 1.5 W cm^{-2} NIR laser for 10 min) [66]. **b** and **c** are reprinted with permission from [66]. Copyright © 2017, American Chemical Society

In summary, this shows the usefulness of diazonium chemistry in a variety of drug delivery applications. Many of the studies reviewed were focused on cancer treatment using carbon-based nanomaterials, and diazonium chemistry was used to control the nanomaterial wettability and to conjugate biomolecules for cancer treatment or diagnostic. In other examples, including both protein-based drug delivery systems and surface modification of inorganic substrates, diazonium chemistry was used to conjugate drugs and to control pharmacokinetics. These works emphasize the diversity of drug delivery applications, including theranostic, where diazonium modifications are useful.

5 Summary and Perspective

Diazonium chemistry has been successfully used to modify the surface of a wide range of implant materials including metals, alloys, metal oxides, and polymers used as stents, orthopedic, and dental devices. Implants were modified to control their wettability, produce a bioactive surface, reduce abrasion, and link different sections of the implants with a strong permanent bond. All these applications take advantage of the versatility of the diazonium salts, the mild reaction conditions for their conjugation, and robustness of the bond with the surfaces. Similarly, surfaces of scaffold materials have been modified with diazonium chemistry to change wettability, conjugate biomolecules, and modulate immune response. Hydrophobic polymers such as PCL and PDLLA were treated with hydrophilic functional groups to reduce non-specific protein adsorption and reduce fibrous tissue formation. Many studies used diazonium chemistry to change silk fibroin surface to conjugate peptides, hormones, and growth factors.

In drug delivery, diazonium has been extremely useful to covalently modify a variety of substrates with layers that are used for three main goals: (i) to change hydrophilicity or surface charge of the substrate, and thus increase water stability and interactions with drug molecules; (ii) to bind polymerization initiators that are then used to graft polymeric layers that include the drug molecules to be released; or (iii) to stably bind dye molecules to be used to detect the location of the drug delivery vessel in theranostic systems. The key reason for using diazonium chemistry in these applications is the stability of the modification introduced on the substrates.

Future work could exploit diazonium chemistry to systematically investigate the relationship between surface properties, protein adsorption profiles, and cellular response in different tissues, leading to a new generation of biomaterials with predictable and controlled body response. The tyrosine-based reaction used to modify silk and viral nanoparticles could be exploited to modify many more protein-based nanomaterials (films, coatings, nanoparticles), both as drug delivery systems and as ECM coatings on implants. Also, drug-eluting coatings anchored through diazonium chemistry could be synthesized on many more biomaterials beyond stainless steel stents [24]. Two interesting applications come to mind in this regard: the inclusion of anti-inflammatory drugs as coatings to decrease foreign-body reaction around inert implants such as PEEK, which we have shown can be easily modified through diazonium chemistry [22]; and the inclusion of anti-microbial or antibiotic compounds to prevent infection around dental or osteointegrated percutaneous implants, which often fail due to lack of good sealing with the epidermis. Since these implants are usually made in metallic alloys that can be easily modified by diazonium chemistry, this technique could be used to anchor drug-loaded layers on their surface and decrease risk of failure due to infection.

A few challenges lie ahead before diazonium chemistry can be a routinely used technique in clinical applications. One of them is the fine control over the thickness and homogeneity of the coating layer. This is a crucial challenge that needs to be addressed for successful translation of the technique to the clinic, where reproducibility is a significant factor considered for FDA or Health Canada approval [67]. Another challenge is that much more work will need to be done to understand long-term stability and biocompatibility of diazonium modifications in physiological environment. This is especially important for coatings of implants that are designed to be permanent devices in the body. In fact, only 7 studies out of the studies mentioned in this chapter assessed the in-vivo response of the fabricated materials [22, 25, 40, 44, 45, 52, 66]. To the best of our knowledge drug eluting stents from Sinomed are the only devices that have undergone clinical trials and have reached the market. This highlights the need of more long-term preclinical and clinical examination of diazonium-modified materials to have real translation into the healthcare market.

References

1. Qiu Z-Y, Chen C, Wang X-M, Lee I-S (2014) Regenerative Biomater 1(1):67–79.https://doi.org/10.1093/rb/rbu007
2. Ratner BD, Hoffman AS, Schoen FJ, Lemons JE (2004) Biomaterials science: an introduction to materials in medicine. Elsevier. https://doi.org/10.1557/mrs2006.17
3. Kurup A, Dhatrak P, Khasnis N (2021) Mater Today: Proc 39:84–90. https://doi.org/10.1016/j.matpr.2020.06.163
4. Qi P, Maitz MF, Huang N (2013) Surf Coat Technol 233:80–90. https://doi.org/10.1016/j.surfcoat.2013.02.008
5. Chouirfa H, Bouloussa H, Migonney V, Falentin-Daudré C (2019) Acta Biomater 83:37–54. https://doi.org/10.1016/j.actbio.2018.10.036
6. Meyers SR, Grinstaff MW (2012) Chem Rev 112(3):1615–1632. https://doi.org/10.1021/cr2000916
7. Schreiber F (2000) Prog Surf Sci 65(5):151–257. https://doi.org/10.1016/S0079-6816(00)00024-1
8. Park J-W, Kim H, Han M (2010) Chem Soc Rev 39(8):2935–2947. https://doi.org/10.1039/B918135K
9. Vericat C, Vela M, Benitez G, Carro P, Salvarezza R (2010) Chem Soc Rev 39(5):1805–1834. https://doi.org/10.1039/B907301A
10. Cerruti M, Fissolo S, Carraro C, Ricciardi C, Majumdar A, Maboudian R (2008) Langmuir 24(19):10646–10653. https://doi.org/10.1021/la801357v
11. Ohl A, Schleinitz W, Meyer-Sievers A, Becker A, Keller D, Schröder K et al (1999) Surf Coat Technol 116–119:1006–1010. https://doi.org/10.1016/S0257-8972(99)00139-5

12. Chu PK, Chen JY, Wang LP, Huang N (2002) Mater Sci Eng R Rep 36(5):143–206. https://doi.org/10.1016/S0927-796X(02)00004-9
13. Wu G, Li P, Feng H, Zhang X, Chu PK (2015) J Mater Chem B 3(10):2024–2042. https://doi.org/10.1039/C4TB01934B
14. Lu T, Wen J, Qian S, Cao H, Ning C, Pan X et al (2015) Biomaterials 51:173–183. https://doi.org/10.1016/j.biomaterials.2015.02.018
15. Sandomierski M, Voelkel A (2020) J Inorg Organomet Polym Mater 1–21. https://doi.org/10.1007/s10904-020-01725-0
16. Pinson J (2012) Aryl diazonium salts. 1–35. https://doi.org/10.1002/9783527650446
17. Mahjoubi H, Kinsella JM, Murshed M, Cerruti M (2014) ACS Appl Mater Interfaces 6(13):9975–9987. https://doi.org/10.1021/am502752j
18. Chehimi MM (2012) Aryl diazonium salts: new coupling agents in polymer and surface science. Wiley
19. Li W, Wang X, Wu Y, Chen M, Li Z (2021) Surf Coat Technol 407:126802. https://doi.org/10.1016/j.surfcoat.2020.126802
20. Oweis Y, Algizani S, Mezour MA, Alageel O, Abdallah MN, Mahjoubi H et al (2020) J Prosthodont 29(3):251–60. https://doi.org/10.1111/jopr.13129
21. Abdallah M-N, Tran SD, Abughanam G, Laurenti M, Zuanazzi D, Mezour MA et al (2017) Acta Biomater 54:150–63. https://doi.org/10.1016/j.actbio.2017.02.044
22. Mahjoubi H, Buck E, Manimunda P, Farivar R, Chromik R, Murshed M et al (2017) Acta Biomater 47:149–58. https://doi.org/10.1016/j.actbio.2016.10.004
23. Viel P, Le XT, Huc V, Bar J, Benedetto A, Le Goff A et al (2008) J Mater Chem 18(48):5913–5920. https://doi.org/10.1039/B811299A
24. Shaulov Y, Okner R, Levi Y, Tal N, Gutkin V, Mandler D et al (2009) ACS Appl Mater Interfaces 1(11):2519–2528. https://doi.org/10.1021/am900465t
25. Levy Y, Tal N, Tzemach G, Weinberger J, Domb AJ, Mandler D (2009) J Biomed Mater Res Part B: Appl Biomater 91B(2):819–830. https://doi.org/10.1002/jbm.b.31461
26. Alageel O, Abdallah M-N, Luo ZY, Del-Rio-Highsmith J, Cerruti M, Tamimi F (2015) Dent Mater 31(2):105–114. https://doi.org/10.1016/j.dental.2014.11.002
27. Mezour M, Oweis Y, El-Hadad A, Algizani S, Tamimi F, Cerruti M (2018) RSC Adv 8(41):23191–23198. https://doi.org/10.1039/C8RA02634C
28. Shimizu K, Malmos K, Spiegelhauer S-A, Hinke J, Holm AH, Pedersen SU et al (2014) Int J Adhes Adhes 51:1–12. https://doi.org/10.1016/j.ijadhadh.2014.02.002
29. Sandomierski M, Buchwald T, Patalas A, Voelkel A (2020) Sci Rep 10(1):1–10. https://doi.org/10.1038/s41598-020-76360-3
30. Buck E, Lee S, Stone LS, Cerruti M (2021) ACS Appl Mater Interfaces 13(6):7021–7036. https://doi.org/10.1021/acsami.0c16509
31. Bureau C, Haroun F, Henault E (2018) Drug eluting stent with a biodegradable release layer attached with an electro-grafted primer coating. Google Patents
32. Rothfuss LG, Hokett SD, Hondrum SO, Elrod CW (1998) J Prosthet Dent 79(3):270–272. https://doi.org/10.1016/S0022-3913(98)70236-7
33. Matsuda Y, Yanagida H, Ide T, Matsumura H, Tanoue N (2010) J Adhes Dent 12(3):223–229. https://doi.org/10.3290/j.jad.a17550
34. Pan L, Duan L, Zheng Z, Hu Y, Aamir A, Bhuwan S et al (2016) Mater Des 95:555–562. https://doi.org/10.1016/j.matdes.2016.01.076
35. Briem D, Strametz S, Schröoder K, Meenen NM, Lehmann W, Linhart W et al (2005) J Mater Sci: Mater Med 16(7):671–677. https://doi.org/10.1007/s10856-005-2539-z
36. Freed LE, Engelmayr GC Jr, Borenstein JT, Moutos FT, Guilak F (2009) Adv Mater 21(32–33):3410–3418. https://doi.org/10.1002/adma.200900303
37. Dhandayuthapani B, Yoshida Y, Maekawa T, Kumar DS (2011) Int J Polym Sci. https://doi.org/10.1155/2011/290602
38. Jiao Y-P, Cui F-Z (2007) Biomed Mater 2(4):R24. https://doi.org/10.1088/1748-6041/2/4/R02
39. Murphy AR, John PS, Kaplan DL (2008) Biomaterials 29(19):2829–2838. https://doi.org/10.1016/j.biomaterials.2008.03.039

40. Bai S, Zhang M, Huang X, Zhang X, Lu C, Song J et al (2021) Chem Eng J 413:127512. https://doi.org/10.1016/j.cej.2020.127512
41. Vidal G, Blanchi T, Mieszawska AJ, Calabrese R, Rossi C, Vigneron P et al (2013) Acta Biomater 9(1):4935–4943. https://doi.org/10.1016/j.actbio.2012.09.003
42. Wenk E, Murphy AR, Kaplan DL, Meinel L, Merkle HP, Uebersax L (2010) Biomaterials 31(6):1403–1413. https://doi.org/10.1016/j.biomaterials.2009.11.006
43. Chernozem R, Guselnikova O, Surmeneva M, Postnikov P, Abalymov A, Parakhonskiy B et al (2020) Appl Mater Today 20:100758. https://doi.org/10.1016/j.apmt.2020.100758
44. Qian Y, Li L, Song Y, Dong L, Chen P, Li X et al (2018) Biomaterials 164:22–37. https://doi.org/10.1016/j.biomaterials.2018.02.038
45. Song Y, Li L, Zhao W, Qian Y, Dong L, Fang Y et al (2021) Bioact Mater 6(9):2983–2998. https://doi.org/10.1016/j.bioactmat.2021.02.020
46. Mikulska A, Filipowska J, Osyczka AM, Nowakowska M, Szczubiałka K (2015) Front Chem 2:117. https://doi.org/10.3389/fchem.2014.00117
47. Zhang W, Zhang Z, Zhang Y (2011) Nanoscale Res Lett 6(1):1–22. https://doi.org/10.1186/1556-276X-6-555
48. Liu J, Cui L, Losic D (2013) Acta Biomater 9(12):9243–9257. https://doi.org/10.1016/j.actbio.2013.08.016
49. Wei G, Yan M, Dong R, Wang D, Zhou X, Chen J et al (2012) Chem Eur J 18(46):14708–14716. https://doi.org/10.1002/chem.201200843
50. Liu J, Cui L, Kong N, Barrow CJ, Yang W (2014) Eur Polymer J 50:9–17. https://doi.org/10.1016/j.eurpolymj.2013.10.015
51. Lucherelli MA, Raya J, Edelthalhammer KF, Hauke F, Hirsch A, Abellán G et al (2019) Chem A Eur J 25(57):13218–13223. https://doi.org/10.1002/chem.201903165
52. Lucherelli MA, Yu Y, Reina G, Abellán G, Miyako E, Bianco A (2020) Angew Chem 132(33):14138–14143. https://doi.org/10.1002/ange.201916112
53. Wu J, Paudel KS, Strasinger C, Hammell D, Stinchcomb AL, Hinds BJ (2010) Proc Natl Acad Sci 107(26):11698–11702. https://doi.org/10.1073/pnas.1004714107
54. Schoonen L, van Hest JCM (2014.) Nanoscale 6(13):7124–7141. https://doi.org/10.1039/C4NR00915K
55. Hooker JM, Kovacs EW, Francis MB (2004) J Am Chem Soc 126(12):3718–3719. https://doi.org/10.1021/ja031790q
56. Kovacs EW, Hooker JM, Romanini DW, Holder PG, Berry KE, Francis MB (2007) Bioconjug Chem 18(4):1140–1147. https://doi.org/10.1021/bc070006e
57. Schlick TL, Ding Z, Kovacs EW, Francis MB (2005) J Am Chem Soc 127(11):3718–3723. https://doi.org/10.1021/ja046239n
58. Bruckman MA, Kaur G, Lee LA, Xie F, Sepulveda J, Breitenkamp R et al (2008) ChemBioChem 9(4):519–523. https://doi.org/10.1002/cbic.200700559
59. Kaur G, Wang C, Sun J, Wang Q (2010) Biomaterials 31(22):5813–5824. https://doi.org/10.1016/j.biomaterials.2010.04.017
60. Marín-Caba L, Chariou PL, Pesquera C, Correa-Duarte MA, Steinmetz NF (2019) Langmuir 35(1):203–211. https://doi.org/10.1021/acs.langmuir.8b03337
61. Ma D, Chen Z, Yi L, Xi Z (2019) RSC Adv 9(50):29070–29077. https://doi.org/10.1039/C9RA05630K
62. Coburn JM, Na E, Kaplan DL (2015) J Control Release 220:229–238. https://doi.org/10.1016/j.jconrel.2015.10.035
63. Gam-Derouich S, Ngoc Nguyen M, Madani A, Maouche N, Lang P, Perruchot C et al (2010) Surf Interface Anal 42(6–7):1050–6. https://doi.org/10.1002/sia.3210
64. Del Castillo GF-D, Hailes R, Adali-Kaya Z, Robson T, Dahlin A (2020) Chem Commun 56(44):5889–92.https://doi.org/10.1039/D0CC01250E

65. Di Martino A, Guselnikova OA, Trusova ME, Postnikov PS, Sedlarik V (2017) Int J Pharm 526(1–2):380–390. https://doi.org/10.1016/j.ijpharm.2017.04.061
66. Zhao Y, Tong L, Li Z, Yang N, Fu H, Wu L et al (2017) Chem Mater 29(17):7131–7139. https://doi.org/10.1021/acs.chemmater.7b01106
67. Hetemi D, Noël V, Pinson J (2020) Biosensors 10(1):4. https://doi.org/10.3390/bios10010004

On the Use of Diazonium Salts in the Design of Catalytic Hybrid Materials and Coatings

Fatima Mousli, Youssef Snoussi, Mohamed M. Chehimi, and Robert Wojcieszak

Abstract We summarize the existing knowledge on the use of diazonium salts as a new generation of surface modifier and coupling agents to functionalize substrates and nanostructures for applications in photo/catalysis. More specifically, we tackle arylation of carbon allotropes to immobilize organometallic nanocatalyst and monometallic and bimetallic nanoparticles. We also discuss the role of arylation or arylation in situ polymerization sequential steps to tune and enhance the photocatalytic performances of TiO_2 (active under UV) and RuO_2-TiO_2 mixed oxide photocatalysts (active under sunlight). This chapter clearly stresses the important role of diazonium salts in controlling the interfacial and catalytic performances of heterogeneous and immobilized catalysts.

1 Introduction

Sustainable development as well as preservation of the environment is a major issue of the twenty-first century. The concept of green chemistry was proposed over two decades by Anastas and Warner [1]. It is a response from the chemist community to minimize the impact of chemicals and processes on the environment and to balance the negative public opinion on chemistry. The twelve founding principles include limitation of the quantities of solvents, improvement of energy efficiency, use of less harmful chemical syntheses, and use of renewable raw materials and catalysts.

In this context, several researchers have focused their studies on the development of new materials with interesting catalytic performances in order to promote beneficial reactions for the environment such as oxygen reduction reactions (ORR)

F. Mousli
Université Mouloud Mammeri, Tizi-Ouzou, Algeria

F. Mousli · Y. Snoussi · M. M. Chehimi (✉)
Université Paris Cité, CNRS, ITODYS, F-75013 Paris, France
e-mail: mohamed.chehimi@cnrs.fr

R. Wojcieszak
Université de Lille & CNRS, Lille, France

M. M. Chehimi et al. (eds.), *Aryl Diazonium Salts and Related Compounds*,
Physical Chemistry in Action, https://doi.org/10.1007/978-3-031-04398-7_15

[2] and electro-reduction of CO_2 and to reduce the impact of pollution on the environment [3]. Semiconductor metal oxides, noble metals, and some transition metal nanoparticles are interesting catalysts for numerous reactions; their activity depends strongly on several parameters, such as particle size, dispersion [4], method of preparation [5], and materials management [6]. However, the mediocre performances of these catalysts in terms of selectivity, activity, and colloidal stability in organic and aqueous media or in polymeric matrices slow down their development and limit their application [3, 6–8].

To address these issues, research focused on the surface modification of the catalysts by (i) doping with noble metals (Pt, Pd, Au, and Ag) [9–11], metal di/oxide (RuO_2, CuO, Cu_2O, NiO_2...) [12, 13], and organic compounds and polymers [14, 15], or by (ii) surface functionalization with, e.g., alkyl amines. The majority of these studies have led to an improvement of the catalytic performances, particularly in the field of energy storage and heterogeneous photocatalysis which is widely regarded as an interesting route for solving environmental problems [16].

Numerous recent studies have confirmed that the dispersion and electrical conductivity of catalysts are significantly improved after modification with diazonium compounds, resulting in enhanced electrocatalytic performances [17–19]. In addition, modification via arylation allows to provide uniform attachment of functional groups to the surface under test [20].

Diazonium salts present very interesting solution chemistry. In 2007, Taylor and Felpin [21] highlighted the role of diazonium salt in catalysis, especially for selective coupling of Suzuki–Miyaura type in the presence of Pd catalyst supported on carbon (Pd^0/C). The use of tetrafluoroborate diazonium salts made possible a selective coupling on two positions of the aromatic ring also comprising a bromine-containing functional groups (Fig. 1a). This process demonstrates the strong reactivity of the diazonium salt in solution. The bromine function being much less reactive, a base, is necessary for the realization of the second coupling. This double regioselective Suzuki coupling is carried out one-pot, under mild conditions. The use of heterogeneous catalysis, low-polluting solvents, and reaction conditions requiring low energy lead to more practical and environment-friendly processes [22]. This procedure [23] was subsequently adapted using another type of supported catalyst: palladium on barium carbonate, and the coupling took place between a boronic acid and the diazonium salt (Fig. 1b). It is a simple reaction which can be realized under mild conditions without any ligands and base, and it constitutes an eco-compatible alternative compared to the classic reaction.

Commercially available carbon-supported metallic catalysts find less applications due to the corrosion of amorphous carbon during the catalytic process leading to poor durability and blocking of the catalyst activity [24]. Overcoming these problems and improving the performance of the carbonaceous support as well as that of the materials linked to its surface remains challenging task. Nevertheless, carbonaceous materials are widely used as effective catalyst support for numerous reactions, thanks to their high specific surfaces, physicochemical properties, thermal and chemical stability, and electrical conductivity [25]. In addition, diazonium salts can modify the surface of such materials while preserving the inherent, basic properties of each

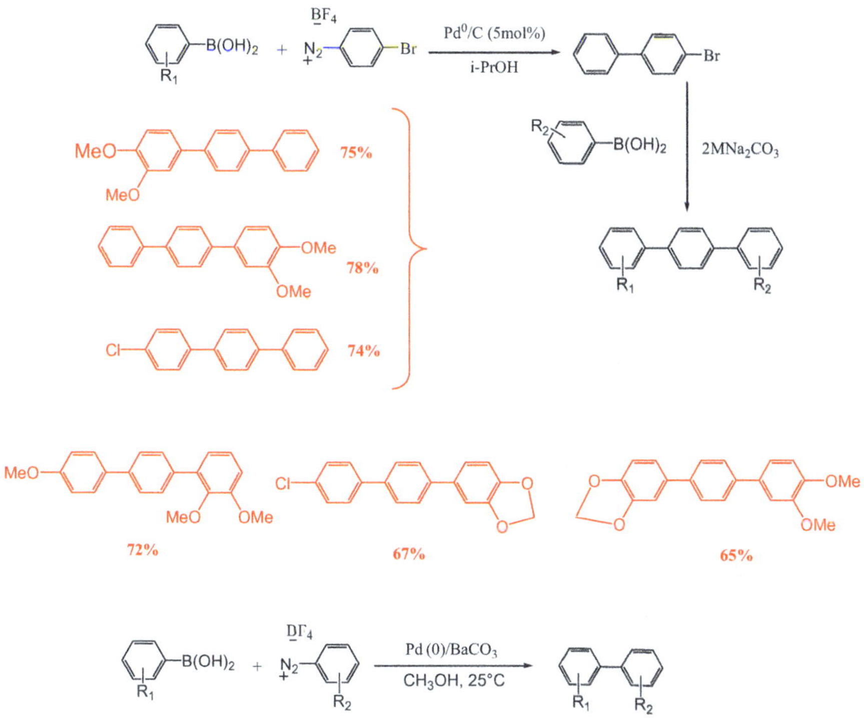

Fig. 1 **a** Double selective one-pot arylation using diazonium salts for the synthesis of terphenyls, **b** C–C coupling using palladium on barium carbonate

component. This provides the basis of research carried out on the use of diazonium salts in the field of heterogeneous catalysis.

In this chapter, we will discuss the recent progress of diazonium salt interface chemistry in design and application of hybrid catalysts.

2 Immobilization of Catalysts on Arylated Surfaces

The immobilization of enzymes onto materials surfaces is known for a long time in the domain of affinity chromatography [26, 27]. The structure of the support materials used for catalyst immobilization has major effect on the catalytic activity [6], i.e., supports with high-surface area promote the dispersion of the active phase on the catalyst surface, thus improving its properties.

Diazonium salts are particularly suited for the modification of carbon allotropes such as carbon nanotubes (CNTs) as they react spontaneously with sp^2 carbon atoms. For this reason, diazonium coupling agents are now routinely used for the robust attachment of nanoparticles to carbon surface [28, 29]. The major drawback of this approach is the resulting sp^3 hybridization which alters the aromaticity of the carbon support and induces low electrical conductivity of CNTs.

Non-covalent modifications of CNTs are also widely exploited; they do not alter the CNT structure and, therefore, do not affect their conduction properties. In addition, the functionalization is easily controlled and the yields are often much superior to that of covalent bonding. However, non-covalent interactions are reversible and therefore less stable than covalent bonds [30].

The modification of CNTs with diazonium salts is a radical reaction; it provides a covalent modification of nanotube sidewalls. The diazonium salts can be prepared (in an ice-bath) in acetonitrile, in concentrated sulfuric acid, or oleum [31], or water [32] or even without solvent [33]; by vigorous stirring in the presence of isoamylnitrite and a substituted aniline. The in situ generated diazonium reacts with a suspension of CNTs in, e.g., acetonitrile [34]. In this way, several substituents can be inserted on the surface CNTs; in addition to simple functions such as nitro, carboxylic acid, chlorine, and thiol, it is also possible to introduce polyamines, diphenylamines, and pyridines [35–37].

Catalysts integration on carbonaceous surface can find applications in various catalyzed processes; some interesting examples are reported below.

Pt is one of the most used but also the most limiting components of fuel cells, due to its high price, which represents 15% of the total cost of the cell [38]. Several studies focused on reducing the Pt mass [39, 40], others on the exploitation of other non-noble metals [41], but the proposed materials must present matching catalytic performances to be competitive with Pt. For example, diazotized catalysts of the general formulae $[Ni(P_2{}^RN_2{}^R)_2]^{2+}(BF_4{}^-)_2$ are interesting candidates for battery applications since they combine the features of the three main families of catalysts: inorganic materials, organic, and bio-molecules [42]. This catalyst can be used to replace platinum in fuel cells. For this reason, several strategies have been developed in order to immobilize Dubois catalyst $\left[Ni\left(P_2^{R_2}N_2^{R_1}\right)_2\right]$ on CNTs to serve as hybrid electrocatalyst (see [43] for the synthesis of Dubois catalysts). In this regard, the collaborative work conducted by Vincent Artero designed CNT-supported Dubois type nickel catalysts via covalent bonding [44], π stacking [45], and electrostatic interactions [46]. The general procedure is simplified and schematically displayed in Fig. 2.

Le Goff et al. [44] covalently grafted $[Ni(P_2{}^RN_2{}^{ester})_2]$ Dubois type catalyst onto MWCNTs previously functionalized via reduction of a 4-(2-aminoethyl)benzenediazonium salt. The amino function allows the formation of a amide linkage with the ester $[Ni(P_2{}^RN_2{}^{ester})_2]$ carrier of an active ester (Fig. 2a). This strategy enabled to achieve anodic currents of 2 mA cm^{-2}, at 500 mV over-voltage, in the study of hydrogen oxidation reaction. In addition, by applying a lower over-voltage of 300 mV, stable currents were recorded for 10 h and which corresponded to a Turn Over Number of 20,000 (±30%), after one hour with a quantity of catalyst estimated to $1.5 \pm 0.5.10^{-9}$ mol cm^{-2}.

The low amounts of catalyst limit the catalytic performances of the final material. In order to address this issue, the method of non-covalent grafting was implemented. Tran et al. [45] immobilized, by non-covalent grafting, a catalyst bearing a pyrene unit

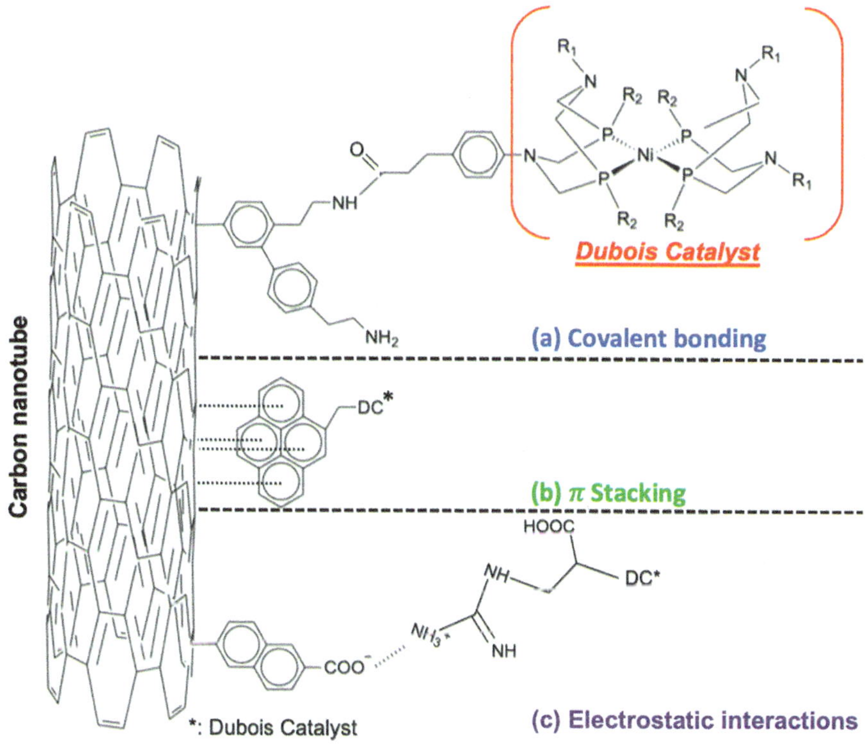

Fig. 2 Schematic representation of the covalent (**a**), non-covalent π stacking (**b**), and electrostatic (**c**) immobilization of Dubois catalyst on the surface of carbon nanotubes. In (**a**), diazonium chemistry permitted covalent grafting by interfacial amidation, whereas in (**c**) immobilization of the Dubois catalyst was permitted by electrostatic interactions between carboxylate group of the arylated CNTs and ammonium moiety borne by the catalyst

on nitrogen via π interactions with MWCNT (Fig. 2b). This led to the same currents but which were obtained at lower deposition extent of catalyst on the sidewall of the MWCNT, i.e., 1.1 ± 1.10^{-8} mol cm^{-2}. This method is therefore particularly interesting for controlling and maximizing the catalyst concentration at the electrode surface. It allows higher surface concentrations to be obtained, but the intrinsic structure of the ligand carrying the pyrene function also constrains the catalytic center due to the small distance between the pyrene function and the amine of the macrocycle. This steric constraint could explain the low currents obtained despite a higher catalyst concentration compared to previous work. It is thus important to stress that although diazonium salts are unique coupling agents for immobilizing large extents of nanocatalysts, they might block the surface and hinder electrocatalysis performance. For this reason, surface arylation remains tricky and might not be the panacea for particular applications where non-covalent modification is preferred.

In 2017, Gentil et al. deposited $[Ni(P_2^{cy}N_2^{Arg})_2]^{8+}$ on MWCNTs functionalized with a diazonium salt bearing a naphthoate unit [46]. These naphthoate functions allow electrostatic interactions with the guanidinium residues of the catalyst (Fig. 2c). The electrostatic interactions allow the catalyst to retain certain flexibility, which is essential for catalysis. The electrochemical characterization provided currents of 16 mA cm^{-2} with a low overvoltage of about 200 mV for a maximum metal charge of the order of $0.2 \pm 0.1 \times 10^{-9}$ mol cm^{-2}. The immobilization via electrostatic interaction is "dynamic," allowing the catalyst to be in favorable conformations to catalyze hydrogen oxidation reaction. The amount of the catalyst is lower compared to the work of Tran et al. [45] due to the covalent modification of the MWCNTs.

Despite these excellent results, the catalytic performances are insufficient to envisage real applications. Several increments remain to be made in order to improve the catalytic performances. This is why new electrode architecture has been developed through a strategy based on the double non-covalent modification of MWCNTs.

The immobilization of the catalyst can also be done by post-functionalization of reactive surfaces through electrochemical reduction of diazonium compounds. This procedure has been shown to be very effective in immobilizing incompatible molecules by electrografting, making it possible to obtain mixed layers. This very simple method is performed by reacting a mixture of catalytic molecules with a surface previously functionalized by a reactive layer [47] as schematically illustrated in Fig. 3.

The final material obtained by this technique can have enhanced catalytic properties due to a synergistic effect [48], or they can generate a succession of catalytic reactions through various functional sites [49].

Zhang et al. [49] have developed a reactive material containing two groups that are prone to post-functionalization by co-grafting vinylbenzene diazonium and azidobenzene diazonium in order to immobilize two catalysts. The first one is the biomacromolecule NAD (D-sorbitol dehydrogenase), and the other, molecular $[Cp^*Rh(bpy)Cl]^+$, allowing NADH to be regenerated by electrochemical reduction of NAD^+ (Fig. 4). The electrode modified by these two catalysts allowed a series of

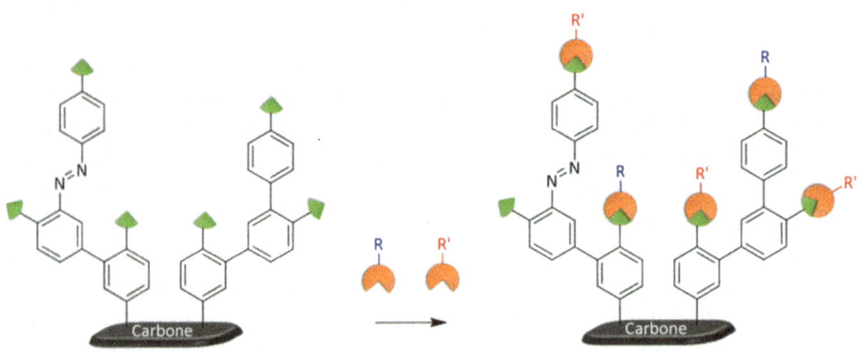

Fig. 3 Mixed post-functionalization reaction between a reactive platform and a mixture of two catalytic reagents

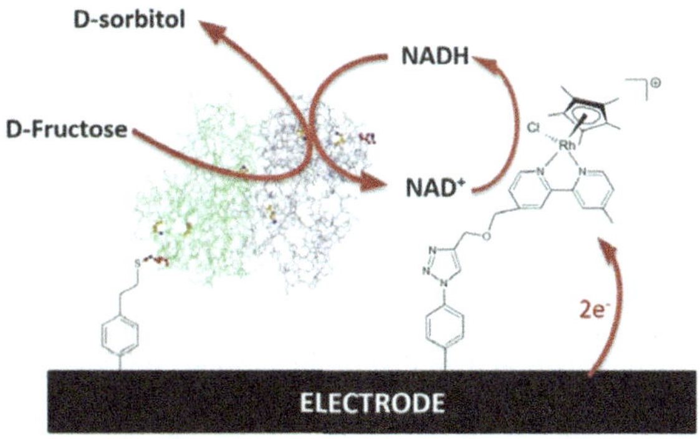

Fig. 4 Electrocatalytic reduction of D-fructose by an electrode functionalized by a mixed layer containing two centers. Reproduced with permission from [49] Copyright 2018 John Wiley & Sons

reactions necessary for the reduction of D-fructose in the presence of NAD^+ in solution. Actually, the electrodes compatible with biological media, obtained by post-functionalization for enzymatic catalysis, require a high resistance of the organic layers to nonspecific adhesion and must contain active centers allowing detection of analytes.

Following the same method, Binding and his collaborators developed a heterogeneous catalyst consisting of two catalytic materials (complexes of rhodium and iridium) for hydroamination [48]. They have shown that mixed layers lead to more efficient catalysts than their monometallic equivalents.

A good distribution of catalytic nanoparticles is a very important factor in improving photocatalytic processes. Although CuO nanoparticles can be directly synthesized on the surface of MWCNTs, Bhakta et al. [50] proposed to use MWCNT surface functionalized with 5-amino-1,2,3-benzenetricarboxylic acid in order to ensure good and homogeneous catalyst distribution.

Deposition of NPs on the surface of the CNTs could be facilitated by their functionalization with oxygen-containing groups such as quinonyl, carboxyl, or hydroxyl. This treatment can generally be carried out with strong oxidizing acidic solutions which, however, can damage the structure of the CNTs. To overcome this problem, Guo and Li have chemically deposited Pt [51] and Pd [52] nanoparticles on CNTs for electrocatalytic application. Their research was carried out with a dual electrochemical treatment consisting in the oxidization of single-walled carbon nanotubes (SWCNTs) with concomitant introduction of C=O, OH, and COOH groups followed by adsorption of metal salts and their in situ reduction to metallic Pt nanoparticles [51]. Instead, the same procedure was repeated with arylation of the surface followed by in situ deposition of palladium salt and in situ reduction to Pd NPs [52]. Arylation permitted to functionalize the 2D carbon material without damaging them. This work has been discussed at length previously [53, 54] and opened new horizons for immobilizing nanocatalysts on arylated carbon allotropes. This could be done by immobilizing pre-fabricated catalysts or chemisorption of metal ions followed by in situ reduction.

Vila et al. [55] reported no synergistic effect in the Cu-Pt bimetallic system supported on a carbon surface grafted with sulfonate groups. The charged sulfonate groups and metallic ions in solution undergo electrostatic interactions. Nevertheless, the catalytic performance was excellent for the electro-reduction reaction of nitrates and thanks to the presence of the organic groups that inhibited platinum aggregation.

Mirzaei et al. [56] exploited the surface chemistry of diazonium salts for the modification of High-surface-area graphite (HSAG) particles. They obtained reactive and well-dispersed bimetallic Cu100-xRhx nanocatalysts on HSAG arylated with thiophenyl groups by in situ metal ion reduction (Fig. 5A). The authors demonstrated that the average size of the nanocatalyst is narrower under an increased grafting rate. Ungrafted composites $Cu_{48}Rh_{52}$/HSAG and $Cu_{65}Rh_{35}$/HSAG present larger size distribution than that of the grafted composite, with average size of 2.2 nm for the ungrafted samples (1.7 nm for $Cu_{48}Rh_{52}$/HSAG and 5.5 for $Cu_{65}Rh_{35}$/HSAG). The size distribution narrows with increasing grafting rate, the standard deviation decreases by a factor of 2 (0.4 instead of 0.8 nm for non-grafted sample) (Fig. 5B) displays the size of the NPs determined from TEM images.

Moreover, during the catalytic study of nitrate electro-reduction by cyclic voltammetry, they demonstrated that the increase in the grafting rate results in lower current intensities and a shift of the reduction peak potential toward the cathodic values. In contrast, at a low rate of grafting, the electrochemical activity increases by 40% owing to significant increase in the catalytic activity of the materials (Fig. 5C).

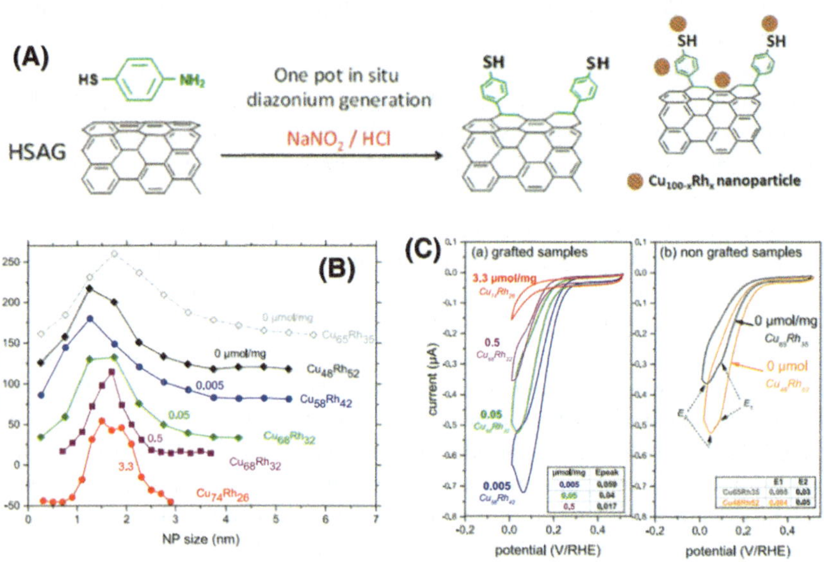

Fig. 5 **A** Synthesis route for grafting benzenethiol groups on HSAG and formation of RhCu/HSAG nanocomposite **B** NP size histograms determined from TEM images, **C** voltammograms recorded with a cavity microelectrode (CME) at 10 mV s^{-1} in 10^{-1} mol L^{-1} KOH and 10^{-2} mol L^{-1} KNO_3 for the $Cu_{100-x}Rh_x$/HSAG composites (**a**) for different grafting and (**b**) with no grafting materials. Adapted with permission from [56], Copyright 2019, American Chemical Society

In a simpler way, and given the commercial availability of gold nanoparticle suspensions, one could disperse gold NPs over an arylated carbon support. Toward this end, Bensghaïer et al. [57] have designed arylated CNT@Au NPs as electrocatalyst for the direct methanol oxidation (DMO). Congo Red (CR) dye was diazotized and reacted in water at RT with CNTs. The resulting CNT-CR nanohybrid, coined "painted" CNTs, was coated on glassy carbon electrode and further decorated with gold NPs (Fig. 6a). Figure 6c shows chronoamperometric curves with a forward anodic wave corresponding to methanol oxidation (the current density of which is J_f), whereas the backward cathodic is assigned to the oxidation of side products, mainly CO. Under optimized conditions, the J_f/J_b ratio was found to be as high as 1.68. Moreover, this electrocatalyst hybrid exhibited remarkable durability as judged from current density (*j*)-versus-time (*t*) chronoamperometric curves.

Mesoporous carbons are widely used in various applications such as water and air purification, chromatography, nanotechnology, energy storage, and catalysis [58–60]. This is due to their pore structure, electrical conductivity, chemical inertness, and high thermal and mechanical stability. However, their inert and hydrophobic nature is unfavorable from the point of view of catalytic processes, which requires a functional and reactive surface with a high specific affinity [61]. Toward this end, Wang et al. [62] demonstrated that the functionalization of mesoporous carbons with sulfonic

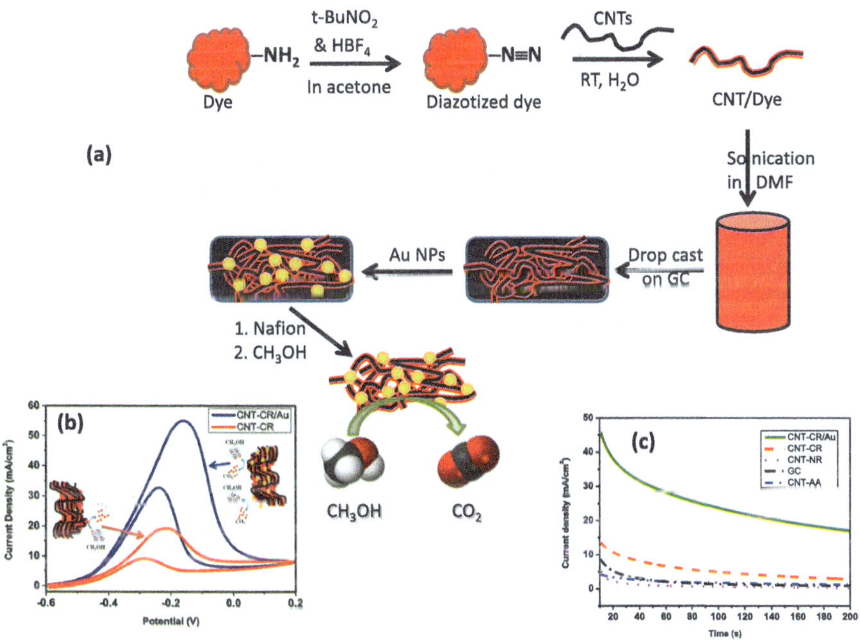

Fig. 6 Arylated CNT@Au NPs for DMO application: **a** design of the electrocatalyst and its deposition on GC, **b** methanol oxidation over CNT-CR/Au and CNT-CR and determination of J_f/J_b, **c** current density-versus-time chronoamperometric curves of DMO over CNT-CR/Au and reference materials. Reprinted with permission from [57] Copyright 2021 Springer Nature

groups provides highly catalytically active carbonaceous solid catalysts. This system is characterized by an ordered structure tolerant to H_2O with a strong acidity which is due to the presence of the sulfonated groups. Goscianska and Malaika [63] reported the preparation of acid catalysts based on mesoporous carbons through modification with the diazonium salt of 4-aminobenzenesulfonic acid (SO_3H) for glycerol acetylation reaction. The design of this material was carried out in two stages: the first one was the modification of carbon with concentrated sulfuric acid and the second step included the reaction of the carbon with the aryl diazonium cation obtained by in situ generation from the sulfanilic acid. More than 95% of the glycerol was converted to di- and triacetins in the presence of modified mesoporous carbon. It turned out that the catalytic efficiency of the material strongly depends on the number of SO_3H groups introduced into the carbon structure; this is an important parameter in the design of this kind of catalytic material. Besides grafting with diazonium salts, functionalization of mesoporous carbon by sulphonic groups can be achieved by built-in and post-grafting. The former seems the easiest method for mesoporous carbon modification; it consists in treating the surface directly with concentrated sulfuric acid. It is a very simple method which allows adjusting the amount of SO_3H with the concentration of sulfuric acid. This process of surface functionalization is difficult to combine with soft-template synthesis of mesoporous carbons, due to the low thermal stability of the C-SO_3H fragments, and the effect of the precursor S in the design of the micellar system mesophase [64]. The post-grafting route for modifying the carbon surface with sulfonic groups involves the use of various sulfonation reagents such as concentrated sulfuric acid, H_2SO_4, $ClSO_3H$, 4-benzenediazonium sulfonate, SH, or SO_3H terminated organosilanes and phenylvinylsulfonate [65–67]. This approach permits to modify the carbon surface without compromising its mesostructure [63].

Several other studies which have been carried out on the process for immobilizing nanoparticles and enzymes on surfaces grafted with diazonium are gathered in Table 1.

Catalyst immobilization via the surface diazonium salt chemistry permits to bind the catalytic layer on the functionalized surface. Although diazonium salts provide catalysts with better dispersion, size distribution, stability, and excellent catalytic activity, the opposite effect is not excluded, e.g., a thick electrochemically grafted aryl layer leads to electrode passivation; hence, the interest in optimizing all parameters thus makes it possible to improve the architecture of the final nanocomposite for the targeted application.

3 Diazonium Modification of Catalysts

As mentioned above, diazonium salts are very good coupling agents making it possible to bind several organic and inorganic materials to the catalysts, in order to improve the performances of the catalytic system.

Metals and metallic oxides are the most promising materials for heterogeneous photo/catalysis. TiO_2 is one of these oxides known for its photoactivity under UV. It

Table 1 Nanoparticles immobilized on aryl group-grafted surfaces

Modified catalyst/substrate	Grafted diazonium salt	Final catalytic material	Catalytic application	References
Pt	Aryl-R (R = $-CH_3$, $-F$, $-Cl$ $-OCF_3$, and $-CF_3$)	Pt-Ar-R	Oxygen reduction reactions (ORR)	[68]
MWCNTs	Anthraquinone-1-diazonium chloride 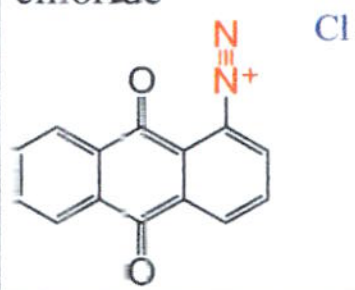	AQ-MWCNTs	Oxygen reduction reactions (ORR)	[69]
Olive Pit (OP)	Thiol (–SH), amine ($-NH_2$) and carboxylic acid (–COOH)	OP-SH, OP-COOH, and OP-NH_2	Reduction of nitrophenol reaction	[70]
RuO_2-TiO_2	4-diphenylamine diazonium tetrafluoroborate 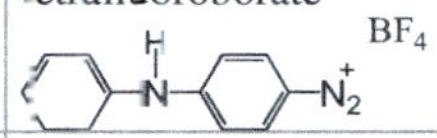	Cotton fabric/RuO_2-TiO_2/DAP/PANI	Methyl orange degradation	[71]
Pt	4-chlorophenyl diazonium	Pt-ArCl	Oxygen reduction reactions (ORR)	[72]
TiO_2	4-sulfanilic acid diazonium 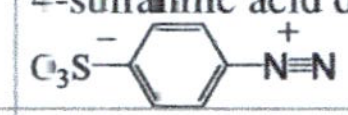	TiO_2/diazonium/graphene oxide	Methylene blue degradation	[73]
Graphene	*p*-aminophenyl	Pd/functionalized graphene (fG) (PdNCs/fG)	Formic acid oxidation	[74]

(continued)

Table 1 (continued)

Modified catalyst/substrate	Grafted diazonium salt	Final catalytic material	Catalytic application	References
Reduced graphene	4-aminothiophenol 	Pt@SC_6H_4–rGO	Oxidation of methanol and ORR	[19]
SWCNT	Aminoethyl/ethylenediamine 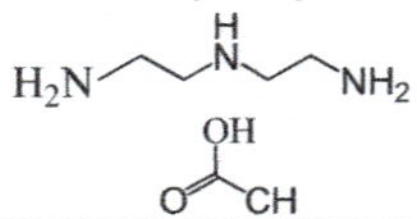	SWCNT-CO-NH-$(CH_2)_2$-NH_2/laccase bioelectrode	ORR	[75]
MWCNTs	Diazotized Congo red (CR) dye 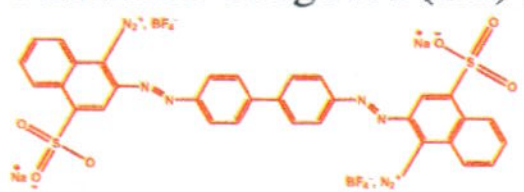	CNT-CR with immobilized Au NPs	Direct methanol oxidation (DMO)	[57]

has indeed raised much interest due to its chemical stability [76], photostability [77], high dielectric constant [78]; physical, optical, and electrical properties [79]; low cost, and nontoxicity [80]. However, the use of the majority of metallic oxides, in particular TiO_2, is limited due to the high rate of recombination of the photo-induced electron–hole pairs produced during photocatalytic processes, few active sites, light scavenging capacity, and negligible absorbance of visible light. Efforts have been made to develop materials with a lower rate of electron–hole pair recombination and high absorption of visible light. This involves the heterostructures that combine metallic oxides with other inorganic (metals, metal oxides) and organic materials (e.g., polymers) in order to achieve a narrow bandgap [10].

Photocatalysts based on TiO_2 heterostructure have been the subject of numerous studies in the last decade. They have shown great potential for application in the degradation of a large variety of organic pollutants and the fractionation of water for the generation of hydrogen due to the excellent photocatalytic performances which are often superior to those of pristine TiO_2 [81]. This activity depends sometimes on the amount of material employed to modify titania.

In previous works [16, 82], some of us have identified the spectacular effect of diazonium salt modification of nanocatalysts on their dispersion, catalytic properties, and stability. Nanocomposite catalysts based on TiO_2 nanoparticles (NPs) were prepared by sol gel route and functionalized with diphenyl amino groups (DPA) from the 4-diphenylamine diazonium salt precursor (reduced with ascorbic acid) and then topped with polyaniline (PANI) [16]. The resulting nanocomposites catalyzed the degradation and the mineralization of Methyl Orange (MO), a model organic pollutant. The analysis of these materials revealed a strong adhesion of PANI on NPs surface thanks to the diazonium salt which provides adhesive layer to the TiO_2 surface and a unique coupling agent for the polymer through the oxidation of the diphenyl amino groups. Figure 7 depicts the proposed mechanism of covalent bond formed between the oxide NPs and DPA; it describes the initiation and growth of the polymer chain at TiO_2 surface. The mechanism is based on the conclusions of Aït Atmane et al. [83], who demonstrated by XPS and ToF-SIMs—Time-of-Flight Secondary Ion Mass Spectroscopy that modification of metal oxides with diazonium salts leads to the formation of an interfacial Metal–O–C(aryl) bond.

These materials exhibit superior catalytic performances in MO degradation in aqueous media under UV light. The degradation of the dye is higher in the presence of TiO_2-PANI than with unmodified TiO_2 nanoparticles as demonstrated by Salem et al. [84] and Reddy et al. [85]. This is due to the conductive polymer loaded on the nanocatalyst surface and its role during the photocatalyzed degradation of allura red, quinoline yellow [84], Rhodamine B (RhB), Methylene Blue (MB), and phenol [85] under UV light. The degradation percentage is twice higher than that obtained with TiO_2-PANI (36.2% after 5 min under irradiation compared to 18% for TiO_2-PANI). The degradation rate constant (Kapp) was 0.133 min^{-1} for TiO_2-DPA-PANI nanocomposite, much higher than 0.059 and 0.085 min^{-1} found for TiO_2 and TiO_2-PANI, respectively (Fig. 8A). Total discoloration of MO solutions was achieved without any spectral evidence of organic byproducts (Fig. 8B(a)), thus suggesting complete mineralization of the dye (Fig. 8B(b)).

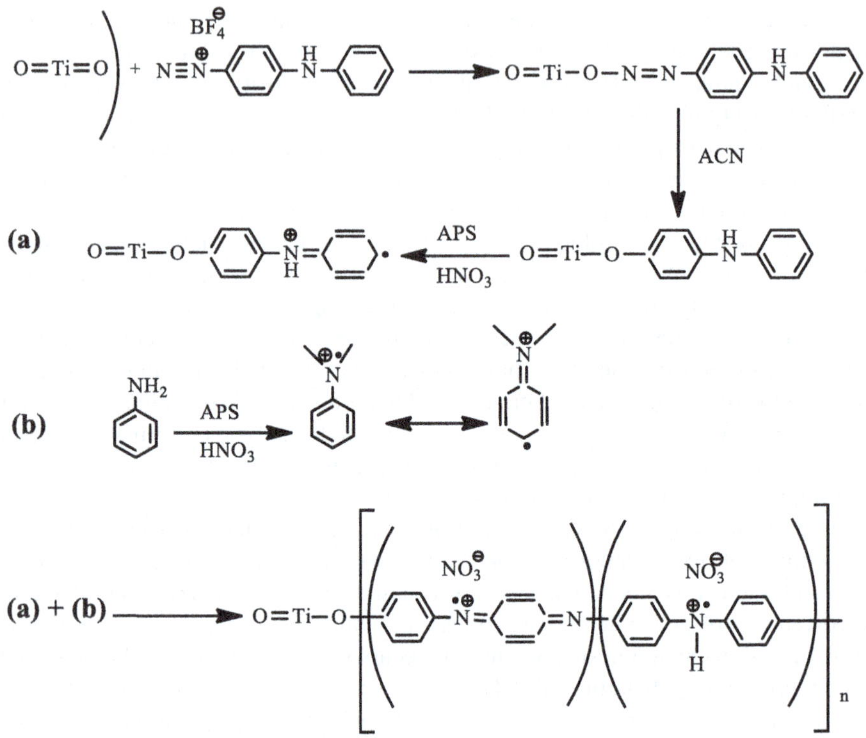

Fig. 7 Chemical route for preparing TiO_2-DPA-PANI nanocomposite adapted with permission from [16] Copyright 2018 Elsevier

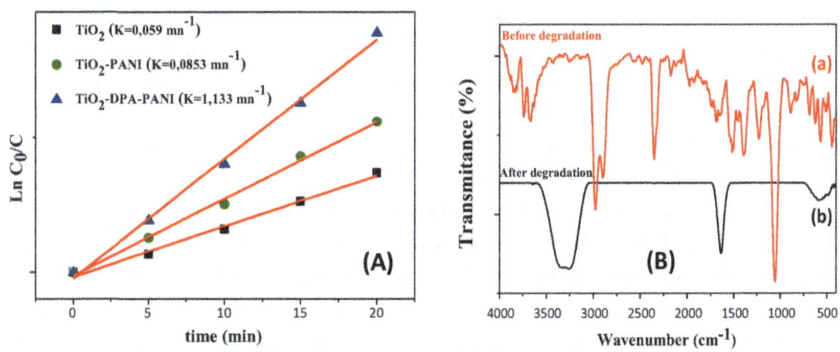

Fig. 8 **A** First order linear transforms of the degradation of methyl orange under UV irradiation, **B** FTIR spectra of methyl orange solution before (**a**), and after (**b**) photocatalyzed degradation using TiO_2-DPA-PANI. Reproduced with permission from [16] Copyright 2018 Elsevier

TiO_2 NPs were modified with RuO_2 in order to obtain a material with better physicochemical properties and a super catalytic power. The RuO_2-TiO_2 (metal-n-semiconductor) heterostructure was prepared by a sol gel route. Functionalization of the heterostructure surface and the preparation of RuO_2-TiO_2/DPA/PANI nanocomposite were carried out by the same method and under the same conditions as in the presence of TiO_2 (Fig. 9A). Functionalization with the diazonium salt provides a thick layer of PANI as shown in the digital photographs of pristine heterostructure (Fig. 9B(a)) and modified surfaces (Fig. 9B(b–d)). Diazonium grafting is manifested by a color change to a pale green for RuO_2-TiO_2 mixed oxide (Fig. 9B(b)). A slight color change was observed for the nanocomposite RuO_2-TiO_2/PANI and RuO_2-TiO_2/DPA, despite the striking color of PANI (Fig. 9B(c)). In the case of PANI polymerized on the functionalized surfaces of the mixed oxides, the original color of the latter is no longer visible and tends toward a dark green due to PANI which covers the entire surface previously modified with DPA (Fig. 9B(d)). This reflects the spectacular role of DPA salt in the polymerization process of PANI on TiO_2 and RuO_2-TiO_2 surface, as well as its effect on the optical properties of the nanocomposite.

The in situ polymerization of PANI on the TiO_2 surface allowed shifting the TiO_2 absorption zone from UV to visible light and from visible light complete light absorption for RuO_2 TiO_2 heterojection. The DPA layer causes a displacement of the electrons from the valence band (HOMO) to the conductive strip (LUMO) of the PANI and increases its conductance. The separation of the electron–hole pairs is at the origin of the improvement of the catalytic properties of the NPs and the acceleration of the photodegradation and mineralization of MO dye. In addition, the functionalization with the diazonium salt leads to a thick layer of PANI that constitutes a protective coating on the oxides nanoparticles; this permits recycling the catalyst several times: 2, 4, and 6 times for TiO_2, TiO_2/PANI and TiO_2-DPA-PANI, respectively, under visible light (Fig. 10A). No decrease in the MO degradation rate could be observed until the 6th, 8th, and 10th test, respectively, in the presence of RuO_2-TiO_2, RuO_2 TiO_2/PANI and RuO_2-TiO_2/DPA/PANI and in the absence of irradiation which shows a promising potential for practical applications.

4 Conclusion

This chapter reviews the surface modification strategies of a wide range of nanomaterials including metals, semi-conductors, simple, mixed, and complex oxides, carbon nanostructure, clays, and even organic polymers using aryl diazonium salts to obtain materials with high catalytic performances active under and without irradiation. The catalyst immobilization on surface functionalized with diazonium salts is a very effective strategy; in particular, it permits the study the impact of the anchoring point on the catalytic performances. The size and distribution of the catalysts are very important factors in photo/catalytic processes; they require controlling the grafting density of the aryl groups. Actually, increasing the grafting surface concentrations

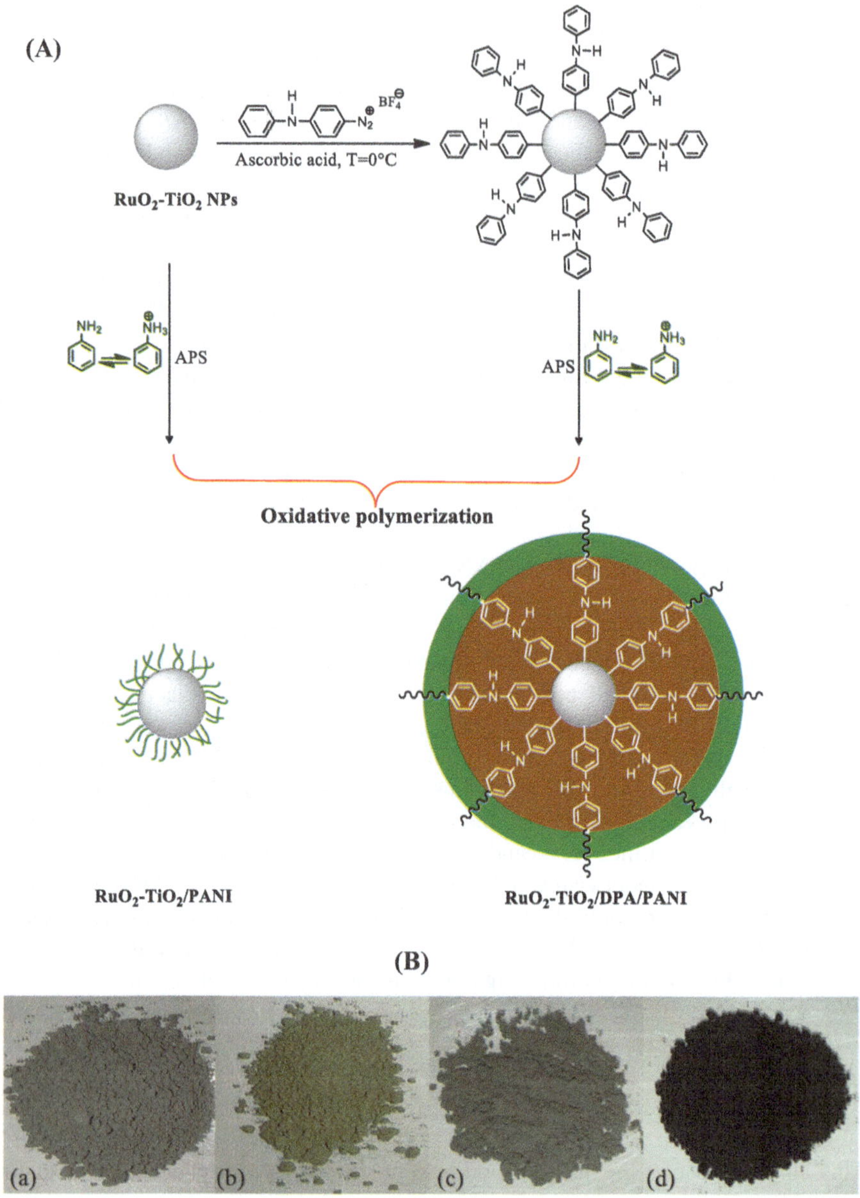

Fig. 9 Synthesis of RuO_2-TiO_2/PANI nanocomposite (upper panel: Fig. 2A) and digital photographs of pristine RuO_2-TiO_2 (**a**), RuO_2-TiO_2-DPA (**b**), RuO_2-TiO_2-PANI (**c**), and RuO_2-TiO_2-DPA-PANI (**d**) (lower panel: Fig. 2B). Adapted from [82, open access]

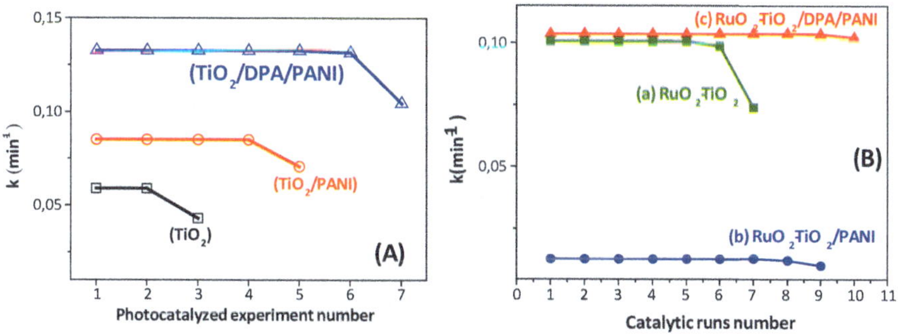

Fig. 10 Plots of kinetic constants-versus-photocatalyzed experiment number for **A** TiO_2, TiO_2-PANI and TiO_2-DPA-PANI, and **B** RuO_2-TiO_2 (**a**), RuO_2-TiO_2-PANI (**b**) and RuO_2-TiO_2-DPA-PANI (**c**). Adapted from [82]

leads to a decrease in the size distribution of the final catalyst; but in electrochemistry, a thick layer of diazonium leads to the passivation of the electrode surface and blocks its electroactivity. Therefore, the optimization of the surface concentration through the conditions under which the functionalization process is carried out (grafting method, time, temperature, pH, nature of the surface to be grafted, and the choice of the aryl salt) is crucial. It permits to obtain materials with spectacular properties.

The literature survey conclusively indicates the growing interest of the surface chemistry of diazonium salts in the field of catalysis either by dispersing nanocatalysts on arylated surfaces or by tuning surface properties of nanocatalysts via arylation using diazonium compounds. In this sense, catalyst arylation option permitted to design catalytic RuO_2-TiO_2/DPA/PANI nanocomposite that is active in darkness, without any UV or visible light assistance. Moreover, considerable stability was achieved enabling the recovery and the reuse of the same catalyst up to 9 times. At the time of finalizing this chapter, a new study reported on the fine tuning of electrocatalytic performances of gold nanoparticles via dual arylation. The mixed aryl layer on gold nanoparticles efficiently permitted to electrocatalyze ethanol oxidation reaction for fuel cell application [86]. One thus should expect new directions in structuring interfaces with aryl diazonium salts in the general domain of catalysis.

References

1. Anastas PT, Warner JC (1998) Principes de la chimie verte. Chimie verte: théorie et pratique, pp 29–56
2. Saada TN, Pang L, Sravan KK, Dourado AH, Germano LD, Vicentini ED, Batista APL, de Oliveira-Filho AGS, Dumeignil F, Paul S, Wojcieszak R, Melinte S, Sandu G, Petretto G, Rignanese GM, Braga A H, Rosado TF, Meziane D, Boukherroub R, de Torresi SIC, da Silva AGM, Szunerits S (2021) L'importance de la forme des nanocristaux de Cu_2O sur la réaction de dégagement d'oxygène sous plasmon en milieu alcalin. Electrochim Acta 390:138810

3. Zouaoui N (2017) Fonctionnalisation de la surface de plomb pour l'électroréduction du CO_2. Doctoral dissertation, Université du Québec, Institut National de la Recherche Scientifique. http://espace.inrs.ca/id/eprint/6956/1/Zouaoui%2C%20Nidhal.pdf
4. Azizullah A, Khattak MNK, Richter P, Häder DP (2011) Water pollution in Pakistan and its impact on public health—a review. Environ Int 37:479–497
5. Bazargan MH, Byranvand MM, Kharat AN (2012) Preparation and characterization of low temperature sintering nanocrystalline TiO_2 prepared via the sol gel method using titanium (IV) butoxide applicable to flexible dye sensitized solar cell. Int J Mat Res 103:347–351
6. Guo DJ, Li HL (2005) Highly dispersed Ag nanoparticles on functional MWNT surfaces for methanol oxidation in alkaline solution. Carbon 43(6):1259–1264
7. Griffete N, Herbst F, Pinson J, Ammar S, Mangeney C (2011) Preparation of water-soluble magnetic nanocrystals using aryl diazonium salt chemistry. J Am Chem Soc 133:1646–1649
8. Mesnage A, Magied MA, Simon P, Herlin-Boime N, Jégou P, Deniau G, Palacin S (2011) Grafting polymers to titania nanoparticles by radical polymerization initiated by diazonium salt. J Mater Sci 46:6332–6338
9. Gesesse GD, Wang C, Chang BK, Tai SH, Beaunier P, Wojcieszak R, Remita H, Colbeau-Justin C, Ghazzal MN (2020) A soft-chemistry assisted strong metal–support interaction on a designed plasmonic core–shell photocatalyst for enhanced photocatalytic hydrogen production. Nanoscale 12(13):7011–7023
10. Ferraz PC, Costa NJ, Teixeira-Neto E, Teixeira-Neto ÂA, Liria CW, Thuriot-Roukos J, Machini MT, Froidevaux R, Dumeignil F, Rossi LM, Wojcieszak R (2020) Oxydation sans base de 5-hydroxyméthylfurfural et de furfural sur des nanoparticules bimétalliques intégrées AuPd. Catalyseurs 10(1):75
11. Silvester L, Ramos F, Thuriot-Roukos J, Heyte S, Araque M, Paul S, Wojcieszak R (2019) Fully integrated high-throughput methodology for the study of Ni- and Cu-supported catalysts for glucose hydrogenation. Catal Today 338:72–80
12. Chen S, Ciotonea C, Ungureanu A, Dumitriu E, Catrinescu C, Wojcieszak R, Dumeignil F, Royer S (2019) Preparation of nickel (oxide) nanoparticles confined in the secondary pore network of mesoporous scaffolds using melt infiltration. Catal Today 334:48–58
13. Shi D, Yang Q, Peterson C, Lamic-Humblot AF, Girardon JS, Griboval-Constant A, Lorenzo S, Moulay TS, Valérie B, Paul AJB, Wojcieszak R, Paul S, Eric M, Marceau E (2019) Bimetallic Fe-Ni/SiO_2 catalysts for furfural hydrogenation: identification of the interplay between Fe and Ni during deposition-precipitation and thermal treatments. Catal Today 334:162–172
14. Alwin E, Kočí K, Wojcieszak R, Zieliński M, Edelmannová M, Pietrowski M (2020) Influence of high temperature synthesis on the structure of graphitic carbon nitride and its hydrogen generation ability. Materials 13:2756
15. Muhd Julkapli N, Bagheri S, Bee Abd Hamid S (2014) Recent advances in heterogeneous photocatalytic decolorization of synthetic dyes. Sci World J
16. Mousli F, Chaouchi A, Hocine S, Lamouri A, Vilar MR, Kadri A, Chehimi MM (2019) Diazonium-modified TiO_2/polyaniline core/shell nanoparticles. Structural characterization, interfacial aspects and photocatalytic performances. Appl Surf Sci 465:1078–1095
17. Lomeda JR, Doyle CD, Kosynkin DV, Hwang WF, Tour JM (2008) Diazonium functionalization of surfactant-wrapped chemically converted graphene sheets. J Am Chem Soc 130:16201–16206
18. Zeb G, Gaskell P, Le XT, Xiao X, Szkopek T, Cerruti M (2012) Decoration of graphitic surfaces with Sn nanoparticles through surface functionalization using diazonium chemistry. Langmuir 28:13042–13050
19. Ensafi AA, Jafari-Asl M, Rezaei B (2014) A new strategy for the synthesis of 3-D Pt nanoparticles on reduced graphene oxide through surface functionalization, application for methanol oxidation and oxygen reduction. Electrochim Acta 130:397–405
20. Park J, Yan M (2013) Covalent functionalization of graphene with reactive intermediates. Acc Chem Res 46:181–189
21. Taylo RH, Felpin FX (2007) Suzuki−Miyaura reactions of arenediazonium salts catalyzed by Pd(0)/C. One-pot chemoselective double cross-coupling reactions. Org Lett 9:2911–2914

22. Felpin FX, Fouquet E (2010) A useful, reliable and safer protocol for hydrogenation and the hydrogenolysis of O-benzyl groups: the in situ preparation of an active Pd^0/C catalyst with well-defined properties. Chem Eur J 16:12440–12445
23. Felpin FX, Fouquet E (2008) Efficient and practical cross-coupling of arenediazonium tetrafluoroborate salts with boronic acids catalyzed by palladium (0)/barium carbonate. Adv Synth Catal 350:863–868
24. Wang Y-J, Wilkinson DP, Zhang J (2011) Noncarbon support materials for polymer electrolyte membrane fuel cell electrocatalysts. Chem Rev 111:7625–7651
25. Li H, Zhang X, Pang H, Huang C, Chen J (2010) PMo12-functionalized graphene nanosheet-supported PtRu nanocatalysts for methanol electro-oxidation. J Solid State Electrochem 14:2267–2274
26. Gübitz G (1988) In selective sample handling and detection in high-performance liquid chromatography. J Chromatogr Libr A 39A:145 (Frei RW, Zech K (eds))
27. Mohamed AA, Salmi Z, Dahoumane SA, Mekki A, Carbonnier B, Chehimi MM (2015) Functionalization of nanomaterials with aryldiazonium salts. Adv Colloid Interface Sci 225:16–36
28. Ossonon BD, Belanger D (2017) Functionalization of graphene sheets by the diazonium chemistry during electrochemical exfoliation of graphite. Carbon 111:83–93
29. Chehimi MM (ed) (2012) Aryl diazonium salts: new coupling agents in polymer and surface science. Wiley
30. Donck S (2015) Catalyse supportée sur nanotubes de carbone. Doctoral dissertation, Université Paris-Saclay. https://tel.archives-ouvertes.fr/tel-01285212/document
31. Hudson JL, Casavant MJ, Tour JM (2004) Water-soluble, exfoliated, nonroping single-wall carbon nanotubes. J Am Chem Soc 126(36):11158–11159
32. Price BK, Tour JM (2006) Functionalization of single-walled carbon nanotubes. J Am Chem Soc 128(39):12899–12904
33. Dyke A, Tour JM (2004) Overcoming the insolubility of carbon nanotubes through high degrees of sidewall functionalization. Chem Eur J 10(4):812–817
34. Bahr JL, Tour JM (2001) Highly functionalized carbon nanotubes using in situ generated diazonium compounds. Chem Mater 13(11):3823–3824
35. Singh P, Samorì C, Toma FM, Bussy C, Nunes A, Al-Jamal KT, Menard-moyon C, Prato M, Kostarelos K, Bianco A (2011) Polyamine functionalized carbon nanotubes: synthesis, characterization, cytotoxicity and siRNA binding. J Mater Chem 21(13):4850–4860
36. Floch FL, Thuaire A, Bidan G, Simonato JP (2009) The electrochemical signature of functionalized single-walled carbon nanotubes bearing electroactive groups. Nanotechnology 20(14):145705
37. Mann JA, Dichtel WR (2013) Improving the binding characteristics of tripodal compounds on single layer graphene. ACS Nano 7(8):7193–7199
38. Thompson ST, Papageorgopoulos D (2019) Platinum group metal-free catalysts boost cost competitiveness of fuel cell vehicles. Nat Catal 2(7), Art no 7
39. Mora C, Spirandelli D, Franklin EC, Lynham J, Kantar MB, Miles W, Hunter CL (2018) Broad threat to humanity from cumulative climate hazards intensified by greenhouse gas emissions. Nat Clim Change 8(12):1062–1071
40. Zappa W, van den Broek M (2018) Analysing the potential of integrating wind and solar power in Europe using spatial optimisation under various scenarios. Renew Sustain Energy Rev 94:1192–1216
41. Jaouen F, Jones D, Coutard N, Artero V, Strasser P, Kucernak A (2018) Toward platinum group metal-free catalysts for hydrogen/air proton-exchange membrane fuel cells. Technol Rev 62:231–255
42. Wilson AD, Newell RH, McNevin MJ, Muckerman JT, Rakowski DuBois M, DuBois DL (2006) Hydrogen oxidation and production using nickel-based molecular catalysts with positioned proton relays. J Am Chem Soc 128(1):358–366. https://doi.org/10.1021/ja056442y
43. Schild J (2020) Catalyse supportée sur nanotubes de carbone pour la production d'énergie bas carbone. Doctoral dissertation, Université Paris-Saclay. https://tel.archives-ouvertes.fr/tel-03103776/document

44. Le Goff A, Artero V, Jousselme B, Tran PD, Guillet N, Métayé R, Fontecave MAL (2009) From hydrogenases to noble metal-free catalytic nanomaterials for H_2 production and uptake. Science 326:1384–1387
45. Tran PD, Le Goff A, Heidkamp J, Jousselme B, Guillet N, Palacin S, Artero V (2011) Noncovalent modification of carbon nanotubes with pyrene-functionalized nickel complexes: carbon monoxide tolerant catalysts for hydrogen evolution and uptake. Angew Chem Int Ed 50(6):1371–1374
46. Gentil S, Lalaoui N, Dutta A, Nedellec Y, Cosnier S, Shaw WJ, Le Goff A (2017) Carbon-nanotube-supported bio-inspired nickel catalyst and its integration in hybrid hydrogen/air fuel cells. Angew Chem Int Ed 56(7):1845–1849
47. Cesbron M (2019) Vers l'élaboration de surfaces modifiées par des monocouches organiques mixtes issues de la réduction de précurseurs diazonium. Doctoral dissertation, Université d'Angers. https://tel.archives-ouvertes.fr/tel-03079022/document
48. Binding SC, Pernik I, Gonçales VR, Wong CM, Webster RF, Cheong S, Tilley RD, Garcia-Bennett AE, Gooding JJ, Messerle BA (2019) Simultaneous functionalization of carbon surfaces with rhodium and iridium organometallic complexes: hybrid bimetallic catalysts for hydroamination. Organometallics 38:780
49. Zhang L, Vilà N, Walcarius A, Etienne M (2018) Molecular and biological catalysts coimmobilization on electrode by combining diazonium electrografting and sequential click chemistry. Chem Electro Chem 5:2208
50. Bhakta AK, Kumari S, Hussain S, Belkhiri S, Lo M, Mascarenhas RJ, Delhalle J, Mekhalif Z (2021) Simultaneous formation of CuO nanoflowers and semi-spherical nanoparticles onto MWCNT surface. Emergent Mater 4:403–411
51. Guo DJ, Li HL (2004) High dispersion and electrocatalytic properties of Pt nanoparticles on SWNT bundles. Electroanal Chem 573:197
52. Guo DJ, Li HL (2004) Electrochemical synthesis of Pd nanoparticles on functional MWNT surfaces. Electrochem Commun 6:999
53. Mahouche-Chergui S, Gam-Derouich S, Mangeney C, Chehimi MM (2011) Aryl diazonium salts: a new class of coupling agents for bonding polymers, biomacromolecules and nanoparticles to surfaces. Chem Soc Rev 40:4143–4166
54. Guo DJ, Mirkhalaf F (2012) Modification of nano-objects by aryl diazonium salts. In: Chehimi MM (ed) Aryl diazonium salts. Wiley-VCH, Weinheim, Germany, pp 103–124
55. Vila N, Brussel MV, D'Amours M, Marwan J, Buess-Herman C, Belanger D (2007) Metallic and bimetallic Cu/Pt species supported on carbon surfaces by means of substituted phenyl groups. J Electroanal Chem 609(2):85–93
56. Mirzaei P, Bastide S, Aghajani A, Bourgon J, Leroy E, Zhang J, Snoussi Y, Bensgheir A, Hamouma O, Chehimi MM, Cachet-Vivier C (2019) Bimetallic Cu–Rh nanoparticles on diazonium-modified carbon powders for the electrocatalytic reduction of nitrates. Langmuir 35(45):14428–14436
57. Bensghaïer A, Bhullar V, Kaur N, Lo M, Bdiri M, Mahajan A, Chehimi MM (2021) "Painted CNT"@ Au nanoparticles: a nanohybrid electrocatalyst of direct methanol oxidation. Emergent Mater 4:515–524
58. Maluta JR, Machado SAS, Chaudhary U, Manzano JS, Kubota LT, Slowing II (2018) Development of a semigraphitic sulfur-doped ordered mesoporous carbon material for electroanalytical application. Sens Actuators B Chem 257:347–353
59. Gao X, Du D, Li S, Yan X, Xing W, Bai P, Xue Q, Yan Z (2018) Outstanding capacitive performance of ordered mesoporous carbon modified by antraquinone. Electrochim Acta 259:110–121
60. Goscianska J, Pietrzak R, Matos J (2018) Catalytic performance of ordered mesoporous carbons modified with lanthanides in dry methane reforming. Catal Today 301:204–216
61. Wu Z, Webley PA, Zhao D (2010) Comprehensive study of pore evolution, mesostructural stability, and simultaneous surface functionalization of ordered mesoporous carbon (FDU-15) by wet oxidation as a promising adsorbent. Langmuir 26:10277–10286

62. Wang P, Zhao Y, Liu J (2018) versatile design and synthesis of mesoporous sulfonic acid catalysts. Sci Bull 63:252–266
63. Goscianska J, Malaika A (2020) A facile post-synthetic modification of ordered mesoporous carbon to get efficient catalysts for the formation of acetins. Catal Today 357:84–93
64. Zhong R, Sels BF (2018) Sulfonated mesoporous carbon and silica-carbon nanocomposites for biomass conversion. Appl Catal B Environ 236:518–545
65. Yuan Z, Zhang Z, Zheng J, Lin J (2015) Efficient synthesis of promising liquid fuels 5-ethoxymethylfurfural from carbohydrates. Fuel 150:236–242
66. Kaper H, Grandjean A, Weidenthaler C, Schüth F, Goettmann F (2012) Surface diels–alder reactions as an effective method to synthesize functional carbon materials. Chem A Eur J 18(13):4099–4106
67. Muylaert I, Verberckmoes A, Spileers J, Demuynck A, Peng L, De Clippel F, Sels B, Van Der Voort P (2013) Synthesis of sulphonated mesoporous phenolic resins and their application in esterification and asymmetric aldol reactions. Mater Chem Phys 138:131–139
68. Zhou ZY, Kang X, Song Y, Chen S (2012) Ligand-mediated electrocatalytic activity of Pt nanoparticles for oxygen reduction reactions. J Phys Chem C 116(19):10592–10598
69. Banks CE, Wildgoose GG, Heald CGR, Compton RG (2005) Oxygen reduction catalysis at anthraquinone centres molecularly wired via carbon nanotubes. J Iran Chem Soc 2(1):60–64
70. Belbekhouche S, Kebe SI, Mahouche-Chergui S, Guerrouache M, Carbonnier B, Jaziri M, Chehimi MM (2017) Aryl diazonium-modified olive waste: a low cost support for the immobilization of nanocatalysts. Colloids Surf A Physicochem Eng Aspects 529:541–549
71. Mousli F, Khalil AM, Maurel F, Kadri A, Chehimi MM (2020) Mixed oxide-polyaniline composite-coated woven cotton fabrics for the visible light catalyzed degradation of hazardous organic pollutants. Cellulose 27(13):7823–7846
72. Zhou ZY, Kang X, Song Y, Chen S (2012) Enhancement of the electrocatalytic activity of Pt nanoparticles in oxygen reduction by chlorophenyl functionalization. Chem Commun 48(28):3391–3393
73. Hoan NTV, Minh NN, Nhi TTK, Van Thang N, Tuan VA, Nguyen VT, Khieu DQ (2020) TiO_2/diazonium/graphene oxide composites: synthesis and visible-light-driven photocatalytic degradation of methylene blue. J Nanomater
74. Kankla P, Limtrakul J, Green ML, Chanlek N, Luksirikul P (2019) Electrooxidation of formic acid enhanced by surfactant-free palladium nanocubes on surface modified graphene catalyst. Appl Surf Sci 471:176–184
75. Nazaruk E, Karaskiewicz M, Żelechowska K, Biernat JF, Rogalski J, Bilewicz R (2012) Powerful connection of laccase and carbon nanotubes: material for mediator free electron transport on the enzymatic cathode of the biobattery. Electrochem Commun 14(1):67–70
76. Griffete N, Herbst F, Pinson J, Ammar S, Mangeney C (2011) Preparation of water-soluble magnetic nanocrystals using aryl diazonium salt chemistry. J Am Chem Soc 133:1646–1649
77. Reddy KR, Hassan M, Gomes VG (2015) Hybrid nanostructures based on titanium dioxide for enhanced photocatalysis. Appl Catal A Gen 489:1–16
78. Saleh A, Abd Rasin F, Ali Ameen M (2009) TiO_2 nanoparticles prepared by sol-gel. Mater Sci Eng 3:81–83
79. Byranvand MM, Nemati Kharat A, Fatholahi L, Beiranvand ZM (2013) A review on synthesis of nano-TiO_2 via different methods. J Nanostruct 3:1–9
80. Mital GS, Manoj T (2011) A review of TiO_2 nanoparticles. Rev Sci Bull 56:1639–1657
81. Zhao Z, Tian J, Sang J, Cabot A, Liu H (2015) Structure, synthesis, and applications of TiO_2 nanobelts. Adv Mater 27(16):2557–2582
82. Mousli F, Chaouchi A, Jouini M, Maurel F, Kadri A, Chehimi MM (2019) Polyaniline-grafted RuO_2-TiO_2 heterostructure for the catalysed degradation of methyl orange in darkness. Catalysts 9(7):578
83. Aït Atmane Y, Sicard L, Lamouri A, Pinson J, Sicard M, Masson C, Nowak S, Decorse P, Piquemal JY, Galtayries A, Mangeney C (2013) Functionalization of aluminum nanoparticles using a combination of aryl diazonium salt chemistry and iniferter method. J Phys Chem C 117:26000–26006

84. Salem MA, Al-Ghonemiy AF, Zaki AB (2009) Photocatalytic degradation of allura red and quinoline yellow with polyaniline/TiO_2 nanocomposite. Appl Catal B Environ 91:59–66
85. Reddy KR, Karthik KV, Prasad SB, Soni SK, Jeong HM, Raghu AV (2016) Enhanced photocatalytic activity of nanostructured titanium dioxide/polyaniline hybrid photocatalysts. Polyhedron 120:169–174
86. Aghajani A, Santoni M-P, Mirzaei P, Mohamed AA, Chehimi MM, Jouini M, Tuning diazonium modification of gold nanoparticles for fuel cell application. ChemRxiv https://doi.org/10.33774/chemrxiv-2021-cmbr3

Aryldiazonium Salts as Photoinitiators for Cationic and Free Radical Polymerizations

Yusuf Yagci and Mohamed M. Chehimi

Abstract This chapter discusses the use of aryl diazonium salts in the activation of cationic and free radical polymerizations which can be triggered by specific responses to light, with possible combination of additives. Diazonium salts can be activated with additives in a thermodynamically favorable way owing to their ease of reduction. The current literature clearly indicates that these salts would find wider application in the near future in polymer science.

1 Introduction

Aryldiazonium salts are the class of onium salts used as the first efficient photoinitiators for cationic and free radical polymerizations [1]. Although they have favorable redox potential and can be synthesized by simple procedures, they suffer from thermal instability. On the other hand, thermal stability can be improved [2] with alkyloxy substitution in certain solvents such as acetonitrile, methylene chloride, and dichlomethane. The conventional method applied for their synthesis involves treatment of aniline derivatives with sodium nitrite and a Brønsted acid with nonnucleophilic counter anion. The polymerizations by diazonium salts are initiated after an external stimulation such as irradiation or heating. In this chapter, we will focus on the light-activated initiation processes, specifically for cationic polymerizations. Initiation of free radical polymerization under certain conditions will also be addressed. Such polymerizations are considered in two main approaches, namely direct and indirect initiating systems.

Y. Yagci (✉)
Department of Chemistry, Istanbul Technical University, Maslak, Istanbul 34469, Turkey
e-mail: yusuf@itu.edu.tr

M. M. Chehimi
Université Paris Cité, CNRS, ITODYS, F-75013 Paris, France

M. M. Chehimi et al. (eds.), *Aryl Diazonium Salts and Related Compounds*,
Physical Chemistry in Action, https://doi.org/10.1007/978-3-031-04398-7_16

2 Cationic Polymerization

2.1 Direct Photolysis

Depending on the aryl moiety, these salts are photosensitive at different wavelengths (Table 1) and undergo photolysis with high decomposition quantum yield [3] ($\O_{dec} = 0.3 - 0.6$) to form aryl halide, nitrogen gas, and Lewis acid. The latter can initiate cationic polymerization of epoxides and vinyl ethers directly or by reacting with a hydrogen donating compounds such as water or alcohols.

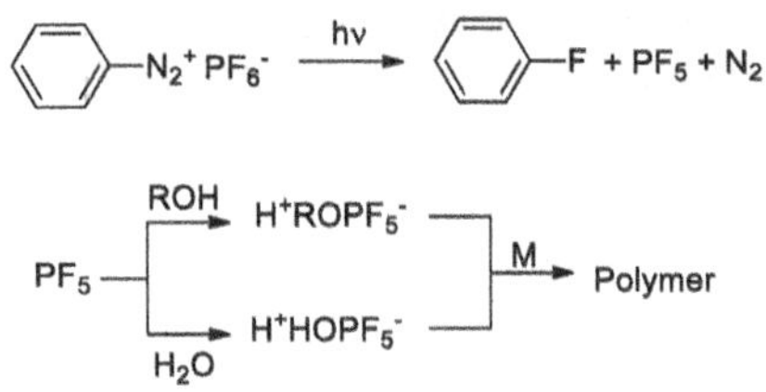

Table 1 Absorption maxima of aryldiazonium salts ($ArN_2^+PF_6^-$)

Aryl group (Ar-)	Absorption max, λ_{max}(nm)
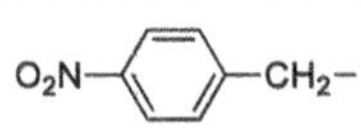	258
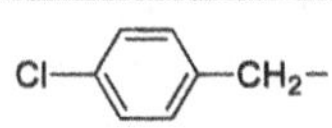	273
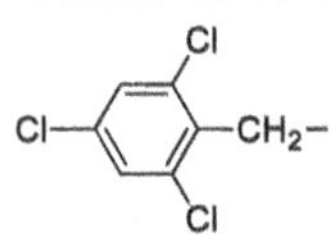	294
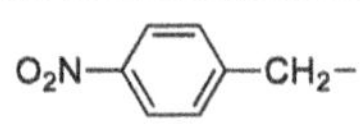	310
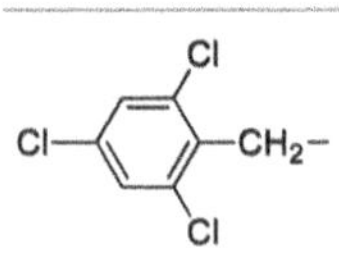	337
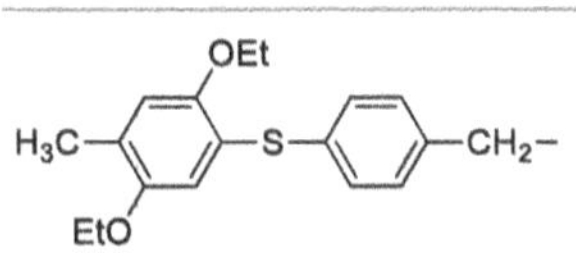	357
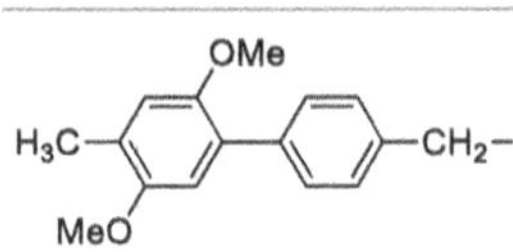	405

The efficiency of the polymerization depends on both the cationic and anionic portions of the salt structure [4, 5].

2.2 *Indirect Acting Photoinitiating Systems*

As stated above, in the direct initiation the salt is responsible for the absorption of the light and subsequent decomposition. However, with indirect-acting systems, the light energy is absorbed by an additional component [6]. After absorbing the energy, the excited state of the additives can undergo several reactions such as electron transfer with the diazonium salt, thus producing initiating species, or generating radicals that can be oxidized. Depending on the type of the additive, polymerizations can be conducted thermally or photochemically at various temperature wavelength ranges, respectively.

2.2.1 Oxidation of Free Radicals

This method, also called as free radical promoted cationic polymerization [7], is based on the oxidation of electron donor radicals according to the following reaction, thus generating reactive cations.

$$R^{\cdot} + ArN_2PF_6^- \rightarrow R^+PF_6^- + Ar^{\cdot} + N_2$$

Wide range of thermally or photochemically active free radical sources in combination with various onium salts can efficiently be used [8–10] to promote the cationic polymerization of vinyl and cyclic ethers. Benzoin derivatives [11], acyl phosphine oxides [12] acyl germanes [13], acyl silanes [14], and organo tellurium compounds [15] are the most widely used cleavage type free radical photoinitiators. Typical reactions for so-called free radical promoted cationic polymerization are presented below in the example of benzoin which proceed with high quantum yields [16].

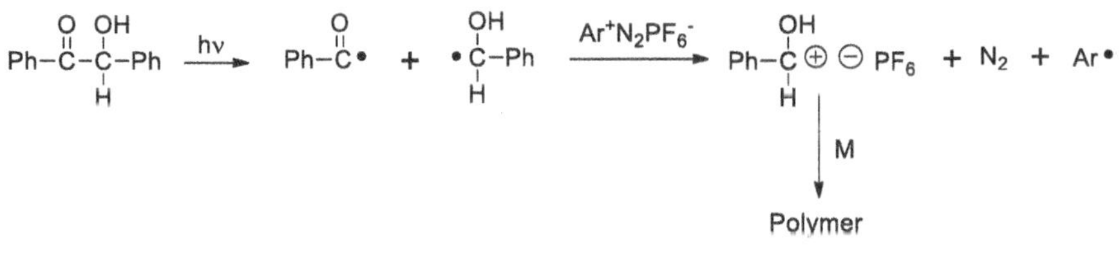

Oxidizable radicals can also be generated by using *Type II* photoinitiators such as benzophenone [11] and thioxanthone [17] derivatives (Scheme 3).

Table 2 Reduction potentials of common onium salts

Onium salts	Reduction potential, $E_{1/2}^{red}$ (V)	References
Diphenyliodonium	−0.2	[18]
Triphenylsulphonium	−1.1	[19]
N-Ethoxy pyridium	−0.7	[20]
p-Cl-Diazonium	0.35	[21]

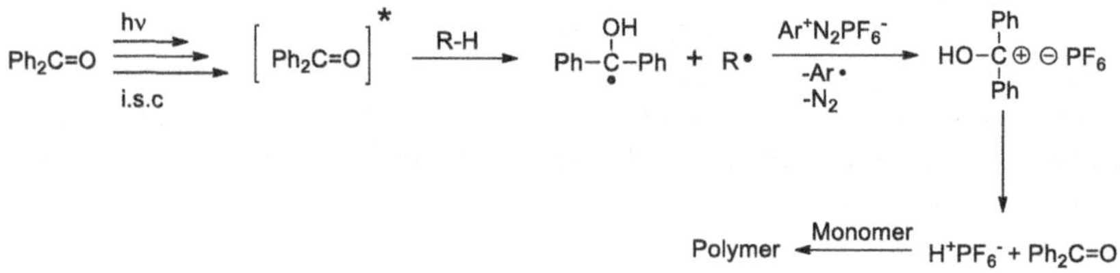

The efficiency of the process is strongly related to the oxidation and reduction potentials of the free radical and the onium ion, respectively. The possibility of successful electron transfer reaction can be estimated by the Rehm-Weller equation.

$$\Delta G = F\left[E_{1/2}^{ox}(R.) - E_{1/2}^{red}(\mathrm{On}+)\right]$$

As can be seen from Table 2, among various onium salts, diazonium salts have the most favorable reduction potential for the radical oxidation process.

2.2.2 Photoinduced Electron Transfer Reaction with Sensitizers

Certain polynuclear aromatic compounds such as anthracene, pyrene, perylene, and phenothiazine in their excited state undergo electron transfer reactions with onium salts to generate reactive cations according to the following general scheme [22].

$$PS \xrightarrow{h\upsilon} PS^*$$
$$PS^* + \mathrm{On}^+X^- \rightarrow PS^{*+}X^- + \mathrm{On}^{\cdot}$$
$$PS = \text{Photosensitizer}$$

The sensitized radical cations can themselves initiate the polymerization of appropriate polymers or abstract hydrogen from solvent or monomer resulting in the release of Brønsted acid.

$$PS^{+\cdot}X^- + R - H \rightarrow PS + R^{\cdot} + H^+X^-$$
$$H^+X^- + \text{Monomer} \rightarrow \text{Polymer}$$

The electron transfer reaction may successfully proceed providing ΔG determined according to the extended Rehm-Weller equation is negative

$$\Delta G = F\left[E_{1/2}^{ox}(S) - E_{1/2}^{red}(\mathrm{On}+)\right] - E\left(S^*\right)$$

Iodonium [6] and alkoxypyridinium [22] salts were shown to be successfully activated by various photosensitizers. As diazonium salts have more favorable reduction potential compared to these salts and their combinations with sensitizers are expected to initiate cationic polymerizations more efficiently.

3 Free Radical Polymerization

In principle, all the initiating systems described for cationic polymerization are applicable for the corresponding free radical polymerizations. Direct and indirect-acting photoinitiating systems are based on the electron transfer process in which diazonium salt is reduced to the diazoradical which essentially forms an aryl radical capable of initiating free radical polymerization. The feasibility of this route has recently been evidenced in electrochemical Radical Addition Fragmentation Transfer (e-RAFT) polymerization by Matyjaszewski and co-workers [23]. Electrochemical reduction of *p*-bromodiazonium resulted in effective generation of aryl radicals to initiate a RAFT polymerization. Homopolymers and block copolymers with well-defined structures and functionalities were prepared by applying either fixed potential or fixed current conditions. The photochemical application of the concept was recently demonstrated by our group. We have recently shown [24] that free radical polymerization of various monomers can be initiated under visible and near-IR (NIR) light irradiation by using two-dimensional (2D) few-layer black phosphorus (BP) exfoliated in a solvent as photocatalyst. Upon the light exposure, BP generates excited electrons and holes, which undergo electron transfer reactions with the onium salts to form free radicals capable of initiating free radical polymerization. Among the onium salts tested, aryldiazonium salt was found to be the most efficient in the photopolymerization process owing to its favorable reduction potential with the conduction edge potential of BP.

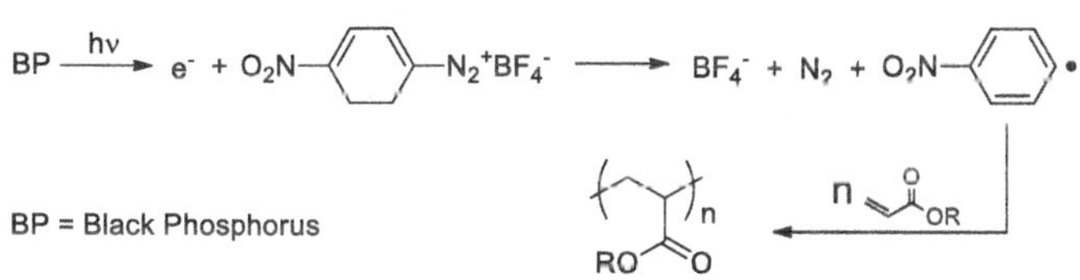

The presented strategy also provides possibility for the in situ preparation of BP-polymer composite materials.

4 Conclusions

In conclusion, photopolymerization is a continuously growing and diverse area that has been broadly exploited for various applications including coating, 3D printing, dental fillings, composites, adhesives, inks, printing plates, optical waveguides, surface patterning, and microelectronics. In this chapter, we have discussed the activation of cationic and free radical polymerizations by diazonium salts which can be triggered by specific responses to light at different wavelengths directly or in combination with additives. Compare to the other onium salts, diazonium salts can be activated with additives in a more thermodynamically favorable manner facilitated by their reduction potentials. Despite the thermally unstable nature of diazonium salts, the possibility of photoinduced polymerizations in the low-energy region of the electromagnetic spectrum indicates that they would find wider application in the near future. Typical applications for the preparation of clay-polymer nanocomposites are covered in another chapter.

References

1. Lohse F, Zweifel H (1987) Photocrosslinking of epoxy resins. Adv Polym Sci 78:61
2. Müller U, Utterodt A, Mörke W, Deubzer B, Herzig C (2001) New insights about diazonium salts as cationic photoinitiators. J Photochem Photobio A: Chem 140:53–66
3. Smets G, Aerts A, van Erum J (1980) Photochemical initiation of cationic polymerization and its kinetics. Polym J 12:539
4. Schessinger SI (1974) Photopolymerisation o f epoxides. Photo Sci Eng 18:387
5. Dreyfuss P, Dreyfuss MP (1967) Polytetrahydrofuran. Adv Polym Sci 4:528
6. Yagci Y, Reetz I (1998) Externally stimulated initiator systems for cationic polymerization. Prog Polym Sci 23:1485–1538
7. Tasdelen MA, Lalevée J, Yagci Y (2020) Photoinduced Free Radical Promoted Cationic Polymerization 40 Years After Its Discovery. Polym Chem 11:1111–1121
8. Abdul-Rasoul FA, Ledwith A, Yagci Y (1978) Photo-chemical and thermal cationic polymerizations. Polymer 19:1219–1222
9. Ledwith (1978) Possibilities for promoting cationic polymerization by common sources of free radicals. Polymer 19:1217–1219
10. Yagci Y, Ledwith A (1988) Mechanistic and kinetic-studies on the photoinitiated polymerization of tetrahydrofuran. J Polym Sci Part A, Polym Chem 26:1911–1918
11. Yagci Y, Schnabel W (1992) New aspects on the photoinitiated free-radical promoted cationic polymerization. Makromol Chem Macromol Symp 60:133–143
12. Yagci Y, Schnabel W (1987) Acylphosphine oxides as free-radical promoters in cationic polymerizations. Makromol Chem-Rapid Commun 8:209–213
13. Durmaz YY, Moszner N, Yagci Y (2008) Visible light initiated free radical promoted cationic polymerization using acylgermane based photoinitiator in the presence of onium salts. Macromolecules 41:6714–6718
14. Sari E, Mitterbauer M, Liska R, Yagci Y (2019) Visible light induced free radical promoted cationic polymerization using acylsilanes. Prog Org Coat 132:139–143
15. Kaya K, Seba M, Fujita T, Yamago S, Yagci Y (2018) visible light-induced free radical promoted cationic polymerization using organotellurium compounds. Polym Chem 9:5639–5643
16. Ptitsyna OA, Levashova TW, Butin KP (1971) Dokl Akad Nauk 201:372

17. Yilmaz G, Beyazit S, Yagci Y (2011) Visible light induced free radical promoted cationic polymerization using thioxanthone derivatives. J Polym Sci Part A, Polym Chem 49:1591–1596
18. Grimshaw J (1981) Electrochemistry of the sulfonium group. In: Stirling CJM, Patai S (eds) The chemistry of sulphonium groups Wiley, New York
19. Bottcher A, Hasebe K, Hizal G, Yagci Y, Stellberg P, Schnabel W (1991) Initiation of cationic polymerization via oxidation of free-radicals using pyridinium salts. Polymer 32:2289
20. Elofson FM, Gadallah FF (1969) Substituent effects in the polarography of aromatic diazonium salts. J Org Chem 94:854
21. Baumann H, Müller U, Pfeifer D, Timpe H-J (1982) Lichtinitiierte Polymer-und Polymerisationsreaktionen. III. Photoinduzierte Zersetzung von Aryldiazoniumsalzen durch Benzoinderivate. J Prakt Chem 324:217
22. Dossow D, Zhu QQ, Hizal G, Yagci Y, Schnabel W (1996) Photosensitized cationic polymerization of cyclohexene oxide: a mechanistic study concerning the use of pyridinium-type salt. Polymer 37:2821–2826
23. Wang Y, Fantin M, Park S, Gottlieb E, Fu L, Matyjaszewski K (2017) Electrochemically mediated reversible addition–fragmentation chain-transfer polymerization. Macromolecules 50:7872–7879
24. Kocaarslan Z, Eroglu G, Yilmaz O, Metin Y, Yagci (2021) ACS Macro Letters 10:679–683

Polymer Surface Science and Adhesion Using Diazonium Chemistry

Nebewia Griffete, Khouloud Jlassi, Ahmed M. Khalil, Hatem Ben Romdhane, Mohamed M. Chehimi, and Yusuf Yagci

Abstract Polymer adhesion to modified surfaces is a hot, ever-progressing topic. It concerns the attachment of prefabricated polymers or polymers grown by in-situ polymerization. Herein, we summarize the recent progress achieved in 2012–2021 in polymer adhesion to arylated surfaces bearing functional groups able to initiate polymerization or to favorably interact with prefabricated or precipitating (pre)polymers. We focus first on radical polymerization techniques with emphasis on UV or sunlight-triggered polymerization processes. The latter enables obtaining patterned polymer coatings. We also discuss the design of imprinted polymer and antibacterial coatings grown on arylated flat, flexible, or particulate surfaces, as well the making of adhesive layers of vinylic polymers bearing redox groups in their repeat units. Diazonium salts permit also attach "clickable" groups for click polymerization or for making layered sol-gel coatings. In a second important section, we demonstrate that diazonium salts are unique coupling agents for obtaining adhesive conjugated polymer layers of major importance in the design of flexible electrochemical sensors or in the development of electronic materials. The Chapter finishes with eye-catchy new trends in surface-confined polymerization such as plasmon-triggered nitroxide mediated polymerization, and the design of covalently bonded biopolymer to alloy surface for corrosion control applications in simulated Dead Sea water.

N. Griffete
Sorbonne Université, Paris, France

K. Jlassi
Qatar University, Doha, Qatar

A. M. Khalil
National Research Centre, Giza, Egypt

H. B. Romdhane
Université Tunis El-Manar, Manar II, Tunisia

M. M. Chehimi (✉)
Université Paris Cité, CNRS, ITODYS, F-75013 Paris, France
e-mail: mohamed.chehimi@cnrs.fr

Y. Yagcı
Istanbul Technical University, Maslak, Istanbul, Turkey

M. M. Chehimi et al. (eds.), *Aryl Diazonium Salts and Related Compounds*, Physical Chemistry in Action, https://doi.org/10.1007/978-3-031-04398-7_17

1 Introduction and Scope

The surface chemistry of aryl diazonium for the attachment of polymers has been developed in the late 20th Century when Jean Pinson and co-workers have considered the grafting of an epoxy coating to aminoarylated carbon fibers [1]. It was the first example of modern surface chemistry of diazonium salts for the attachment of prepolymers and polymers by surface-confined polymerization (termed "*grafting from*" or "*grafting through*" techniques) or by reaction of prefabricated polymers with reactive aryl layers (termed "*grafting to*") as depicted in Fig. 1.

The topic of polymer grafting to diazonium-modified flat surfaces and 0-3D nano-objects has been summarized in detail in the literature in books and review articles [2–8]. In this chapter, we will focus on important developments achieved in the period 2012–2021, subsequent to the publication of the 2012 "Aryl diazonium salts" book [3].

The Chapter is organized as follows:

- a brief summary of achievements in polymer grafting covering 1997–2011
- an overview of the methods of coating polymers on arylated surfaces with emphasis on the mechanisms of polymer growth and potential applications
- new trends and conclusions.

It should be noted that polymer surface modification for the attachment of polymers is not covered as most of the salient features of this topic are discussed in Chap. 10 (Vieillard et al.). Concerning conductive polymers, polyaniline modification photocatalysts will be out of scope; the reader is referred to Chap. 15 by Mousli et al. Despite their important use, diazo resins, diazotized reactive oligomers, or prepolymers, are kept out of the scope, since their relevance to surface science is limited.

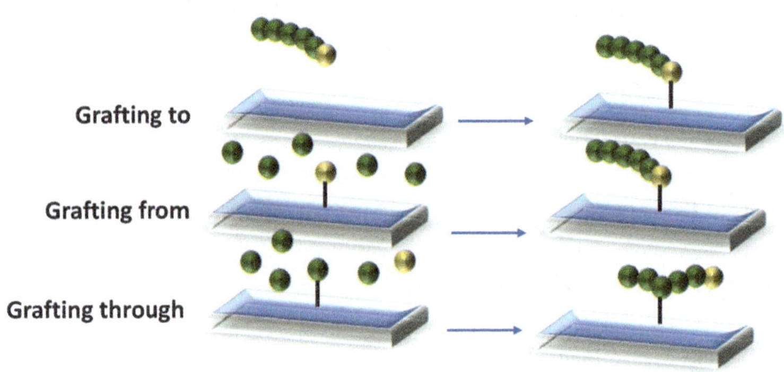

Fig. 1 Methods of attachment of polymers to surfaces via "grafting to", "grafting from" and "grafting through". "Grafting to" could also be achieved with polymers bearing reaction or "clickable" groups

2 Earlier Developments (1997–2011)

The realization that arylation of surfaces with strong surface-carbon adhesive bonding, together with the possibility to equip aniline precursors with numerous functional groups, made of diazonium salts unique coupling agents for bonding polymers to the surfaces. Pinson and co-workers [1] published a seminal work on the reaction of epoxy molecules to aminated carbon fiber obtained by reductive grafting of nitrophenyl groups which were further reduced to amino groups. This strategy has opened new avenues for the preparation of wide range of polymer composites. In a follow-up paper, polystyrene was photografted to surfacebound 4-benzoylphenyl groups, whereas poly(1,2-propanediyl fumarate) was attached to the phenylcarboxylate functions through ionic bonds with Mg(II) ions leading to the formation of anticorrosive protective coatings on iron surface [9]. In the years 2004–2005, with the advent of atom transfer radical polymerization (ATRP), it was soon realized that this controlled radical polymerization (CRP) technique could also be applied to grow numerous vinylic polymers on arylated flat surfaces; such as iron [10, 11], glassy carbon [12], ultrananocrystalline diamond (UNCD) [13], gold [14, 15] and stainless steel [16]; as well as particulate materials such as carbon black [17], carbon nanotubes [18, 19] and nanodiamonds [20]. Radical addition-fragmentation transfer (RAFT) polymerization is another CRP technique employed for tethering polymers to materials surfaces [21], as well as photo-iniferter, and plasmon-induced nitroxy mediated polymerization (discussed in Sect. 3).

In this period, the main objectives were the demonstration of proof of concept of grafting polymer chains to surfaces by a variety of surface-confined polymerization approaches. Main issues concerned were related to the preparation of polymer brushes with hydrophilic or hydrophobic properties, or switchable from hydrophobic to hydrophilic by cleavage of *tert*-butyl pendant groups, molecularly imprinted or biomimetic stimuli-responsive. Attachment of polyaniline and other conjugated polymers to electrode surfaces by grafting strategy has also been investigated [2, 8]. This first major period of polymer grafting to arylated surfaces was summarized and discussed in depth [2, 3, 22, 23].

3 Recent Progress in Polymer Grafting to Arylated Surfaces: 2012–2021

Following the publication of *Aryl Diazonium Salts* book [3], the last decade, witnessed important developments in the field of polymer grafting to arylated surfaces and attachment of diazotized oligomers. In addition, and as for silane modification of surfaces, adhesion does not necessarily occur through covalent bonding. It could also be the result of strong interfacial van der Waals or electrostatic interactions. We will consider CRP, photopolymerization, and oxidative polymerization techniques, which could be triggered chemically, photochemically, or electrochemically, respectively. CRP and photopolymerization processes provide additional advantages. CRP is a versatile method to facilitate surfaceinitiated polymerization yielding hybrid materials with well-defined structures [24]. The advantages of photopolymerization include attractive productivity, energy-saving potential, and favorable environmental impact and the possibility of spatial and temporal control is particularly important for surface coating applications [25]. Each section will also cover applications involving multilayered and patchy coatings, metal-coating adhesive joints, and sensors.

3.1 Surface-Initiated ATRP

By virtue of its mechanism, ATRP is perfectly adapted CRP to surface-initiation on arylated surfaces. Initiating sites could be incorporated onto the surface directly by using diazonium salts such as N_2^+-C_6H_4–$CH(CH_3)Br$ or indirectly through hydroxylated attached aryl groups followed by post-functionalized with 2-bromoisobutyryl bromide. The general surface-initiated ATRP mechanism is displayed in Fig. 2. The method has dramatically improved and can be conducted even in the presence of air [26–28] or externally triggered by either light [29] or electrochemically [30].

For the electrochemical route, two attractive strategies have been devised; the group of Daasbjerg [31] designed brushes of poly(ferrocenyl methyl methacrylate), P(FcMMA) at mild conditions. Attached 2-hydroxyethylphenyl groups (–C_6H_4–CH_2–CH_2–OH) were activated by 2-bromoisobutyryl bromide (BrIB), resulting in the formation of surface-bound –O–C(=O)–$C(CH_3)_2Br$ ATRP initiating sites (Fig. 3). The thickness of the aryl layer (0.5–2.5 nm) depended on the scan rate of diazonium reduction (1, 5, 10 V/s; lower scan rate induced higher aryl thickness) with a result of more densely packed chains at high aryl grafting density. The thickness of the PFcMMA films was almost the same ($\geq$25 nm). The more densely packed brushes exhibited less average roughness (0.9 nm).

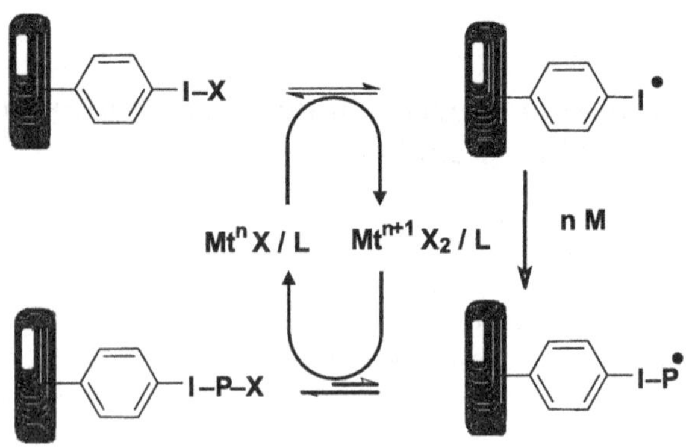

Fig. 2 Principle of surface-confined ATRP using initiator attached by reductive grafting of aryl diazonium salt. L: ligand, X = mostly Br, but also Cl, M: metal usually copper, P/ polymer chain

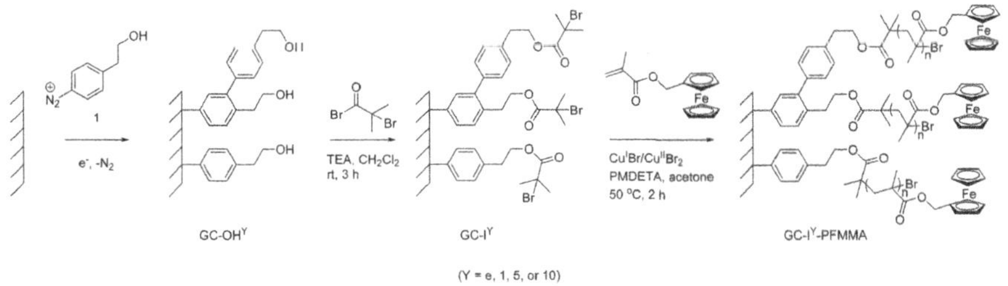

Fig. 3 Method to electrografting ATRP initiator and growth of poly(ferrocenyl methyl methacrylate), (PFcMMA). Repoduced from [31] with permission of ACS

The electron transfer through the aryl initiator layer to the redox ferrocenyl of the PFcMMA was investigated and the apparent rate constant of electron transfer was found to exponentially decrease with the thickness of the aryl layer in the dry state.

In another electrochemical application of polymer brushes, e-ATRP was used to graft PGMA chains to carbon fibers (CFs) [30]. First, the isolated diazonium salt is equipped with $-CH(CH_3)Br$ groups were electrochemically reduced at -0.3 V/SCE for 5 min and the CF-Br fibers were dipped in the polymerization medium with $CuBr_2$ and the bipyridine ligand (Bpy). Cu(II)/Bpy was electrochemically reduced to Cu(I)/Bpy (30 cycles from 0 to -1.5 V). Thus formed Cu(I) started then to catalyze the ATRP process leading to CF-PGMA hybrids which were further activated using iminodiacetic for recovering Ni(II) from electroless nickel plating baths (Fig. 4). A -0.7 V potential was applied to CF-PGMA-IDA immobilized Ni(II) species (30 s in 0.5 M H_2SO_4) in order to immobilize nano-nickel on the modified carbon fibers. About 53 mg.g^{-1} (0.918 mM g^{-1}) were recovered within 40 min at 50 °C and optimal pH = 5.2. Recovery followed pseudo-second-order kinetics and adsorption isotherm fitted Langmuir model.

Another way to trigger ATRP under air tolerant conditions is to use a reducing agent for converting Cu(II) to Cu(I) catalyst. The method was coined ARGET ATRP for Activator ReGenetaed by Electron Transfer ATRP. In this method, the reducing agent (e.g., Vitamin C) converts Cu(II) to Cu(I), but remaining oxygen reoxidizes Cu(I) to Cu(II) and the cycle restarts until complete consumption of oxygen. At this stage, the newly formed Cu(I) is no longer oxidized to Cu(II) and ATRP is triggered. Interestingly, the group of Daasbjerg has employed surface-initiated ARGET ATRP to grow PGMA brushes as molecular glue for the adhesive bonding of ethylene-propylene-diene M class (EPDM) rubber to stainless steel [28]. Addition of benzoyl peroxide (BPO) at the rubber/SS-PGMA interface, heating (up to 170 °C), and compression led to crosslinking of PGMA-Rubber and thus to remarkable adhesion, as strong as that obtained with commercially available glues. It should be pointed out that in this case, excellent adhesion was achieved with nanoscale adhesive joint.

In an interesting study, Lee et al. [32] compared the growth of brushes of poly(oligo (ethylene glycol) methacrylate), POEGMA, on gold by SI-ATRP and SI-ARGET ATRP and found that the latter yield thickness tenfold higher.

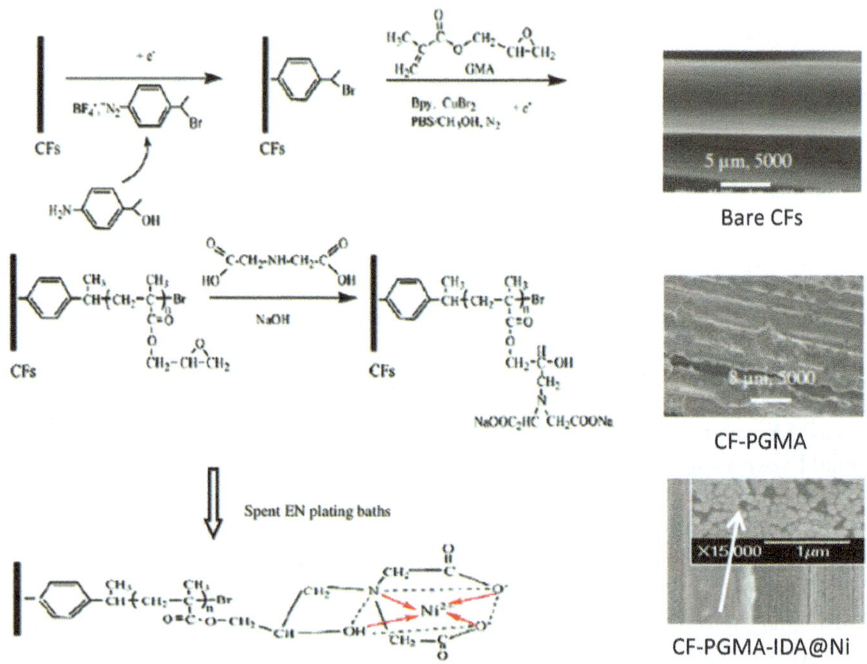

Fig. 4 Preparation of IDA-functionalized CF-PGMA for the recovery of nickel: **a** sequential electroreduction of aryl diazonium salt, e-ATRP of GMA, capture, and reduction of Ni(II); **b** SEM images of pristine CFs, **c** CF-PGMA, and **d** CF-PGMA-IDA@Ni. Reproduced from [30] with permission of Springer

3.2 RAFT and (Photo)iniferter

Diazonium salts can be used as iniferter agent for controlled polymerization. An iniferter is a chemical compound that simultaneously acts as initiator, transfer agent, and terminator in RAFT polymerization. Developed in the late 1990s by the team of Professor E. Rizzardo [33], RAFT polymerization method consists of a succession of reversible addition-fragmentation reactions of polymer chains growing on a chain transfer agent.

General mechanism of RAFT polymerization is presented in Fig. 5a. In this process, the radicals produced from thermal or photoinitiators initiate the polymerization. The propagating radicals add to the RAFT agent. In the chain equilibration step, unstable polymeric radicals formed undergo rapid fragmentation to yield polymers with RAFT functionality and a new radical capable of reinitiating.

A rapid balance addition-fragmentation between growing radical species and thiocarbonylthio species allows all chains to grow at the same rate. The method can be confined to surfaces [34] and complies with diazonium surface chemistry for attaching RAFT agents [21]. When the RAFT agent is equipped with photochromic

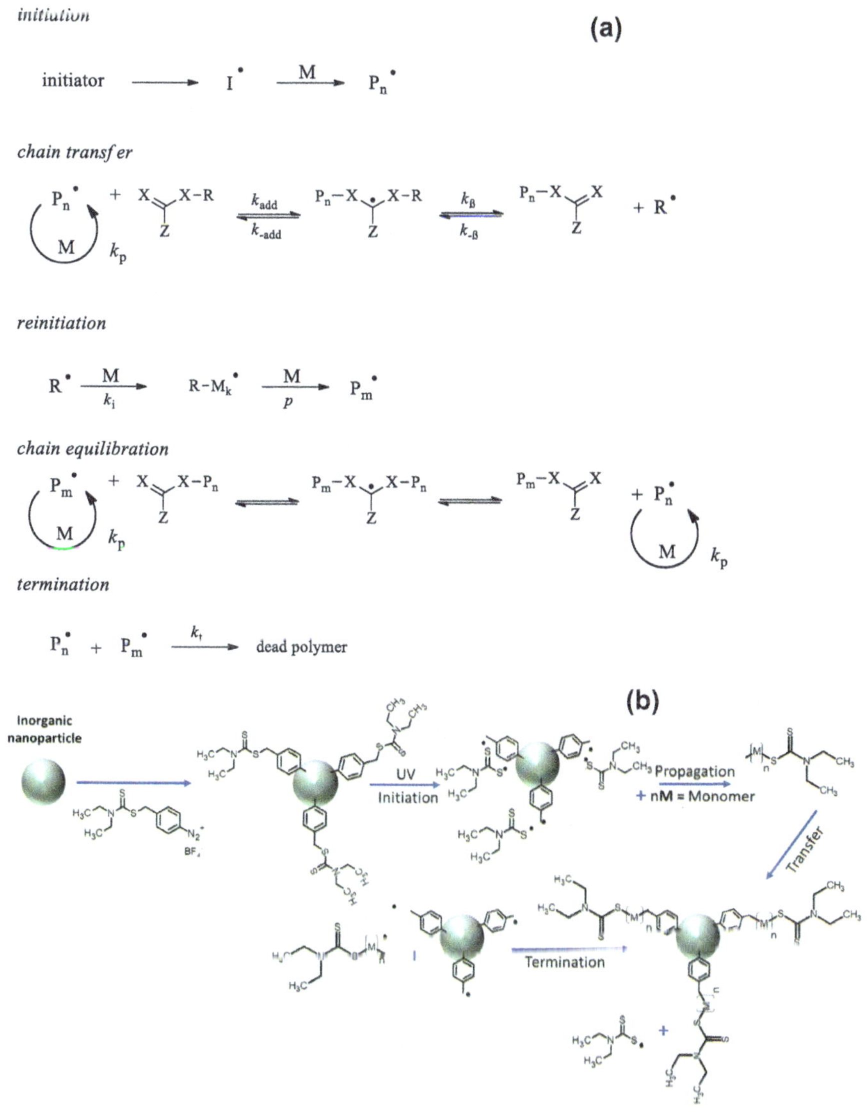

Fig. 5 **a** General mechanism of RAFT polymerization, **b** Schematic illustration of the inorganic nanoparticle surface modification using a diazonium salt for fabrication of nanocomposites by iniferter

groups, the process can be applied photochemically without requirement of an additional initiator [35]. In this case, the polymerization can also be considered as iniferter process since the polymers possess RAFT structural units at the chain ends. Photoiniferter technique can easily be adapted surface coating of nanoparticles using *N*,*N*-diethyldithiocarbamate modified benzyl-diazonium tetrafluoroborate (DEDTC) as RAFT agent (Fig. 5b). This approach offers several advantages over conventional methods: (i) ease and rapidity of inorganic nanoparticles' surface functionalization using diazonium salts; (ii) stability of the covalent link between the inorganic core and the organic coating; (iii) formation of small and dispersed nanoparticles.

Shortlisted applications of arylated surface-confined photoiniferter are discussed below.

(a) ***Selective proteins adsorption using magnetic molecularly imprinted polymer***

A novel synthetic pathway for protein specific molecularly imprinted polymer coatings on magnetic nanoparticles was developed by combining diazonium modification of magnetic iron oxide nanoparticles and UV-light induced surfaceconfined photoiniferter copolymerization of acrylamide and *N, N*-methylene-bis-acrylamide [36] (Fig. 6a). The resulting crosslinked polymer-grafted nanoparticles exhibited high adsorption capacity toward template proteins, and it was observed that affinity constants matched with those obtained with antibodies (Fig. 6b). The synthesis method proved to be robust toward proteins of various MW and isoelectric points (bovine serum albumin, BSA; ovalbumin, OVA; and lysozyme, Lyz) which makes it applicable for a whole range of proteins of interest in nanomedicine (for example green fluorescent protein, GFP).

(b) ***Selective pollutants adsorption using magnetic MIP nanoparticles***

First, ultra-small dispersed magnetic nanocrystals coated by a thin layer of a bisphenol A (BPA) imprinted polymer was synthesized [37]. Magnetic particles were grafted with BPA-imprinted poly(methacrylic acid-co-ethylene glycol dimethacrylate). The obtained biomimetic nanoparticles were dispersed in basic medium due to the ionization of the COOH groups and were efficiently attracted by magnet, therefore, facilitating recovery of BPA (Fig. 7a). The imprinting technique provided to impart numerous artificial receptor sites to the polymer layer compared to the non-imprinted polymer, resulting in four-fold uptake of BPA by MIP-grafted nanoparticles (Fig. 7b), which is an important imprinting factor.

(c) ***pH sensitive polymer embedded magnetic nanoparticles***

An original and simple route combining aryl diazonium salt chemistry and the iniferter method to elaborate individually dispersed, highly soluble, and pH-sensitive poly(methacrylic acid)-coated magnetic nanoparticles was developed [38]. The response to pH stimulus is due to the presence of carboxylic groups all along the poly(methacrylic acid)-grafted chains, which are fully protonated or deprotonated below or above pKa, respectively.

(d) ***Molecularly imprinted polymer-coated gold nanorods as folic acid sensor***

Individually dispersed hybrid materials composed of gold nanorod cores and molecularly imprinted polymer shells were successfully used as folic acid sensors [39]. The 2-(phenoxy)ethyldiethylcarbamadithionate diazonium chloride was reacted with gold nanorods in water and at RT to provide surface-bound initiators to act as photoiniferter. Subsequent UV-light induced copolymerization of methacrylic acid

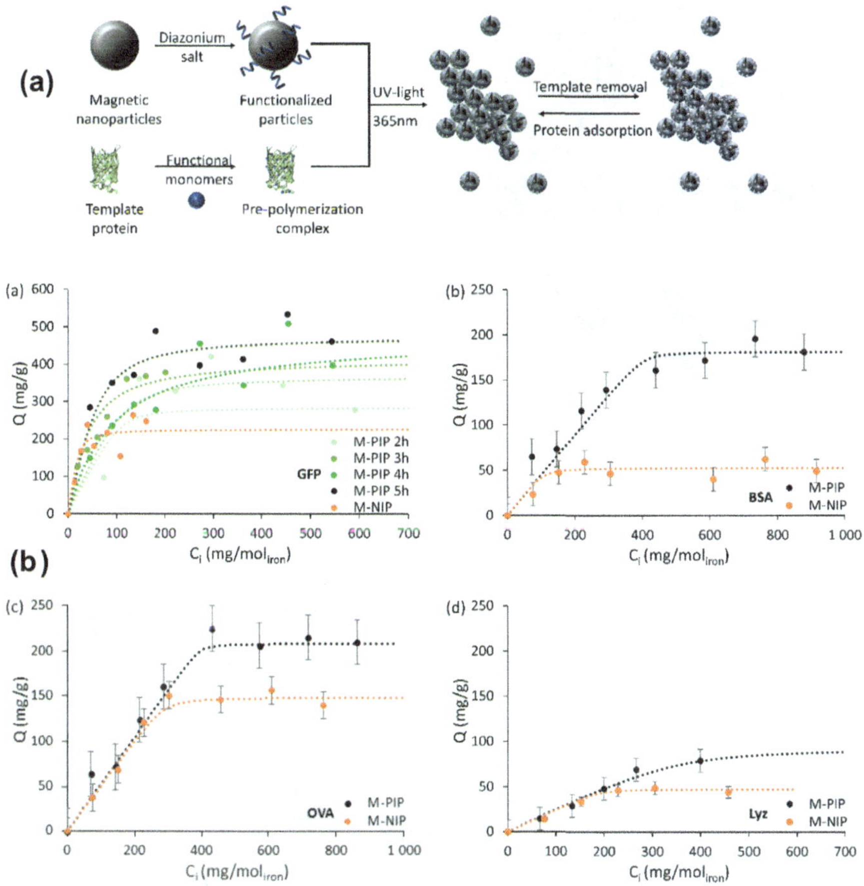

Fig. 6 **a** Synthetic pathway for the preparation of functional nanoparticles using the diazonium salt as iniferter agent. **b** Adsorption isotherms of different proteins on the corresponding magnetic molecularly imprinted polymer (M-PIP) and non-imprinted polymer (M-NIP). **a** GFP, m = 0.2 mg; **b** (BSA), m = 0.8 mg; **c** OVA, m = 0.8 mg; **d** Lyz, m = 0.9 mg. All experiments V = 2 mL, 24 h. Reproduced from [36] with permission of ACS

and *N,N*-bisacrylamide in the presence of folic acid permitted to obtain folic acid-imprinted MIP polymer-coated nanorods (Fig. 8A). The MIP thickness was controlled by polymerization time (Fig. 8B). The plasmonic properties of the nanorods enabled the direct detection of folic acid (FA) capture and release.

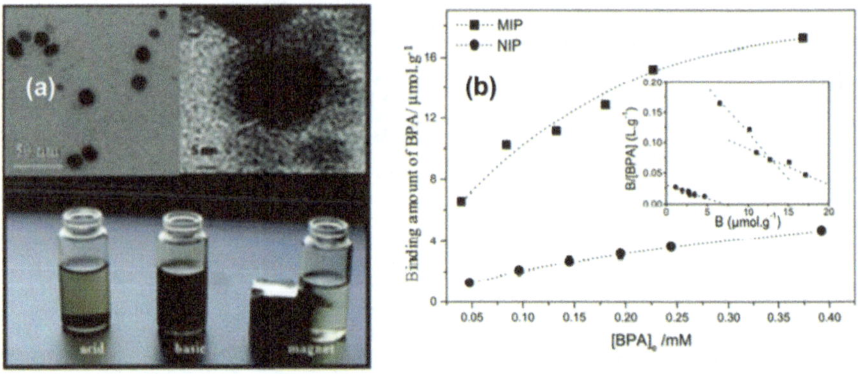

Fig. 7 **a** TEM micrographs of the magnetic MIP nanoparticles and their dispersion in different media. **b** Binding isotherms of magnetic MIP (squares) and magnetic NIP (circles) for Bisphenol A. The inset shows the Scatchard plot for the BPA binding. Reproduced from [37] with permission of the Royal Society of Chemistry

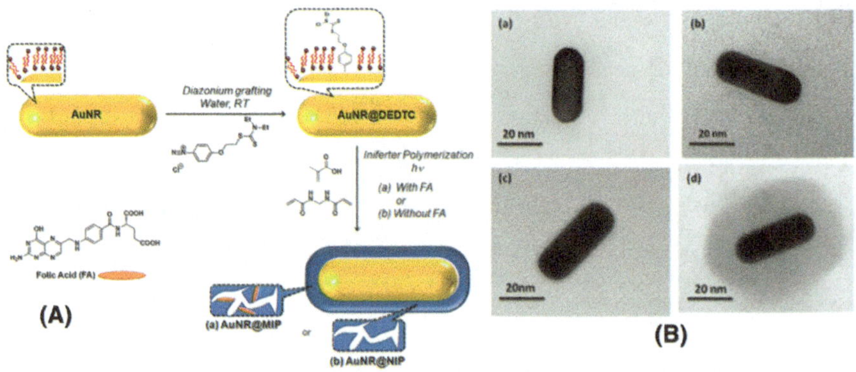

Fig. 8 **A** Schematic illustration of the surface modification strategy for coating gold nanorods (AuNRs) with a MIP layer by combining the diazonium salt chemistry and the iniferter photopolymerization method. **B** TEM micrographs of gold nanorods: **a** CTAB-coated AuNRs, **b** AuNRs@DEDTC, AuNR@MIP$_{4h}$ after **c** 4 h and **d** 8 h polymerization leading to AuNR@MIP nanocomposites. Reproduced from [39] with permission of the Royal Society of Chemistry

3.3 *Radical Photopolymerization Using* Type II *Photoinitiators*

Radical photopolymerization using *Type II* photoinitiators is an interesting technique for the synthesis of polymers; it is particularly suited for the modification of surfaces. It has the following features: (i) does not require drastic purification of the monomer, (ii) can be conducted at near UV and visible range of electromagnetic spectrum at RT, and (iii) efficient and rapid polymerization. This approach was successfully applied to the surface modification through diazonium chemistry. Attachment of typical *Type II* photoinitiator, benzophenone to the surface and using dimethylaminoaniline (DMA) as co-initiator led to grafted surface layers with remarkable properties such as controlled thickness and hydrophilic/hydrophobic character [40, 41]. The method could be extended to PHEMA-co-PMAA grafts which served for complexing Ag(I) ions; the latter was switched to immobilized Ag nanoparticles. The copolymer/Ag coatings grafted to flexible ITO substantially resisted *L. monocytogenes* ATCC 19,115 and *S. auerus* biofilm formation [42]. However, the system consisting of surface-bound DMA (from the reduction of diazotized DMA) serving as hydrogen and free benzophenone as hydrogen abstractor is better option for growing polymer chains [43]. Indeed, excess of free DMA in solution leads to turbid solution due to the synthesis of free polymer chains in solution (unpublished observations). In contrast, free benzophenone leads to semi-pinacol radical which is less prone to trigger polymerization in bulk solution.

Figure 9 depicts model surface-initiated radical photopolymerization with *Type II* initiator system based on diazonium chemistry [43]. Diazonium salt was prepared and isolated and further electro reduced at the surface of gold electrode. In the presence of the arylated gold surface, monomer and solvent, and under UV-light irradiation, benzophenone strips a hydrogen from the surfacebound dimethylamino group, therefore yielding radical species at the surface ($-N(CH_3)CH_2^{\bullet}$) and a semi-pinacol in solution, poorly active in triggering bulk solution polymerization (Fig. 9a). Hence, polymerization is confined to the surface and could be coined *"grafting from"*. Increase of benzophenone in solution permitted to control PHEMA thickness (Fig. 9b) without any observation of turbidity. The OH groups from PHEMA grafts were activated with carbonyldiimidazole to provide imidazole-functionalized PHEMA grafts for the covalent immobilization of bovine serum albumin (BSA). The extent of imidazole functionalization was ~35% (Fig. 9c) sufficient to graft quasi-one monolayer of BSA (Fig. 9d). This approach was further extended to crosslinked biomimetic coating for the picomolar electrochemical selective recognition of melamine [44]. In the case of the radical scavenger quercetin (flavonoid type molecule), isopropylthioxanthone was employed as hydrogen abstractor from the diazonium-derived 2-hydroxyethylamino groups for the attachment of quercetin imprinted polymer grafts [45].

Radical photopolymerization using *Type II* initiators can also be triggered using visible light in a sunlight simulator device or even under exposure to sunlight which is "greener" and more appealing process. In this way, camphorquinone (CQ) was used to abstract hydrogen from the surfacebound DMA groups grafted to gold (Fig. 10a) [46]. The method is versatile and could adapted to either gold or flexible ITO electrode for the attachment of homopolymer or crosslinked polymer grafts (PHEMA and PMMA, or P(HEMA-co-EGDMA) and P(MMA-co-EGDMA)). Densely packed polymer coatings were in the nanometer scale, ~ 4 to 11.2 nm for PHEMA and P(HEMA-co-EGDMA), respectively. This is most probably due to dense and thick DMA aryl layer (~8.8 nm). Using masks, it is possible to fabricate thin polymer patches (Fig. 10b). The nature of the polymer permits to control hydrophilic/hydrophobic character (Fig. 10c); PHEMA coating is relatively more hydrophilic (contact angle $\theta = 55°$, Fig. 10c) than PMMA ($\theta = 74°$), XPS permits to account for the surface chemical

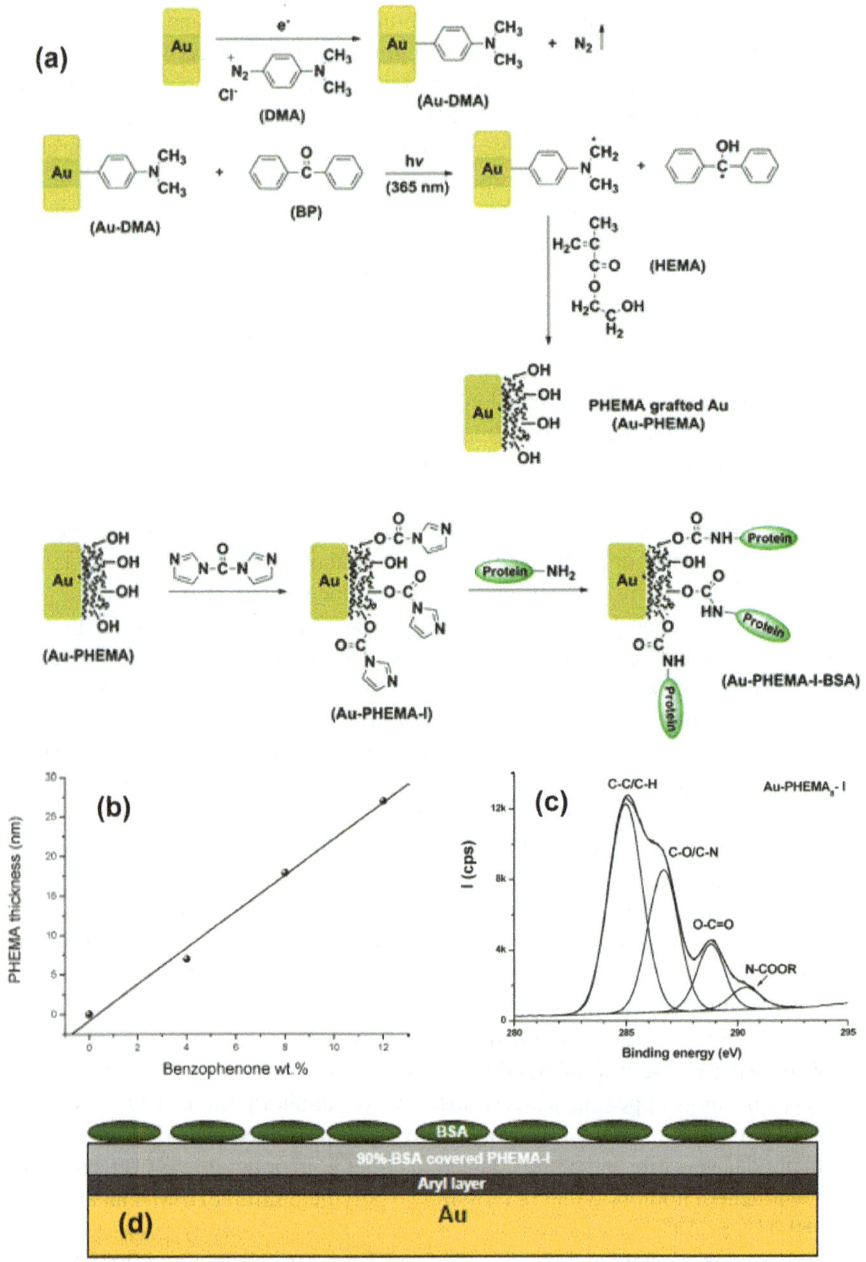

Fig. 9 Preparation of PHEMA grafts on DMA-modified gold plates: **a** mechanisms of polymer brush growth and post-activation by carbonyldiimidazole for protein covalent attachment, **b** control of polymer thickness with concentration of benzophenone photosensitizer, **c** high-resolution XPS C1s spectrum of imidazole-functionalized PHEMA brush, and **d** 90% coverage of BSA grafted to imidazole-modified PHEMA brush. Reproduced from [43] with permission of ACS

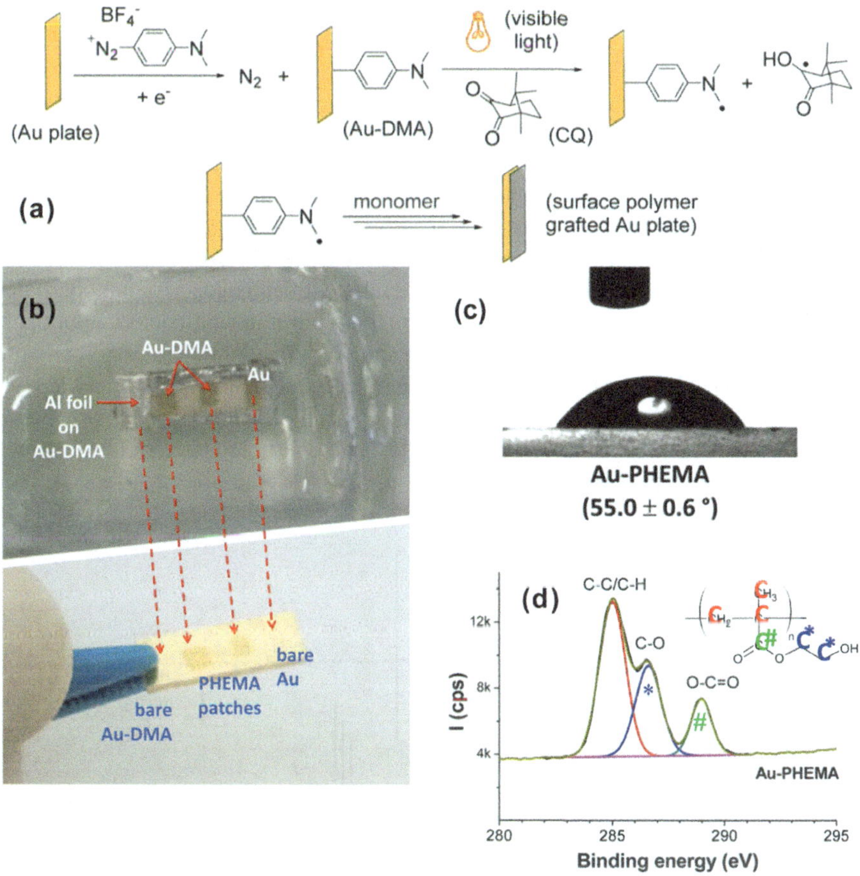

Fig. 10 **a** Visible light radical photopolymerization initiated on dimethylaminophenyl-modified gold surface, **b** design of polymer patches, **c** relatively low contact angle of a water drop on Au-PHEMA, and **d** high-resolution C1s region from the PHEMA graft. Reproduced from [46] with permission of John Wiley & Sons

composition: Fig. 10d displays peak-fitted C1s region from Au–PHEMA; the C–C/C–H, C–O, and O–C = O peak component relative intensity are in the 3:2:1 ratio, in line with the chemical structure.

This general approach using aryl diazonium salts to conduct radical photopolymerization with *Type II* initiators has been applied for developing ion-imprinted clay-polymer nanocomposites. Both studies were conducted with DMA-intercalated clay and radical photopolymerization was carried out in the presence of the hydrogen abstracting photoinitiators, benzophenone [47] or CQ [47] under UV or visible light, respectively. The resulting imprinted nanocomposites were highly selective adsorbents of Pb(II) [47] and Cu(II) [48] when Pb(II) and Cu(II) were used as template ions in the photopolymerization process.

3.4 Concurrent Radical Photopolymerization and Metal Nanoparticle Formation

This is an interesting technique originally developed by Yagci and Sangermano [49, 50], particularly suitable for incorporating metallic nanostructures in vinylic polymer grafts. It is based on the redox reactions between the radicals generated upon light exposure and the metal ion employed in the polymerization medium. The technique has been tested with vinylic monomers in the presence of silver nitrate in view of designing antibacterial hydrophilic polymer grafts with embedded Ag nanoparticles [51]. Figure 11 illustrates the mechanism of making ultrathin, antibacterial polymer-silver nanocomposites coating. Both HEMA and polyglycidol macromonomer were photopolymerized resulting in PHEMA or comb poly(α-*tert*-butoxy-ω-vinylbenzyl-polyglycidol) (PPGL) grafts. Either linear or comb hydrophilic polymers resist to protein or bacterial adhesion; with embedded Ag nanoparticles, bactericidal coatings were obtained with zero immobilized bacteria.

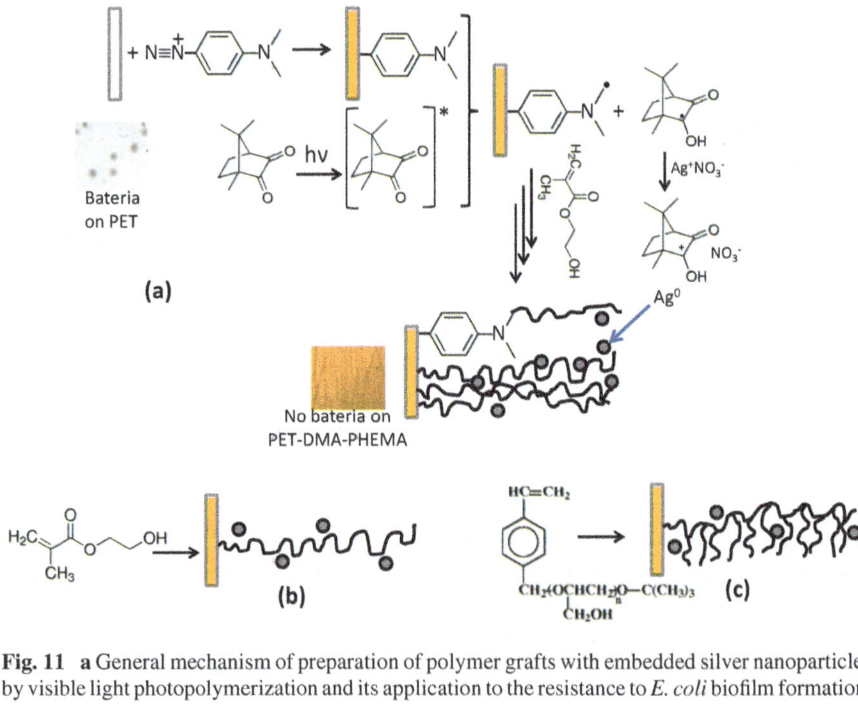

Fig. 11 **a** General mechanism of preparation of polymer grafts with embedded silver nanoparticles by visible light photopolymerization and its application to the resistance to *E. coli* biofilm formation. Bacteria stick to pristine PET but PET-DMA-PHEMA is antibacterial. Fabrication of **b** linear polymer grafts of PHEMA, and **c** comb PPGL polymer grafts. See [51] for article details

3.5 Azide-Alkyne Click Polymerization

Huisgen 1,3-cycloaddition is probably the most investigated "click" chemical reaction in modern materials chemistry [52]. It has been particularly explored for synthesizing linear, graft, hyperbranched, cyclic, star, and dendritic polymeric architectures [53–55] as well as main chain- and side-chain poly(1,2,3-triazolium)s [56]. Bensghaïer et al. [57] have prepared CNTs grafted with main chain polytriazole starting from the arylation of the nanotubes with 4-azidobenzenediazonium (Fig. 12a). The clickable CNTs were reacted with two co-monomers, diazido and dialkynyl monomers, to yield very stable MWCNT-polytriazole (MWCNT-PTAz) nanohybrids. The chemical composition of the nanohybrids was verified by XPS. Survey regions (Fig. 12b) demonstrate stepwise increase of N1s/C1s and O1s/C1s intensity ratios resulting from arylation and click polymerization. The N1s narrow region from MWCNT-N_3 testifies for azidophenyl grafting to MWCNTs and for the existence of N=N azo bonds within the tethered aryl layer (Fig. 12c). Complex C1s region from MWCNT-PTAz (Fig. 12d) exhibits intense C–N and C–O C1s components in line with the chemical structure of the polytriazole shown in Fig. 12a. Triazole moieties in the grafted chains yield complete change in the N1s region of MWCNT-PTAz, i.e., the peak is fitted with two components assigned to N = N and > N– in the 2/1 ratio as expected for triazole groups (Fig. 12e).

3.6 Oxidative Polymerization of Conjugated Monomers

Conjugated polymers such as polypyrrole (PPy), polyaniline (PANI), polythiophene (PTh), or poly(3,4-ethylenedioxythiophene) (PEDOT) or their diazotized oligomers have been the subject of numerous studies and concerned electropolymerization, chemical precipitation polymerization and photopolymerization. Numerous substrates have been investigated; glassy carbon [58], gold [59], ITO-coated glass [60] and flexible ITO electrodes [61, 62]; shape memory nitinol [63], carbon nanotubes [64, 65] and graphene oxide [66]. Polymers or oligomers were attached to the surface via non-covalent interactions or covalent bonds, depending on the attached aryl groups. Note however that in the case of diazotized oligomers, substrate-oligoaryl adhesion is expected to occur via covalent bonding. The team of Lacroix paved the way for the attachment of conjugated polymers [58] and oligomers [67] as summarized in an excellent Chapter with emphasis on molecular electronics [68]. An interesting paper published in 2010 reported the study of molecular weight effect of high diazotized oligoaniline on the grafting to single-walled carbon nanotubes [64]. Previous studies on in-situ polymerization of pyrrole and aniline and considering covalent and non-covalent attachment of the corresponding conductive polymers are summarized below.

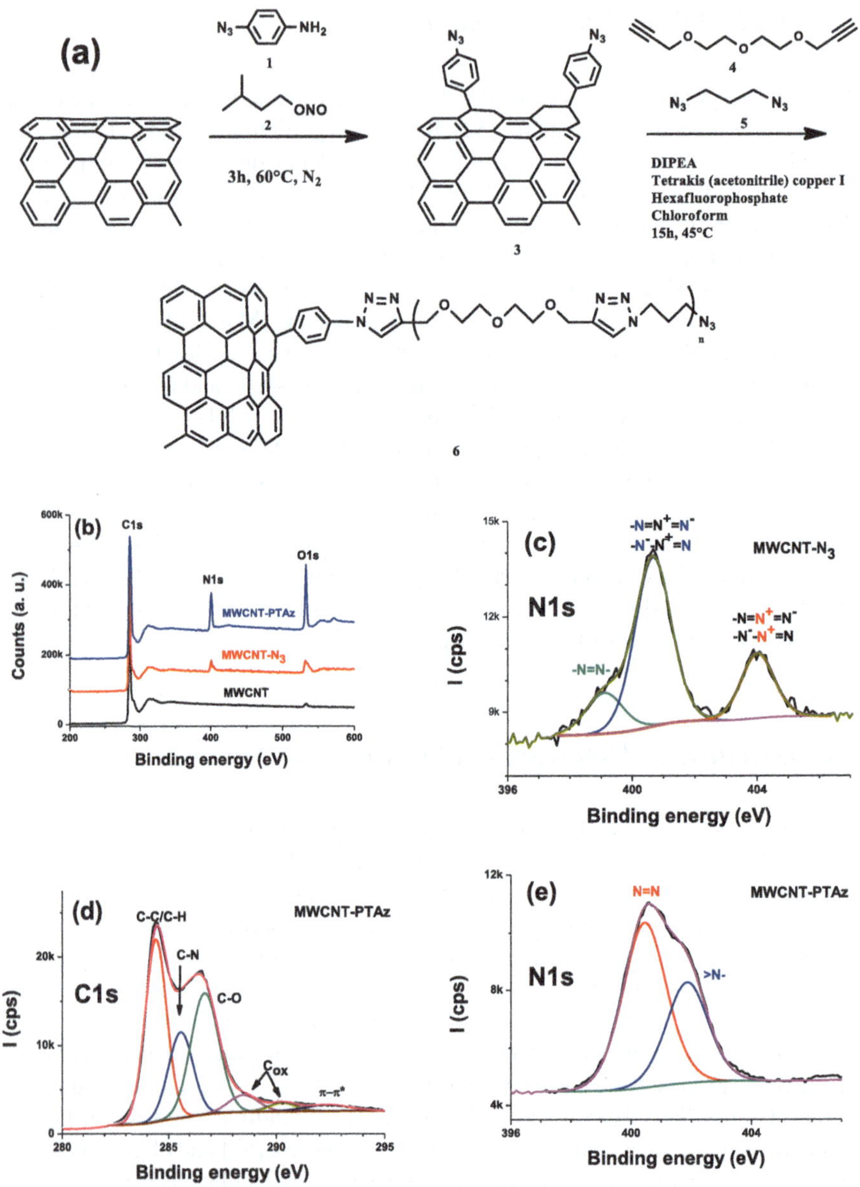

Fig. 12 Preparation and XPS surface analysis of MWCNT-polytriazole nanohybrid: **a** Synthetic route for the in-situ grafting of azidoaryl groups to MWCNTs followed by surface-confined 1,3-cycloaddition click polymerization. XPS spectra of pristine and functionalized MWCNTs: **b** survey spectra of MWCNT, MWCNT-N_3, and MWCNT-PTAz; **c** high-resolution N1s narrow region of MWCNT-N_3; **d** peak-fitted high-resolution C1s spectrum of MWCNT-PTAz; **e** peak-fitted high-resolution N1s spectrum of MWCNT-PTAz. Reproduced from [57] with permission from John Wiley & Sons

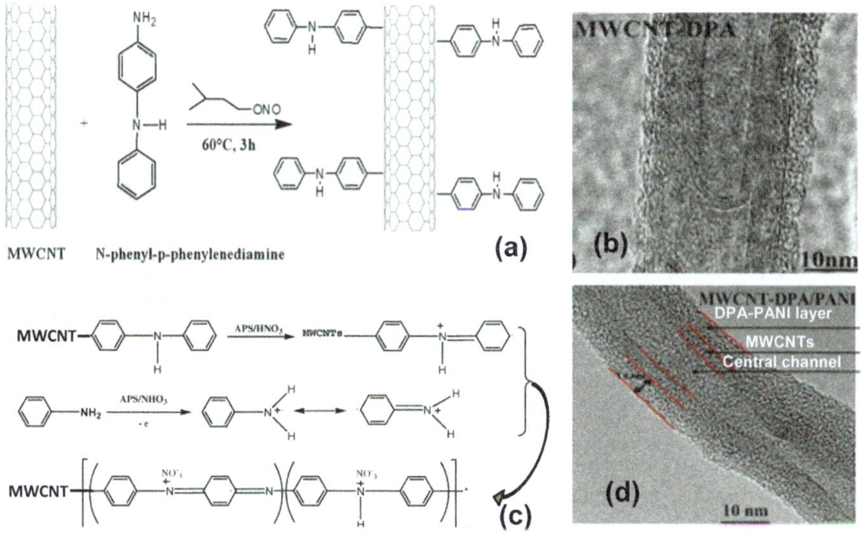

Fig. 13 Preparation of core/double-shell MWCNT/DPA/PANI nanocomposite: **a** arylation of CNTs and its TEM image (**b**); **c** in-situ oxidative polymerization of aniline in the presence of MWCNT-DPA and the resulting nanocomposite as imaged by TEM **d** Reproduced from [65] with permission of Elsevier

3.6.1 Covalent Grafting of Conductive Polymers by in-Situ Polymerization

Chemical Polymerization

MWCNT/polyaniline was prepared by oxidative polymerization of aniline in the presence of CNTs grafted with diphenylamine groups (DPA) [65]. D/G Raman peak intensity ratios were 1.01 and 1.33 for MWCNTs and MWCNT-DPA, respectively, an indication of effective covalent grafting of DPA to the CNT sidewall. Densely packed, protuberance-free PANI chains layer was found to evenly wrap the underlying CNTs, resulting in unique core/shell structure (Fig. 13). Arylation played a significant role in this regard since without any pretreatment, PANI was coated on pristine CNTs but the coating was not uniform. This has also significant effect on the electric conductivity of the CNT/Aryl/PANI hybrid which was found to be 12.4 S. cm^{-1}, higher than that of MWCNT/PANI (7.3 S. cm^{-1}), for 50/50 wt.% initial aniline/MWCNT-DPA ratio.

Covalent Attachment of Polypyrrole by Photopolymerization

Covalent attachment of polypyrrole was achieved using the diazonium compound of 4-(1H-pyrrol-1-yl)aniline, electrochemically reduced to electrode surfaces as

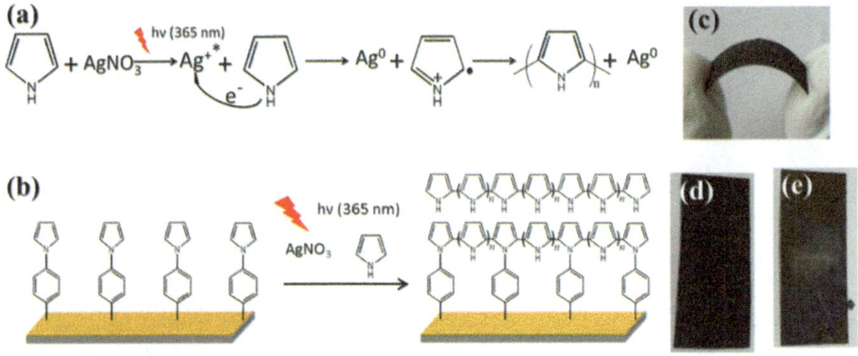

Fig. 14 Photopolymerization of pyrrole on flexible ITO using $AgNO_3$ photosensitizer: **a** mechanism of the photopolymerization process, **b** pyrrole photopolymerization ITO surface-bound pyrrolyl, (**c**, **d**) digital photographs of the PPy-Ag nanocomposite (ITO/Py-D/PPy-Ag) under bent and flat conditions. **e** Digital photograph of the PPy/Ag nanocomposite (ITO/PPy-Ag) on bare ITO. Reproduced from [61] with permission of ACS

described by the team of Zineb Mekhalif [69]. This compound provides surface-bound pyrrolyl groups attached to the underlying electrode via interfacial C-N = N–O-M (observation of negatively charged ToF–SIMS fragments of $NiCN_2O^-$, $NiC_2N_2O^-$ and $NiC_4N_2O^-$) and C-N = N-M groups (detection of $NiC_2N_2^-$, $NiC_4N_2^-$ and $NiC_6N_2^-$). This is very important because covalent bonds ensure excellent adhesion, but the existence of the interfacial diazoether –N = N–O group is surprising since diazoethers are known to be highly unstable in solution [70]. Their existence could also be confirmed by XPS analysis of diazonium intercalated montmorillonite as well as Nitinol grafted aminophenyl and pyrrolylphenyl groups [63]. With surfacebound pyrrolyl groups in hand, photopolymerization of pyrrole was adapted from the 2001 work of Breimer et al. [71] in order to prepare silver-polypyrrole nanocomposites on glass-coated ITO electrodes [72], flexible ITO electrodes [61], and shape memory Nitinol alloy [63]. In these studies, $AgNO_3$ was tested as photosensitizer in order to provide silver-polypyrrole nanocomposite coatings (Fig. 14). Silver imparts conductivity [61] and bactericidal property [73].

Adhesion to ITO is particularly weak and requires an adhesive layer, ideally provided by the aryl groups. Figure 14a displays the mechanism of photopolymerization of pyrrole using AgNO3 photosensitizer under UV-light. Figure 14b shows arylation of flexible ITO using 4-the in-situ generated diazonium compound of (1H-pyrrol-1-yl)aniline. Figure 14c demonstrates that despite the bending stress, the polypyrrole film did not suffer any adhesion failure as for flat substrate. However, without any arylation, PPyAg adhesion was found to be weak and the film could be removed with an adhesive tape. Similarly, Jacques et al. [63] noted excellent adhesion of PPyAg nanocomposite on Nitinol grafted with pyrrolyl groups. In contrast, PPyAg nanocomposite film was delaminated from bare Nitinol alloy upon simple cleaning the film after synthesis using water wash bottle.

3.6.2 Non-covalent Attachment of Conductive Polypyrrole

Molecular interactions at interfaces play major role in polymer adhesion, harnessed with covalent bonds. However, other interactions of lower energies could also be sufficient for adhesive joints. This is the reason why polypyrrole was grown on a variety of surfaces bearing functional groups that undergo electrostatic interactions or hydrogen bonding with polypyrrole on top of van der Waals interactions including the ubiquitous London dispersion forces (instantaneous dipole–induced dipole forces). In this regard, aryl diazonium compounds can provide a range of functional groups, such as alkyl, perfluorinated, and the most employed strong hydrogen bonding COOH and NH_2 groups. In line with the well-known remarkable adhesion of polypyrrole on aminosilane-modified glass plates [74], Lo et al. [62] electrochemically deposited benzene sulfonic acid-doped polypyrrole (PPyBSA) on aminophenyl-modified flexible ITO which led to design high-performance electrochemical sensors of metal ions in aqueous solutions prepared in laboratory and in wastewater [75]. Figure 15 depicts the strategy of the fabrication of the films (Fig. 15a). Arylation of ITO yields excellent adhesion of PPy coating, however, adhesion failure was noted on bare ITO by simple drying with compressed air (Fig. 15b). Also washing of the electrodes after electropolymerization removes PPy from bare ITO, but not arylated ITO (see supplementary video in [76]). The pretreatment time to graft the aryl layer by chronoamperometry was also found to be important to control in order to avoid passivation of ITO (Fig. 15c). The ideal time was ~ 30 to 45 s at −0.8 V/saturated calomel. For lower pretreatment time, PPyBSA could be deposited but exhibits some charge transfer resistance, which is the same situation for 60 s chronoamperometric aryl grafting. For longer treatment, i.e., 120 s, no film was formed as shown in Fig. 15d.

Another important parameter is the nature of the functional group in para position of the diazonium. As several groups have reported, they impart varying hydrophilic/hydrophobic characters and exhibit charge transfer resistance as probed by the $Fe(CN)_3{}^{-}/_4{}^{-}$ redox probe (Fig. 15e). The relative peak intensity I_{rel} ($I_{rel} = I_{ap}(ITO–X)/I_{ap}(bare\ ITO) \times 100\%$) correlated very well with the water contact angle on modified ITO [76]. The most hydrophobic surfaces gave the highest I_{rel} values, i.e., weak barrier, whereas the most hydrophilic surfaces interact more favorably with water rather than with the probe. Figure 15e justifies the use of para-aminobenzenediazonium salt as precursor for grafting aminophenyl groups to ITO in the design of metal ion sensors.

Concerning metal ion sensing performances, differential pulse voltammetry (DPV) was employed to detect Pb^{2+}, Cd^{2+}, and Cu^{2+} ions in single, or in metal ion mixture solutions. The limits of detection (LODs) were ~1.0, 8.95, and 11.1, nmol.L^{-1}, respectively. For similar initial concentrations, no competitive adsorption was noted and heavy metal ions could be recovered without any significant loss. In contrast, recovery was lower than 50% for Pb^{2+} in the presence of 100 fold more concentrated metal ions Cu^{2+} or Cd^{2+}. When these ions were spiked in wastewater, similar flexible electrodes ITO-NH_2-PPyBSA were permitted to reach 81 and 95% recovery of Cd^{2+} (good) and Cu^{2+} (excellent recovery score), respectively [75].

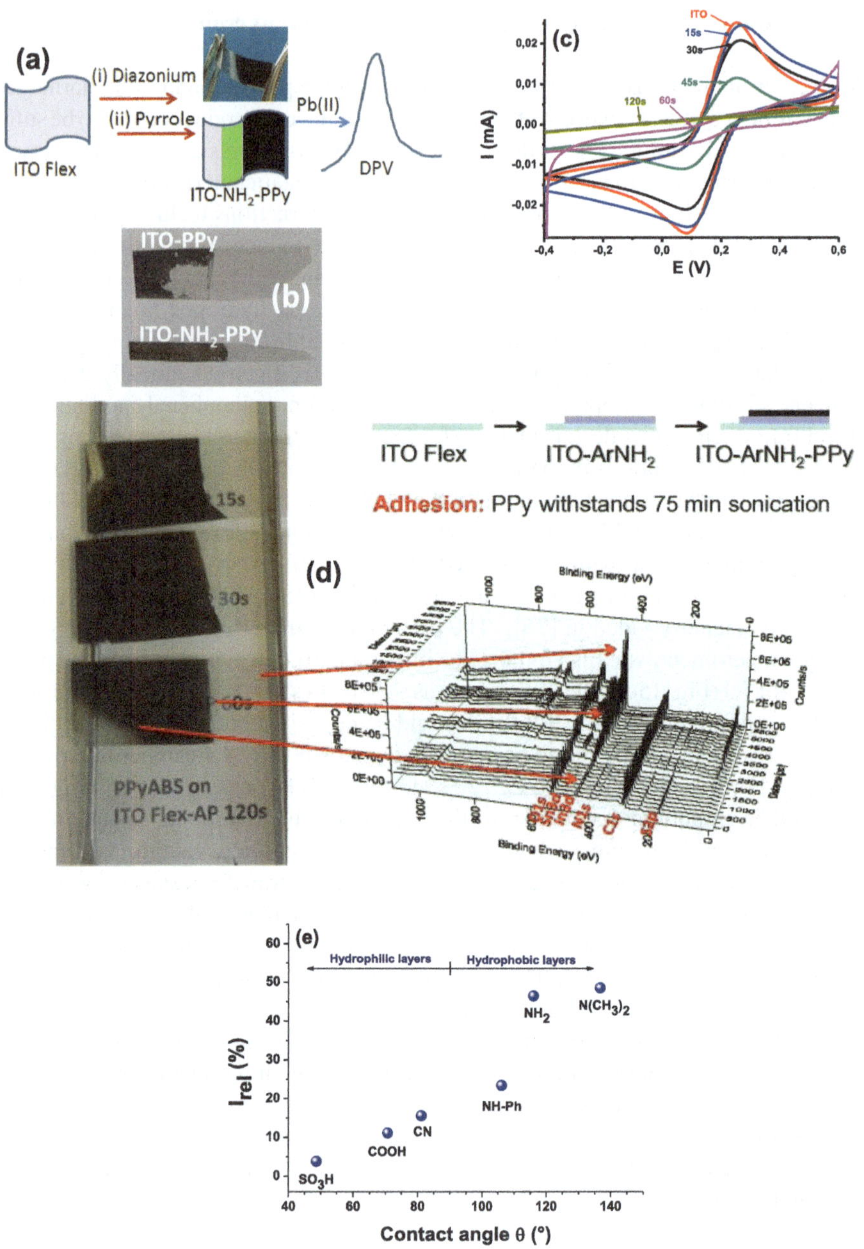

Fig. 15 Adhesion of PPyBSA to aryl-modified flexible ITO: **a** general strategy and target sensing application, **b** effect of drying with compressed air of ITO-PPy (adhesion failure) and ITO-NH_2-PPy (good adhesion), **c** cyclic voltammograms recorded for 1 mM $Fe(CN)_3^-/_4^-$ redox probe in 0.1 M KCl on ITO and ITONH_2 (t), at indicated electrografting time (t in s), **d** effect of pretreatment time with aminophenyl groups on the electrodeposition of PPyBSA, and **e** plot of Irel-vs-water contact angles on various aryl-modified flexible ITO surfaces. **a** and **c** reproduced from [62] with permission of Elsevier; **d** reproduced by permission of Springer Nature from [75]; **e** reproduced from [76]. Figure 15b: previously unpublished

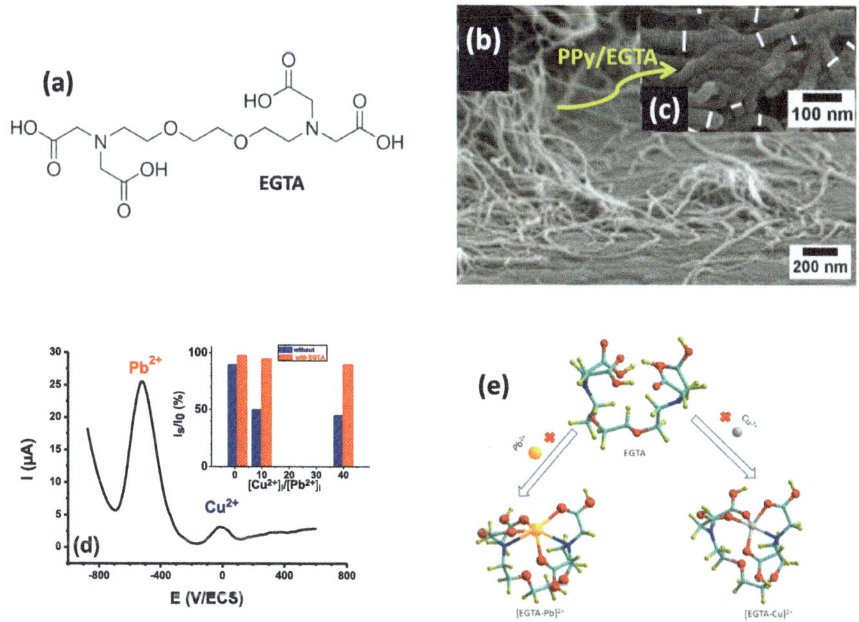

Fig. 16 **a** EGTA chemical structure, **b** CNTs immobilized on ITO-NH_2, **c** PPy/EGTA-wrapped CNTs on ITO, **d** selectivity of the sensor with total recovery using PPy/EGTA even in the presence of Cu^{2+}, and **e** structures of stable conformers of EGTA and, $(EGTA\text{-}Pb)^{2+}$ and $(EGTA\text{-}Cu)^{2+}$ complexes. Reproduced from [77]

To enhance selectivity and improve conductivity of the sensing films, PPyBSA was prepared in the presence of the EGTA chelator, ethylene glycolbis(2-aminoethylether)-tetraacetic acid on aminophenyl-modified ITO decorated with MWCNTs (Fig. 16a–b) [77]. CNTs exhibited remarkable adhesion on ITO NH_2 surface but were easily removed from bare ITO by simple washing, indicating the importance of aryl diazonium pretreatment. Electrodeposition of PPyBSA on CNT-decorated ITO-NH_2 led to PPyBSA-wrapped nanotubes (Fig. 16c). The average diameter was 18 ± 2.6 nm for bare CNTs and increased to 27 ± 4.8, 35.6 ± 5.9, and 175 ± 20.1 nm after 1, 5 and 10 cycles of pyrrole electropolymerization, respectively. EGTA-containing PPyBSA could recognize Pb^{2+} and Cu^{2+} simultaneously, however, no recovery loss was noted (Fig. 16d) since, for an initial Cu^{2+} concentration, 40-fold that of Pb^{2+}, the response of the sensor to the presence of Pb^{2+} was not altered, that is no significant competitive adsorption of copper occurred. DFT calculation permitted us to understand the excellent selective behavior of the EGTA-containing sensor; the interaction energy of the EGTA–metal ion pairs (with solvation effects), was −374.6 kJ/mol for Pb^{2+}, more than threefold that computed for Cu^{2+} (−116.4 kJ/mol). This clear difference is due to the hexacoordination of Pb^{2+} compared to the pentacoordiantion of Cu^{2+}.

3.7 New Trends

In this chapter, it is not intended to systematically citation of the numerous studies on polymer grafting to arylated surfaces. Instead, we concentrated on the most eye-catchy strategies for gluing polymers to the surfaces. While above-shortlisted strategies proved to be robust, there are new trends that are worth further investigation. For example, Fig. 17 shows the power of photo-induced thiol-yne reaction between clickable gold substrates and mercaptosilanes to provide silanized aryl layer able to react further with aminosilane [78] leading to form of final sol-gel polymer coating (Fig. 17a). UV-light induced thiol-yne click reaction permits to obtain patterned sol-gel coatings which could be monitored by XPS. S/Au atomic ratio was determined along the main axis of the photo-patterned mercaptosilane; the S/Au was highest in the arylated regions exposed to light-driven click reaction (Fig. 17a, right). As silanes undergo sol-gel polymerization, there is clear interest in this approach which could be for example exploited to design unbreakable capillary microextraction devices [79].

Another way to conduct surface-initiated polymerization is via plasmon-induced process. This is convenient for nitroxide mediated polymerization (NMP), a well-known CRP. The C-ON bond from the NMP initiator is usually cleaved at 100 °C; but on plasmonic surfaces such as gold gratings, shining 785 nm laser beam excites plasmons which in turn cleave surface-bound C-ON group from the NMP initiator thus enabling the polymerization to proceed (Fig. 17b) [80]. It was interesting to show that poly(N-isopropyl acrylamide), PNIPAM, could be grown and verified to be thermoresponsive. Due to the living character of NMP, vinyl boronic acid (VBA) could also be polymerized on Au-PNIPAM terminated by NMP initiator moieties. The polyBVA grafts served for specific recognition of cis-diol compounds (catechol and glycoprotein).

Recently, the year 2020 marked the 100th anniversary of the first ever published article on polymerization by Hermann Staudinger. Editors of Macromolecular Chemistry and Physics [81] celebrated the event and discussed the milestones of polymer chemistry and its benefit to the Society. However, they also underlined the concerns raised by plastic pollution and the need for a sustainable future for polymers. Toward this end, we were delighted to see that not only synthetic polymers are concerned with adhesion to arylated surfaces, but also naturally occurring polyelectrolytes such as polyalginate. This natural polymer can also be prepared by graft polymerization to provide barrier to corrosion of AA-5083 aluminum-magnesium alloy immersed in Dead Sea water, the saltiest water on Earth (21.8% of chloride species) [82].

4 Conclusion

We have summarized the essential developments of polymer attachment to surfaces by various polymerization techniques conducted in the presence of arylated

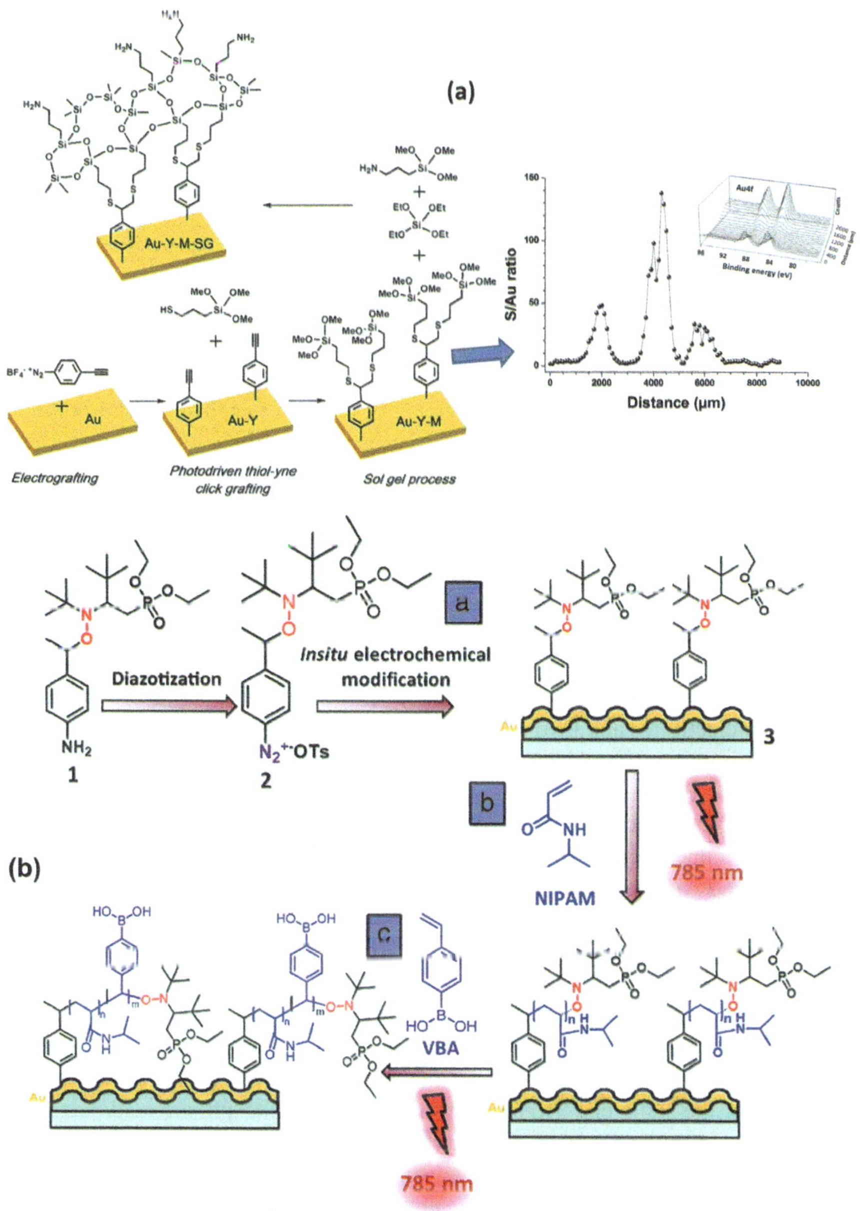

Fig. 17 New trends of surface-confined polymerization on arylated surfaces: **a** sol-gel polymerization on alkynylated gold photo clicked with mercaptosilane, **b** three-step plasmon-induced nitroxide mediated polymerization of NIPAM followed by VBA on arylated gold gratings, and **c** alginate poly merization on 4-hydroxybenzenediazonium modified AA-5083 alloy. **a** reproduced from [78] with permission of ACS, **b** reproduced from [80], **c** adapted from [82] with permission of Springer Nature

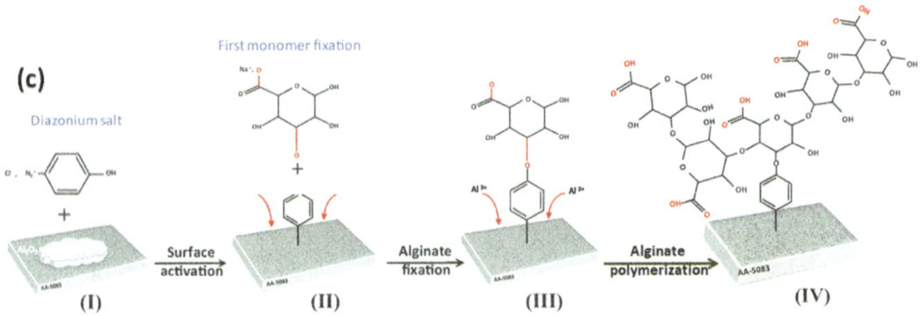

Fig. 17 (continued)

surfaces. Radical polymerization techniques offer wide panel of vinylic polymers which offer several properties: controlled hydrophilic/hydrophobic character, chemical reactivity, redox reactions, biomimetism, sensing, and modification by post-polymerization in the case of CRP. Surface patterning is possible via radical photopolymerization triggered by either UV or sunlight. We have introduced first ever examples of simultaneously proceeding radical polymerization and electron transfer reactions conducted on aryl-base surfacebound initiator in view of designing bactericidal silver-polymer nanocomposite thin films. Patterning via photo thiol-yne reaction opens new avenues for sol-gel polymer coatings. Concerning conjugated polymers, we discussed the essential parameter of near aryl monolayer in order not to passivate electrodes and favor electropolymerization. Tuning aryl layer thickness is not only important for electrochemical reactions, but also it is a key factor in conductive polymer adhesion, particularly on ITO electrodes. We have finished the chapter with new trends relevant to pure surface polymer chemistry and first report on plasmon-induced nitroxide mediated polymerization (NMP) conducted on gold grating grafted with diazonium-based NMP initiator to provide diblock copolymer brushes. Polymerization of alginate on hydroxyphenyl-modified Al-Mg alloy is an important step toward sustainable surface polymer chemistry.

Overall, diazonium salts have much to offer in the general domain of surface polymer chemistry and polymer adhesion. Interestingly, Amory Jacques[1] has compared the cost of Nitinol surface modification with various diazonium salts (bearing NH_2, pyrrolyl, SO_3H groups) to the same process conducted with phosponic acids bearing the same end groups and found that the phosphonic acid/diazonium salt cost ratio is in the 18–6066 range, which advocates the development of diazonium surface modification for polymer adhesion. Although it has come of age, one can anticipate new and surprising, highly challenging findings in the coming years. Diazonium modification of substrates will certainly find a niche in surface polymer chemistry as they are unique coupling agents for polymers and provide robust substrate-aryl-polymer interphase region, i.e., highly resistant to hydrolysis,

[1] Jacques A. PhD Thesis, University of Namur, Namur, Belgium (2016). https://researchportal.unamur.be/fr/studentTheses/les-sels-de-diazonium-une-bo%C3%AEte-%C3%A0-outils-pour-la-modification-et, last accessed 28 September 2021.

a phenomenon that provokes polymer leaching and fast drop of performances of devices exposed to humid conditions.

Declaration of Interest The authors declare no conflict of interest.

References

1. Delamar M, Desarmot G, Fagebaume O, Hitmi R, Pinson J, Savéant J-M (1997) Carbon 35:801–807
2. Mahouche-Chergui S, Gam-Derouich S, Mangeney C, Chehimi MM (2011) Chem Soc Rev 40:4143–4166
3. Chehimi MM (2012) In: Aryl diazonium salts: new coupling agents in polymer and surface science. Wiley
4. Chehimi MM, Pinson J, Salmi Z (2013) Appl Surf Chem Nanomater 95–143
5. Bakas I, Salmi Z, Gam-Derouich S, Jouini M, Lépinay S, Carbonnier B, Khlifi A, Kalfat R, Geneste F, Yagci Y (2014) Surf Interface Anal 46:1014–1020
6. Cao C, Zhang Y, Jiang C, Qi M, Liu G (2017) ACS Appl Mater Interfaces 9:5031–5049
7. Sandomierski M, Voelkel A (2020) J Inorganic Organometallic Polym Mater 1–21
8. Bensghaïer A, Mousli F, Lamouri A, Postnikov PS, Chehimi MM (2020) Chem Africa 1–35
9. Adenier A, Cabet-Deliry E, Lalot T, Pinson J, Podvorica F (2002) Chem Mater 14:4576–4585
10. Matrab T, Chehimi MM, Perruchot C, Adenier A, Guillez A, Save M, Charleux B, Cabet-Deliry E, Pinson J (2005) Langmuir 21:4686–4694
11. Matrab T, Save M, Charleux B, Pinson J, Cabet-Deliry E, Adenier A, Chehimi MM, Delamar M (2007) Surf Sci 601:2357–2366
12. Matrab T, Chehimi M, Pinson J, Slomkowski S, Basinska T (2006) Surface Interf Anal: Int J Devoted Developm Appl Techniq Anal Surf Interf Thin films 38:565–568
13. Matrab T, Chehimi M, Boudou J-P, Benedic F, Wang J, Naguib N, Carlisle J (2006) Diam Relat Mater 15:639–644
14. Gam-Derouich S, Ngoc Nguyen M, Madani A, Maouche N, Lang P, Perruchot C, Chehimi MM (2010) Surface Interface Anal 42:1050–1056
15. Mrabet B, Mejbri A, Mahouche S, Gam-Derouich S, Turmine M, Mechouet M, Lang P, Bakala H, Ladjimi M, Bakhrouf A (2011) Surf Interface Anal 43:1436–1443
16. Iruthayaraj J, Chernyy S, Lillethorup M, Ceccato M, Røn T, Hinge M, Kingshott P, Besenbacher F, Pedersen SU, Daasbjerg K (2011) Langmuir 27:1070–1078
17. Liu T, Casado-Portilla R, Belmont J, Matyjaszewski K (2005) J Polym Sci Part A: Polym Chem 43:4695–4709
18. Matrab T, Chancolon J, L'hermite MM, Rouzaud J-N, Deniau G, Boudou J-P, Chehimi MM, Delamar M (2006) Colloids and Surfaces A: Physicochem Eng Aspects 287:217–221
19. Wu W, Tsarevsky NV, Hudson JL, Tour JM, Matyjaszewski K, Kowalewski T (2007) Small 3:1803–1810
20. Dahoumane SA, Nguyen MN, Thorel A, Boudou J-P, Chehimi MM, Mangeney C (2009) Langmuir 25:9633–9638
21. Wang G-J, Huang S-Z, Wang Y, Liu L, Qiu J, Li Y (2007) Polymer 48:728–733
22. Salmi Z, Gam-Derouich S, Mahouche-Chergui S, Turmine M, Chehimi M (2012) Chem Pap 66:369–391
23. Mohamed AA, Salmi Z, Dahoumane SA, Mekki A, Carbonnier B, Chehimi MM (2015) Adv Coll Interface Sci 225:16–36
24. Braunecker WA, Matyjaszewski K (2007) Prog Polym Sci 32:93–146
25. Yagci Y, Jockusch S, Turro NJ (2010) Macromolecules 43:6245–6260

26. Matyjaszewski K, Dong H, Jakubowski W, Pietrasik J, Kusumo A (2007) Langmuir 23:4528–4531
27. Szczepaniak G, Fu L, Jafari H, Kapil K, Matyjaszewski K (2021) Acc Chem Res 54:1779–1790
28. Buhl KB, Møller RK, Heide-Jørgensen S, Kolding AN, Kongsfelt M, Budzik MK, Hinge M, Pedersen SU, Daasbjerg K (2018) ACS Omega 3:17511–17519
29. Słowikowska M, Chajec K, Michalski A, Zapotoczny S, Wolski K (2020) Materials 13:5139
30. Jin G-P, Fu Y, Bao X-C, Feng X-S, Wang Y, Liu W-H (2014) J Appl Electrochem 44:621–629
31. Lillethorup M, Torbensen K, Ceccato M, Pedersen SU, Daasbjerg K (2013) Langmuir 29:13595–13604
32. Lee BS, Kim JY, Park JH, Cho WK, Choi IS (2016) J Nanosci Nanotechnol 16:3106–3109
33. Chiefari J et al (1998) Macromolecules 31:5559–5562
34. Zoppe JO, Ataman NC, Mocny P, Wang J, Moraes J, Klok H-A (2017) Chem Rev 117:1105–1318
35. Tasdelen MA, Durmaz YY, Karagoz B, Bicak N, Yagci Y (2008) J Polym Sci Part A: Polym Chem 46:3387–3395
36. Boitard C, Lamouri A, Ménager C, Griffete Nbw (2019) ACS Appl Polym Mater 1:928–932
37. Griffete N, Li H, Lamouri A, Redeuilh C, Chen K, Dong C-Z, Nowak S, Ammar S, Mangeney C (2012) J Mater Chem 22:1807–1811
38. Griffete N, Lamouri A, Herbst F, Felidj N, Ammar S, Mangeney C (2012) RSC Adv 2:826–830
39. Ahmad R, Félidj N, Boubekeur-Lecaque L, Lau-Truong S, Gam-Derouich S, Decorse P, Lamouri A, Mangeney C (2015) Chem Commun 51:9678–9681
40. Gam-Derouich S, Carbonnier B, Turmine M, Lang P, Jouini M, Ben Hassen-Chehimi D, Chehimi MM (2010) Langmuir 26:11830–11840
41. Gam-Derouich S, Gosecka M, Lepinay S, Turmine M, Carbonnier B, Basinska T, Slomkowski S, Millot M-C, Othmane A, Ben Hassen-Chehimi D (2011) Langmuir 27:9285–9294
42. Ben Slama R, Mrabet B, Bakhrouf A, Salmi Z, Bakas I, Chehimi MM (2015) Surface Innovations 3:103–114
43. Gam-Derouich S, Lamouri A, Redeuilh C, Decorse P, Maurel Fo, Carbonnier B, Beyazıt S, Yilmaz G, Yagci Y, Chehimi MM (2012) Langmuir 28:8035–8045
44. Bakas I, Salmi Z, Jouini M, Geneste F, Mazerie I, Floner D, Carbonnier B, Yagci Y, Chehimi MM (2015) Electroanalysis 27:429–439
45. Salmi Z, Benmehdi H, Lamouri A, Decorse P, Jouini M, Yagci Y, Chehimi MM (2013) Microchim Acta 180:1411–1419
46. Bakas I, Yilmaz G, Ait-Touchente Z, Lamouri A, Lang P, Battaglini N, Carbonnier B, Chehimi MM, Yagci Y (2016) J Polym Sci, Part A: Polym Chem 54:3506–3515
47. Msaadi R, Ammar S, Chehimi MM, Yagci Y (2017) Eur Polymer J 89:367–380
48. Msaadi R, Yilmaz G, Allushi A, Hamadi S, Ammar S, Chehimi MM, Yagci Y (2019) Polymers 11:286
49. Yagci Y, Sangermano M, Rizza G (2008) Polymer 49:5195–5198
50. Yagci Y, Sangermano M, Rizza G (2008) Macromolecules 41:7268–7270
51. Samanta S, Bakas I, Yilmaz G, Cabet E, Lilienbaum A, Sun X, Gosecka M, Basinska T, Slomkowski S, Singh A (2014) J Colloid Sci Biotechnol 3:58–67
52. Singh MS, Chowdhury S, Koley S (2016) Tetrahedron 72:5257–5283
53. Qin A, Lam JW, Tang BZ (2010) Chem Soc Rev 39:2522–2544
54. Arslan M, Tasdelen MA (2019) Chemistry Africa 2:195–214
55. Neumann S, Biewend M, Rana S, Binder WH (2020) Macromol Rapid Commun 41:1900359
56. Obadia MM, Drockenmuller E (2016) Chem Commun 52:2433–2450
57. Bensghaïer A, Salmi Z, Le Droumaguet B, Mekki A, Mohamed AA, Beji M, Chehimi MM (2016) Surf Interface Anal 48:509–513
58. Santos LM, Ghilane J, Fave C, Lacaze P-C, Randriamahazaka H, Abrantes LM, Lacroix J-C (2008) J Phys Chem C 112:16103–16109
59. Ait-Touchente Z, Sakhraoui HEEY, Fourati N, Zerrouki C, Maouche N, Yaakoubi N, Touzani R, Chehimi MM (2020) Appl Sci 10:7010

60. Villemin E, Lemarque B, Vũ TT, Nguyen VQ, Trippé-Allard G, Martin P, Lacaze P-C, Lacroix J-C (2019) Synth Met 248:45–52
61. Samanta S, Bakas I, Singh A, Aswal DK, Chehimi MM (2014) Langmuir 30:9397–9406
62. Lo M, Diaw AK, Gningue-Sall D, Aaron J-J, Oturan MA, Chehimi MM (2017) Electrochem Commun 77:14–18
63. Jacques A, Saad A, Chehimi MM, Poleunis C, Delcorte A, Delhalle J, Mekhalif Z (1808) ChemistrySelect 3(2018):11800–11801
64. Chiang LY, Anandakathir R, Hauck TS, Lee L, Canteenwala T, Padmawar PA, Pritzker K, Bruno FF, Samuelson LA (2010) Nanoscale 2:535–541
65. Mekki A, Samanta S, Singh A, Salmi Z, Mahmoud R, Chehimi MM, Aswal DK (2014) J Colloid Interface Sci 418:185–192
66. Guettiche D, Mekki A, Lilia B, Fatma-Zohra T, Boudjellal A (2021) J Mater Sci: Mater Electron 32:10662–10677
67. Santos L, Ghilane J, Martin P, Lacaze P-C, Randriamahazaka H, Lacroix J-C (2010) J Am Chem Soc 132:1690–1698
68. Lacroix JC, Ghilane J, Santos L, Trippe-Allard G, Martin P, Randriamahazaka H (2012) Electrografting of conductive oligomers and polymers, Wiley-VCH Verlag GmbH & Co. KGaA: Weinheim, Germany
69. Jacques A, Devillers S, Delhalle J, Mekhalif Z (2013) Electrochim Acta 109:781–789
70. Boyle WJ, Broxton TJ, Bunnett J (1971) J Chem Soc D: Chem Commun 1469–1470
71. Breimer MA, Yevgeny G, Sy S, Sadik OA (2001) Nano Lett 1:305–308
72. Bouktit B, Salmi Z, Decorse P, Lecocq H, Jouini M, Aswal D, Singh A, Chehimi MM (2013) J Colloid Sci Biotechnol 2:200–210
73. Saad A, Cabet E, Lilienbaum A, Hamadi S, Abderrabba M, Chehimi MM (2017) J Taiwan Inst Chem Eng 80:1022–1030
74. Perruchot C, Chehimi M, Delamar M, Cabet-Deliry E, Miksa B, Slomkowski S, Khan M, Armes S (2000) Colloid Polym Sci 278:1139–1154
75. Lo M, Diaw AK, Gningue-Sall D, Aaron J-J, Oturan MA, Chehimi MM (2018) Environ Sci Pollut Res 25:20012–20022
76. Lo M, Pires R, Diaw K, Gningue-Sall D, Oturan MA, Aaron J-J, Chehimi MM (2018) Surfaces 1:43–58
77. Lo M, Seydou M, Bensghaïer A, Pires R, Gningue-Sall D, Aaron J-J, Mekhalif Z, Delhalle J, Chehimi MM (2020) Sensors 20:580
78. Bengamra M, Khlifi A, Ktari N, Mahouche-Chergui S, Carbonnier B, Fourati N, Kalfat R, Chehimi MM (2015) Langmuir 31:10717–10724
79. Bagheri H, Bayat P, Piri-Moghadam H (2013) J Chromatogr A 1318:58–64
80. Guselnikova O, Marque SR, Tretyakov EV, Mares D, Jerabek V, Audran G, Joly J-P, Trusova M, Svorcik V, Lyutakov O (2019) J Mater Chem A 7:12414–12419
81. Abd-El-Aziz AS, Antonietti M, Barner-Kowollik C, Binder WH, Böker A, Boyer C, Buchmeiser MR, Cheng SZ, D'Agosto F, Floudas G (2020) Macromol Chem Phys 221:2000216
82. Sassi W, Zrelli R, Hihn J-Y, Berçot P, Rezrazi E-M (2020) J Polym Res 27:1–18

Diazonium-Modification of Plasmonic Surfaces Formed by Laser Ablation

Roman Elashnikov, Elena Miliutina, Vaclav Svorcik, and Oleksiy Lyutakov

Abstract In this chapter, the utilization of diazonium chemistry for surface modification of plasmon-active surface(s) and the creation of analyte-targeted SERS substrates is discussed. Such substrates are able to selectively entrap and recognize the targeted analyte from solution by SERS measurement, ensuring its detection at very low concentrations even at dominant background of interfering molecules. The stability of diazonium-created coatings allows performing several steps of plasmonic surface modification and their subsequent utilization in various, often challenging conditions. In the first part of the chapter, the main principle of SERS method and functional SERS substrates is described. In the second part, the preparation and diazonium-functionalization of plasmon-active grating surface aimed at selective SERS detection are discussed. In the last part of the chapter, several examples of diazonium chemistry utilization for SERS-based recognition of enantiomers, environmental contaminants, dangerous compounds, and disease markers are given.

1 Surface Enhanced Raman Spectroscopy

A surface plasmon is a collective oscillation of electrons, excited on the metal surface under illumination [1, 2], which can exist in two forms—localized surface plasmon (LSP) typical for confined metal nanoparticles and traveled surface plasmon-polariton wave (SPP) [3–6], observed on roughed or periodically patterned metal films. Both LSP and SPP ensure an intense focusing of light energy near the metal surface and found several applications in the field of sensors and triggering of chemical transformations [7–14], starting from water splitting up to classical organic reactions [15–18].

R. Elashnikov · E. Miliutina · V. Svorcik · O. Lyutakov (✉)
Department of Solid State Engineering, University of Chemistry and Technology, 16628 Prague, Czech Republic
e-mail: lyutakoo@vscht.cz

E. Miliutina · O. Lyutakov
Research School of Chemistry and Applied Biomedical Science, Tomsk Polytechnic University, 634050 Tomsk, Russian Federation

M. M. Chehimi et al. (eds.), *Aryl Diazonium Salts and Related Compounds*,
Physical Chemistry in Action, https://doi.org/10.1007/978-3-031-04398-7_18

In this chapter, the utilization of plasmon(s)-related phenomena in the detection field, in particular in Surface Enhanced Raman Spectroscopy (SERS) will be described [19]. The Raman spectrum is produced due to the inelastic scattering of photons by matter and contains information about the structure of "probed" matter [20, 21]. The Raman signal is intrinsically weak, but the plasmon-assisted light energy focusing (i.e., SERS) increases its intensity by many orders of magnitude [22]. Actually, the SERS can be considered as a common and beneficial analytical technique for the detection/analysis of relevant organic molecules, including drugs, disease markers, pesticides, as well as for discrimination of various microorganisms, cells, and bacteria strains [23–26]. The initial research in the SERS field was focused mostly on the increasing of high SERS enhancement factor (EF) and determination of the mechanism of SERS enhancement [27]. In the case of SERS EF, the absolute limit of detection—single-molecule identification was reached several years ago [28, 29]. However, the implementation of SERS for analysis of real samples, usually containing various molecules, has made no significant progress since the Raman signal from all molecules is enhanced [30]. Thus, if the detection of a targeted analyte can be possible up to very low concentration, it real detection near a concentration limit of detection (LOD) is not possible since the response of background molecules are dominant in the spectra [31–35]. For the SERS implementation in the real system, the functional SERS surfaces were proposed, which make use plasmon-active surface(s) grafting with specific anchors to ensure the selective entrapping of targeted analyte from the complex sample [36–38]. As a result, only targeted analytes are kept near the SERS active surface and produce the resonant SERS signal.

So, the preparation of targeted SERS substrates, able to selectively entrap and detect the targeted compounds, is a hot topic in the SERS development [39, 40]. The main question is how to perform the functionalization of SERS substrate, which should remain stable during several steps of surface decoration as well as under conditions of subsequent substrates utilization [41]. The most common and widely used way of plasmon metals (Me) grafting is based on thiol chemistry [42, 43]. However, the Me-S bonds strength is relatively weak and can be easily disrupted under temperature treatment, pH variation, or due to substitution reactions [44]. Thus, utilization of thiol chemistry at the first stage of functional and analyte-targeted SERS substrate preparation significantly restricts the available chemical routes and subsequent samples utilization under specific conditions or in bio-liquids [41, 45]. In this regard, the diazonium-based approach becomes advantageous. First, the diazonium grafting results in the formation of covalent bonds, which are stable [46]. This fact ensures the ability of safe subsequent steps of surface modification as well as the utilization of functionalized SERS substrates in various conditions, without the risk of the functionality loss [47, 48]. The diazonium grafting is often the first step for introducing reactive sites on the surface (in a more simple case—the carboxy or amino groups), which can be further chemically linked with anchor molecules in one or several steps of chemical transformation (Fig. 1a).

Additionally, diazonium grafting can be performed under various activation mechanisms, with the formation of organic layers with desired thickness and density [46, 49–53]. For example, spontaneous metal surface grafting or application of external

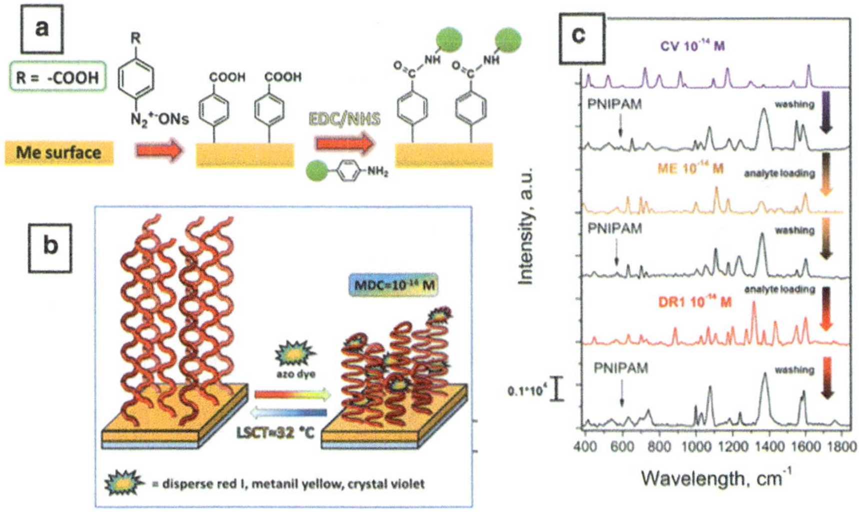

Fig. 1 **a**—diazonium surface grafting—introduction of NH_2 or -COOH chemical groups on metal surface for their subsequent EDC/NHS coupling with anchors; **b**—schematic representation of PNIPAm-based reusable SERS substrate; **c**—demonstration of SERS reusability in the entrapping/detection of dyes molecules. Reprinted with permission from reference [64] Copyright 2017 Wiley

triggering, such as UV-light, heating, external electric potential, or addition of reactive chemical agents can lead to the desired level of anchors density on each particular SERS substrate [54–56]. An additional point should also be mentioned—almost all created SERS substrates are disposable [57, 58]. The diazonium-functionalization of SERS substrates is based on cheap and wide available materials and reagents, provided by diazonium chemistry [59, 60]. Last but not least, the diazonium chemistry was demonstrated to introduce the smart surface coatings, able to reversibly entrap/release the targeted molecules from the surface of SERS substrates (Fig. 1b, c), making their re-utilization possible [61–64].

2 Functional SERS Substrates Based on Diazonium Surface Modification

LSP-based substrates for SERS measurements commonly provide the higher local SERS EF but, due to the inhomogeneous distribution of plasmonic hot spots, it is accompanied by a higher uncertainty in obtained analytical signal [65, 66]. SPP-based substrates ensure more homogeneous smearing of plasmon field intensity across the SERS active substrate area, making the semiquantitative SERS measurements possible (Fig. 2a) [67, 68]. Somewhat weaker SPP-induced SERS enhancement can

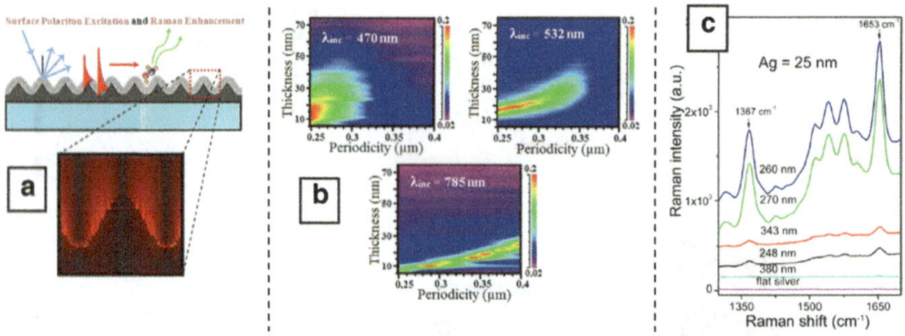

Fig. 2 **a**—schematic representation of surface plasmon-polariton excitation and propagation, **b**—theoretical fit of optimal combination of grating period and silver thickness for different SERS excitation wavelength (amplitude −50 nm), **c**—R6G SERS spectra for gratings with different periods(excitation wavelength—532 nm). Reprinted with permission from [75]. Copyright 2015 American Chemical Society

be enhanced by the creation of spatially modulated, grating-like surface, or by fabrication of LSP-SPP coupled structures, which combines the homogeneity of SERS signal, characteristic for SPP and SERS enhancement, typical for LSP case [69–73].

The creation of SPP-supported metal-covering periodical substrates is a relatively simple technique, based on the preparation of patterned polymer substrates and subsequent metal deposition [74–76]. Several methods of polymer template patterning allow the generation of periodical surface structures with various amplitude and periodicity [77–84], which further determine the strength and excitation wavelength of SPP. As a plasmon-active metal, silver and gold are commonly used, but utilization of less expensive copper or aluminum is also possible [85–87]. The gold is more stable against oxidation and is commonly preferred [88]. The thickness of deposited metal, periodicity, and amplitude of the initial template (Fig. 2b–c) determine the wavelength position of SPP (which should correlate with SERS excitation wavelength) and its strength (which should be as high as possible). To ensure the SERS active grating selectivity, the Au surface should be grafted with various organic moieties, ideally forming stable covalent bonds [89, 90]. In these regards, the two typical examples can be mentioned—introduction of partial selectivity through the changing of surface hydrophobicity/lipophilicity [45, 76, 91] or several steps of grafting with the introduction of single (or multi) analyte-targeted affinity. As an example of the first approach, the surface grafting with alkyl aryl moieties of various carbon chain lengths via diazonium chemistry was proposed in [92] (Fig. 3a). The grafting, performed using a one-step procedure, allows entrapping of the lipophilic disease markers from a water-based environment and their SERS detection with close to 10 μM LOD. Indeed, such low LOD would be unattainable without previous grating surface "hydrophobization." A similar approach was also introduced with the grafting of the plasmonic surface with folic acid (FA) [93], which ensures the selective entrapping of a range of biomolecules associated with tumor cells activity (Fig. 3b). The utilization of diazonium-modification allows supposing

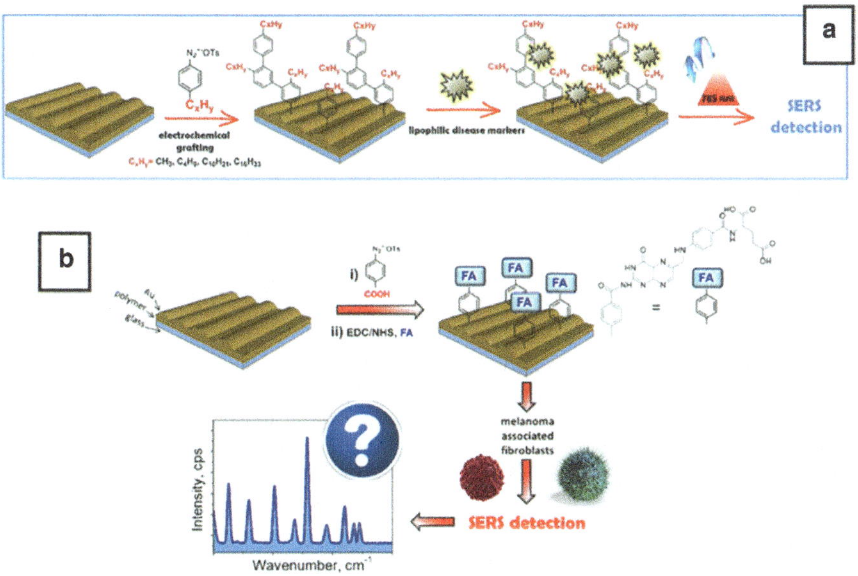

Fig. 3 **a**—schematic representation of Au gratings surface lipophilization [92] Copyright 2018 Elsevier, **b**—FA grafting to Au grating surface, and concept of tumor activity sensing. Reprinted with permission from reference [93] Copyright 2019 Springer

the implementation of SERS substrate for detection in bioenvironment, where the thiol-thiol substitution reaction with peptides can be expected in the case of common thiol-based surface grafting [94].

3 Diazonium Grafted SERS Substrates for Selective Entrapping and Detection

3.1 Heavy Metal Ions and Radicals SERS Detection

Utilization of partial selective entrapping is an interesting way, how to improve the SERS selectivity. However, it still lacks selectivity, and more specific surface anchors should also be considered. The grafting of these anchors can be performed in several steps procedure, where the diazonium chemistry is responsible for the first step [95–97] (Fig. 1a). Depending on the grafted molecules, such functional SERS substrates can be oriented on the SERS detection of a wide range of analytes, from metal ions up to large biomolecules. To detect metal ions, the grating surface was decorated with a chelator, ensuring ions entrapping [98, 99]. As a chelator, the DTPA molecules were used, and the shift of carboxyl group SERS band allows not only metal ions presence detection but also identification of their kind (Fig. 4a–

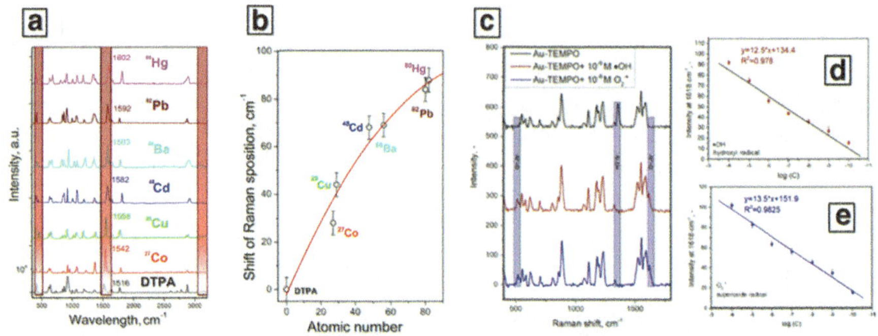

Fig. 4 **a**—SERS spectra of DTPA-grafted Au grating before and after its interaction with metal ions; **b**—dependence of the shift of Raman signal of DTPA carbonyl group on the atomic number of complexed metal ions [99] Copyright 2017 Elsevier; **c**—characteristic SERS spectra of grafted TEMPO before and after reaction with hydroxyl and superoxide radicals; **d–e**—corresponded calibration curves in TEMPO-based radicals detection. Reprinted with permission from reference [100] Copyright 2019 Elsevier

b). The intensity of shifted peak allows for semiquantitative determination of the concentration of metal ions, even in their mixed solution [99]. A similar interesting approach was demonstrated in [100], where the surface immobilization of TEMPO molecules was used. In the presence of hydroxyl or superoxide radicals, the TEMPO molecules undergo a chemical transformation, which was used to estimate these radicals concentrations, through appropriate changes in the SEPC spectra (Fig. 4c–e).

3.2 *MOF-Based Surface Functionalization for Pollutants and Drug SERS Detection*

Among the materials, which can be used for selective SERS detection, the class of metal-organic frameworks (MOFs) deserves special attention [101]. The MOFs are highly porous materials with a great diversity of porous size and physico-chemical properties, making them an ideal candidate for selective entrapping of targeted analyte(s) [102, 103]. The utilization of diazonium grafting will ensure the surface functionalization responsible for further surface-assisted MOF growth. As an example, grafting of the surface with –Ar-COOH moieties allows the strong surface immobilization of the so-called MOF-5, as is shown in Fig. 5 [104]. Such structures were applied for selective entrapping and detection of organophosphorus pesticides in soil with the LODs near 10^{-12} M, significantly exceeding the common analytical approach (Fig. 5b). A similar approach was also demonstrated in [34, 105], where the grafting of homochiral MOFs ensures the enantioselective entrapping of model methylsulfoxide and cysteine as well as more relevant DOPA and pseudoephedrine (Fig. 5c–d). The targeted enantiomers were selectively entrapped by homochiral MOF layer, and the enantioselective SERS detection of enantiomers was performed

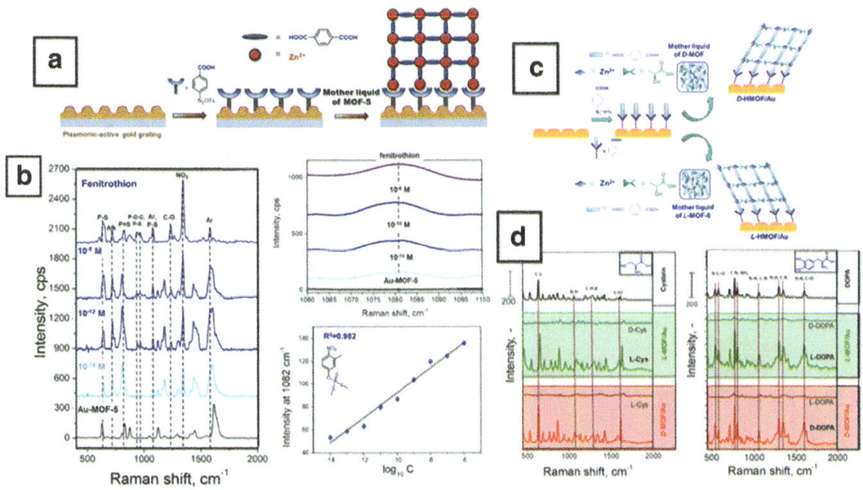

Fig. 5 **a** — Schematic representation of Au grating surface-assisted MOF-5 growth; b—SERS sensing of Paraoxon in different concentrations on Au gratings/MOF-5 substrates, including - magnified area of SERS spectra and calibration curves [104] Copyright 2019 Elsevier. **c**—Schematic representation of preparation of L- or D-HMOF-6 Au grating-assisted surface growth **d**—enantiospecific detection of L -cysteine, and L –DOPA. Reprinted with permission from reference [105] Copyright 2021 Elsevier

without any additional requirement for sample initial separation or sophisticated laboratory equipment with a very low LODs in all cases.

3.3 *Enantiospecific and Large Biomolecules SERS Detection*

Unlike the commonly used Raman Optical Activity or chiral plasmonic approaches [101, 106–109], the utilization of SERS detection through enantioselective analyte capture greatly simplifies the analytical procedure and can be performed with standard (or even portable) Raman spectrometer. The utilization of the MOF-based approach is not the only possible choice. The diazonium chemistry-based grafting of tartaric acid and subsequent enantioselective entrapping/recognition of DOPA enantiomer was demonstrated in [110] (Fig. 6a). Such an approach was implemented on SERS active grating surface and further expanded to the field of plasmon-active optical fibers [111], also able to support SPP excitation (Fig. 6b). In the latter case, the authors used two plasmon metals with wavelengths-separated absorption bands. Their decoration with L- or D-TA results in selective shift of the plasmon absorption maxima in the presence of L- or D-DOPA enantiomers.

Another critical problem, which can be solved using the diazonium-functionalized SERS substrate, is the detection of large biomolecules [26, 112]. The surface-assisted

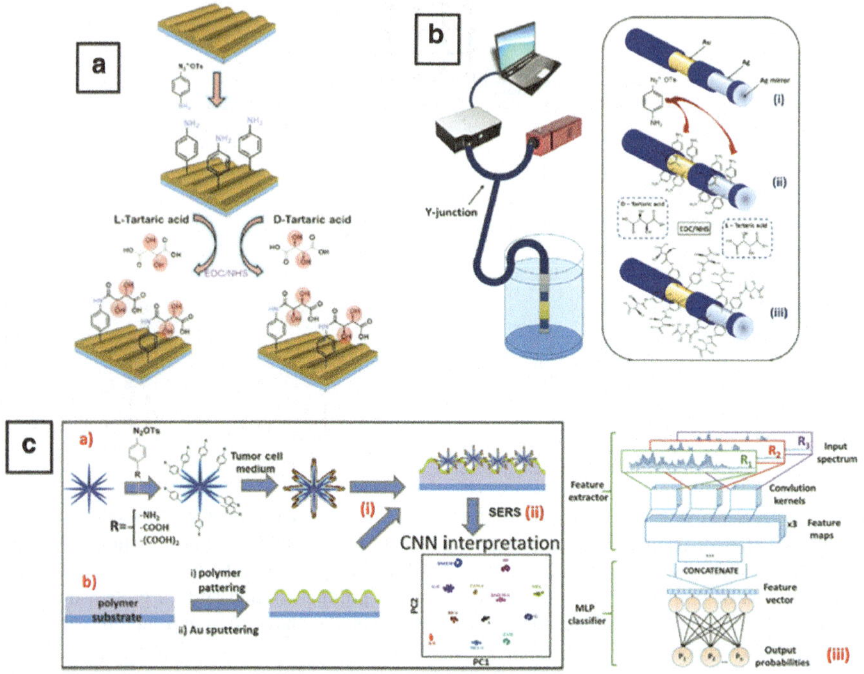

Fig. 6 **a**—preparation of L- and D-TA decorated Au grating surface for selective entrapping and SERS detection of DOPA enantiomers, reprinted with permission from [110]. Copyright 2019 American Chemical Society; **b**—plasmon-fiber assisted approach for enantioselective instant recognition of DOPA enantiomers reprinted with permission from [111]. Copyright 2020 American Chemical Society; **c**—representation of functional SERS-NN system utilization for tumor detection: **a**—preparation of functional Au nanourchins and their interaction with culture medium sample, **b**—simultaneous preparation of gold grating: (i)—deposition of AuMs on the surface of gold grating, (ii)—SERS measurements, (iii)—implementation of convolutional NN for SERS results in interpretation. Reprinted with permission from reference [116]. Copyright 2020 Elsevier

immobilization of poly-4-vinylboronic with the diazonium chemistry allows to selectively entrap the glycoprotein(s) from model solution and their subsequent SERS identification [113]. A similar approach was used for surface immobilization of single strain DNA (ssDNA), which serves as an anchor for its counterpart in solution. In such a way, detection and identification of the presence of ssDNA or even a small mutation in ssDNA structure was demonstrated [114, 115]. However, in this case, the high SERS peaks interference makes detection/recognition significantly more difficult. As a solution, the utilization of machine learning was proposed. The combination of diazonium chemistry, SERS measurements, and neural networks (NN) make the detection and recognition of large biomolecules indeed possible [115, 116]. The ability to create various functional groups on the plasmon-active surface as subsequent data for NN learning, validation and utilization deserves special attention. Recent grafting of plasmon-active nanoparticles with amino, carboxy, and oleophilic chemical moieties provided the quasi-selective entrapping of biomolecules, and significantly increased the reliability of cancer cell metabolic activity [116] (Fig. 6c).

4 Conclusion and Outlook

In summary, the further development of SERS is closely related to the design and creation of analyte-oriented substrates, able to selectively entrap targeted molecules from multi-component solutions. Diazonium chemistry allows stable surface modification of SERS substrate, with the direct introduction of desired surface anchors or creation of available chemical groups for their subsequent surface immobilization. In these regards, the stability of Me-C bonds, reached with diazonium chemistry, ensures the possibility of further SERS surface modification as well as plasmon-active substrates utilization in various environments and "strong" conditions. Functionalization by diazonium-based SERS surface decoration makes possible SERS detection of pesticides, heavy metal ions, and radicals, as well as disease markers, DNA strains, and enantiomers relative concentration. Without a doubt, further implementation of diazonium chemistry for SERS surface decoration will be used for solving many relevant analytical and bioanalytical goals and tune the utilization of SERS substrates in more application-useful areas.

Acknowledgments This work was supported by Czech Science Foundation, project No. 21-06065S and by the OP VVV Project NANOTECH ITI II. No. CZ.02.1.01/0.0/0.0/18_069/0010045.

References

1. Guselnikova O, Kalachyova Y, Elashnikov R et al (2020) Microporous Mesoporous Mater 309. https://doi.org/10.1016/j.micromeso.2020.110577
2. Barnes WL, Dereux A, Ebbesen TW (2003) Nature 424:824–830. https://doi.org/10.1038/nature01937
3. Jain PK, Huang X, El-Sayed IH, El-Sayed MA (2009) ChemInform 40. https://doi.org/10.1002/chin.200914223
4. Zayats AV, Smolyaninov II, Maradudin AA (2005) Phys Rep 408:131–314. https://doi.org/10.1016/j.physrep.2004.11.001
5. Elashnikov R, Lyutakov O, Kalachyova Y et al (2015) React Funct Polym 93:163–169. https://doi.org/10.1016/j.reactfunctpolym.2015.06.010
6. Elashnikov R, Lyutakov O, Ulbrich P, Svorcik V (2016) Mater Sci Eng C 64:229–235. https://doi.org/10.1016/j.msec.2016.03.047
7. Daniel M-C, Astruc D (2004) ChemInform 35. https://doi.org/10.1002/chin.200416213
8. Linic S, Aslam U, Boerigter C, Morabito M (2015) Nat Mater 14:567–576. https://doi.org/10.1038/nmat4281
9. Guselnikova O, Olshtrem A, Kalachyova Y et al (2018) J Phys Chem C 122:26613–26622. https://doi.org/10.1021/acs.jpcc.8b07398
10. Erzina M, Guselnikova O, Miliutina E et al (2021) J Phys Chem C 125:10318–10325. https://doi.org/10.1021/acs.jpcc.0c11535

11. Olshtrem A, Guselnikova O, Postnikov P et al (2020) Nanoscale 12:14581–14588. https://doi.org/10.1039/d0nr02934c
12. Elashnikov R, Zahorjanova K, Miliutina E et al (2020) Nanoscale 12:12068–12075. https://doi.org/10.1039/d0nr00295j
13. Miliutina E, Guselnikova O, Soldatova NS et al (2020) J Phys Chem Lett 11:5770–5776. https://doi.org/10.1021/acs.jpclett.0c01350
14. Guselnikova O, Postnikov P, Kosina J et al (2021) J Mater Chem A 9:8462–8469. https://doi.org/10.1039/d0ta12530j
15. Ingram DB, Linic S (2011) J Am Chem Soc 133:5202–5205. https://doi.org/10.1021/ja200086g
16. Robatjazi H, Zhao H, Swearer DF, et al (2017) Nat Commun 8. https://doi.org/10.1038/s41467-017-00055-z
17. Zabelina A, Zabelin D, Miliutina E et al (2021) J Mater Chem A 9:17770–17779. https://doi.org/10.1039/d1ta04505a
18. Guselnikova O, Trelin A, Miliutina E et al (2020) ACS Appl Mater Interfaces 12:28110–28119. https://doi.org/10.1021/acsami.0c04029
19. Stiles PL, Dieringer JA, Shah NC, Van Duyne RP (2008) Annu Rev Anal Chem 1:601–626. https://doi.org/10.1146/annurev.anchem.1.031207.112814
20. Cheng JX, Xie XS (2004) J Phys Chem B 108:827–840. https://doi.org/10.1021/jp035693v
21. Gouadec G, Colomban P (2007) Prog Cryst Growth Charact Mater 53:1–56. https://doi.org/10.1016/j.pcrysgrow.2007.01.001
22. Kneipp K, Kneipp H, Itzkan I et al (1999) Chem Rev 99:2957–2975. https://doi.org/10.1021/cr980133r
23. Langer J, de Aberasturi DJ, Aizpurua J et al (2020) ACS Nano 14:28–117. https://doi.org/10.1021/acsnano.9b04224
24. Pang S, Yang T, He L (2016) TrAC—Trends Anal Chem 85:73–82. https://doi.org/10.1016/j.trac.2016.06.017
25. Kneipp K, Kneipp H, Itzkan I et al (2002) J Phys Condens Matter 14:R597–R624. https://doi.org/10.1088/0953-8984/14/18/202
26. Zong C, Xu M, Xu LJ et al (2018) Chem Rev 118:4946–4980. https://doi.org/10.1021/acs.chemrev.7b00668
27. Haynes CL, Yonzon CR, Zhang X, Van Duyne RP (2005) J Raman Spectrosc 36:471–484. https://doi.org/10.1002/jrs.1376
28. Lim DK, Jeon KS, Kim HM et al (2010) Nat Mater 9:60–67. https://doi.org/10.1038/nmat2596
29. Kneipp K, Wang Y, Kneipp H et al (1997) Phys Rev Lett 78:1667–1670. https://doi.org/10.1103/PhysRevLett.78.1667
30. Kozik A, Pavlova M, Petrov I, et al (2021) Anal Chim Acta 338978. https://doi.org/10.1016/j.aca.2021.338978
31. Yaseen T, Pu H, Sun DW (2018) Trends Food Sci Technol 72:162–174. https://doi.org/10.1016/j.tifs.2017.12.012
32. Wen S, Miao X, Fan GC et al (2019) ACS Sensors 4:301–308. https://doi.org/10.1021/acssensors.8b00682
33. Kalachyova Y, Erzina M, Postnikov P et al (2018) Appl Surf Sci 458:95–99. https://doi.org/10.1016/j.apsusc.2018.07.073
34. Guselnikova O, Lim H, Na J, et al (2021) Biosens Bioelectron 180:113109. https://doi.org/10.1016/j.bios.2021.113109
35. Fraser JP, Postnikov P, Miliutina E et al (2020) ACS Appl Mater Interfaces 12:47774–47783. https://doi.org/10.1021/acsami.0c11231
36. Liu R, Zhao J, Han G et al (2017) ACS Appl Mater Interfaces 9:38222–38229. https://doi.org/10.1021/acsami.7b10409
37. Huang L, Sun DW, Wu Z, et al (2021) Anal Chim Acta 1167:338570. https://doi.org/10.1016/j.aca.2021.338570
38. Bainova P, Miliutina E, Burtsev V et al (2021) Chem List 115:447–451

39. Hakonen A, Andersson PO, Stenbæk Schmidt M et al (2015) Anal Chim Acta 893:1–13. https://doi.org/10.1016/j.aca.2015.04.010
40. Hu Y, Liao J, Wang D, Li G (2014) Anal Chem 86:3955–3963. https://doi.org/10.1021/ac5002355
41. Burtsev V, Miliutina E, Erzina M et al (2019) J Phys Chem C 123:30492–30498. https://doi.org/10.1021/acs.jpcc.9b06751
42. Xue Y, Li X, Li H, Zhang W (2014) Nat Commun 5. https://doi.org/10.1038/ncomms5348
43. Elashnikov R, Radocha M, Panov I et al (2019) Mater Sci Eng C 102:192–199. https://doi.org/10.1016/j.msec.2019.04.029
44. Demers LM, Mirkin CA, Mucic RC et al (2000) Anal Chem 72:5535–5541. https://doi.org/10.1021/ac0006627
45. Burtsev V, Erzina M, Guselnikova O et al (2021) Analyst 146:3686–3696. https://doi.org/10.1039/d0an02360d
46. Mahouche-Chergui S, Gam-Derouich S, Mangeney C, Chehimi MM (2011) ChemInform 42. https://doi.org/10.1002/chin.201149242
47. Bakas I, Salmi Z, Gam-Derouich S et al (2014) Surf Interface Anal 46:1014–1020. https://doi.org/10.1002/sia.5451
48. Gam-Derouich S, Mahouche-Chergui S, Turmine M et al (2011) Surf Sci 605:1889–1899. https://doi.org/10.1016/j.susc.2011.06.029
49. Li D, Luo Y, Onidas D, et al (2021) Adv Colloid Interface Sci 294:102479. https://doi.org/10.1016/j.cis.2021.102479
50. Belanger D, Pinson J (2011) ChemInform 42. https://doi.org/10.1002/chin.201144274
51. Guselnikova O, Svanda J, Postnikov P et al (2017) Adv Mater Interfaces 4:1600886. https://doi.org/10.1002/admi.201600886
52. Guselnikova O, Elashnikov R, Postnikov P et al (2018) ACS Appl Mater Interfaces 10:37461–37469. https://doi.org/10.1021/acsami.8b06840
53. Idriss H, Guselnikova O, Postnikov P et al (2020) ACS Appl Polym Mater 2:977–986. https://doi.org/10.1021/acsapm.9b01185
54. Jacques A, Chehimi MM, Poleunis C et al (2016) Electrochim Acta 211:879–890. https://doi.org/10.1016/j.electacta.2016.06.060
55. Msaadi R, Yilmaz G, Allushi A et al (2019) Polymers (Basel) 11:286. https://doi.org/10.3390/polym11020286
56. Lei Y, Huang Q, Gan D, et al (2020) J Environ Chem Eng 8:103780. https://doi.org/10.1016/j.jece.2020.103780
57. Wu W, Liu L, Dai Z, et al (2015) Sci Rep 5. https://doi.org/10.1038/srep10208
58. Yu WW, White IM (2010) Anal Chem 82:9626–9630. https://doi.org/10.1021/ac102475k
59. Ahmad AAL, Marutheri Parambath JB, Postnikov PS et al (2021) Langmuir 37:8897–8907. https://doi.org/10.1021/acs.langmuir.1c00884
60. Bernard MC, Chaussé A, Cabet-Deliry E et al (2003) Chem Mater 15:3450–3462. https://doi.org/10.1021/cm034167d
61. Guselnikova O, Postnikov P, Kalachyova Y et al (2017) ChemNanoMat 3:135–144. https://doi.org/10.1002/cnma.201600284
62. Elashnikov R, Mares D, Podzimek T et al (2017) Analyst 142:2974–2981. https://doi.org/10.1039/c7an00419b
63. Erzina M, Guselnikova O, Postnikov P, et al (2018) Adv Mater Interfaces 5. https://doi.org/10.1002/admi.201801042
64. Gehan H, Dos Santos L, Lau S et al (2011) Plasmon. Met Nanostruct Their Opt Prop IX 8096:80961N
65. Wustholz KL, Henry AI, McMahon JM et al (2010) J Am Chem Soc 132:10903–10910. https://doi.org/10.1021/ja104174m
66. Dieringer JA, McFarland AD, Shah NC et al (2006) Faraday Discuss 132:9–26. https://doi.org/10.1039/b513431p
67. Chu Y, Banaee MG, Crozier KB (2010) ACS Nano 4:2804–2810. https://doi.org/10.1021/nn901826q

68. Lal S, Hafner JH, Halas NJ et al (2012) Acc Chem Res 45:1887–1895. https://doi.org/10.1021/ar300133j
69. Kim K, Lee HS (2005) J Phys Chem B 109:18929–18934. https://doi.org/10.1021/jp052665z
70. Orendorff CJ, Gole A, Sau TK, Murphy CJ (2005) Anal Chem 77:3261–3266. https://doi.org/10.1021/ac048176x
71. Kalachyova Y, Mares D, Jerabek V et al (2016) J Phys Chem C 120:10569–10577. https://doi.org/10.1021/acs.jpcc.6b01587
72. Kalachyova Y, Mares D, Jerabek V et al (2017) Phys Chem Chem Phys 19:14761–14769. https://doi.org/10.1039/c7cp01828b
73. Slepička P, Siegel J, Lyutakov O et al (2018) Biotechnol Adv 36:839–855. https://doi.org/10.1016/j.biotechadv.2017.12.011
74. Kalachyova Y, Lyutakov O, Kostejn M et al (2015) Electron Mater Lett 11:288–294. https://doi.org/10.1007/s13391-014-4336-7
75. Kalachyova Y, Mares D, Lyutakov O et al (2015) J Phys Chem C 119:9506–9512. https://doi.org/10.1021/acs.jpcc.5b01793
76. Lyutakov O, Huttel I, Siegel J, Švorčík V (2009) Appl Phys Lett 95:173103.https://doi.org/10.1063/1.3254210
77. Viswanathan NK, Kim DY, Bian S et al (1999) J Mater Chem 9:1941–1955. https://doi.org/10.1039/a902424g
78. Cheng JY, Ross CA, Smith HI, Thomas EL (2006) Adv Mater 18:2505–2521. https://doi.org/10.1002/adma.200502651
79. Schäffer E, Thurn-Albrecht T, Russell TP, Steiner U (2000) Nature 403:874–877. https://doi.org/10.1038/35002540
80. Lyutakov O, Tůma J, Huttel I et al (2013) Appl Phys B Lasers Opt 110:539–549. https://doi.org/10.1007/s00340-012-5291-3
81. Lyutakov O, Hüttel I, Prajzler V et al (2009) J Polym Sci Part B Polym Phys 47:1131–1135. https://doi.org/10.1002/polb.21718
82. Lyutakov O, Tuma J, Prajzler V et al (2010) Thin Solid Films 519:1452–1457. https://doi.org/10.1016/j.tsf.2010.08.019
83. Miliutina E, Guselnikova O, Marchuk V et al (2018) Langmuir 34:12861–12869. https://doi.org/10.1021/acs.langmuir.8b02025
84. Kalachyova Y, Lyutakov O, Slepicka P et al (2014) Nanoscale Res Lett 9. https://doi.org/10.1186/1556-276X-9-591
85. Švanda J, Kalachyova Y, Ajami A et al (2021) Opt Mater (Amst) 112:110695. https://doi.org/10.1016/j.optmat.2020.110695
86. Guselnikova O, Samant R, Postnikov P et al (2019) J Mater Chem C 7:14181–14187. https://doi.org/10.1039/c9tc03690c
87. Chan GH, Zhao J, Hicks EM et al (2007) Nano Lett 7:1947–1952. https://doi.org/10.1021/nl070648a
88. Kalachyova Y, Mares D, Jerabek V et al (2019) Sensors actuators. A Phys 285:566–572. https://doi.org/10.1016/j.sna.2018.11.037
89. Laurentius L, Stoyanov SR, Gusarov S, et al (2011) ACS Nano 5:4219–4227. https://doi.org/10.1021/nn201110r
90. Mesnage A, Lefèvre X, Jégou P et al (2012) Langmuir 28:11767–11778. https://doi.org/10.1021/la3011103
91. Miliutina E, Guselnikova O, Bainova P et al (2018) Adv Mater Interfaces 5. https://doi.org/10.1002/admi.201800725
92. Guselnikova O, Kalachyova Y, Hrobonova K et al (2018) Sensors actuators. B Chem 265:182–192. https://doi.org/10.1016/j.snb.2018.03.016
93. Guselnikova O, Dvorankova B, Kakisheva K et al (2019) Anal Bioanal Chem 411:3309–3319. https://doi.org/10.1007/s00216-019-01801-6
94. Olshtrem A, Chertopalov S, Guselnikova O et al (2021) 2D Mater. https://doi.org/10.1088/2053-1583/ac27c0

95. Guselnikova O, Postnikov P, Elashnikov R et al (2017) Colloids Surfaces A Physicochem Eng Asp 516:274–285. https://doi.org/10.1016/j.colsurfa.2016.12.040
96. Guselnikova O, Postnikov P, Chehimi MM et al (2019) Langmuir 37:8897-8907. https://doi.org/10.1021/acs.langmuir.8b03041
97. Guselnikova O, Váňa J, Phuong LT et al (2021) Chem Sci 12:5591–5598. https://doi.org/10.1039/d0sc05898j
98. Guselnikova O, Svorcik V, Lyutakov O et al (2019) Sensors (Basel) 19:2110. https://doi.org/10.3390/s19092110
99. Guselnikova O, Postnikov P, Erzina M et al (2017) Sensors actuators. B Chem 253:830–838. https://doi.org/10.1016/j.snb.2017.07.018
100. Guselnikova O, Postnikov P, Marque SRA et al (2019) Sensors Actuators, B Chem 300:127015. https://doi.org/10.1016/j.snb.2019.127015
101. Miliutina E, Guselnikova O, Chufistova S et al (2019) ACS Sensors 4:3133–3140. https://doi.org/10.1021/acssensors.9b01074
102. Chen B, Xiang S, Qian G (2010) Acc Chem Res 43:1115–1124. https://doi.org/10.1021/ar100023y
103. Furukawa H, Cordova KE, O'Keeffe M, Yaghi OM (2013) Science (80):341. https://doi.org/10.1126/science.1230444
104. Guselnikova O, Postnikov P, Elashnikov R et al (2019) Anal Chim Acta 1068:70–79. https://doi.org/10.1016/j.aca.2019.03.058
105. Guselnikova O, Postnikov P, Kolska Z et al (2020) Appl Mater Today 20. https://doi.org/10.1016/j.apmt.2020.100666
106. Ding SY, You EM, Tian ZQ, Moskovits M (2017) Chem Soc Rev 46:4042–4076. https://doi.org/10.1039/c7cs00238f
107. Schuller JA, Barnard ES, Cai W et al (2010) Nat Mater 9:193–204. https://doi.org/10.1038/nmat2630
108. Kalachyova Y, Guselnikova O, Elashnikov R et al (2019) ACS Appl Mater Interfaces 11:1555–1562. https://doi.org/10.1021/acsami.8b15520
109. Miliutina E, Zadny J, Guselnikova O et al (2021) Sensors Actuators, B Chem 343:130122.https://doi.org/10.1016/j.snb.2021.130122
110. Guselnikova O, Postnikov P, Trelin A et al (2019) ACS Sensors 4:1032–1039. https://doi.org/10.1021/acssensors.9b00225
111. Miliutina E, Guselnikova O, Kushnarenko A et al (2020) ACS Sensors 5:50–56. https://doi.org/10.1021/acssensors.9b01328
112. Doering WE, Piotti ME, Natan MJ, Freeman RG (2007) Adv Mater 19:3100–3108. https://doi.org/10.1002/adma.200701984
113. Guselnikova O, Marque SRA, Tretyakov E V et al (2019) J Mater Chem A 7:12414–12419. https://doi.org/10.1039/c9ta01630a
114. Guselnikova O, Postnikov P, Pershina A et al (2019) Appl Surf Sci 470:219–227.https://doi.org/10.1016/j.apsusc.2018.11.092
115. Guselnikova O, Trelin A, Skvortsova A et al (2019) Biosens Bioelectron 145. https://doi.org/10.1016/j.bios.2019.111718
116. Erzina M, Trelin A, Guselnikova O, et al (2020) Sensors actuators B Chem 308. https://doi.org/10.1016/j.snb.2020.127660

Diazonium Salts and the Related Compounds for the Design of Biosensors

Guozhen Liu and Cheng Jiang

Abstract Aryldiazonium salt chemistry, as a versatile surface decoration and functionalization method, has been intensively employed in construction of sensing interface. The *para*-position (–*R*) of aryldiazonium salt bearing different functional moieties endows the sensing surface with different purposes. Typically, bioreceptor linkage groups (e.g. –COOH, $-NH_2$) are essential, and inert groups (e.g. –H, $-CH_3$)-based spacer and antifouling moieties can be incorporated according to the actual requirement. Introduction of molecular wires, or nanomaterials (e.g. carbon based or noble metal based) can further enhance signal and/or electron transfer helping to increase the sensitivity. This chapter focuses on the design of the sensing interface by applying the aryldiazonium salt surface chemistry with the format of forming single/mixed aryldiazonium salt layers, and their translations into biosensing of different target analytes.

1 Introduction

A biosensor is an analytical device used to measure and transmit the presence or concentration of a biological analyte (e.g. microorganism, biomolecules or biological structures) [1]. Typically, there are three key components for construction of a biosensor: a bioreceptor for recognizing and binding the target bioanalytes, a transducer for converting the biorecognition events into a measurable signal, and a signal processor for reporting the analyte signal [2]. The popular choices for bioreceptors are usually enzymes, cells, antibodies, aptamers, molecularly imprinted polymers, et al. For signal reporting, electrochemical, optical, mass or mechanical signal transmission approaches have been developed. The biosensor, combining with the associated

G. Liu (✉)
School of Medicine, The Chinese University of Hong Kong, Shenzhen 518172, P.R. China
e-mail: liuguozhen@cuhk.edu.cn

C. Jiang
Department of Chemistry, University of Oxford, Oxford OX1 3QZ, UK

M. M. Chehimi et al. (eds.), *Aryl Diazonium Salts and Related Compounds*,
Physical Chemistry in Action, https://doi.org/10.1007/978-3-031-04398-7_19

receptors and signal processors, aims to display the bioinformation into a recognizable and user-friendly way. Biosensors have become highly versatile platforms for a wide range of applications in different research fields (e.g. biomedical diagnostics, environment monitoring, food safety and quality, biosecurity, etc.) because of their simplicity of use and capability of operation in complex fluids like serum. With the advances in nanotechnology, biotechnology, and biomedical engineering, biosensors have developed to be a cross-disciplinary field. Since 2017, CRISPR/Cas systems have been demonstrating great potentials in biosensing due to the super sensitivity and specificity [3–5], offering a valuable toolbox for its cooperation with the aryldiazonium salt chemistry. The fast growth of biosensing technologies has narrowed the gap of biosensing and clinical diagnostics. With the development of advanced materials, nanobiotechnology and electronic engineering, biosensing devices provide the possibility of real-time monitoring of individual multiple health conditions through wearable and implantable biosensors [6].

The good coordination of signal transduction with selective recognition and binding of a biological or chemical target of interest is always the leading factor for a sensory device to perform the detection. This can be achieved by modification of the transducer interface with a chemical layer serving as linking agent that can guarantee the anchoring of the biorecognition unit. Therefore, the intellectual decoration of the recognition domain is of paramount significance in defining the sensitivity, selectivity, specificity, and reproducibility for the final sensory device [7]. With dedicated and precise manipulation on the sensing interfacial design, e.g. incorporation with nanomaterials, biotechnology or careful selection of a signal amplification module, the sensitivity of sensors can reach the single-molecule level [8–11].

The formation of organized monolayers on a sensing interface to which the recognition molecules are immobilized, is crucial to improve the performance of the sensors. The gold-thiol chemistry is the most classic method to endow gold surface with self-assembled monolayers (SAMs) for biosensing applications [12]. To date, there are a number of other facile surface chemistries available including aryldiazonium salts [13] and silane-based chemistries on metal or oxide surfaces [14], the hydrosilylation or chemical grafting of alkenes on hydrogen-terminated silicon surface [15] or on hydrogen terminated diamond surfaces [16], respectively. As a facile strategy for growth of molecular system on certain substrates, aryldiazonium salt chemistry is a well-recognized competitor to the existing SAMs-based systems, attracting dramatic research interests since Pinson and coworkers firstly demonstrated the reaction mechanism using modified carbon electrodes in 1992 [17]. Aryldiazonium salt can generate corresponding reactive aryl radicals under the reductive condition, which can effectively attack and covalently anchor to the substrates. The versatile properties of aryldiazonium salts with wide choices of $-R$ functional groups in *para*-position of benzene have attracted chemists for a whole century for exploring and understanding both the underlying fundamentals and the application aspects for aryldiazonium salts. This radical-based surface grafting methodology makes it possible to efficiently modify a large panel of substrates for diverse applications, e.g. molecular electronic junctions [18], corrosion control, [19] energy storage

[20], removal of heavy metals by complexation layers [21], and sensing application [22]. Notably, among these applications, the aryldiazonium salt chemistry has been advancing at an astonishing pace in sensing filed. The merits of using aryldiazonium salts in sensing are mainly attributed to their simple preparation, fast surface grafting at time scale of seconds to minutes, wide choices of *para*-position functionalities, compatibility with broad spectra of substrates and materials, and strong aryl-carbon/metal covalent bond, etc. As a coupling agent, aryldiazonium salts can attach a recognition unit (i.e. bioreceptor) on the interfaces for sensing applications. In 2008, Gooding group have systematically reviewed the aryldiazonium salts chemistry for developing biosensors, they highlighted the attachment of single-component-based phenyl layers on electrodes (e.g. metal and carbon) [23]. Then we summarized existing methods for modification of interfaces with aryldiazonium salts for biosensing (optical-, electrochemical-) applications [24]. A recent comprehensive review on grafting of diazonium salts on surfaces for biosensing has been reported by Pinson [25]. Last decade has witnessed tremendous advancement of aryldiazonium salt derived surface chemistry in regard to (1) preparation of aryl-diazonium derivatives with diverse functionalities; (2) design and fabrication of multipurpose interface by introducing mixed aryl layers [26]; (3) controllable growth of thinner aryl layers for sensing applications; (3) coupling of different nanomaterials via aryl layers to enhance sensing performance [27]. This chapter is to discuss advances in diazonium salts and the related compounds for the design of biosensors.

2 Strategies for Generation of the Aryldiazonium Salt Chemistry Decorated Substrates

2.1 *Preparation of Aryldiazonium Salts*

Typically, aryldiazonium salts can be synthesized through the diazotization of the corresponding aromatic compounds under acidic condition ($pH < 3$) with addition of $NaNO_2$ [28] despite the popular reaction in acetonitrile solution with tertbutylnitrite [29] or in neutral [30] and alkaline [31] aqueous media. Due to the stability limitation, majority aryldiazonium salts are home-made. This makes hinders for constructing new multicomponent materials to generate robust interfaces. Thus, it is more favorable for researchers to synthesize the aryldiazonium salts with particular functional groups meeting their specific requirements [32]. Moreover, the clear limitations with aryldiazonium salts, e.g. their chemical instability at room temperature, the difficulty for isolation purification, and their explosive characteristics have motivated scientists to pursue their targets using the same medium without need of pre-isolated aryldiazonium salts. Bélanger and coworkers reported a simpler procedure consisting of the in situ generation of a aryldiazonium salt from an benzeneamine (i.e. aniline) by standard diazotization procedures which eliminated the purification step and avoided the inherent instability of selected diazonium salts [33]. Therefore, there are mainly two pathways for obtaining aryldiazonium salts-based interfacial functionalization: (i) using pre-isolated aryldiazonium salts, (ii) using in-situ generation and grafting aryldiazonium salts from aniline using reductive medium.

2.2 Compatible Substrates for Aryldiazonium Salt Chemistry

Many studies have proved that aryldiazonium salts can be used to decorate all types of carbon derived materials [34], metal [35], semi-conductors [36, 37]. Additionally, aryldiazonium salts can be used to link different materials, such as polymers [38], adhesive resins [39], biomacromolecules [40], and nanoparticles [41], to various of surfaces [42]. It is reported that chemically exfoliated molybdenum disulfide (MoS_2) was decorated with 4-methoxyphenyl diazonium salts through the carbon−sulfur bond [43], which is an energetically favorable process. Moreover, insulin-modified cell surfaces have been explored using aryldiazonium salt chemistry [44]. In the study, human mesenchymal stem cells that grow on the insulin decorated diazoresin and pectin surfaces presented an elevated proliferation and higher osteogenic activity. This surface chemistry is highly adaptive for different materials, making aryldiazonium salts an efficient coupling reagent in sensing applications.

2.3 Methods for Modification of Substrates Using Aryldiazonium Salts

As illustrated in Fig. 1, a wide range of anilines can be used to endow substrate with diverse functions like biomolecule linkage, inert spacer, fouling resistance, electron transfer enhancer, etc. using aryldiazonium salt chemistry. To anchor them to a specific substrate, there are a couple of widely studied methods available. The most popular method to fabricate aryldiazonium salt to a surface is the electrochemical reduction [35], which can be realized in aqueous, organic, and also ionic liquid media [45, 46]. Pinson and his team have summarized different grafting methods of diazonium salts in his recent review [25]. Besides the common electrochemical reduction, chemical reductive grafting of aryldiazonium salts has also been reported [47–49]. In a recent work, Zeb et al., have demonstrated the chemical modification of titanium nitride with aryldiazonium salt chemistry [50]. Aryldiazonium salts-based ion exchange of sodium was also explored as anther efficient method to obtain clay-based nanocomposites [51]. Furthermore, radiation-based strategies like microwave irradiation [52], UV or visible light assisted route [53, 54], mechanochemical method, [55], and ultrasonication [56], have also been employed for aryldiazonium salt fabrication. Aryldiazonium salt reduction happens on a substrate as soon as the substrate is capable to reduce such as carbon materials or gold (without the need of chemical

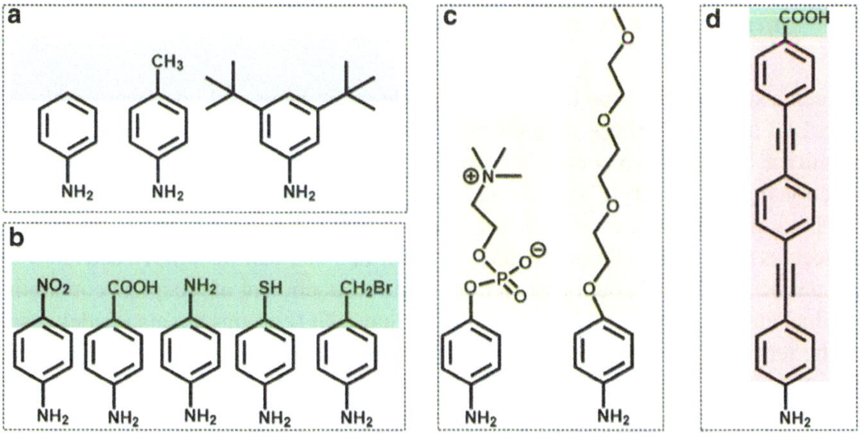

Fig. 1 The typical anilines with papa-position groups bearing different functional moieties, **a** inert spacer, **b** biomolecule linkage groups, **c** non-ionic or zwitterionic antifouling groups, and **d** conductive molecular wires. Their cooperative surface chemistry would be favorable for construction of diverse sensing interfaces. Note that multiple functions can also be integrated into a single aryl layer

reductants or electrochemical/radiation-induced radical processes) [57, 58]. Aryldiazonium salt layers can be spontaneously grafted to diverse surfaces, which highlight the advantages in functionalization of non-conductive substrates. Aryl groups were normally attached to a substrate by the covalent binding [59] indicating the elegant success in aryldiazonium salt chemistry. More importantly, aryldiazonium salt chemistry is able to yield an ultrathin grafted film of 3–5 nm by adjusting the underlying support and the experimental parameters [60]. Combellas et al., have demonstrated aryl monolayer can be achieved when conducting the grafting experiments with ionic liquids or using a radical scavenger. Moreover, aryl monolayers can also be achieved through a steric hindrance-based method by using the aryldiazonium salts bearing two bulky tert-butyl groups at *meta*-positions of the aryl group [61].

3 Applications of Aryl Diazonium Salts and Related Compounds in Biosensing Applications

As the Chap. 18 mainly focuses on electrochemical sensor using aryl layer bearing polymers, in the current chapter, we will specifically emphasize the sensory devices with non-polymer-based aryldiazonium molecular systems, e.g. shorter chemical groups or oligomers on the *para*-position. According to a single aryl layer or mixed aryl layers used for sensing interfaces, this section will be divided into **Single aryl layer-based biosensors** and **Mixed aryl layers-based biosensors**.

3.1 Single Aryl Layer-Based Sensors

In such a sensing interface format, only single-component aryl layer was employed (Fig. 2). Functional groups on *para*-position of aryldiazonium cations are needed to permit the linkage of a receptor that can bind target analyte. The most frequently used functional groups in aryldiazonium salt-based biosensing interfaces are –COOH and $–NH_2$, both of them are easy to form the chelating bonds with metals or charged molecules due to the charged nature. The former one can be simply activated by the classic EDC/NHS coupling reaction to allow attachment of receptors containing residual amines by the amide bond. Besides that $-NO_2$ groups are also widely used as the terminal groups on aryldiazonium salts since $-NO_2$ group can be reduced to $-NH_2$, reaching a more controlled mode. The –SH terminated aryldiazonium salts are studied for sensing purpose with benefit from gold-thiol chemistry [62–64]. In a typical example, Mousavisani et al., have reported label-free impedance

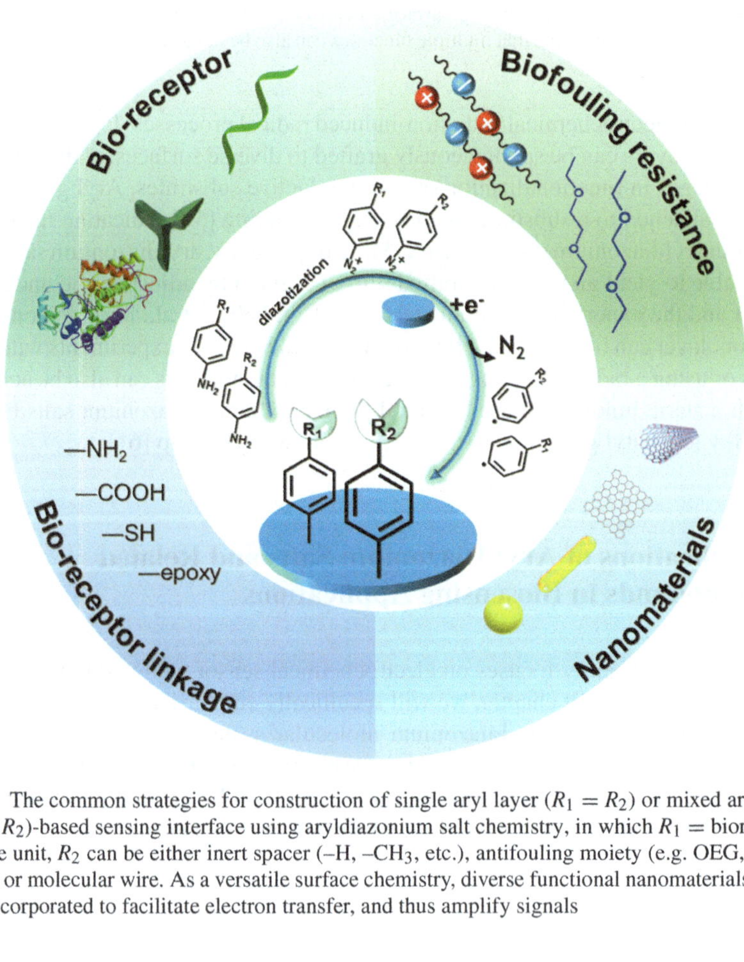

Fig. 2 The common strategies for construction of single aryl layer ($R_1 = R_2$) or mixed aryl layer ($R_1 \neq R_2$)-based sensing interface using aryldiazonium salt chemistry, in which R_1 = bioreceptor linkage unit, R_2 can be either inert spacer (–H, $–CH_3$, etc.), antifouling moiety (e.g. OEG, zwitterions), or molecular wire. As a versatile surface chemistry, diverse functional nanomaterials can be also incorporated to facilitate electron transfer, and thus amplify signals

biosensor for detection of breast cancer 1 (BRCA1) gene [65]. The 4-aminobenzoic acid was used to allow electrochemical grafting of –COOH groups on the screen-printed carbon electrode (SPCE), followed by linkage of DNA probe targeting the BRCA1gene. The sensor was capable of detection of BRCA1 gene and also studying the protective effect of glutathione over the DNA damage. Similar label-free DNA sensor was reported by Torréns et al. [66]. They explored detection of Francisella tularensis genomic sequence by attaching corresponding DNA probe to the 3,5-bis(4-diazophenoxy)benzoic acid coated surface. The lowest detection limit (LoD) of 0.5 nM and 0.9 nM was obtained using amperometric detection on the electrochemical gold and carbon grafted surfaces, respectively. In another study, Sharma et al., utilized glutaraldehyde to modify glucose oxidase (GO_x) to benzeneamine terminated pyrolytic carbon surface, which was generated by reduction of $-NO_2$ to $-NH_2$ on *para*-position of aryl layer [67]. A LOD of 0.4 μM and a dynamic range of 0.001–1 mM were achieved, respectively. Similarly, a glucose biosensor was obtained through the immobilization of 4-nitrophenyl diazonium salt on a glassy carbon (GC) electrode and the subsequent attachment of GO_x (Fig. 3a) [68]. The proposed biosensor was successfully used to detect glucose based on by GO_x catalyzed glucose oxidation with the consumption of O_2. The glucose sensor presented a broad linear

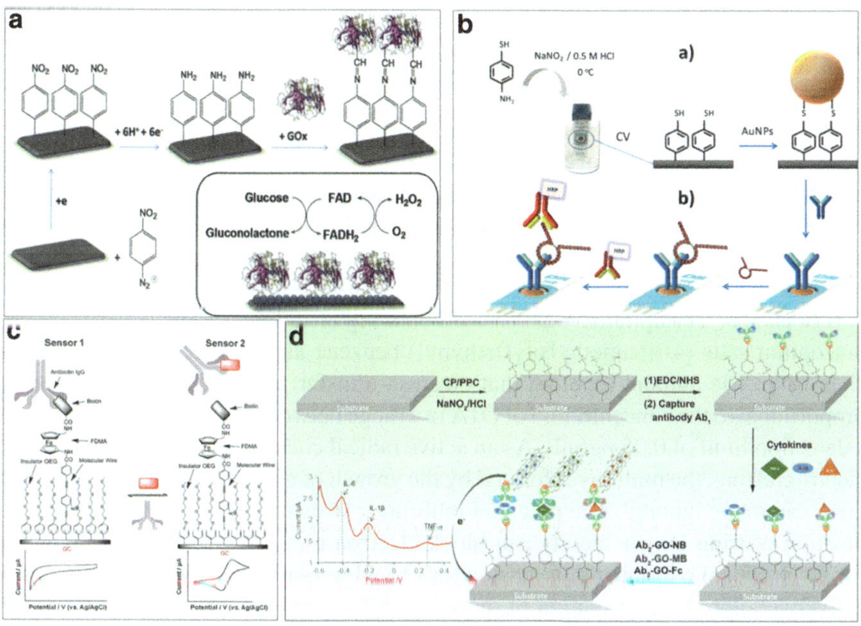

Fig. 3 The typical aryldiazonium salt modified biosensors with a single aryl layer of nitrophenyl (**a**), single aryl layer of thiophenyl with AuNPs (**b**), mixed aryl layers bearing OEG moiety and molecular wire (**c**), and mixed aryl layers containing zwitterionic PC and CP-based multiplexed sensing with decoration of three antibody pairs and GO loaded electroactive probes (**d**). Reused with permission from Refs. [68–71] Copyright (2013, 2018, 2011) Elsevier, and (2018) American Chemical Society, respectively

range (0.05–4.5 mM) and low detection limit (~10 μM). The great stability of this biosensor, and technically simple preparation at short period of time make this method suitable for fabrication of low-cost glucose biosensors.

Stepping further, as a versatile surface functionalization methodology for sensor development, the aryldiazonium salts-based surface chemistry can coordinate with numerous nanomaterials like gold nanoparticles (AuNPs), carbon nanotubes (CNTs), graphene oxide (GO), etc., to form composite architectures for sensing interfaces [20, 72]. The inclusion of conductive nanomaterials can particularly boost electron transfer for compensation of the potential multilayer structure of aryldiazonium chemistry for the electrochemical sensors. In a typical example, Serafín et al., have fabricated a sandwich immunosensor for sensing brain natriuretic peptide (BNP) (Fig. 3b) [69]. AuNPs were attached with the electrografted 4-thiol-benzene surface via Au–S bond, followed by adsorption of capture antibody to the AuNP surfaces. Then, with addition of standards or samples, the horse radish peroxidase (HRP) conjugated detection antibody was finally added. The LoD of 4 pg mL^{-1} with capability of working in serum sample was demonstrated. With displacement of AuNPs with CNTs and GO, the same group extended the strategy for detection of adiponectin cytokine [73] and obesity-related hormones [74] (ghrelin and peptide YY), respectively.

3.2 Mixed Aryl Layer-Based Sensors

In the Sect. 3.1, single aryl layer bearing individual functionality, usually the functional groups for linkage of bioreceptor, was used. To further extend the scope and achieve multifunctional interfaces, mixed aryl layers bearing different functionalities have been investigated (Fig. 2). Typically, the additional functional modules can be a molecular spacer, an antifouling layer, a molecular wire, etc. [75]. Hayat et al., reported an electrochemical impedimetric aptasensor for the detection of ochratoxin A (OTA). [76] The aptamer was attached onto organized mixed layers derived from diazonium salts (4-((trimethylsilyl)ethynyl) benzene and *p*-nitrobenzene) via click chemistry. The authors reported that electron-transfer resistance (R_{et}) values were proportional to the concentration of OTA in a range between 1.25 and 500 pg/mL with a detection limit of 0.25 pg/mL. As an active radical chemistry-based aryldiazonium electrografting, the multilayer formed by the growth of aryl layer on the as-deposited layer cannot be ignored. The potential multilayer structure and aryl layer with *para*-position bearing longer moiety can block electron transfer, which is not favorable for electrochemical sensing in most cases. To address the problem, Liu et al., have actively explored the formation of monolayers on interfaces for aryldiazonium salt chemistry in sensing applications by control the input concentration and deposition time [77, 78]. They have thoroughly studied the mixed aryl layers containing either oligoethyleneglycol (OEG) or 4-carboxyphenyl with oligo(phenylethynylene) molecular wire (MW) to switch on electrochemistry thus permit electron transfer between model redox proteins like horseradish peroxidase (HRP) or GO_x and underlying electrode [32, 35, 79]. The rigid MWs with 20 Å length can interact directly with the protein through penetrating the outer shell of proteins and reduce the tunneling distance to facilitate efficient electron transfer behavior despite the redox active centers being embedded, which can undoubtedly extend the scope of using aryldiazonium salt chemistry for diverse protein electrochemistry and biosensing and not limited to the short aryl layers [80]. The diluent OEG molecules with three ethylene glycol units are able to resist non-specific adsorption of proteins in blood serum, which is extremely important for clinical applications [81, 82]. This is a very important fundamental study that provides opportunity for an electrochemical interface that does not block electron transfer even in the presence of long/high-impedance aryl layers. Benefit from the mixed aryl layers bearing OEG and MWs, Gooding and co-worker further explored the real sensing application for detection of biotin (Fig. 3c), enrofloxacin, and HbA1c in complex sample matrix (e.g. serum, milk) by attaching relevant receptors onto the interface [70, 83, 84].

Since the synthesis of MWs needs complex procedure, in such regard, Liu's group explored using employment of conductive nanomaterials with mixed aryl layers to develop hybrid multifunctional sensing interfaces which can also permit electron transfer and increase sensitivity. In one example, mixed layers of 4-nitrophenyl/OEG were deposited on GC surface, followed by conversion $-NO_2$ to $-NH_2$ to allow adsorption of AuNPs, carboxylphenyl was then deposited onto the AuNPs for linkage of methyl parathion hydrolase (MPH) [85]. The obtained biosensor can achieve detection of methyl parathion in the range of 0.2 to 100 ppb with the detection limit of 0.07 ppb. Alternatively, CNTs can be also used to form composite structures with the mixed aryl layers and was investigated for sensing of paraoxon [86] and hydrogen peroxide [87] on GC and gold electrode surface, respectively. Besides the "bottom-up" approach that grafting of mixed aryl layers on the electrode, Liu et al., further explore the "top-down" method by modification of mixed aryl layers tailoring (OEG-/amino-/carboxyl-)phenyl onto AuNPs first, and then deposit them on a GO sheets, the resultant GO-Ab/AuNPs/mixed aryl layer nanocomposite was modified with capture antibody, detection antibody-ferrocene separately to obtain the capture and detection probes. The nanocomposite-based system was applied for detection of cardiac marker troponin-I with a detectable concentration of 0.05 ng mL^{-1} using square wave voltammetry (SWV) [88].

It should be noted that OEG layer suffers some limitations, e.g. their low tolerance toward transition metals and reactive oxygen species which are common in biological fluids [89, 90]. In such regard, aryl layer directly bearing zwitterions-based antifouling system was explored. Particularly, the phenyl phosphoryl choline (PPC) was electrografted on to GC, gold, and indium tin oxide (ITO) electrode surfaces and reported effective antifouling performance with high stability compared to OEG [37, 91, 92]. Taking such advantages and the benefits that conjugation of short aryldiazonium salt with biomolecules like proteins or antibodies can still permit electron transfer, [93–95] Jiang et al., have explored the use of shorter (~0.8 nm) mixed aryl layers consisting of PPC and phenylbutyric acid (PBA) to detect TNF-alpha in whole blood using the sandwich format on the mixed-layer system [96].

Benefit from the low-impedance and antifouling PPC layer, and conductive nanomaterials enhanced electron transfer, Liu's group constructed several composite sensing interfaces using mixed aryl layers containing PPC and carboxyl phenyl (CP) with graphene oxide (GO) for detection of cytokines in live cells [97]. Wei et al., further explored the multiplexing capability by attach three capture antibodies to the mixed PPC/CP/GC surface, followed by addition of samples and the redox probes-detection antibodies conjugated GO (Fig. 3d). Simultaneous detection of three cytokines (IL-6, IL-1β, and TNF-α) can be achieved using the distinct redox potentials of the redox probes comprising of nile blue (NB), methyl blue (MB), and ferrocene (Fc) [71]. More details about the single-/mixed-aryl layer-based biosensor are summarized in Table.1.

Besides the in vitro sensing, the in vivo sensing potential of aryldiazonium salt chemistry was also explored. Encouragingly, Liu's group pioneerly extended the potential of aryldiazonium surface chemistry for in vivo electrochemical monitoring of cytokines. For example, they modified Fc tagged drug (aspirin) loaded aptamer (targeting IFN-γ) onto the carboxyl phenyl deposited GC electrodes, it was implanted into the subcutaneous pockets on the back of rat model [119]. The aspirin release can be triggered by the immunoregulatory cytokine IFN-γ due to conformational change of the aptamer, generating electron communication barrier for the Fc and the underlying electrode, which can be quantitatively monitored in real time. In another example, the incorporation of GO into the mixed aryl layers (i.e. phenyl phosphorylcholine and carboxyl phenyl) can further boost electron transfer by decoration of capture antibody/detection antibody to GO sheet, and permit the in vivo detection of IL-1β, IL-6, and TNF-α in mouse brain [120, 121].

In the above-discussed examples, the concept of incorporation of aryldiazonium salts-based antifouling layers into a biorecognition layer to permit effective resistance against non-specific protein adsorption is mainly used. Jiang et al., have critically reviewed the antifouling strategies for different surfaces and materials in biosensing [89]. However, the reality is that no surface chemistry can provide absolute antifouling, thus joint effort with incorporation of different supportive measures can be applied to enhance the antifouling performance. Tymoczko and co-workers have reported that external electric field can help mitigate electrostatics to permit DNA hybridization [122] and modulation of protein adsorption behaviors on zwitterion (i.e. phosphorylcholine) decorated surface [123]. It is known that the non-specific and abundant proteins are usually charged at physiological pH. Thus, it might be possible to resisting these interferences from accessing the interfacial moieties by using an external electric field to allow manipulation of the charge state for the overall biosensing system. For example, positively/negatively charged interface can repel positively/negatively charged proteins accordingly with negligible side effect on binding event of target analyte. The proposed method is undoubtedly worthy to be employed to support antifouling performance of aryl layers functionalized interface.

Table.1 The construction of aryldiazonium salt modified biosensors and their sensing performance

Sensing interface composition	Detection method	Sample	Analyte	Dynamic range	Sensitivity	Reference
GO_x-Ph-AuNP/GO	CA	CCM	glucose	0.3–20 mM	42 μA mM^{-1} cm^{-2}	[98]
Ab1/(FPC-PBA)/ITO	CA	Blood	TNF-α	(0.01–500) ng mL^{-1}	10 pg mL^{-1}	[96]
Ab/(Nitro-/Carboxyl-/OEG)Ph/AuNPs/GC	CV	Milk	BoNT/A	(4–35) pg mL^{-1}	1 pg mL^{-1}	[99]
MPH/CP/AuNPs/(OEG-Nitro-)Ph/GC	SWV	Green tea	MP	(0.2–100) ppb	0.07 ppb	[85]
Ab2-GO-Ph-Fc/analyte/Ab1/PPC(-ph-COOH))/rGO-AuNP/Ph/Au	SWV	CCM	TNF-α	(0.1–150) pg mL^{-1}	0.1 pg mL^{-1}	[97]
Ab2-GO-Fc(NB/MB)/analyte/Ab1/PPC-CP/GC	SWV	Serum	IL-6, IL-1β, TNF-α	(5–150) pg mL^{-1} (5–200) pg mL^{-1} (5–200) pg mL^{-1}	5 pg mL^{-1} 5 pg mL^{-1} 5 pg mL^{-1}	[71]
Apt-streptavidin MB/GO-Ph/GC	SWV	Serum	VEGF	10–500 pg mL^{-1}	10 pg mL^{-1}	[100]
HRP-anti-BNP-BNP-anti-BNP-AuNPs-S-Ph/SPCE	Amperometry	Serum	BNP	(0.014–15) ng mL^{-1}	4 pg mL^{-1}	[69]
Ab1/CP/AgNPs/SiO_2/GO/GC	Amperometry	Urine	EE2	(0.14–50) ng mL^{-1}	65 pg mL^{-1}	[101]
Apt/graphene/S-Ph–S/Au	IDE-SPQC	Milk	Staphylococcus aureus	$4.1 \times 10^1 - 4.1 \times 10^5$ cfu mL^{-1}	41 cfu mL^{-1}	[102]
Ab1-PEA-CP/SPCE	DPV	Serum	ACTH	(0.025–1.0) pg mL^{-1}	0.018 pg mL^{-1}	[103]
Apt/CP/SPCE	EIS	Milk	AFM1	(2–150) pg mL^{-1}	1.15 pg mL^{-1}	[104]
DNA probe/AP/SPCE	DPV	PB	Human gene MRP3	(1.0–7.3) μg mL^{-1}	210 ng mL^{-1}	[105]
Apt/CP/Au	EIS	Apple juice	S. typhimurium	$(1 \times 10^1–1 \times 10^8)$ CFU mL^{-1}	6 CFU mL^{-1}	[106]
Ab1-Bp/CP/Au	EIS	PB	Methotrexate	$(3 \times 10^{-12}–3 \times 10^{-4})$ M	7×10^{-12} M	[107]
Ab1/GA/AP/Gr/SPE	DPV	Milk	β-lactoglobulin	(0.001–100) ng mL^{-1}	0.85 pg mL^{-1}	[108]

(continued)

Table.1 (continued)

Sensing interface composition	Detection method	Sample	Analyte	Dynamic range	Sensitivity	Reference
Apo-A4-ABA/ZIF-8@N-Gr/GC	DPV	Serum	Apo-A4	$(1.47 \times 10^{-10}$–$3.0 \times 10^{-7})$ g/mL	8.33×10^{-11} g/mL	[109]
Ab2-HRP@SWCNT/TGF-β1/Ab1/CP/SPCE	Amperometry	Saliva	TGF-β1	(2.5–1000) pg mL^{-1}	0.95 pg mL^{-1}	[110]
Apt/AuNPs-S-Ph-GO/GC	CA	Serum	HER2	(0.5–2) ng mL^{-1}	0.16 nM	[111]
Ab1/AP/Gr/AuNPs/Nafion/GC	EIS	Cereal	DON	(6–30) ng mL^{-1}	0.3 μg mL^{-1}	[112]
Ab1/CP/CNT/FET	FET	Buffer	OPN	–	30 fM	[113]
Ab2-ALP@AuNPs/BoNT/E/Ab1/Gr/AP-Ph/GC	SV	Juice/milk	BoNT/E	(0.01–10) ng mL^{-1}	5.0 pg mL^{-1}	[114]
Ab1/CP/CNT/SPE	SWV	Blood	SMN protein	$(10^{-3}$–100) ng mL^{-1}	0.75 pg mL^{-1}	[115]
HRP@Ab2/AFP/Ab1/GA/AP/Gr/GC	CA	PBS	AFP	(0.1–2) ng mL^{-1}	0.01 ng mL^{-1}	[116]
Ab1/VTA2/ITO	EIS	Cell media	HL-60 cells	$(270$–$2.7 \times 10^{7})$ cells mL^{-1}	90 cells mL^{-1}	[117]
Tyrosinase/CNT/AP/BDD	CV	River water	BPA	(0.01–100) nM	0.01 nM	[118]

Ph: phenyl, GO_x: glucose oxidase, CCM: cell condition media, GC: glassy carbon, PPC: phenyl phosphorylcholine, EIS: electrochemical impedance spectroscopy, CP: 4-carboxylic phenyl, PBA: phenylbutyric acid, ITO: indium tin oxide, TNF-α: tumor necrosis factor-α, IL-6: interleukin-6, IFN-γ: interferon-γ, VEGF: vascular endothelial growth factor, BRCA1: breast cancer 1 gene, SWV: square wave voltammetry, MPH: methyl parathion hydrolase, MP: methyl parathion, BoNT/A: botulinum neurotoxin type A, CV: cyclic votammetry, DPV: differential pulse voltammetry, Ab: antibody, CA: chronoamperometry, rGO: reduced graphene oxide, BNP: B-type natriuretic peptide, NB: Nile blue, MB: methylene blue, EE2: ethinylestradiol, AgNPs: silver nanoparticles, Apt: aptamer, IDE-SPQC: interdigitated electrode-series electrode piezoelectric quartz crystal, ACTH: adrenocorticotropin hormone, AFM1: aflatoxin M1, APBA: phenylboronic acid, SPCE: screen-printed carbon electrodes, PB: phosphate buffer, AP: aminophenyl, Bp: blocking polymer, Ab1: capture antibody, Ab2: detection antibody, GA: glutaraldehyde, Apo-A4: apolipoprotein A4, Gr: graphene, HER2: human epidermal growth factor receptor 2, DON: deoxynivalenol, OPN: osteopontin, BoNT/E: botulinum neurotoxin-E, ALP: alkaline phosphatase, SV: stripping voltammetry, SMN: survival motor neuron, AFP: Alpha-fetoprotein, TP: theophylline, AD: adenosine, AT: 2,4-diamino-1,3,5-triazine, VTA2: 2-p-aminophenyl-1, 3, 2-dithiarsenolane, BDD: boron doped diamond electrode, BPA: Bisphenol A

4 Conclusions

Comparing to existing SAMs-based surface coupling strategies for biosensing, the clear advantage of aryldiazonium salt derived biosensors is the robustness and stability through the formation of covalent bonds (C−C, C−Au, C−S, etc.) and their ability to introduce different functionalities and form composite with numerous nanomaterials. Aryldiazonium salt surface chemistry derived biosensors have been widely investigated for detection of wide range of analytes in body fluids and biological media (as summarized in Table 1), from in vitro diagnostics [124] to in vivo monitoring [121]. It is worth noting that, besides biosensing applications, the versatile and feasible aryldiazonium salt chemistrybased sensors have also been widely applied for sensing a variety of analytes (metal ions, acrylamide, pesticides, hydrogen peroxide etc.) in food and environmental samples with attachment of appropriate receptors [86, 87, 125–132]. Although aryldiazonium salts have achieved great achievements in sensing applications, the challenges and obstacles are still existing, such as how to precisely control the thickness and surface organization of aryldiazonium salt to enhance the sensitivity of the sensors, and how to make the fabrication of aryldiazonium salt in situ under in vivo conditions to expand the application of aryldiazonium salt chemistry for in vivo detection. With the fast development with respect to advanced micro-/nano-fabrication, fabrication of the switch-based biosensors133, 134], soft electronics [136], active exploration of aryldiazonium salts derivatives (e.g. diaryliodonium salt) [137], automation, and simulation for the new molecular design and synthesis [6, 135] their combination with aryldiazonium salts with undoubtedly generate fruitful scientific sparks and step into a new era. It is expected to see that continuous efforts will be devoted in further exploring aryl diazonium salt chemistry for biosensing with broad detection methods (electrochemical-, optical-, or their combination) [24, 138, 139] and molecular diagnostics applications.

References

1. Turner AP (2013) Biosensors: sense and sensibility. Chem Soc Rev 42(8):3184–3196
2. Liu G (2021) Grand Challenges in Biosensors and Biomolecular Electronics. Frontiers in Bioengineering and Biotechnology, 707615, 1–5
3. Li Y, Li S, Wang J et al (2019) CRISPR/Cas systems towards next-generation biosensing. Trends Biotechnol 37(7):730 743
4. Li Y, Liu L, Liu G (2019) CRISPR/Cas multiplexed biosensing: a challenge or an insurmountable obstacle? Trends Biotechnol 37(8):792–795
5. Dai Y, Wu Y, Liu G et al (2020) CRISPR mediated biosensing toward understanding cellular biology and point-of-care diagnosis. Angew Chem Int Ed 132(47):20938–20950
6. Qiao L, Benzigar MR, Subramony JA et al (2020) Advances in sweat wearables: sample extraction, real-time biosensing, and flexible platforms. ACS Appl Mater Interfaces 12(30):34337–34361
7. Ni S, Shen Z, Zhang P et al (2020) Enhanced performance of an electrochemical aptasensor for real-time detection of vascular endothelial growth factor (VEGF) by nanofabrication and ratiometric measurement. Anal Chim Acta 1121:74–82

8. Gooding JJ, Gaus K (2016) Single-molecule sensors: challenges and opportunities for quantitative analysis. Angew Chem Int Ed 55(38):11354–11366
9. Lei J, Ju H (2012) Signal amplification using functional nanomaterials for biosensing. Chem Soc Rev 41(6):2122–2134
10. Su Q, Jiang C, Gou D et al (2021) Surface plasmon-assisted fluorescence enhancing and quenching: from theory to application. ACS Appl Bio Mater 4(6):4684–4705
11. Liu G, Jiang C, Lin X et al (2021) Point-of-care detection of cytokines in cytokine storm management and beyond: significance and challenges. VIEW 2(4):20210003
12. Love JC, Estroff LA, Kriebel JK et al (2005) Self-assembled monolayers of thiolates on metals as a form of nanotechnology. Chem Rev 105(4):1103–1170
13. Leroux YR, Hapiot P (2013) Nanostructured monolayers on carbon substrates prepared by electrografting of protected aryldiazonium salts. Chem Mater 25(3):489–495
14. Meiners F, Ross JH, Brand I et al (2014) Modification of slicon oxide surfaces by monolayers of an oligoethylene glycol-terminated perfluoroalkyl silane. Colloids Surf A 449:31–41
15. Wang J, Wu F, Watkinson M et al (2015) "Click" patterning of self-assembled monolayers on hydrogen-terminated silicon surfaces and their characterization using light-addressable potentiometric sensors. Langmuir 31(35):9646–9654
16. Lud SQ, Steenackers M, Jordan R et al (2006) Chemical grafting of biphenyl self-assembled monolayers on ultrananocrystalline diamond. J Am Chem Soc 128(51):16884–16891
17. Delamar M, Hitmi R, Pinson J et al (1992) Covalent modification of carbon surfaces by grafting of functionalized aryl radicals produced from electrochemical reduction of diazonium salts. J Am Chem Soc 114:5883–5884
18. McCreery RL (2004) Molecular electronic junctions. Chem Mater 16(23):4477–4496
19. Adenier A, Cabet-Deliry E, Lalot T et al (2002) Attachment of polymers to organic moieties covalently bonded to iron surfaces. Chem Mater 14:4576–4585
20. Assresahegn BD, Brousse T, Bélanger D (2015) Advances on the use of diazonium chemistry for functionalization of materials used in energy storage systems. Carbon 92:362–381
21. Cui Y, Hu Z-J, Yang J-X et al (2012) Novel phenyl-iminodiacetic acid grafted multiwalled carbon nanotubes for solid phase extraction of iron, copper and lead ions from aqueous medium. Microchim Acta 176(3–4):359–366
22. Liu GZ, Zhang Y, Guo WQ (2014) Covalent functionalization of gold nanoparticles as electronicbridges and signal amplifiers towards an electrochemical immunosensor for botulinumneurotoxin type A. Biosens Bioelectron 61:547–553
23. Gooding JJ (2008) Advances in interfacial design for electrochemical biosensors and sensors: aryl diazonium salts for modifying carbon and metal electrodes. Electroanalysis 20(6):573–582
24. Cao C, Zhang Y, Jiang C et al (2017) Advances on aryldiazonium salt chemistry based interfacial fabrication for sensing applications. ACS Appl Mater Interfaces 9(6):5031–5049
25. Hetemi D, Noël V, Pinson J (2020) Grafting of diazonium salts on surfaces: application to biosensors. Biosensors 10(1):4
26. Jiang C, Silva SM, Fan S et al (2017) Aryldiazonium salt derived mixed organic layers: from surface chemistry to their applications. J Electroanal Chem 785:265–278
27. Pilan L (2020) Tailoring the performance of electrochemical biosensors based on carbon nanomaterials via aryldiazonium electrografting. Bioelectrochemistry 107697
28. Pazo-Llorente R, Bravo-Diaz C, Gonzalez-Romero E (2004) pH effects on ethanolysis of some arenediazonium ions: evidence for homolytic dediazoniation proceeding through formation of transient diazo ethers. Eur J Org Chem 15:3221–3226
29. Sv B, Bélanger D (2008) In situ generation of diazonium cations in organic electrolyte for electrochemical modification of electrode surface. Electrochim Acta 53:6961–6967
30. Ceccato M, Bousquet A, Hinge M et al (2011) Using a mediating effect in the electroreduction of aryldiazonium salts to prepare conducting organic films of high thickness. Chem Mater 23(6):1551–1557
31. Podvorica FI, Kanoufi F, Pinson J et al (2009) Spontaneous grafting of diazoates on metals. Electrochim Acta 54:2164–2170

32. Liu GZ, Gooding JJ (2006) An interface comprising molecular wires and poly(ethylene glycol) spacer units self-assembled on carbon electrodes for studies of protein electrochemistry. Langmuir 22:7421–7430
33. Baranton S, Bélanger D (2005) Electrochemical derivatization of carbon surface by reduction of in situ generated diazonium cations. J Phys Chem B 109(51):24401–24410
34. Liu GZ, Liu J, Böcking T et al (2005) The modification of glassy carbon and gold electrodes with aryl diazonium salt: the impact of the electrode materials on the rate of heterogeneous electron transfer. Chem Phys 319(1):136–146
35. Liu G, Paddon-Row MN, Justin GJ (2007) A molecular wire modified glassy carbon electrode for achieving direct electron transfer to native glucose oxidase. Electrochem Commun 9(9):2218–2223
36. El Hadj FA, Amiar A, Cherkaoui M et al (2012) Study of organic grafting of the silicon surface from 4-nitrobenzene diazonium tetrafluoroborate. Electrochim Acta 70:318–324
37. Jiang C, Tanzirul Alam M, Parker SG et al (2015) Zwitterionic phenyl phosphorylcholine on indium tin oxide: a low-impedance protein-resistant platform for biosensing. Electroanalysis 27(4):884–889
38. Salmi Z, Gam-Derouich S, Mahouche-Chergui S et al (2012) On the interfacial chemistry of aryl diazonium compounds in polymer science. Chem Pap 66(5):369–391
39. Delamar M, Desarmot G, Fagebaume O et al (1997) Modification of carbon fiber surfaces by electrochemical reduction of aryl diazonium salts: application to carbon epoxy composites. Carbon 35(6):801–807
40. Flavel BS, Gross AJ, Garrett DJ (2010) A simple approach to patterned protein immobilization on silicon via electrografting from diazonium salt solutions. ACS Appl Mater Interfaces 2:1184–1190
41. Liu GZ, Luais E, Gooding JJ (2011) The fabrication of stable gold nanoparticle-modified interfaces for electrochemistry. Langmuir 27:4176–4183
42. Mahouche Chergui S, Gam Derouich S, Mangeney C et al (2011) Aryl diazonium salts: a new class of coupling agents for bonding polymers, biomacromolecules and nanoparticles to surfaces. Chem Soc Rev 40(7):4143–4166
43. Knirsch KC, Berner NC, Nerl HC et al (2015) Basal-plane functionalization of chemically exfoliated molybdenum disulfide by diazonium salts. ACS Nano 9(6):6018–6030
44. Mikulska A, Filipowska J, Osyczka AM et al (2015) Osteoinductive activity of insulin-functionalized cell culture surfaces obtained using diazonium chemistry. Front Chem 2:1–7
45. Shul G, Ruiz CAC, Rochefort D et al (2013) Electrochemical functionalization of glassy carbon electrode by reduction of diazonium cations in protic ionic liquid. Electrochim Acta 106:378–385
46. Padilha JC, Noël JM, Bergamini JF et al (2016) Functionalization of carbon materials by reduction of diazonium cations produced in situ in a brønstedt acidic ionic liquid. ChemElectroChem 3:572–580
47. Mesnage A, Magied MA, Simon P et al (2011) Grafting polymers to titania nanoparticles by radical polymerization initiated by diazonium salt. J Mater Sci 46(19):6332–6338
48. Mahouche Chergui S, Ledebt A, Mammeri F et al (2010) hairy carbon nanotube@nano-pd heterostructures: design, characterization, and application in suzuki c–c coupling reaction. Langmuir 26(20):16115–16121
49. Grondein A, Bélanger D (2011) Chemical modification of carbon powders with aminophenyl and aryl-aliphatic amine groups by reduction of in situ generated diazonium cations: applicability of the grafted powder towards CO_2 capture. Fuel 90(8):2684–2693
50. Zeb G, Viel P, Palacin S et al (2015) On the chemical grafting of titanium nitride by diazonium chemistry. RSC Adv 5(62):50298–50305
51. Salmi Z, Benzarti K, Chehimi MM (2013) Diazonium cation-exchanged clay: an efficient, unfrequented route for making clay/polymer nanocomposites. Langmuir 29(44):13323–13328

52. Karousis N, Economopoulos SP, Iizumi Y et al (2010) Microwave assisted covalent functionalization of C60@ SWCNT peapods. Chem Commun 46(48):9110–9112
53. Pan Y, Tong B, Shi J et al (2010) Fabrication, characterization, and optoelectronic properties of layer-by-layer films based on terpyridine-modified MWCNTs and ruthenium (III) ions. J Phys Chem C 114(17):8040–8047
54. Bouriga M, Chehimi MM, Combellas C et al (2012) Sensitized photografting of diazonium salts by visible light. Chem Mater 25(1):90–97
55. Pandurangappa M, Ramakrishnappa T, Compton RG (2009) Functionalization of glassy carbon spheres by ball milling of aryl diazonium salts. Carbon 47(9):2186–2193
56. Dahoumane SA, Nguyen MN, Thorel A et al (2009) Protein-Functionalized hairy diamond nanoparticles. Langmuir 25(17):9633–9638
57. Jayasundara DR, Cullen RJ, Colavita PE (2013) In situ and real time characterization of spontaneous grafting of aryldiazonium salts at carbon surfaces. Chem Mater 25:1144–1152
58. Mesnage A, Lefèvre X, Jégou P et al (2012) Spontaneous grafting of diazonium salts: chemical mechanism on metallic surfaces. Langmuir 28:11767–11778
59. Laurentius L, Stoyanov SR, Gusarov S et al (2011) Diazonium-derived aryl films on gold nanoparticles: evidence for a carbon-gold covalent bond. ACS Nano 5:4219–4221
60. Mohamed AA, Salmi Z, Dahoumane SA et al (2015) Functionalization of nanomaterials with aryldiazonium salts. Adv Colloid Interface Sci 225:16–36
61. Combellas C, Jiang D-E, Kanoufi F et al (2008) Steric effects in the reaction of aryl radicals on surfaces. Langmuir 25(1):286–293
62. Liu GZ, Liu JQ, Davis TP et al (2011) Electrochemical impedance immunosensor based on gold nanoparticles and aryl diazonium salt functionalized gold electrodes for the detection of antibody. Biosens Bioelectron 26:3660–3665
63. Taufik S, Barfidokht A, Alam MT et al (2016) An antifouling electrode based on electrode–organic layer–nanoparticle constructs: electrodeposited organic layers versus self-assembled monolayers. J Electroanal Chem 779:229–235
64. Liu G, Iyengar SG, Gooding JJ (2012) An electrochemical impedance immunosensor based on gold nanoparticle-modified electrodes for the detection of HbA1c in human blood. Electroanalysis 24(7):1509–1516
65. Mousavisani SZ, Raoof J-B, Turner APF et al (2018) Label-free DNA sensor based on diazonium immobilisation for detection of DNA damage in breast cancer 1 gene. Sens Actuators B Chem 264:59–66
66. Torréns M, Ortiz M, Turner APF et al (2015) Amperometric detection of Francisella tularensis genomic sequence on Zn-mediated diazonium modified substrates. Electrochem Commun 53:6–10
67. Sharma D, Lim Y, Lee Y et al (2015) Glucose sensor based on redox-cycling between selectively modified and unmodified combs of carbon interdigitated array nanoelectrodes. Anal Chim Acta 889:194–202
68. Nasri Z, Shams E (2013) A glucose biosensor based on direct electron transfer of glucose oxidase immobilized onto glassy carbon electrode modified with nitrophenyl diazonium salt. Electrochim Acta 112:640–647
69. Serafín V, Torrente-Rodríguez RM, González-Cortés A et al (2018) An electrochemical immunosensor for brain natriuretic peptide prepared with screen-printed carbon electrodes nanostructured with gold nanoparticles grafted through aryl diazonium salt chemistry. Talanta 179:131–138
70. Khor SM, Liu G, Fairman C et al (2011) The importance of interfacial design for the sensitivity of a label-free electrochemical immuno-biosensor for small organic molecules. Biosens Bioelectron 26(5):2038–2044
71. Wei H, Ni S, Cao C et al (2018) graphene oxide signal reporter based multifunctional immunosensing platform for amperometric profiling of multiple cytokines in serum. ACS Sens 3(8):1553–1561
72. Yáñez-Sedeño P, Campuzano S, Pingarrón JM (2018) Integrated affinity biosensing platforms on screen-printed electrodes electrografted with diazonium salts. Sensors 18(2):675

73. Ojeda I, Barrejón M, Arellano LM et al (2015) Grafted-double walled carbon nanotubes as electrochemical platforms for immobilization of antibodies using a metallic-complex chelating polymer: application to the determination of adiponectin cytokine in serum. Biosens Bioelectron 74:24–29
74. Martínez-García G, Agüí L, Yáñez-Sedeño P et al (2017) Multiplexed electrochemical immunosensor for obesity-related hormones using grafted graphene-modified electrodes as platforms for antibodies immobilization. Procedia Tech 27:187–189
75. Jiang C, Alam MT, Parker SG et al (2016) Strategies to achieve control over the surface ratio of two different components on modified electrodes using aryldiazonium salts. Langmuir 32(10):2509–2517
76. Hayat A, Sassolas A, Marty J-L et al (2013) Highly sensitive ochratoxin A impedimetric aptasensor based on the immobilization of azido-aptamer onto electrografted binary film via click chemistry. Talanta 103:14–19
77. Xia Z, Leonardi F, Gobbi M et al (2016) Electrochemical functionalization of graphene at the nanoscale with self-assembling diazonium salts. ACS Nano 10(7):7125–7134
78. Nguyen VQ, Sun X, Fdr L et al (2016) Unprecedented self-organized monolayer of a Ru (II) complex by diazonium electroreduction. J Am Chem Soc 138(30):9381–9384
79. Liu G, Paddon-Row MN, Justin GJ (2008) Protein modulation of electrochemical signals: application to immunobiosensing. Chem Commun 33:3870–3872
80. Polsky R, Harper JC, Dirk SM (2007) Diazonium-functionalized horseradish peroxidase immobilized via addressable electrodeposition: direct electron transfer and electrochemical detection. Langmuir 23:364–366
81. Fu Y, Jiang C, Tofaris GK et al (2020) Facile impedimetric analysis of neuronal exosome markers in Parkinson's disease diagnostics. Anal Chem 92(20):13647–13651
82. Jiang C, Hopfner F, Katsikoudi A et al (2020) Serum neuronal exosomes predict and differentiate Parkinson's disease from atypical Parkinsonism. J Neurol Neurosurg Psychiatry 91:720–729
83. Khor SM, Liu G, Peterson JR et al (2011) An electrochemical immunobiosensor for direct detection of veterinary drug residues in undiluted complex matrices. Electroanalysis 23(8):1797–1804
84. Liu G, Khor SM, Iyengar SG et al (2012) Development of an electrochemical immunosensor for the detection of HbA1c in serum. Analyst 137(4):829–832
85. Liu G, Guo W, Yin Z (2014) Covalent fabrication of methyl parathion hydrolase on gold nanoparticles modified carbon substrates for designing a methyl parathion biosensor. Biosens Bioelectron 53:440–446
86. Liu G, Song D, Chen F (2013) Towards the fabrication of a label free amperometric immunosensor using SWNTs for direct detection of paraoxon. Talanta 104:103–108
87. Guo W, Jiang F, Chu J et al (2013) A stable interface based on aryl diazonium salts/SWNTs modified gold electrodes for sensitive detection of hydrogen peroxide. J Electroanal Chem 703:63–69
88. Liu G, Qi M, Zhang Y et al (2016) Nanocomposites of gold nanoparticles and graphene oxide towards an stable label-free electrochemical immunosensor for detection of cardiac marker troponin-I. Anal Chim Acta 909:1–8
89. Jiang C, Wang G, Hein R et al (2020) Antifouling strategies for selective in vitro and in vivo sensing. Chem Rev 120(8):3852–3889
90. Ostuni E, Chapman RG, Holmlin RE et al (2001) A survey of structure–property relationships of surfaces that resist the adsorption of protein. Langmuir 17(18):5605–5620
91. Gui AL, Luais E, Peterson JR et al (2013) Zwitterionic phenyl layers: finally, stable, anti-biofouling coatings that do not passivate electrodes. ACS Appl Mater Interfaces 5(11):4827–4835
92. Parviz M, Darwish N, Alam MT et al (2014) Investigation of the antifouling properties of phenyl phosphorylcholine-based modified gold surfaces. Electroanalysis 26(7):1471–1480
93. Polsky R, Harper JC, Wheeler DR et al (2008) Electrically addressable diazonium-functionalized antibodies for multianalyte electrochemical sensor applications. Biosens Bioelectron 23(6):757–764

94. Polsky R, Harper JC, Dirk SM et al (2007) Diazonium-functionalized horseradish peroxidase immobilized via addressable electrodeposition: direct electron transfer and electrochemical detection. Langmuir 23(2):364–366
95. Polsky R, Harper JC, Wheeler DR et al (2008) Electrically addressable cell immobilization using phenylboronic acid diazonium salts. Angew Chem Int Ed 120(14):2671–2674
96. Jiang C, Alam M, Silva S et al (2016) Unique sensing interface that allows the development of an electrochemical immunosensor for the detection of tumor necrosis factor α in whole blood. ACS Sens 1(12):1432–1438
97. Qi M, Zhang Y, Cao C et al (2016) Decoration of reduced graphene oxide nanosheets with aryldiazonium salts and gold nanoparticles toward a label-free amperometric immunosensor for detecting cytokine tumor necrosis factor-α in live cells. Anal Chem 88(19):9614–9621
98. Qi M, Zhang Y, Cao C et al (2016) Increased sensitivity of extracellular glucose monitoring based on AuNP decorated GO nanocomposites. RSC Adv 6(45):39180–39187
99. Liu G, Zhang Y, Guo W (2014) covalent functionalization of gold nanoparticles as electronic bridges and signal amplifiers towards an electrochemical immunosensor for botulinum neurotoxin type A. Biosens Bioelectron 61:547–553
100. Ni S, Qiao L, Shen Z et al (2020) Physical absorption versus covalent binding of graphene oxide on glassy carbon electrode towards a robust aptasensor for ratiometric electrochemical detection of vascular endothelial growth factor (VEGF) in serum. Electrochim Acta 331:135321
101. Cincotto FH, Martínez-García G, Yáñez-Sedeño P et al (2016) Electrochemical immunosensor for ethinylestradiol using diazonium salt grafting onto silver nanoparticles-silica–graphene oxide hybrids. Talanta 147:328–334
102. Lian Y, He F, Wang H et al (2015) A new aptamer/graphene interdigitated gold electrode piezoelectric sensor for rapid and specific detection of Staphylococcus aureus. Biosens Bioelectron 65:314–319
103. Moreno-Guzmán M, Ojeda I, Villalonga R et al (2012) Ultrasensitive detection of adrenocorticotropin hormone (ACTH) using disposable phenylboronic-modified electrochemical immunosensors. Biosens Bioelectron 35(1):82–86
104. Istamboulié G, Paniel N, Zara L et al (2016) Development of an impedimetric aptasensor for the determination of aflatoxin M1 in milk. Talanta 146:464–469
105. Revenga-Parra M, García-Mendiola T, González-Costas J et al (2014) Simple diazonium chemistry to develop specific gene sensing platforms. Anal Chim Acta 813:41–47
106. Bagheryan Z, Raoof J-B, Golabi M et al (2016) Diazonium-based impedimetric aptasensor for the rapid label-free detection of Salmonella typhimurium in food sample. Biosens Bioelectron 80:566–573
107. Phal S, Shatri B, Berisha A et al (2018) Covalently electrografted carboxyphenyl layers onto gold surface serving as a platform for the construction of an immunosensor for detection of methotrexate. J Electroanal Chem 812:235–243
108. Eissa S, Tlili C, L'Hocine L et al (2012) Electrochemical immunosensor for the milk allergen β-lactoglobulin based on electrografting of organic film on graphene modified screen-printed carbon electrodes. Biosens Bioelectron 38(1):308–313
109. Sun B, Li D, Hou X et al (2020) A novel electrochemical immunosensor for the highly sensitive and selective detection of the depression marker human apolipoprotein A4. Bioelectrochemistry 135:107542
110. Sánchez-Tirado E, Arellano LM, González-Cortés A et al (2017) Viologen-functionalized single-walled carbon nanotubes as carrier nanotags for electrochemical immunosensing. Application to TGF-β1 cytokine. Biosens Bioelectron 98:240–247
111. Saeed AA, Sánchez JLA, O'Sullivan CK et al (2017) DNA biosensors based on gold nanoparticles-modified graphene oxide for the detection of breast cancer biomarkers for early diagnosis. Bioelectrochemistry 118:91–99
112. Sunday CE, Masikini M, Wilson L et al (2015) Application on gold nanoparticles-dotted 4-nitrophenylazo graphene in a label-free impedimetric deoxynivalenol immunosensor. Sensors 15(2):3854–3871

113. Lerner MB, D'Souza J, Pazina T (2012) Hybrids of a genetically engineered antibody and a carbon nanotube transistor for detection of prostate cancer biomarkers. ACS Nano 6:5143–5149
114. Narayanan J, Sharma MK, Ponmariappan S et al (2015) Electrochemical immunosensor for botulinum neurotoxin type-E using covalently ordered graphene nanosheets modified electrodes and gold nanoparticles-enzyme conjugate. Biosens Bioelectron 69:249–256
115. Eissa S, Alshehri N, Rahman AMA et al (2018) Electrochemical immunosensors for the detection of survival motor neuron (SMN) protein using different carbon nanomaterials-modified electrodes. Biosens Bioelectron 101:282–289
116. Wang R, Xue C (2013) A sensitive electrochemical immunosensor for alpha-fetoprotein based on covalently incorporating a bio-recognition element onto a graphene modified electrode via diazonium chemistry. Anal Methods 5(19):5195–5200
117. Xu Y, Wu H, Huang C et al (2015) Sensitive detection of tumor cells by a new cytosensor with 3D-MWCNTs array based on vicinal-dithiol-containing proteins (VDPs). Biosens Bioelectron 66:321–326
118. Zehani N, Fortgang P, Saddek Lachgar M et al (2015) Highly sensitive electrochemical biosensor for bisphenol A detection based on a diazonium-functionalized boron-doped diamond electrode modified with a multi-walled carbon nanotube-tyrosinase hybrid film. Biosens Bioelectron 74:830–835
119. Cao C, Jin R, Wei H et al (2020) Adaptive in vivo device for theranostics of inflammation: real-time monitoring of interferon-γ and aspirin. Acta Biomater 101:372–383
120. Qi M, Huang J, Wei H et al (2017) Graphene oxide thin film with dual function integrated into a nanosandwich device for in vivo monitoring of interleukin-6. ACS Appl Mater Interfaces 9(48):41659–41668
121. Shen Z, Huang J, Wei H et al (2020) Validation of an in vivo electrochemical immunosensing platform for simultaneous detection of multiple cytokines in Parkinson's disease mice model. Bioelectrochemistry 134:107532
122. Tymoczko J, Schuhmann W, Gebala M (2014) Electrical potential-assisted DNA hybridization. How to mitigate electrostatics for surface DNA hybridization. ACS Appl Mater Interfaces 6(24):21851–21858
123. Xie Y, Pan Y, Zhang R et al (2015) Modulating protein behaviors on responsive surface by external electric fields: a molecular dynamics study. Appl Surf Sci 326:55–65
124. Kaur J, Jiang C, Liu G (2019) Different strategies for detection of HbA1c emphasizing on biosensors and point-of-care analyzers. Biosens Bioelectron 123:85–100
125. Ait-Touchente Z, Falah S, Scavetta E et al (2020) Different electrochemical sensor designs based on diazonium salts and gold nanoparticles for pico molar detection of metals. Molecules 25(17):3903
126. Puy-Llovera J, Pérez-Ràfols C, Serrano N et al (2017) Selenocystine modified screen-printed electrode as an alternative sensor for the voltammetric determination of metal ions. Talanta 175:501–506
127. Serrano N, González-Calabuig A, del Valle M (2015) Crown ether-modified electrodes for the simultaneous stripping voltammetric determination of Cd(II), Pb(II) and Cu(II). Talanta 138:130–137
128. Liu G, Wang S, Liu J et al (2012) An Electrochemical immunosensor based on chemical assembly of vertically aligned carbon nanotubes on carbon substrates for direct detection of the pesticide endosulfan in environmental water. Anal Chem 84(9):3921–3928
129. Liu G, Zhang Y, Qi M et al (2015) Covalent anchoring of multifunctionized gold nanoparticles on electrodes towards an electrochemical sensor for the detection of cadmium ions. Anal Methods 7(13):5619–5626
130. Lau PY, Ng KL, Yusof NA et al (2019) A sample pre-treatment-free electrochemical immunosensor with negative electro-pulsion for the quantitative detection of acrylamide in coffee, cocoa and prune juice. Anal Methods 11(33):4299–4313
131. Gervais E, Aceta Y, Gros P et al (2018) Study of an AuNPs functionalized electrode using different diazonium salts for the ultra-fast detection of Hg(II) traces in water. Electrochim Acta 261:346–355

132. Nguyen TTK, Nguyen TN, Anquetin G et al (2018) Triggering the electrolyte-gated organic field-effect transistor output characteristics through gate functionalization using diazonium chemistry: application to biodetection of 2,4-dichlorophenoxyacetic acid. Biosens Bioelectron 113:32–38
133. Cao C, Zhang F, Goldys EM et al (2018) Advances in structure-switching aptasensing towards real time detection of cytokines. TrAC Trends Anal Chem 102:379–396
134. Cao C, Jin R, Wei H et al (2018) Graphene oxide based recyclable in vivo device for amperometric monitoring of interferon-γ in inflammatory mice. ACS Appl Mater Interfaces 10(39):33078–33087
135. Benzigar MR, Dasireddy VDBC, Guan X et al (2020) Advances on emerging materials for flexible supercapacitors: current trends and beyond. Adv Funct Mater 30(40):2002993
136. Steeno R, Rodríguez González MC, Eyley S et al (2020) Covalent functionalization of carbon surfaces: diaryliodonium versus aryldiazonium chemistry. Chem Mater 32(12):5246–5255
137. Angelone D, Hammer AJS, Rohrbach S et al (2021) Convergence of multiple synthetic paradigms in a universally programmable chemical synthesis machine. Nat Chem 13(1):63–69
138. Zheng Y, Jiang C, Ng SH et al (2016) Unclonable plasmonic security labels achieved by shadow-mask-lithography-assisted self-assembly. Adv Mater 28(12):2330–2336
139. Wang Z, Wang J, Xiao Z et al (2013) A paper indicator for triple-modality sensing of nitrite based on colorimetric assay, Raman spectroscopy, and electron paramagnetic resonance spectroscopy. Analyst 138(24):7303–7307

Reinforced Polymers: The Emerging Role of Diazonium Modification of Fillers

Arvind K. Bhakta, Khouloud Jlassi, Beata Strzemiecka, Karim Benzarti, and Mohamed M. Chehimi

Abstract This chapter summarizes the knowledge of filler modification with diazonium salts, with a view to designing high-performance reinforced polymer composites. Arylation of fillers serves for attaching prepolymers or for triggering polymerization processes. The level of modification is molecular or macromolecular in nature. Both arylated and macromole-modified fillers served to design reinforced polymer composites with unique mechanical and dielectric properties. One important feature is the rheology of formulations prior to cure. Interfacial chemical compositions could be investigated with surface-specific techniques such as XPS and ToF–SIMS. Hand-picked cases concern the design of reinforced polymers with silica, zeolite, clays, and carbon allotropes. We demonstrate that fine control over the surface chemical composition of fillers has profound effect on mechanical, thermal, viscoelastic, and electrical properties of end polymer composites.

1 Introduction

Owing to their nanoscale size, high surface-to-volume aspect ratio, and high surface energy, fillers impart to polymers enhanced mechanical properties such as yield strength (or elastic limit) and modulus [1]. Such mechanical properties of filled

---

A. K. Bhakta
St. Joseph's College, Bangalore, Karnataka, India

A. K. Bhakta · M. M. Chehimi (✉)
Université Paris Cité, CNRS, ITODYS, F-75013 Paris, France
e-mail: mohamed.chehimi@cnrs.fr

K. Jlassi
Qatar University, Doha, Qatar

B. Strzemiecka
Poznan University of Technology, Poznań, Poland

K. Benzarti
Univ Gustave Eiffel, Marne la Vallée, France

M. M. Chehimi et al. (eds.), *Aryl Diazonium Salts and Related Compounds*, Physical Chemistry in Action, https://doi.org/10.1007/978-3-031-04398-7_20

polymer composites depend to a very large extent on the filler polymer matrix interactions [2]. If the latter is sufficiently strong at polymer-filler interface, the filler polymer composite withstands the applied mechanical stress. Mechanical stress is transferred to the interface/interphase region which will either resist or fail depending primarily on the strength of molecular interactions. For example, strong molecular interactions at polymer-filler interface (e.g., covalent bonding) result in effective stress transfer from the polymer matrix to nanofillers [3]. One key issue is to chemically modify fillers with appropriate coupling agents that interact covalently or non-covalently with the fillers. While silanes are known to readily react with silica and other ceramics, onium salts are routinely used to intercalate layered silicates (clays) to make organophilic fillers for making them compatible with organic polymer matrices [4]. In the recent decade, we have introduced diazonium compounds as intercalating agents and/or reactive compounds to design organo-fillers, namely clays [5, 6], silicas [7], and carbon nanomaterials [8]. Diazonium modification of reinforcing materials is incipient but has much to offer therefore justifying this contribution to the book.

In this chapter we will review methods of making organo-fillers using diazonium salts and the use of the said modified fillers to prepare high-performance polymer composites. The chapter is organized as follows:

- Brief introduction to diazonium modification of fillers
- Method of testing mechanical properties
- Surface-specific tools to probe filler-aryl-polymer interfaces
- Case studies of polymer composites prepared using diazonium-modified fillers
- Conclusion and prospects

In this chapter, we concentrate on comprehensive case studies starting from filler arylation to mixing with polymer matrix and then mechanical testing. Studies that are related to composites but for other purposes than structural composites are not covered. Herein, we intentionally consider only structural composite materials the interface/interphase region of which has effect on mechanical, thermal, or dielectric properties.

2 Fillers: Definition, Features and Properties Imparted to Polymer Matrices

Fillers are substances, with various shapes (particles, nanofibers, nanorods, nanotubes, layered porous, hollow spheres), which are mixed with polymers in order to reduce their cost or improve their mechanical or electric, or magnetic properties. They usually have a size less than 100 μm, but much hope has been raised with those with at least on dimension in the "nano" regime that is less than 100 nm. Table 1 summarizes the properties imparted by the shortlisted fillers to polymer matrices (only handpicked fillers that were subjected to arylation are reported).

Table 1 Shortlisted frequently employed fillers and corresponding features and properties imparted to polymer matrices

Fillers	Size, Surface Area	Mass loading range	Main expected improved performances	References
Silicas	Nano (from about 2 nm) and micro, from tens to 500 m^2/g	10–70% wt	thermomechanical properties, abrasion resistance, heat stability, optical properties	[7, 9, 10]
Zeolites	Nano to 50 μm, ~20 m^2/g	2–50% wt	flexural strength, durability, impact strength	[11–13]
Clays	Nano to micro, to max. 200 m^2/g	To max 10%	Gas barrier properties, tensile strength, flexural strength, flexural modulus	[4, 14]
CNTs	Nano, up to 1000 m^2/g	From below 1% to several percents	mechanical strength (greater than 85 MPa), conductivity, thermal properties	[15, 16]
Graphene	Nano, up to over 2000 m^2/g	From below 1–2%	Mechanical strength, conductivity, tear strength	[17, 18]
Carbon black	Nano to micro, smaller than 155 m^2/g	5–30%wt	Mechanical strength, conductivity	[19–21]

One key issue is to modify the filler surface in order to enhance molecular interactions at filler-matrix interactions with resulting drastic effect on the macroscopic behavior of the composites and their resistance to stress. fillers such as silica or carbon could be arylated with in situ generated diazonium salts [7], whereas prefabricated diazonium salts could be employed to intercalate clays by cation exchange reactions with sodium [5, 22].

To understand the effect of filler modification on the mechanical properties of the end polymer composites, it is important to characterize surface chemical composition of the filler. X-ray photoelectron spectroscopy (XPS) [5, 7, 13, 23] and time of flight-secondary ion mass spectroscopy (ToF–SIMS) [13] are surface sensitive techniques tracking changes within a 1–5 nm depth, while Fourier transforms infrared spectroscopy (FTIR) and Raman spectroscopy characterizes functional groups at the surface of fillers [16]. The latter is particularly suited for carbon materials and probes the degree of covalent modification. To these analytical tools, thermogravimetric analysis permits to determination of the mass loading of organic matter grafted to the fillers [24], while nitrogen adsorption measurements provide Brunauer, Emmett,

Table 2 Basic principles of experimental determination of tensile strength, flexural strength, and storage and loss moduli

Tensile strength	Flexural strength	Dynamic mechanical analysis
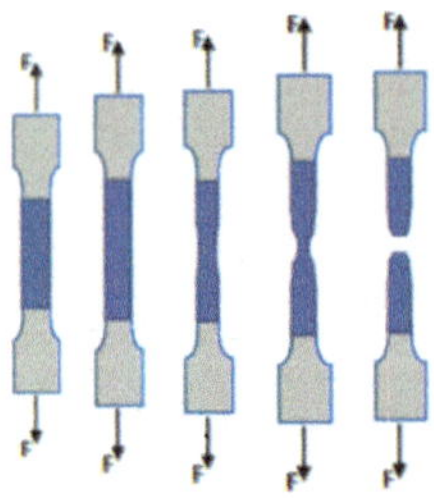	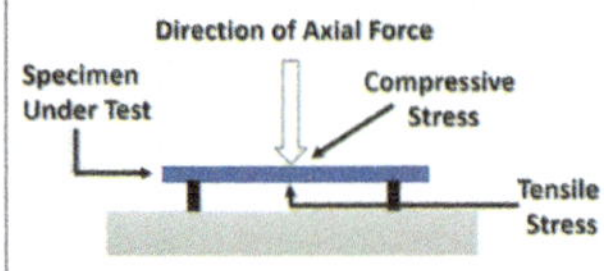	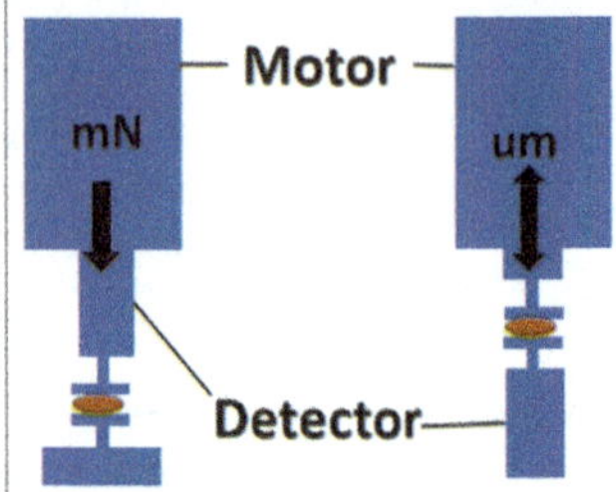

and Teller (BET) specific surface area of the fillers, and pore volume and area [21]. Scanning electron microscopy (SEM) is important for analyzing dramatic morphological changes of the fracture surface induced by the arylation of the filler under test [14], while high-resolution transmission electron microscopy (HRTEM) permits to directly image the surface aryl layer and determine its thickness [23], or to measure the interlayer spacing in clays upon intercalation or exfoliation [6]. Organic modification of crystalline fillers results in an amorphous layer easily distinguishable from that of the filler (see for example arylated CNTs [23]).

As far as the main mechanical testing is concerned on bulk composites, tensile strength, flexural strength, and dynamic mechanical analysis are routinely employed and permit shed a light on the effect of filler nature, mass loading, surface modification, and other parameters on the mechanical properties. Experimental procedures are displayed in Table 2.

Viscoelastic materials, such as polymers, exhibit features of both liquid and solid. When a sample is subjected to sinusoidal strain or stress, their analogous developed stress or strain can be estimated by dynamic mechanical analysis [25]. Thus, knowledge about the loss modulus (G''), storage modulus (G'), and loss factor ($\tan\delta = G''/G'$) within test temperature range can be obtained from dynamic mechanical analysis (DMA) [26]. The loss factor relates to the mobility of the macromolecular chains in the polymer matrix and is very sensitive to any factor affecting this mobility in the composite (such as the crosslink density of the polymer network, the filler content, the surface treatment of fillers using coupling agents, organosilanes or surfactants), especially in the glass transition domain. At fixed oscillation amplitude and frequency, 3-point bending mode was used for dynamic mechanical analysis utilizing rectangular species. It is very useful to study viscoelastic properties/long-term mechanical properties. Tensile strength can be measured via universal testing machine equipped with load cell; room temperature is favored [14]. In a flexure test, material stress prior to bend is described as flexural strength. Their estimation of phenolic composites is important. It is an essential material property and higher flexural strength is generally considered as a signature of better durability. There are three important parameters while carrying out 3-point flexural tests such as distance of supports, speed test, and sample destruction moment [7].

3 Case Studies of Diazonium Modification of Fillers and Uses Thereof

In the following, we will tackle case studies of preparation, characterization, and use of organo-modified fillers for polymer composites. The mechanical properties were determined for composites in terms of, e.g., flexural strength, fracture toughness, storage, and loss moduli. Epoxy composite curing could also be monitored by measuring viscosity of the composite formulation.

3.1 Diazonium Modification of Nanosilicas and Zeolites

Silicas remain among the most employed (nano)materials for reinforcing polymer composites and coatings. In order to avoid phase separation and improve polymer-silica adhesion, surface modification is essential [27] and is routinely achieved with silane coupling agents the chemical structure of which depends on the nature of the polymer matrix, e.g., polymethacrylates, polyaniline, and phenolic resins such as novolac (Scheme 1). The latter is used as binder for several purposes such as abrasive materials and tools [28, 29].

However, novolac is thermoplastic polymer and requires crosslinking process. The crosslinking agent for novolac is most often hexamethylenetetramine (HMTA); but this compound is hazardous, inducing irritation to skin and mucous membranes, and upon thermal decomposition produces NOx, NH_3, CO, and CO_2 [7]. Thus, some efforts are made to replace these crosslinkers with more eco-friendly ones. Toward this end, the team of Strzemiecka [7] prepared diazonium-modified silicas to serve as the active fillers of phenolic resins (used, e.g., as binders in abrasive products) at 10–25 wt. % (Fig. 1a) that can act as crosslinking agent. The diazonium-modified silica nanoparticles with surface-bound $–CH_2–OH$ groups reacted with phenolic resins causing their crosslinking (hardening) (Fig. 1b) and changing their viscosity so that the flow distance (see Fig. 1c) was shorter than that measured for unfilled resin

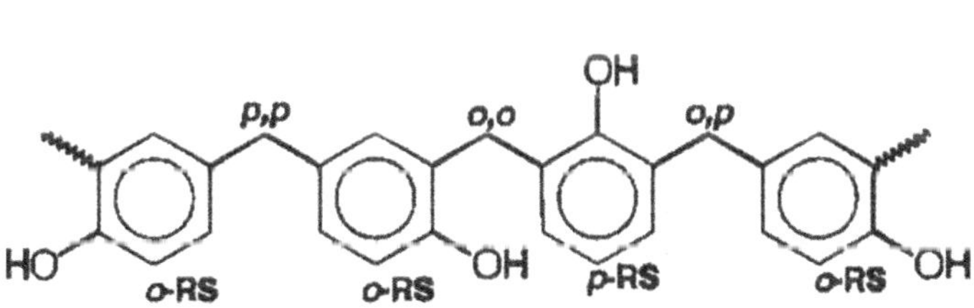

Scheme 1 Novolac chemical structure

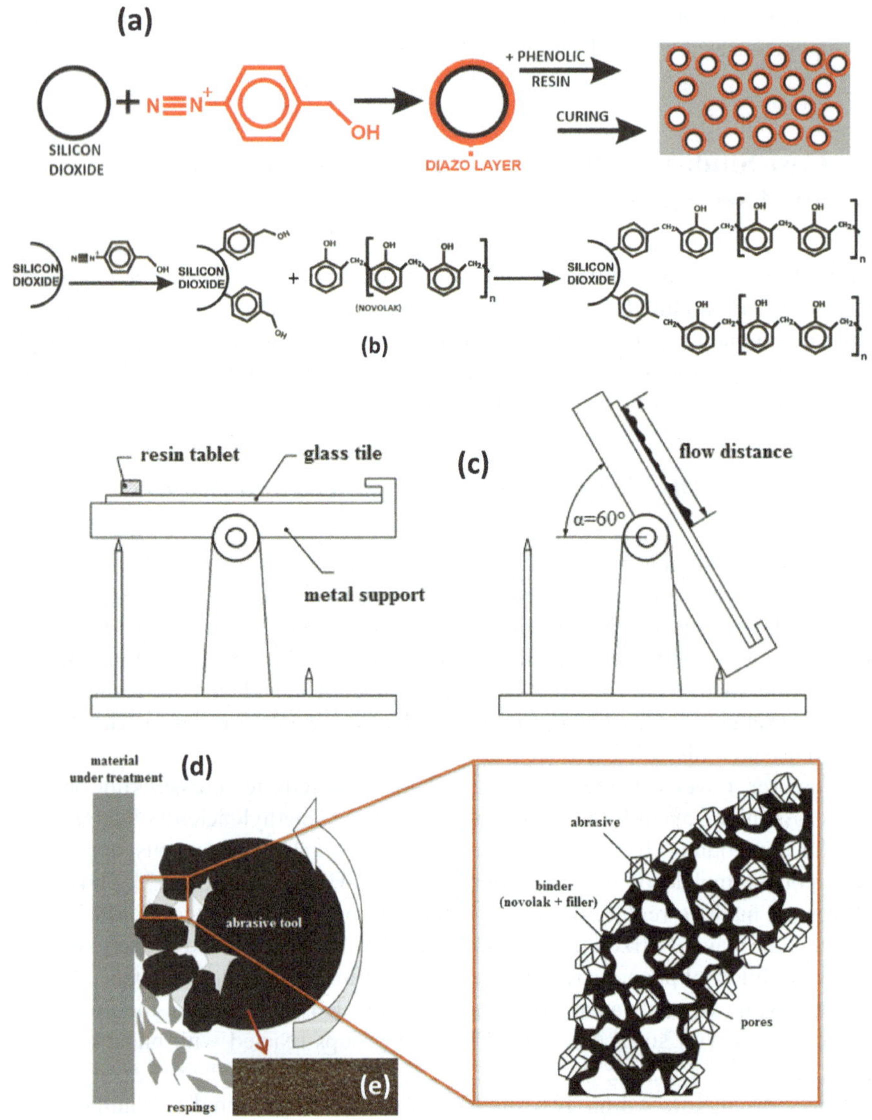

Fig. 1 Fabrication of arylated filler for abrasive tool: **a** Three-step making cured phenolic resin filled with diazonium-modified silica nanoparticles, **b** chemical reaction at arylated silica-novolac interface, **c** determination of flow distance for filled resin prior to cure, **d** schematic illustration of an abrasive tool composite interface, and **e** digital picture of a real abrasive tool sample 50(l) × 10(w) × 4(t) mm. Figures **a** and **b** are reproduced with permission from [7] Copyright 2016 American Chemical Society, and **e** is reproduced with permission [30] Copyright 2019 Elsevier

(Fig. 1c). After annealing at 180 °C, the diazonium-modified silica/phenolic resin composites were found to be mechanically robust and the three-point flexural test indicated 35% increase in the flexural strength compared to pristine Syloid type silica or other standard fillers for phenolic resins.

Diazonium-modified silica with surface-bound $-CH_2-OH$ groups are thus ideal reactive filler for phenolic resins. Such filler ensures interfacial chemical reactions with the matrix and imparts robust mechanical properties to the final composites.

Interestingly, arylated silica of the Syloid 244 type served to design model abrasive tool as shown in Fig. 1d, e. The resin was filled with arylated silica and mixed with alumina particles [30]. The arylated silica imparted 68% improvement of flexural strength of novolac but an impressive quantum jump of 246% for the model abrasive tool compared to the same materials containing untreated silica. Therefore, fabrication of a mechanically robust binder (robustness due to covalent silica filled novolac) has major effect on flexural strength of the ternary abrasive composite tool. In contrast, arylation of silica imparts higher Young modulus to novolac but not to the abrasive tool.

Similar surface modification was performed on zeolites with two diazonium isomers, one was sterically hindered (Fig. 2) [13]:

XPS results show relatively more intense C1s peak compared to Si2p in the decreasing trend, Z-4CH_2OH > Z-2CH_2OH > pristine zeolite [13]. This was confirmed using ToF–SIMS, a very powerful surface analytical technique that probes near 1 nm thickness in the static regime (low sputtering rate of the surface). Figure 2c demonstrates the existence of a chemical reaction between the diazonium via its $-CH_2-OH$ group and the silanol from silica: $-CH_2OH + HO\text{-}Si\equiv \rightarrow -CH_2O\text{-}Si\equiv$.

Reaction of the CH_2OH group in para position is much more favorable than the one with the diazonium bearing CH_2OH in ortho position of the diazonium group, due to steric hindrance. No peak from an aluminum-containg fragment was detected by ToF–SIMS, an important result indicating that for the aluminosilicate type material zeolite, reaction of the diazonium occurs via the silicon chemical environment.

Following the works on intercalation of clay for the making of clay/PGMA [5] and clay/ion-imprinted polymer nanocomposites [22, 31], zeolites were also modified with $N\equiv N^+\text{-}C_6H_4\text{-}N(CH_3)_2$ to provide $-N(CH_3)_2$ surface-bound group [32], a Type II photoinitiator of radical polymerization under visible light.[1] For dental medicine purposes, camphorquinone was used as photosensitizer to strip a hydrogen atom from $-N(CH_3)_2$ surface-bound group resulting in $-N(CH_3)CH_2^{\bullet}$ radical that triggers polymerization. Bis-GMA (2,2-bis{p- (2′-hydroxy-3′ methacryloxypropoxy)phenyl}-propane) 60% by weight was co-polymerized with TEGDMA (triethylene glycol dimethacrylate) (40% by weight). This provides a crosslinked polymer network. Flexural strength (~40 MPa) was maximal for 5% weight modified filler (zeolite-$C_6H_4\text{-}N(CH_3)_2$)

[1] *cf* Chapter 17 "*Polymer surface science and adhesion using diazonium chemistry*" by Griffete et al..

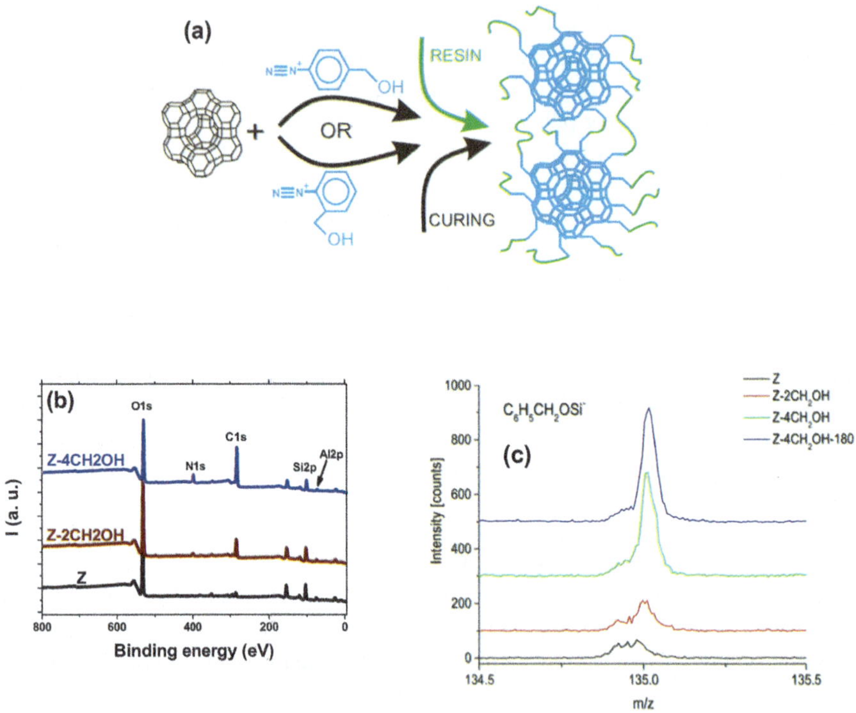

Fig. 2 Three steps of preparation of phenolic resin filled with Z-2CH_2OH or Z-4CH2OH (a), and their surface analysis by XPS (b), and ToF–SIMS (c). Reproduced with permission from [13] Copyright 2018 Elsevier

3.2 Clay Modification with Aryldiazonium Salts

Diazonium salts are unique alternatives to the traditional silanes and ammonium salts. They easily intercalate clays by exchange mechanism followed by the formation of diazoether linkages: Si-O-N = N-C_6H_4-R which dediazonize at 60 °C. The choice of the radical R depends on the selected type of polymer and the polymerization method. R = $N(CH_3)_2$ (DMA) [5, 31], -NO_2 [5], -NH-C_6H_5 (DPA) [6, 14, 33]), and $CH_2SCSN(C_2H_5)_2$ (iniferter) [24].

The N,N-dimethylaminoaniline (DMA) diazonium requires free benzophenone (BP) as photosensitizer (for UV photopolymerization), and BP diazonium necessitates free N,N-dimethylaniline in solution. DMA diazonium was found to be better intercalated compared to BP diazonium, and an exfoliated photochemically prepared clay/PGMA nanocomposite could be synthesized using clay intercalated with DMA diazonium [5]. The iniferter diazonium is too bulky and intercalation required high initial concentration of diazonium [24]. Concerning in situ polymerization of aniline, the diphenylamine (DPA) diazonium was instead used to intercalate bentonite. Combination of DPA diazonium and in situ oxidative polymerization of aniline provided highly exfoliated bentonite-DPA/PANI as conductive nanocomposite filler [6]. The nanocomposite was found to be exfoliated, compared to the corresponding, intercalated clay/PANI, prepared without any diazonium-pretreated clay. Importance of clay diazonium modification of bentonite was reflected in the electrical conductivity of compressed pellets which was found to be ~2 × 10^{-8} S/cm for clay/PANI, five orders of magnitude lower than the clay-DPA/PANI filler prepared using diazonium salt. After reaction with the epoxy resin and DDS hardener, the clay-DPA/PANI/epoxy final composite exhibited improved fracture toughness by 210–220% much higher than 20–30% in the case of the B/PANI (B: bentonite) filler prepared with any diazonium salt [14]. The fracture surfaces of clay-DPA/PANI/epoxy and clay/PANI/epoxy differ markedly; the former has a very particular fibrillar structure indicating ductile failure mode, whereas the clay/PANI filler induces a smooth fracture surface which accounts for brittle failure mode. Figure 3 compares the effects of diazonium salts on the properties the nanocomposite filler and the properties of the final corresponding epoxy composites.

In another study, bentonite-DPA/PANI was mixed with magnetite (Fe_3O_4) by mechanochemistry and the final hybrid B-DPA-PANI@Fe_3O_4 served as filler of DGEBA at 0.1, 0.5, 1, 2, 3 wt. %. The filler was employed for concomitant corrosion protection and oil sensing applications [33]. Figure 4 shows the interfacial groups resulting from the reaction of the (B-DPA-PANI@Fe_3O_4) hybrid filler with the epoxy resin and the hardener. NH groups from either DPA or PANI could react with epoxy groups via ring-opening, therefore, ensuring covalent bonding of the resin to the clay

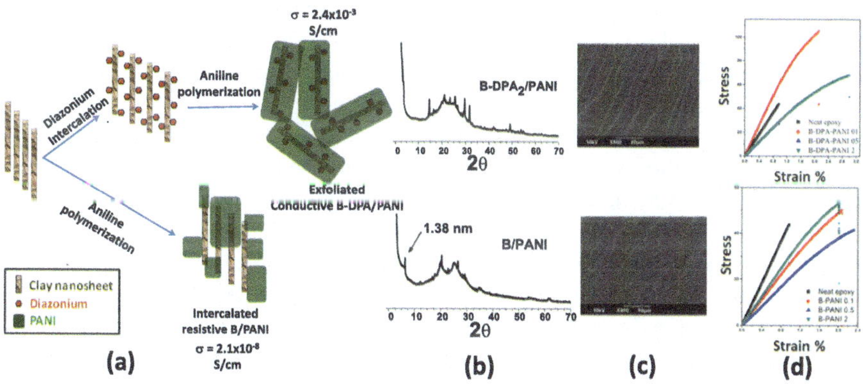

Fig. 3 Upper row: oxidative in situ polymerization of polyaniline in presence of DPA diazonium salt-modified bentonite (B-DPA). Lower row: oxidative in situ polymerization of polyaniline in presence of pure bentonite B. **a** synthesis of hybrid fillers with (upper) and without diazonium (lower); **b**: SAX patterns; **c** SEM image of fracture surface of epoxy matrix filled with B-DPA/PANI (upper row) and B/PANI (lower row); **d** Fracture toughness of expoxy reinforced with B-DPA/PANI and B/PANI (•: neat epoxy, ◆: B-PANI 0.1 wt %, ▲: B-PANI 0.5 wt %, ▼: B-PANI 2 wt %). **b** Adapted with permission from [6] Copyright 2014 The Royal Society of Chemistry; other images reproduced with permission from [14] Copyright 2016 American Chemical Society

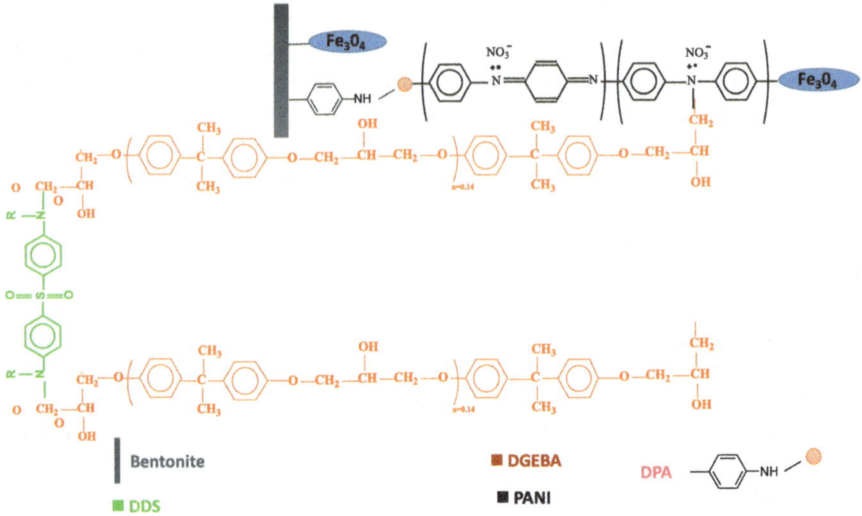

Fig. 4 Molecular view of the epoxy- B-DPA-PANI@Fe_3O_4 interface. Reproduced with permission from [14] Copyright 2016 American Chemical Society, and from [33]

sheets (via PANI and DPA). Similarly, DDS has two amino groups, which react with epoxy by ring-opening leading to crosslinking of the resin.

The magnetic filler improved the mechanical, thermal, and anticorrosion properties compared to unmodified filler (Fig. 5). The addition of only 3 wt. % of filler improved the tensile strength of the composites by 256% (Fig. 5a), and the glass transition temperature Tg by 37% (Fig. 5b). The dielectric spectra indicate a strong interfacial interaction between the hybrid fillers and epoxy matrix (Fig. 5c). Moreover, a network structure within the reinforced matrix was noted, in comparison to neat cured epoxy thus confirming very strong adhesion between the nanofillers and the matrix induced by the diazonium coupling agent and Fe_3O_4 nanoparticles. The addition of the filler to the DGEBA epoxy resin showed most significant improvement in corrosion inhibition using electrochemical impedance spectroscopy (EIS) in 3.5 wt % NaCl, High charge transfer resistance of $100 \times 106\ \Omega\ cm^2$ using 3 wt.% only of filler was noted compared to $0.35 \times 106\ \Omega\ cm^2$ for the pure epoxy, see Fig. 5d. Moreover, increasing the content of the hybrid filler in the epoxy coating, was found to decrease the hydrophilicity of the nanocomposites and consequently increased the water contact angle (WCA) from $50 \pm 4°$ to $85 \pm 2°$ for the 3 wt.% of filler (Fig. 5f), hence the decreased diffusion of the corrosive ions through the coating. As seen in Fig. 5f, the sample showed significant improvement in oil sensing test, the highest conductivity change in the oil occurred with epoxy filled with 3 wt.% of filler, a behavior attributed to the absorbance and expansion of the coating upon exposure to oil. This expansion leads to a decrease in the effective filler volume fraction, which in turn decreases the overall electrical conductivity.

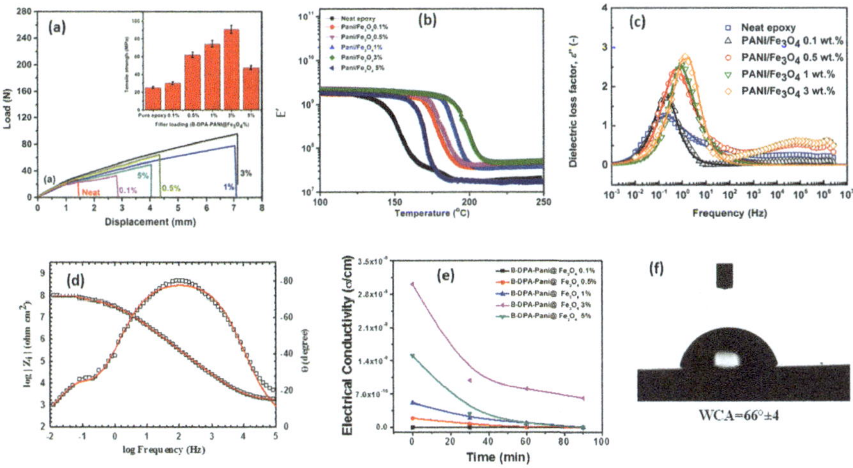

Fig. 5 **a** Tensile load-displacement curves of the cured epoxy and epoxy nanocomposites, **b** Temperature dependence of storage of the cured pure and filled epoxies, **c** Dielectric spectra in the broad frequency range for the cured pure epoxy and filled epoxy with 0.1, 0.5, 1, 2 and 3-wt % B-DPA/PANI-Fe_3O_4 loadings, **d** Bode plot of epoxy coatings with 3 wt % of B-DPA-PANI@Fe_3O_4 filler **e** Sensing characterization of samples. Conductivity (/cm) versus time (min) for DGEBA-B-DPA-PANI@Fe_3O_4 composite films in oil, **f** Water contact angle after the addition of 3 wt % of B-DPA-PANI@Fe_3O_4 filler. Reproduced from [33]

This applied approach highlights new surface and interface chemistry using diazonium salt to prepare efficient and inexpensive bio-based epoxy, for oil sensing and anti-corrosive smart protection, due to the very strong interfacial interactions between filler and host matrix mediating the diazonium salts chemistry.

3.3 *Modification of Carbon Allotropes*

Nanocarbons are of great interest because of their mechanical strength, modifiable surface, electrical and thermal conductivity, catalytic activity, and ultrahigh surface area. Out of all other carbon allotropes, one-dimensional (1D) carbon nanotubes (CNTs) [34] and two-dimensional (2D) graphene have received tremendous attention since their discovery. Carbon nanohorns (CNHs), one of the newest allotropes are also getting much popularity. They have been applied for plenty of applications such as nanofillers for reinforcement of softer and inert polymers matrices or resins [35]. Their low solubility in various solvents is due to the van der Waals forces which hold them together, therefore, resulting in bundles, hence the difficulty to prepare CNT-polymer nanocomposites. In order to exploit real properties of carbon nanomaterials, it should be modified in order to favor their dispersibility and stabilization in polymer matrix. Chemical modification of carbon surface via diazonium chemistry [36] is an elegant way to effectively tune the solubility properties

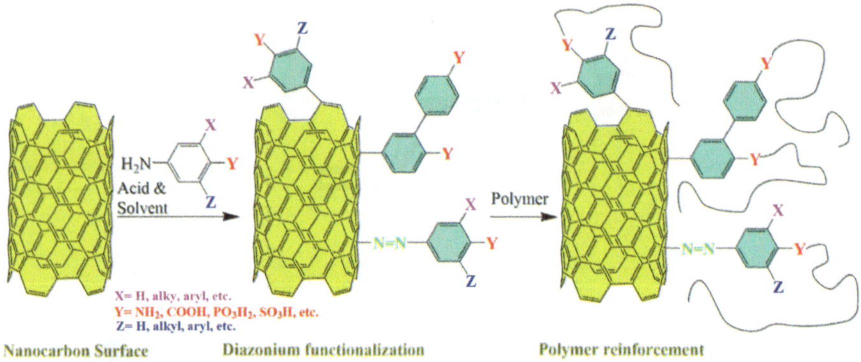

Fig. 6 Steps involved in the synthesis of carbon allotrope-polymer nanocomposite. Example is given for carbon nanotube sidewall

[37]. Figure 6 depicts a general pathway to preparing composite of nanocarbon and polymer utilizing diazonium species (see for example references [35, 36, 38]).

3.3.1 Carbon Nanotubes

Moaseri et al. [39] carried out diazonium salts reaction for functionalizing multi-walled carbon nanotubes (MWCNTs) with amine functions. For this purpose, they have considered 1,6-hexamethylenediamine, aminobenzene, ethylenediamine and 1,4-diaminobenzene. The surface energy and dispersion (in the matrix) and steric stability of the modified MWCNTs increased due to the grafted functional groups. This avoids their aggregation and imparts better distribution of applied shear inside the functionalized MWCNT-reinforced epoxy composite. Consequently, higher mechanical properties have been induced as compared to pristine MWCNTs nanofiller. Trivedi et al. [40] have achieved improved moduli of nanocomposite with incorporation of Ni-coated MWCNTs grafted with aryl diazotized curing agent (reduced bundle size via loosening and shortening) up to 1.3 wt % to the EPON 862 resin matrix. The interaction between resin and curing agent is through covalent bond formation.

A primary arylamine (4,4-diaminodiphenyl sulfone hardener) (DDS) [41] was grafted onto MWCNTs surface via in situ generated diazotization procedure to contrast the unsteadiness (bulk preparation-3 g). Later, it was utilized as a supplementary hardening agent for epoxy curing systems resulting in nanocomposite with self-healing efficiency, and good thermal, mechanical, and electrical performance. In contrast, MWCNTs without diazonium modification do not possess any self-repair ability in the polymer matrix. Covalently interconnected SWCNT and DDS through diazonium salt chemistry along with several strategies of covalent functionalization of ligands based on inbuilt epoxy chemical structure have been discussed [16]. Martinez-Rubi et al. have in situ generated unique diazonium salts of interest for the epoxy composite technology as depicted in Fig. 7. Various functionalization degrees range from 0.2–1.7% carbon substitution; DDS grafting was about four to

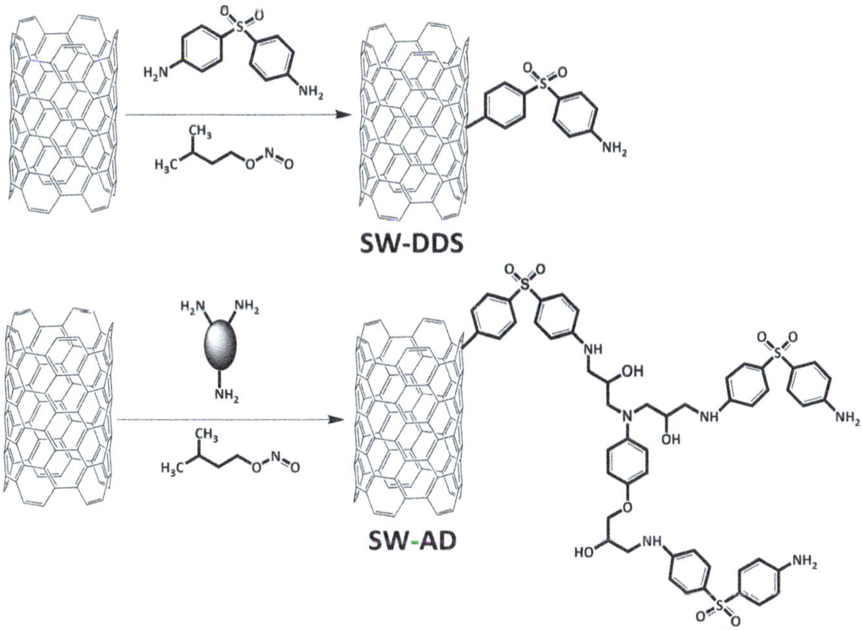

Fig. 7 Arylation of SWCNTs with in situ generated diazotized curing agent DDS and the 1TGAP-3DDS derivative AD. TGAP: epoxy resin triglycidyl p-aminophenol (Araldite MY0510); the hardener 4,40-diaminodiphenyl sulfone: DDS; AD = 1TGAP-3DDS

fivefold higher than that of the bulky AD large molecule as determined by TGA and confirmed by D/G Raman peak intensity ratio (*D*/*G* = 0.03, 0.07 and 0.26 for pristine SWCNT containing 20% metal, SWCNT-AD, and SWCNT-DDS, respectively). SWCNT-DDS and SWCNT-AD dispersed much better in epoxy resin compared to unmodified SWCNTs, and the filled resin had less viscosity which is favorable for processing. Interestingly, the high molecular weight AD induced less grafting, beneficial for retaining the same electrical conductivity order of magnitude. This is important when considering designing electrically conductive epoxy composites.

Work reported by Mammeri et al. [42] on nanoindentation of different methacrylate-grafted CNTs. With higher loads viz. 150 or 200 mN, slight cracking and absence of delamination in hybrid coatings were found.

One-pot aryl diazonium (p-phenylenediamine) reaction on CNTs, followed by its grafting onto carbon fibers for silicon resin strengthening was developed by Wu et al. [43]. Grafting density also determines interfacial strength of nanocomposite. CNTs with terminal -NH_2 act as bridge between fibers and resin matrix, and boost surface roughness of fiber and wettability (Fig. 8), due to mechanical interlocking and enhanced interfacial properties. Previously, carbon nanotube/carbon fiber hybrid reinforced composites exhibited enhancement in interfacial shear strength by 104% [44].

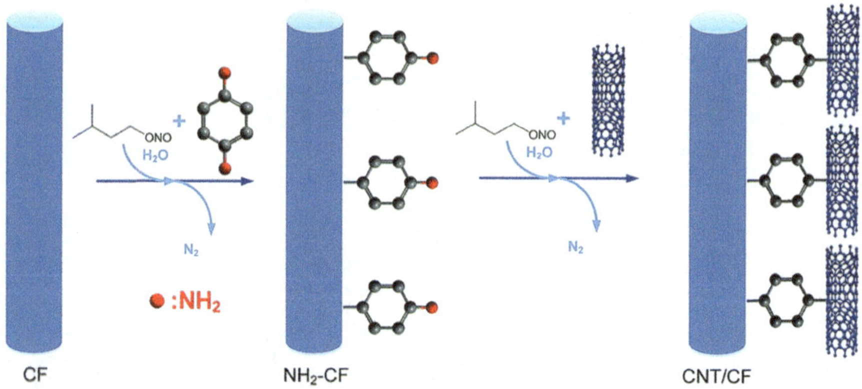

Fig. 8 Two-step diazonium chemical reactions for the design of carbon nanotube-grafted carbon fiber hybrid reinforcement. Reproduced with permission from [44] Copyright 2015 The Royal Society of Chemistry

3.3.2 Graphene Oxide and Graphene Oxide-Based Hybrid Fillers

3-Aminophenyl boronic acid was used to prepare azo-boron grafted reduced graphene oxide (RGO) insertion with sodium metaborate [18]. Incorporation (2 wt %) of this material into poly(lactic acid), composite has shown enhanced tensile strength (49.1%), Young's modulus (34.9%), and high fire safety (flame retardant and smoke/toxic fumes suppression), which can be a potential candidate for engineering applications.

Aryl diazonium (p-aminophenol, a parent aniline) functionalized graphene was incorporated in water-borne epoxy coating formulation as reactive filler (structural fortifier). This treated coating has shown better tribological properties (friction and wear) [45].

A robust polymer-filler interface was developed through diazonium coupling utilizing p-phenylene diamine, graphene nanoplatelets (GNPs), and poly (methyl methacrylate) (PMMA) [46]. Amidation reaction occurs between terminal -NH_2 groups of GNPs with polymer chains resulting in strong adhesion via a chemical bond. Improvement in nanocomposite elastic modulus (24%), strain (~13%), and stress at break (~28%) until optimized loading (2 wt %) was noticed. Further, up to 5 wt % (high concentration), increased glass transition temperature (Tg), and thermal stabilities were noted. In this paper, Vallés et al. [46] have also referred to the theory given by Young et al. [47], modulus of elasticity (E) of a nanocomposite does not gets affected by the *E*' of the filler but relies on their structural parameters: (a) filler/matrix interface- robust interface facilitates stress transfer for high reinforcement degree, (b) aspect ratio, and (c) orientation.

Graphene nanoplatelets-natural rubber composite foam was synthesized after modification (trans-1, 4-cyclohexane diamine) of GNPs by applying diazonium chemistry [48]. Foam product features were found to be influenced by two factors:

structure of microvoid and bubble size. Sluggish sulfur vulcanization kinetics (low cell density and extended expansion time of bubble), more nanofiller loading, and molecular obstacle created by cyclohexyl group cause little bubble diameter, which had occasioned in composite displaying gained tensile strength (68%), lowest thermal expansion coefficient, but low thermal conductivity due to defects formed on GNPs through diazonium functionalization in high acidic medium.

Comparative study by Vicentini et al. [49] has been carried out by considering three-carbon nanostructures (CNSs) namely, RGO, MWCNTs, and CNHs grafted with p-methoxyphenyl group via diazonium chemistry followed by homogeneous composite preparation with biodegradable poly(L-lactic acid). The nanocomposites have shown biocompatibility, which is important for tissue engineering applications. Mechanical behavior study has revealed elastic modulus of 871 MPa for pure polymer but with addition of minute quantity of nanofillers (0.1 wt %), it has decreased to 745, 538, and 458 MPa unexpectedly for MWCNTs, RGO and CNHs (functionalized ones), respectively. This scenario indicates toward modified CNSs more like plasticizers than reinforcement in the polymer matrix. Grafted CNSs in polymer have behaved differently regarding electrical, thermal, and mechanical features owing to various shapes, sizes, and different tendencies to agglomerate (at high concentration) in spite of present functionalities which helps in their dispersion.

3-D network of modified 1-D MWCNTs and 2 D RGO (Fisher indole and diazonium reaction using 4-aminopyridine) in polyimide matrix (In situ polymerization leads to fillers compatibility and homogeneous dispersion) was developed for synergistic toughening [50]. With addition of nanofiller (pyridine-RGO 0.9 wt % + pyridine-MWCNTs 0.1 wt % = 1 wt % nanocarbon), ternary nanocomposite has shown boosted fracture energy (29.7 MJ m^{-3}, 200%), tensile strength (581 MPa, 221%), and modulus (31 GPa, 312%).

3.3.3 Carbon Black

Carbon black (CB) is probably the very first filler that has been arylated by Cabot company for composite and specialty ink purposes [51]. The team of Daniel Bélanger has thoroughly studied its modification by in situ generated diazonium salts [52], while the group of Richard Compton demonstrated its modification by ball milling method consisting of milling CB particles with diazonium compounds (! caution: the method could be hazardous) [53]. The high propensity of CB and other glassy carbons to undergo arylation has been exploited in the modification of glassy carbon-protected cobalt nanorods with diazonium salt of para, N,N-dimethylaminoaniline [54]. For theranostics purposes, carbon-protected magnetic nanoparticles were also arylated to provide COOH, NH_2, and clickable reactive magnetic nanoparticles (https://www.turbobeads.com/, last accessed 8 July 2021). Of relevance to the present chapter, Sandomierski et al. [21] modified CB particles with in situ generated 4-hydroxymethylbenzenediazonium salt to provide reactive CB-C_6H_4-CH_2OH fillers toward phenol–formaldehyde (PF) resins. Although the modification was quite weak, it was sufficient to design arylated CB-reinforced cured PF resin with improved flexural strength. A near one order of magnitude better improvement was even noticed for the abrasive tool containing arylated filled PF compared to unmodified CB corresponding abrasive tool. For this purpose resole (Lerg), filler, novolac, and aluminum oxide (120 mesh, Imerys) were mixed in the ratio of 3:5:12:80 wt.%. Such improvements are due to the reactivity of the arylated CB fillers *vis-à-vis* PF resin, resulting in covalently bonded 3D composite network.

4 Summary of Shortlisted Polymeric Systems Reinforced with Diazonium-Modified Fillers

Table 3 summarizes the experimental conditions for the preparation of arylated fillers and indicates the final composite material therefrom. It also reports salient mechanical features exhibited by the reinforced polymer composite material. With such features in hand, Table 3 conclusively demonstrates the benefits one could achieve using diazonium-modified fillers. For more details about the shortlisted arylated fillers and their corresponding composites, the reader is referred to the references reported in Table 3.

5 Conclusion: What Have We Learned So Far from Diazonium Filler Modification for Polymer Composites

Diazonium chemistry has paved its way in the vast domain of polymer composites and is concerned with numerous combinations of fillers and polymer matrices [35]. From the early years of diazonium surface modification [56] and knowledge of spontaneous reactions of diazonium compounds with carbon allotropes [51], it was not surprising to see numerous examples of modification of CNTs [36], graphene, carbon fibers, and hybrid reinforcements such as carbon fiber/CNTs [44]. In contrast, silica and aluminosilicates were not expected to be prone to surface modification. Not only the reaction turned out to be efficient but also resulted in improved properties of filled binders [7] and even their abrasive tool derivative final product [30]. We have discussed the paramount role of position isomers in the case of zeolites modified with ortho and para-hydroxymethyl groups. Not only, steric hindrance due to ortho position of the reactive $-CH_2OH$ group and less reactivity was noted but the final flexural strength of arylated zeolite-filled phenolic resin was not improved compared to untreated zeolite [13]. Isomery influenced also the rheology of the filled resin; the para position imparted much more viscosity.

The reactivity with clays was completely unexpected as these layered materials react by cation exchange of sodium with the diazonium cation, followed by formation of diazoether and eventually final Si-O-aryl interfaces [5]. Such an intercalation is important as it facilitates obtaining exfoliated clay-polymer nanocomposites [6, 14].

Among the techniques offered to the composite specialist, ToF-SIMMS, though highly expensive, is recommended in order to understand the mechanisms of arylation

Table 3 Design and performances of shortlisted organo-fillers

Filler	Diazonium	Filler modification conditions	Final reinforced polymer system	Performances	Refs
Silica -SYLOID 244	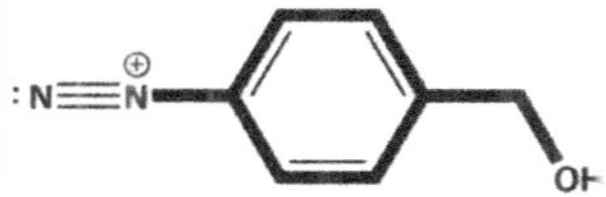	(i) 0.2 g ABA in 20 mL kept at 5 °C; (ii) 0.16 ml HCl added to ABA; (iii) 128 mg $NaNO_2$ in 5 mL water added dropwise, (iv) Syloid was added after 30 min and left 2 h to react	Silica-reinforced phenolic resin composite	Aryl layer resists heating at 110 °C and acetone wash Flexural strength: 40.1 ± 3.0 MPa	[7]
Zeolite—natural zeolite of granulation 20 μm	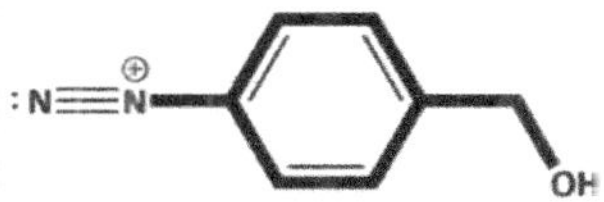 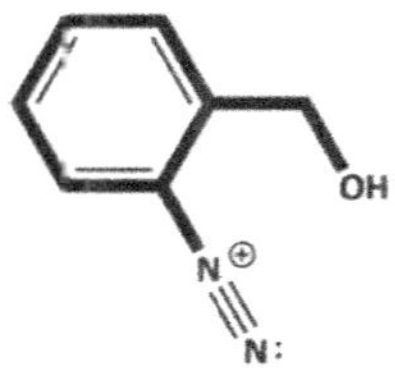	(i) 0.2 g ABA in 20 mL kept at 5 °C; (ii) 0.16 ml HCl added to ABA; (iii) 128 mg $NaNO_2$ in 5 mL water added dropwise, (iv) Syloid was added after 30 min and left 2 h to react	Zeolite-reinforced phenolic resin composite	Thermal stability at range 30–1000 °C of the filler with the coupled aryl layer Influence of the aryl-coupled zeolites on the crosslinking of phenolic resins Comparison of the mechanical properties of the phenolic resin composites with zeolite modified with benzyl alcohol with diazonium group in ortho and para position Flexural strength: 44.8 ± 4.8 MPa	[13, 55]

(continued)

Table 3 (continued)

Filler	Diazonium	Filler modification conditions	Final reinforced polymer system	Performances	Refs
Bentonite clay	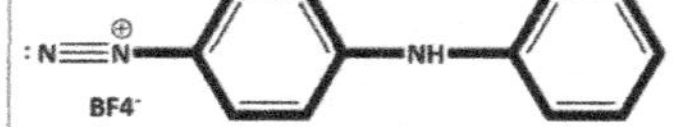	Dropwise addition of 0.15 or 0.3 or 0.45 g diazonium salt in 30 mL of de-ionized water diazonium to 1 g clay suspended in 100 mL of water. Heat up to 60 °C, at reflux in acetonitrile. Arylated clay was reacted with aniline in the PANI synthesis medium using aniline/$(NH_4)_2S_2O_8$ ratio = 4 in HNO_3. The final composite filler is noted: B–DPA_x/PANI (x = fraction of bentonite cation exchange capacity (CEC) = 0.6, 1.3 or 2	Bentonite/PANI-reinforced epoxy composite. Mass loading of B–DPA_x/PANI was 0.1, 0.5, and two mass loadings (wt %)	Best improvement of flexural strength of epoxy composite was reached with 0.1 wt.% hybrid filler B–DPA_2/PANI (53.2 MPa compared to 31.6 MPa for unfilled, crosslinked epoxy) 210 – 220% increase in fracture toughness for B-DPA/PANI-filled epoxy composite compared to 20 – 30% increase with B/PANI filler prepared without any arylation of clay	[6, 14]

(continued)

Table 3 (continued)

Filler	Diazonium	Filler modification conditions	Final reinforced polymer system	Performances	Refs
Carbon nanotube/Carbon fiber hybrid	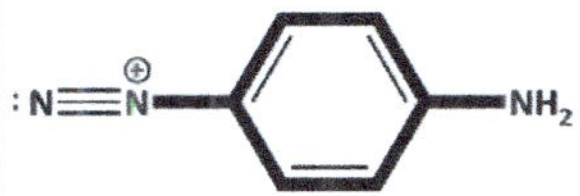	(i) CNTs added to mixture of 0.4 mL isopentyl nitrite, 1 g p-Phenylenediamine and 50 mL H_2O, stirred for 12 h at 353 K (CNT-NH_2); (ii) CFs in acetone/$H_2O$for 0.5 h at 633 K (Untreated CF); (iii) CNT-NH_2 sonicated for 1 h in 50 ml H_2O for various concentration; (iv) Bundled untreated CF on glass frame added to solution of 0.5 mL isopentyl nitrite and CNT-NH_2, stirred for 12 h at 353 K (CF-g-CNT); (v) Compression molding method for composites	Carbon nanotube/Carbon fiber hybrid reinforced methylphenylsilicone resin (MPSR) composites	Increased Interlaminar shear strength (ILSS) by 73.4% and Interfacial shear strength (IFSS) by 66.9%	[43]

(continued)

Table 3 (continued)

Filler	Diazonium	Filler modification conditions	Final reinforced polymer system	Performances	Refs
GO and MWCNTs	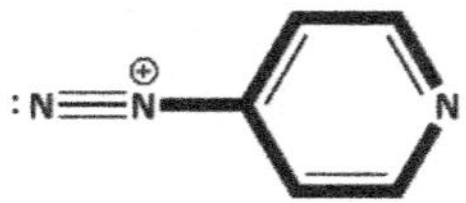	(i) 70 mmol $NaNO_2$ solution added to 50 mL 4 M HCl + 70 mmol 4-aminopyridine and stirred for 30 min. 0.5gOxidized CNTs in 200mLDMF were added. Left at 0 °C and 3 h stirring. Then stirring for 15 h at room temp. Filtered, redispersion in 2M HCl, filtered, neutralized with H_2O. Solid product dispersed in 2 M NaOH, stirred overnight; (ii) GO modification via Fisher indole reaction; (iii) nanocomposite through *in- situ* polymerization and thermal imidization	Functionalized graphene oxide (GO) and multiwalled carbon nanotubes (MWCNTs) reinforced polyimide nanocomposite	Fracture energy (29.7 MJ m^{-3}, 200%), tensile strength (581 MPa, 221%), and modulus (31 GPa, 312%)	[50]

(continued)

Table 3 (continued)

Filler	Diazonium	Filler modification conditions	Final reinforced polymer system	Performances	Refs
MWCNTs, CNHs and RGO	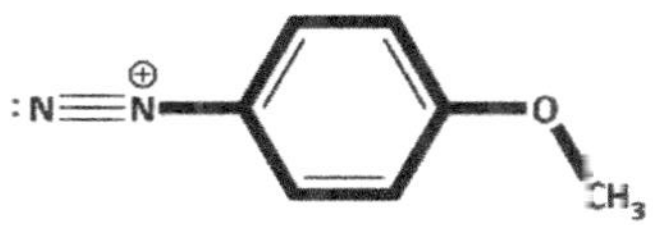	(i) As received CNSs (6.25 mmol in 35 mL CHP sonicated to 10 min; (ii) 4-methoxyaniline solution (3.13 mmol for CNTs & RGO; 6.25 mmol for CNHs) in 15 mL CHP, heated to 80 °C in N_2 (iii) Isopentylnitrite (3.12 mmol for CNTs & RGO; 6.25 mmol for CNHs) addition followed by dilution with 500 mL cold methanol after 15 min stirring (iv) $CHCl_3$ solution of PLLA added with sonication and solvent evaporation at 50 °C	p-methoxyphenyl functionalized MWCNTs, CNHs, and RGO modified poly(L-lactic acid) composite	Elastic modulus of 871 MPa for pure polymer but decreased to 745, 538, and 458 MPa with added functionalized MWCNTs, RGO, and CNHs, respectively (Plasticizers)	[49]

(continued)

Table 3 (continued)

Filler	Diazonium	Filler modification conditions	Final reinforced polymer system	Performances	Refs
RGO	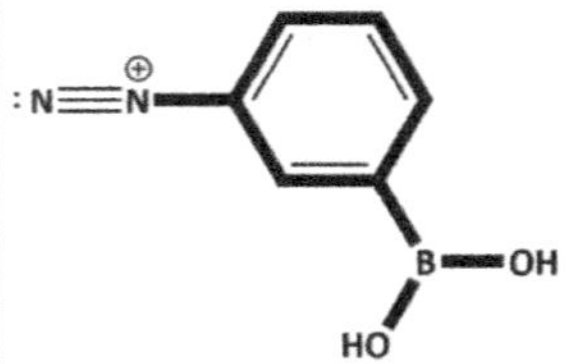	(i) 0.114 mol APBA added to a mixture of 40 mL H_2O and 0.76 mol HCl in ice bath. Then, 0.114 mol $NaNO_2$ solution poured; (ii) 10 g GO dispersed in 20 mL H_2O for 1 h sonication (ii) Mixing both solution with 12 h stirring (iv) 0.38 mol $NaBH_4$ added and stirred for 12 h followed by 12 h aging. washed with ethanol and H_2O, dried for 24 h at 80 °C; (v) PLA nanocomposite through solvent mixing, and compression molding	Azo-boron grafted RGO insertion with *sodium metaborate* poly(lactic acid)composite	Tensile strength >590 MPa and Young's Modulus > 570 MPa	[18]

(continued)

Table 3 (continued)

Filler	Diazonium	Filler modification conditions	Final reinforced polymer system	Performances	Refs
Carbon black	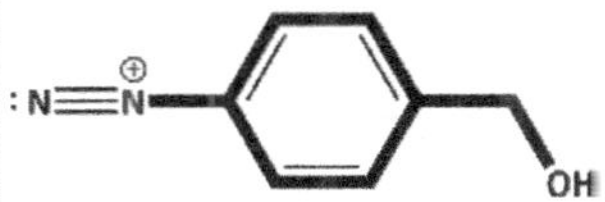	(i) 1 g 4-Aminobenzyl alcohol in 20 mL H_2O, cooled to < 5 °C; (ii) mixed, stirred, 0.8 mL HCl added; (iii)After 30 min,0.64 g $NaNO_2$ in 5 ml H_2O added (iv)after 30 min, 5g Carbon black added, stirred at 25 °C for 2 h, washed, filtered and dried at 110 °C; (v) Novolac with filler(3:1 by weight), placed in mold and annealed	4-hydroxymethylbenzenediazonium modified carbon black reinforced Novolac composite	Increased Flexural strength by 26% and model abrasive tools > 1000%	[21]

of fillers. Reaction via the diazonium group is not systematic as demonstrated by some of us [13]; indeed, -CH_2OH group could also react with the filler (zeolite).

After a decade of frequent investigations of arylation of fillers, clearly, the 2020s years will see systematic research on diazonium modification of reinforcing nanomaterials. One could expect scaling up the preparation of specialty arylated fillers and their marketing as this was done by the company Cabot for diazonium-modified carbon blacks for the making of polymer, rubber, paper, and textile composites [57].

Let's finish by saying that as Edwin Pludemann's silanes were the chief coupling agents in the twentieth century, so aryldiazonium salts will be the coupling agents marking the twenty-first century.

Acknowledgments A.K.B. thanks Wallonie Bruxelles International (WBI) for the provision of post-doc fellowship to conduct research at Université Paris Cité (SOR/2021, N ◦Imputation 101386 and Article Budgétaire 33.01.00.07).

Conflict of Interests Authors declares no conflict of interest.

References

1. Afzal A et al (2019) Polymer-particulate composites with differential interfaces: synthesis, characterization, and mathematical modeling to evaluate interface-yield strength correlations. Colloid Polym Sci 297(4):545–556
2. Rong M, Zhang M, Ruan W (2006) Surface modification of nanoscale fillers for improving properties of polymer nanocomposites: a review. Mater Sci Technol 22(7):787–796
3. Mousa MH, Dong Y, Davies IJ (2016) Recent advances in bionanocomposites: preparation, properties, and applications. Int J Polym Mater Polym Biomater 65(5):225–254
4. Jlassi K, Chehimi MM, Thomas S (2017) Clay-polymer nanocomposites. Elsevier
5. Salmi Z, Benzarti K, Chehimi MM (2013) Diazonium cation-exchanged clay: an efficient, unfrequented route for making clay/polymer nanocomposites. Langmuir 29(44):13323–13328
6. Jlassi K et al (2014) Exfoliated clay/polyaniline nanocomposites through tandem diazonium cation exchange reactions and in situ oxidative polymerization of aniline. RSC Adv 4(110):65213–65222
7. Sandomierski M et al (2016) Reactive diazonium-modified silica fillers for high-performance polymers. Langmuir 32(44):11646–11654
8. Bensghaïer A et al (2018) Dye diazonium-modified multiwalled carbon nanotubes: light harvesters for elastomeric optothermal actuators. Vacuum 155:178–184
9. Zou H, Wu S, Shen J (2008) Polymer/silica nanocomposites: preparation, characterization, properties, and applications. Chem Rev 108(9):3893–3957
10. Rahman IA, Padavettan V (2012) Synthesis of silica nanoparticles by sol-gel: size-dependent properties, surface modification, and applications in silica-polymer nanocomposites—a review. J Nanomater 2012
11. Santoro M et al (2013) High-pressure synthesis of a polyethylene/zeolite nano-composite material. Nat Commun 4(1):1–6
12. Yıldız Yiğit M, Baran ES, Moral ÇK (2021) A polymer–zeolite composite for mixed metal removal from aqueous solution. Water Sci Technol 83(5):1152–1166

13. Sandomierski M et al (2018) Diazonium-modified zeolite fillers. Effect of diazonium substituent position on the filler surface modification and the mechanical properties of phenolic/zeolite composites. Int J Adhes Adhes 85:157–164
14. Jlassi K et al (2016) Clay/polyaniline hybrid through diazonium chemistry: conductive nanofiller with unusual effects on interfacial properties of epoxy nanocomposites. Langmuir 32(14):3514–3524
15. Muralidharan N et al (2018) Carbon nanotube reinforced structural composite supercapacitor. Sci Rep 8(1):1–9
16. Martinez-Rubi Y et al (2012) Tailored SWCNT functionalization optimized for compatibility with epoxy matrices. Nanotechnology 23(28):285701
17. Liu X et al (2018) Research progress of graphene-based rubber nanocomposites. Polym Compos 39(4):1006–1022
18. Tawiah B et al (2019) Highly efficient flame retardant and smoke suppression mechanism of boron modified graphene Oxide/Poly (Lactic acid) nanocomposites. Carbon 150:8–20
19. Islam I et al (2018) Electrical and tensile properties of carbon black reinforced polyvinyl chloride conductive composites. C 4(1):15
20. Spahr ME, Rothon R (2016) Carbon black as a polymer filler. In: Polymers and polymeric composites: a reference series, pp 1–31
21. Sandomierski M et al (2020) Carbon black modified with 4-hydroxymethylbenzenediazonium salt as filler for phenol-formaldehyde resins and abrasive tools. J Appl Polym Sci 137(3):48160
22. Msaadi R et al (2019) Highly selective copper ion imprinted clay/polymer nanocomposites prepared by visible light initiated radical photopolymerization. Polymers 11(2):286
23. Bensghaïer A et al (2017) Efficient covalent modification of multiwalled carbon nanotubes with diazotized dyes in water at room temperature. Langmuir 33(27):6677–6690
24. Salmi-Mani H et al (2016) Diazonium salt-based photoiniferter as a new efficient pathway to clay–polymer nanocomposites. RSC Adv 6(91):88126–88134
25. Menczel JD, Prime RB (2009) Thermal analysis of polymers: fundamentals and applications. Wiley
26. Sepe M (1998) Dynamic mechanical analysis for plastics engineering. William Andrew
27. Martins TD, Ribeiro T, Farinha JPS (2021) Overview of silica-polymer nanostructures for waterborne high-performance coatings. Polymers 13(7):1003
28. Gardziella A, Pilato LA, Knop A (2013) Phenolic resins: chemistry, applications, standardization, safety and ecology. Springer Science & Business Media
29. Strzemiecka B et al (2014) Assessment of the chemical changes during storage of phenol-formaldehyde resins pyrolysis gas chromatography mass spectrometry, inverse gas chromatography and Fourier transform infra red methods. J Chromatogr A 1359:255–261
30. Sandomierski M et al (2019) Improvement of mechanical properties of silica/phenolic composites and abrasive tools by modification of filler using diazonium salt with hydroxymethyl groups. Polym Testing 75:373–379
31. Msaadi R et al (2017) Diazonium-based ion-imprinted polymer/clay nanocomposite for the selective extraction of lead (II) ions in aqueous media. Eur Polymer J 89:367–380
32. Sandomierski M, Okulus Z, Voelkel A (2019) Active diazonium-modified zeolite fillers for methacrylate-based composites. Compos Interfaces 26(7):643–657
33. Jlassi K et al (2018) Anti-corrosive and oil sensitive coatings based on epoxy/polyaniline/magnetite-clay composites through diazonium interfacial chemistry. Sci Rep 8(1):1–13
34. Gan L et al (2019) Wet functionalization of carbon nanotubes and its applications in rubber composites. Carbon-based nanofillers and their rubber nanocomposites. Elsevier, pp 77–108
35. Sandomierski M, Voelkel A (2020) Diazonium modification of inorganic and organic fillers for the design of robust composites: a review. J Inorg Organomet Polym Mater, pp 1–21
36. Bensghaïer A et al (2020) The molecular and macromolecular level of carbon nanotube modification via diazonium chemistry: emphasis on the 2010s years. Chemistry Africa pp 1–35

37. Detriche S et al (2020) Assessment of catalyst selectivity in carbon-nanotube silylesterification. Appl Sci 10(1):109
38. Szabó L et al (2018) Carbon fibre reinforced cellulose-based polymers: intensifying interfacial adhesion between the fibre and the matrix. RSC Adv 8(40):22729–22736
39. Moaseri E et al (2019) Enhancement in mechanical properties of multiwalled carbon nanotube-reinforced epoxy composites: crosslinking of the reinforcement with the matrix via diamines. Polym Eng Sci 59(9):1905–1910
40. Trivedi S et al (2019) Effect of diazotization and magnetic assembly on CNT dispersion observed with hardness and modulus measurement of their epoxy composite of low CNT volume fraction. J Nanopart Res 21(12):1–10
41. Raimondo M et al (2020) Multifunctionality of structural nanohybrids: the crucial role of carbon nanotube covalent and non-covalent functionalization in enabling high thermal, mechanical and self-healing performance. Nanotechnology 31(22):225708
42. Mammeri F et al (2014) Carbon nanotube–poly (methyl methacrylate) hybrid films: Preparation using diazonium salt chemistry and mechanical properties. J Colloid Interface Sci 433:115–122
43. Wu G, Liu L, Huang Y (2019) Grafting of active carbon nanotubes onto carbon fiber using one-pot aryl diazonium reaction for superior interfacial strength in silicone resin composites. Composites Communications 13:103–106
44. Wang Y et al (2015) Preparation and properties of carbon nanotube/carbon fiber hybrid reinforcement by a two-step aryl diazonium reaction. RSC Adv 5(55):44492–44498
45. Liu Y et al (2019) Surface modification of graphene for use as a structural Fortifier in water-borne epoxy coatings. Coatings 9(11):754
46. Vallés C et al (2020) PMMA-grafted graphene nanoplatelets to reinforce the mechanical and thermal properties of PMMA composites. Carbon 157:750–760
47. Young RJ et al (2018) The mechanics of reinforcement of polymers by graphene nanoplatelets. Compos Sci Technol 154:110–116
48. Charoeythornkhajhornchai P, Samthong C, Somwangthanaroj A (2019) Effect of graphene treated with cyclohexyl diamine by diazonium reaction on cure kinetics, mechanical, thermal, and physical properties of natural rubber/graphene nanocomposite foam. Polym Compos 40(S2):E1766–E1776
49. Vicentini N et al (2018) Effect of different functionalized carbon nanostructures as fillers on the physical properties of biocompatible poly (l-lactic acid) composites. Mater Chem Phys 214:265–276
50. Nam K-H et al (2017) Synergistic toughening of polymer nanocomposites by hydrogen-bond assisted three-dimensional network of functionalized graphene oxide and carbon nanotubes. Compos Sci Technol 149:228–234
51. Mohamed AA et al (2015) Functionalization of nanomaterials with aryldiazonium salts. Adv Coll Interface Sci 225:16–36
52. Toupin M, Bélanger D (2007) Thermal stability study of aryl modified carbon black by in situ generated diazonium salt. J Phys Chem C 111(14):5394–5401
53. Pandurangappa M, Ramakrishnappa T, Compton RG (2009) Functionalization of glassy carbon spheres by ball milling of aryl diazonium salts. Carbon 47(9):2186–2193
54. Piquemal J-Y et al Preparation, surface modification and functionalization of anisotropic ferro-magnetic metal nanoparticles: towards multifunctional materials. Appl Surf Chem Nanomater p 205
55. Sandomierski M et al (2019) Mechanically robust and thermally stable abrasive tools from phenolic resins reinforced with diazonium-modified zeolites. Polym Compos 40(8):3209–3219
56. Delamar M et al (1997) Modification of carbon fiber surfaces by electrochemical reduction of aryl diazonium salts: application to carbon epoxy composites. Carbon 35(6):801–807
57. Belmont JA, Amici RM, Galloway CP (2007) Reaction of carbon black with diazonium salts, resultant carbon black products and their uses. Google Patents

Diazonium Salts for the Preparation of Carbon Composites with a Focus on Applications of Carbon Fibers

Melissa K. Stanfield and Luke C. Henderson

Abstract This chapter will provide insight into the use of aryl diazonium salts to functionalize carbon fiber. Examining the production of carbon fiber and its current industry applications. The methods of analysis of functionalized carbon fiber; mechanical tests to evaluate the structural integrity and strength of the fiber will also be covered. Moreover, the chemical analysis that can be used to confirm successful surface grafting has occurred. The literature reports of aryl diazonium salt grafting to carbon fiber over the past decade are explored, examining approaches of functionalization and the increased diversity of applications these chemistries provide.

1 Introduction: With a Focus on Application

The central role of diazonium salts in the modification of carbon fiber is to improve the properties between carbon fiber and polymer matrix they reinforced. This provides better adhesion, and therefore stress transfer, leading to an overall stronger composite. The mechanisms imparting increased adhesion are complex, derived from a combination of physical and chemical interactions, and have never been definitively proven or observed. These chemistries and physical interactions will be further discussed in this chapter.

Carbon-fiber composites are high-strength materials where carbon fibers are embedded in a polymer matrix. These materials have a high strength-to-weight ratio and thus have found application across a vast range of industries though primarily in the aerospace and automotive sectors, replacing traditional materials such as steel and aluminum. When used in transport applications, the lighter composites assist to reduce environmental emissions by increasing fuel efficiency. The demand for these materials has increased significantly over the last decade and an additional compounded annual growth rate of up to 7% is estimated over the coming

M. K. Stanfield · L. C. Henderson (✉)
Deakin University, Geelong, VIC, Australia
e-mail: luke.henderson@deakin.edu.au

M. M. Chehimi et al. (eds.), *Aryl Diazonium Salts and Related Compounds*,
Physical Chemistry in Action, https://doi.org/10.1007/978-3-031-04398-7_21

decade. This is aligned with the increasing desire to increase fuel efficiency for mass transportation and decrease the emission of greenhouse gasses.

Structural materials are the typical application associated with carbon fibers, though their small diameter (~7 μm) and good conductivity have meant that they have found some use as microelectrodes for biological and sensing purposes [1–3] This has been done to monitor various aspects of brain activity as single filaments, or as part of arrays. This is typically due to their combination of high conductivity and flexibility compared to metallic wires. Nevertheless, the primary use of this material is for engineering applications given their lightweight and strong properties and thus, this end-use will be the focus of this chapter.

1.1 Production

The carbon fiber manufacturing process used in industry today has undergone little alteration since the 1960's. The process can be summarized by the complete pyrolysis of an organic fiber precursor, usually primarily composed of polyacrylonitrile (PAN), in addition to several undisclosed co-monomers. PAN-derived carbon fibers are currently the most common polymer used by industry to produce carbon fiber, with a small remainder (<10%) composed of Pitch or Rayon-derived fibers [4].

The PAN precursor fiber, typically produced via wet spinning, is first passed through an oxidation process, which involves heating the fiber (250–300 °C) in an air atmosphere while under tension. This step is intended to stabilize the fibers, making them resistant to thermal degradation thereby preventing melting or burning when exposed to higher temperatures in subsequent steps [5]. Notably, the physical properties of these precursor fibers are decreased after this process, and the exact chemistry and transformations being undertaken within the fiber are still the focus of speculation. The oxidized PAN is then heated in a low-temperature furnace (~600–900 °C) in an inert atmosphere, to initiate the formation of a ladder polymer structure, introducing the development of the desirable physical properties of the final carbon fibers. Further heating at higher temperatures (>1000 °C), entering the high-temperature furnace, that removes a large portion of the non-carbonaceous species (e.g., hydrogen, nitrogen, and oxygen) and sets the crystalline structure of the fibers [5]. At this point, the application of tension, temperature, and duration of exposure to these conditions (and their balance) has installed and set the majority of the physical properties possessed by the final fibers.

The fiber is next passed through an aqueous bath of ammonium bicarbonate and an oxidative potential (up to 20 V) is applied to remove any lubricious carbon species and to install oxygenated functional groups. The electrochemical nature of this step allows the opportunity to apply aryldiazonium chemistry for surface modification and would be introduced into an existing manufacturing process with minimal disruption. This would be facilitated by the reversal of potential (i.e., a reductive potential applied to the fiber rather an oxidative), an operationally simple adjustment to make, and this

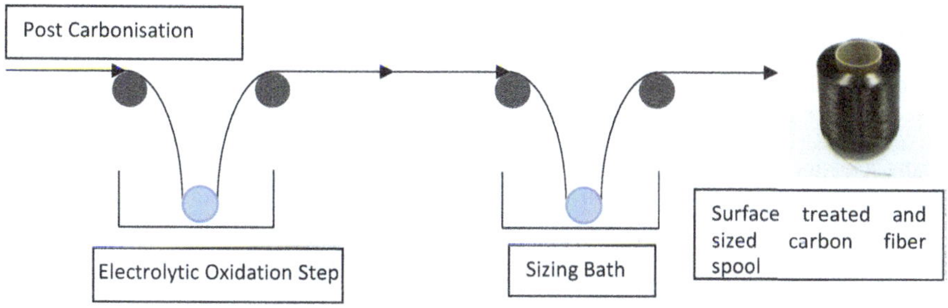

Fig. 1 Schematic of typical carbon fiber surface treatment process

process would be much milder with respect to the applied potentials, considering that aryl diazonium salts are typically reduced around −0.3 V (vs. Ag/AgCl).

The final step in the manufacturing process is referred to as "sizing" (Fig. 1). Here, the fibers are passed through an aqueous emulsion and the contents of the emulsion are deposited onto the fiber surface. The contents of these emulsions are industry secrets, guarded with extreme caution. The role of the sizing agent is to enable downstream processes in which the fiber is involved, such as weaving and braiding, and to protect the fiber from damage.

It is important to note that the interface and interphase of any composite is of critical importance to the final physical properties of the material. This is because the stress transfer from the soft matrix to the strong fiber is facilitated across this juncture, and thus, an interface that consists of complementary chemical interactions will facilitate the translation of stress more efficiently. The ability to modify the carbon-fiber surface, with a known chemistry, and hence the tailor the interface of the final composite material is an incredibly desirable concept. This is due, largely, to the secrecy surrounding the carbon-fiber industry, the processing conditions of manufacture, surface treatments, and sizing constituents are not typically divulged to the end-user and thus, replication of data is notoriously difficult. Therefore, the ability to influence the carbon-fiber surface with a known chemistry removes a significant number of unknowns from the manufacturing process and facilitates the design of composite interfaces, rather than optimization by empirical evaluation.

1.2 *Introduce Interphase/Interface Region*

While CFRP materials are synonymous with high performance, common failure mechanisms such as micro-cracking and delamination typically arise from the weak interface between the fibers and the polymer resin. The interphase is understood to be the principal limitation in the application of the composites, its improvement is vital for manufacture of more reliable composites with enhanced applicability to a wider range of industries. Indeed, a notorious limitation of CFRPs is their poor toughness

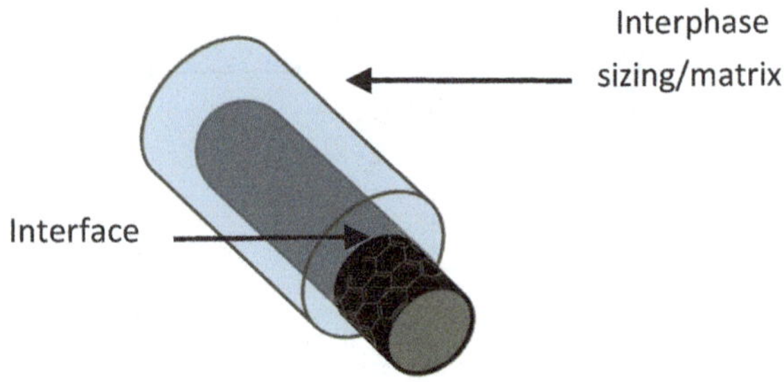

Fig. 2 Schematic of the fiber-matrix interface and interphase regions being where the fiber and matrix physically meet and the localized region around the fiber, respectively

and performance after impact. This is primarily due to any impact causing micro-cracking at the fiber-matrix interface and the onset of catastrophic failure (Fig. 2) [6].

For carbon fibers derived from PAN, the fiber surface largely consists of turbostratic carbon and graphite crystals. When considering the matrices in which these fibers are used to reinforce, such as epoxy, polyamide, polypropylene, poly(ethyleneimine), and others, the native fiber surface offers little molecular complementarity to any of these resins. Methods of improving the fiber-to-matrix adhesion, involve surface treatment during manufacture, as described above, and surface modification of the fiber. These treatments allow the typically inert material to be able to interact with the polymer matrix, through chemical bonding and mechanical interlocking.

Surface treatments involve electrochemical oxidation and the application if sizing—confidential cocktails of proprietary chemicals. Surface modifications include wet chemical procedures, chemical oxidation with acids or chemical oxidants, and gas-phase oxidation. A large portion of the literature focuses on the "activation" of carbon fibers via treatment with concentrated nitric acid, followed by chemical derivatization, usually taking the form of carbodiimide coupling reactions or treatment with alkyl silanes, the latter of which is similar to glass fiber surface treatments. Other common methods of fiber pre-treatment include exposure to ozone and plasma, and electron-beam and gamma radiation have been examined. Of particular interest in this chapter is the electrochemical modification of carbon fibers, which has been used to graft small molecules or polymers to the fiber surface.

Over the years, aryldiazonium salts have been sporadically utilized to modify the carbon-fiber surface, introduce molecules or assist in introducing polymers to the surface. There are numerous methods of achieving their grafting to carbon fiber, these approaches, and their outcomes will be discussed below.

2 Analysis of Functionalised Carbon Fiber

2.1 IFSS and Micro Bond Test

When carrying out the surface modification of carbon fibers it is critical to ensure that the modification process has not impacted their strength-to-weight ratio. Therefore, the determination of physical properties, before and after modification, is commonly carried out. This includes the characterization of tensile strength and tensile modulus of the single carbon-fiber filaments, and it is important to note that a reasonable number, typically more than 50, single filaments should be characterized at each stage of modification to ensure that no damage is or has occurred.

Further to this, evaluation of how well the fibers bond to their respective matrix must also be examined. This can be carried out a multitude of ways and common techniques include the single-fiber pullout test (SFPO), micro-droplet debonding, single fiber push-out, and single fiber fragmentation test (SFFT) (Fig. 3). Each of these techniques has corresponding advantages and disadvantages, though in the authors' experience the SFFT has proven the most reliable.

The SFFT the Kelly-Tyson model, was originally used in metallic systems and later adapted to accommodate ductile resins, Feih et al. [7] to determine the interfacial shear strength (IFSS) a measure of adhesive performance between fiber and matrix. The use of a single filament to determine the interfacial performance in a composite may seem counter-intuitive as the single filament represents a very small portion of the total fibers in a composite. This is typically done as it requires very small amounts of carbon fiber, and resin, and can be carried out with relative ease. The correlation of such micro-scale tests to mesoscale (fiber bundle), and macro-scale tests involving full composite parts has been investigated and reported [8].

Herein, we will provide a brief overview of how to determine IFSS using the SFFT. This process involves layering a single filament of carbon fiber horizontally in a "dog bone" mold, prior to pouring resin in with the fiber and allowing it to cure. To analyze the interfacial adhesion properties an elongation force is applied parallel to the fiber, breaking it into fragments. Small and frequent fragments are typically indicative of a stronger interface. The sample is analyzed under an optical microscope, counting and measuring fiber fragments to determine the average fiber length, l. The average fiber length is then used to calculate the critical fiber length

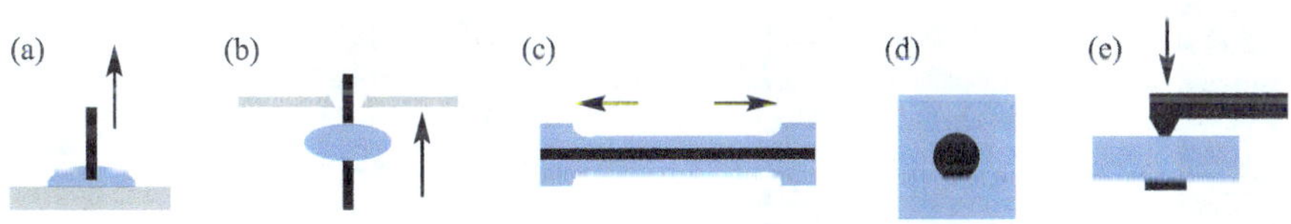

Fig. 3 Commonly used techniques for determining IFSS, arrows show the direction of force applied to the sample to cause interface failure; **a** single fiber pull out; **b** micro droplet debonding; **c** single fiber fragmentation; **d** and **e** single fiber push-out, view from the top and profile, respectively

(l_c) denoted in Eq. 1.

$$l_c = \frac{4}{3}l \tag{1}$$

Finally, using the average fiber diameter d (determined to be 7 μm), the apparent interfacial shear strength τ can be calculated using the following equation:

$$\tau = \frac{\sigma_f(l_c)d}{2l_c} \tag{2}$$

Combined these methods are used to determine mechanical properties of the fibers, however, do not give any information regarding surface chemical composition; and determining fiber functionalization.

Interlaminar shear strength (ILSS) is a control parameter used for laminated advanced composites. It indicates the maximum shear stress existing between the layers of a laminated material. A key distinction between IFSS and ILSS is the scale of the material. IFSS is a single filament in a resin whereas ILSS assists to determine the adhesion between plies in a large-scale structure. Short beam shear (SBS) testing is one method used to determine ILSS of a composite material.

2.2 *Chemical Analysis of Carbon Fiber*

Carbon fibers represent a significant challenge when attempting to characterize them for their surface chemistry, due to their small diameter (~7 μm), conductivity, and their color, resulting in the absorbance of radiation such as IR, etc.

Nevertheless, routine surface characterization techniques include X-ray photoelectron spectroscopy (XPS), Scanning Electron Microscopy (SEM) (Fig. 4), Atomic Force Microscopy (AFM), and Raman spectroscopy. SEM images reveal (Fig. 4), that pristine carbon fiber has a highly striated, uneven surface. The morphology of the surface increases the complexity involved in functionalization and homogeneity.

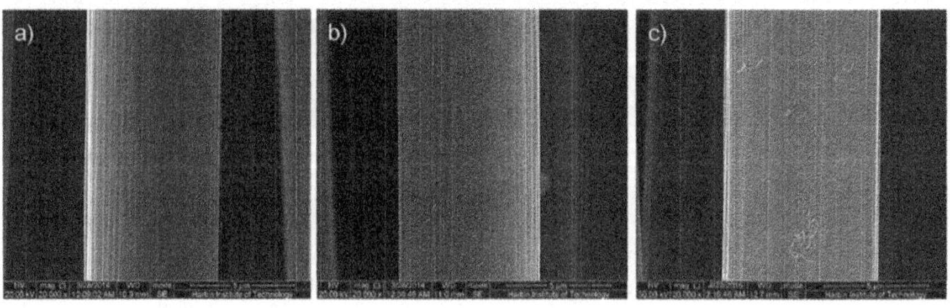

Fig. 4 SEM carbon fiber, Reused with permission from reference [9]. Copyright (2015) Elsevier [9]

Typically, chemical characterization methods involve infrared spectroscopy (IR), and X-ray photoelectron spectroscopy (XPS), these methods are incompatible or difficult to obtain reliable data to relate to surface chemistries on carbon fiber due to their innate inhomogeneity and curved surface. Thus, scanning electron microscopy (SEM) and atomic force microscopy (AFM) are two technologies that involve imaging the carbon-fiber surface, they are useful to determine whether fiber treatment has caused significant damage to the surface. Also, valuable information can be obtained by providing images of polymeric coating and thickness.

3 Diazonium Salts—A Method to Functionalize Carbon Fiber

The mechanisms which lead to increased interfacial adhesion in composites systems are complex and are derived from both physical and chemical interactions. The decoupling of these two influences, and which one is primarily responsible for beneficial macroscopic properties, is exceptionally difficult and still has not been convincingly shown to be possible. The likely scenario is that a balance between micromechanical interlocking between the fiber and resin, based on roughness and surface area, and chemical complementarity and the formation of covalent linkages between the composites phases is required for optimal final properties.

The key objective of modifying carbon fiber via diazonium salt chemistries was initially to improve the adhesion between carbon fiber and its reinforced polymer matrix. The modification of aryl groups to carbon fiber via the reduction of a diazonium salt provides access to chemistries that are able to form covalent linkages between the composite phases. Thus, improving the adhesion of carbon-fiber reinforced composites, although the introduction of these molecules can also increase the roughness and surface area of the fiber, which further augments the interfacial shear strength. Consequently, by providing access to modified carbon fiber, improvements in the adhesion can occur through synergistic improved micromechanical interlocking and chemical bonding. Diazonium salts were first introduced to the surface of carbon fiber in 1997 [10], where 4-nitrophenyldiazonium was introduced to the surface followed by reduction of the nitro group to the corresponding amine functionality (Fig. 5). The amine functionality can participate in the curing mechanism of epoxy resins.

That work opened up a new pathway way for the functionalization of carbon fiber, innovating the surface of the carbon fiber by tailoring its chemistry for best compatibility with supporting resin matrix.

Since this elementary work, many other researchers have adapted the diazonium salt functionalized to carbon fiber. Predominately, the purpose of the functionalization is to further improve interactions at the interface between carbon fiber and its polymer matrix. However, other methods intended for the modified fiber to be used in detected metal ion applications. The various functionalization strategies reported

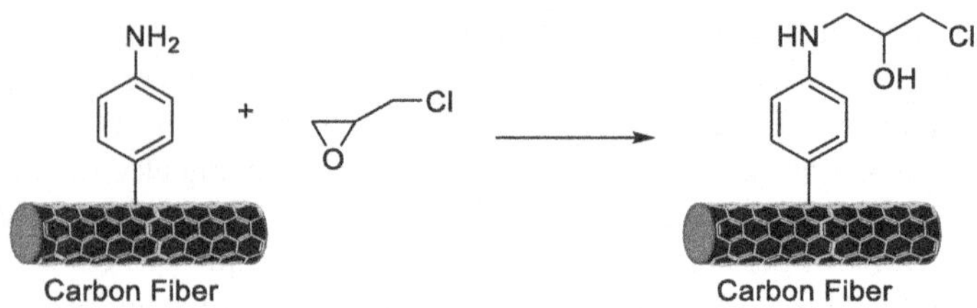

Fig. 5 Aryl amine curing with epoxy resin, adapted from Delamar et al. Adapted with permission from reference [10], Copyright (1997) Elsevier 10

demonstrating the great breadth of diazonium chemistry, using both chemical reduction and electrochemical reduction, introducing both molecules and polymers to the surface of the fiber. Some published methods have utilized the diazonium chemistry to introduce boutique structures to the surface of the fiber.

3.1 Chemically Grafted Approach

While the electrochemical reduction of aryl diazonium salts to conductive surfaces is an incredibly rapid and facile modification strategy, the relative weakness of the aryl-N_2 bond means that stimuli such as heat or light can be used to facilitate reaction with the intended surface. Indeed, even spontaneous grafting of aryl diazonium salts has been observed when modifying both metals and carbonaceous materials. Several publications have explored the chemical grafting of various diazonium salt derivatives to the surface of carbon fiber.

The diazonium species can be generated and functionalized in situ. This involves submerging carbon fiber in a solution of the reactants (corresponding amine, *tert*-butyl nitrite in MeCN) and refluxing the reaction vessel for 24 h, under nitrogen atmosphere. This chemistry can be extended to design and synthesize complex compounds, possessing a pendant amine moiety, to react with an epoxide-based resin upon cure. Servinis et al. [11] achieved functionalized fibers, using this technique. Confirming the functionalization using XPS, via the attachment of a trifluoromethyl groups, as the fluorine unit stands out against the carbonaceous fiber background. The functionalized fibers demonstrated a large increase in coefficient of friction, against stainless steel, and the SFFT indicated an increase in IFFS, attributed to the pendant amine functionality reacting with the epoxy resin (Fig. 6).

In following up on that work another report from the same group expands on the previously established method (above) via a series of reagent concentration studies. This process used a range of concentrations to determine if this had any major ramifications on the interfacial shear strength. The authors found the lowest concentration treatment provided the largest increase in IFSS [12].

Li et al. [13] reported their work on developing an approach to introduce multiple functional groups (amino, hydroxyl, and sulfhydryl) to the surface of CFs using aryl

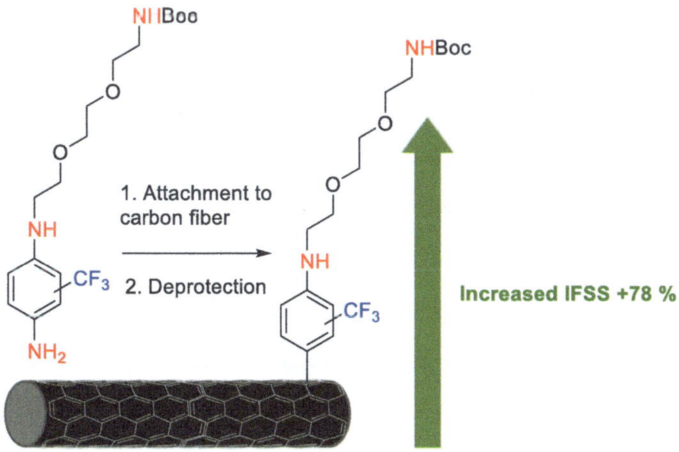

Fig. 6 Functionalisation of carbon fiber and improvement in IFSS. Adapted with permission from reference [11]. Copyright (2015) Royal Society of Chemistry [11]

diazonium salts. The grafting procedure involves stirring in an aqueous solution (with reagents) while undergoing heating at 80 °C for 9 h (Fig. 7). Functionalisation was examined using FT-IR, Raman, XPS, SEM, and SFFT. Results showed the functionalization increased fibers' surface roughness. Composites were prepared using the modified carbon fibers and thermoplastic copoly (phthalazinone ether sulfone)s (PPBES), these were evaluated by ILSS and flexural strength. All functionalized composites achieved increases in ILSS, the composite bearing amine functionality demonstrated the highest increase of 18.6%. The authors attribute the improvement in mechanical properties to the hydrogen bonding formation between the amine and carbonyl and ether groups in the PPBES. Dynamic mechanical analysis (DMA) and hydrothermal aging tests revealed that the CF-NH_2/PPBES confers a resistance that has high-temperature mechanical properties and hydrothermal aging. Notably, the functionalization process did not lead to any discernible decrease in the fiber tensile strength.

Fig. 7 Functionalisation method used by Li et al. Adapted with permission from reference [13]. Copyright (2016) Royal Society of Chemistry [13]

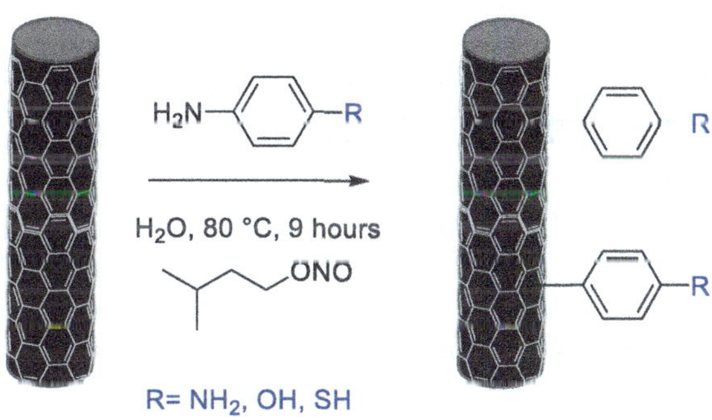

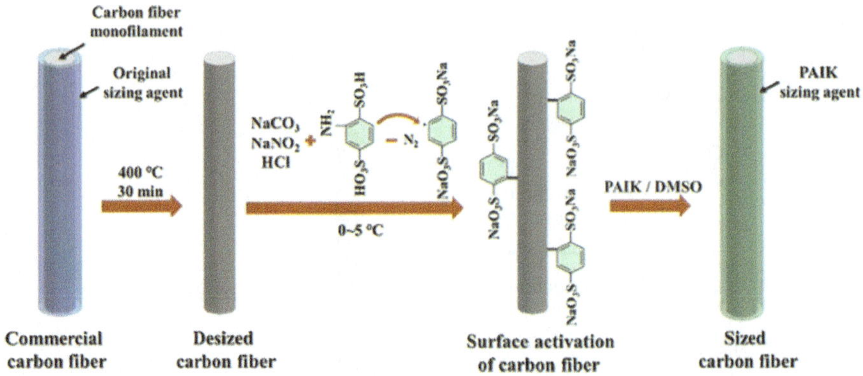

Fig. 8 Schematic of functionalization process published by Wang et al. Adapted with permission from reference [14]. Copyright (2020) Elsevier [14]

The latest reports of chemically grafted treatments by Wang et al. [14] 2020, aryl diazonium treatment was applied to active CF by grafting sodium sulfonate groups to the surface via an aryl diazonium treatment. This modification was confirmed via SEM, AFM, and XPS analysis. Poly(aryl indole ketone) PAIK is a novel heterocyclic derivative of PEEK, was utilized as the sizing agent for CF-PEEK composites and was synthesized via a one-pot random copolymerization. The authors state that a strong cation-π interaction was introduced between CFs and the interfacial layer, resulting in the ILSS of modified CF/PEEK composites increasing significantly (62.7%, from 46.86 MPa to 76.37 MPa). The improvement of mechanical properties was ascribed to the installed cation-π interaction between activated CFs and PAIK, in addition to the enhanced surface roughness of the modified carbon fibers and the excellent compatibility of PAIK and PEEK matrix (Fig. 8) [14].

Reports in the literature of chemical diazonium salt grafting often involve high temperatures and long grafting times. This possess difficulty if the methods we to be used on an industrial scale. To include this process in pre-existing carbon-fiber production lines, these processes could not be performed on the pre-existing continuous line, therefore new infrastructure would need to be manufactured to be able to achieve the desired reaction conditions.

3.2 *ElectroChemically Grafted Approach*

The seminal paper in this area by Delamar et al. [10] took advantage of electrochemistry, allowing for rapid, aqueous grafting to the carbon-fiber surface. Since 1997, other researchers (20+ years later) have also published work investigating the advantageous electrochemical treatment procedures. This work will be further explored below.

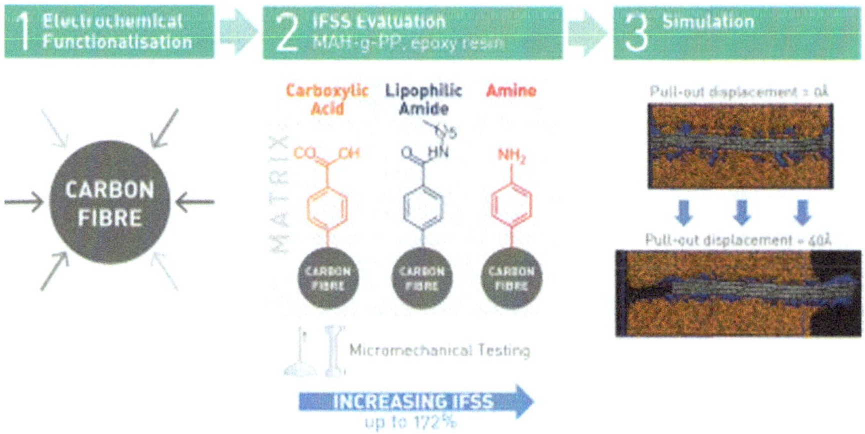

Fig. 9 Introduction of carboxylic acid, lipophilic amide, and amine to the fiber surface. Adapted with permission from reference [15]. Copyright (2015) Elsevier [15]

It is surprising, given the application of electrochemical functionalization of glassy carbon, gold, silicon, and other conductive materials, that the use of electrochemically facilitated aryl diazonium chemistry as a surface modification approach for carbon fibers has not been more thoroughly adopted in the field.

The functionalization has evolved from the initial grafting using the 4-nitro substitution on the diazonium aryl ring and reduction to the corresponding amine moiety. In this same study pendant carboxylic acid and lipophilic amide have since been used in the grafting of diazonium salts to carbon fiber [15]. The choice of aryldiazonium salts used in that study was to examine the effect of a nucleophilic nitrogenous species (amine), a non-nucleophilic nitrogen (amide), and the presence of a carboxylic acid to model that of a freshly oxidized carbon surface (Fig. 9). Interestingly, comparison across the three substitutions reveals amine substitution yields the highest increase in IFSS.

The functionalization saw no detrimental effect on the Young's modulus and tensile strength, achieving IFSS increases of 172% and 30% were achieved relative to control fiber for the amine and amide, respectively. The mechanism by which the unreactive lipophilic amide showed enhanced IFSS proceeded via penetration of the hexane group into the polymeric phase, this was further supported by molecular dynamic simulations. The IFSS was examined in polypropylene grafted with maleic anhydride, revealing only the amine-grafted fibers showed a 67% increase relative to control, attributed to covalent cross-linking between the fiber and maleic anhydride co-monomer (Fig. 10).

Reports in the literature have also explored the grafting conditions, in an attempt to maximize surface attachment at minimum concentration, using diazobenzene [16]. One approach involved holding fibers at a high reductive potential (−1 V vs. Ag/AgCl) for 10, 20, and 30 min. The authors completed this process in two ways (i) reuse the same solution of diazobenzene (ii) use a freshly prepared solution for each

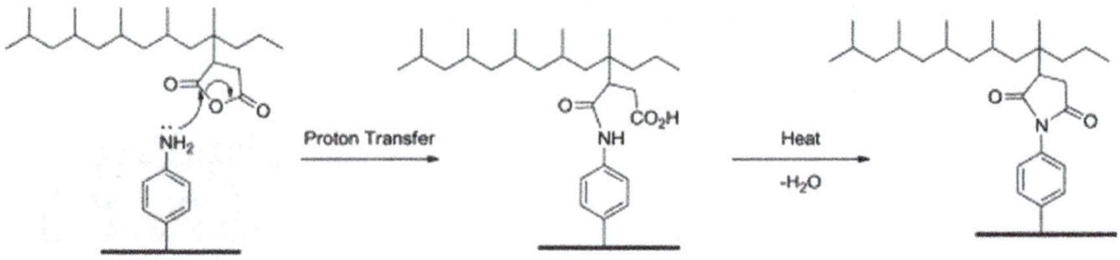

Fig. 10 Scheme of amine reacting with the maleic anhydride. Adapted with permission from reference [15]. Copyright (2015) Elsevier [15]

sample. The authors intended to examine whether there was a substantial difference between each method to understand if this process could be used in a continuous fiber production line, where the same stock of solution would be used for several hours. That work also highlights the functionalization and introduces a passivating layer to the surface of the usually conductive fiber. The authors report a reduced conductivity of the fibers (250%) for a 30 min grafting treatment and IFSS was increased by 19%. This is an interesting outcome as the conductive nature of carbon fibers can result in significant problems for electrical equipment as errant fibers can electrically short circuit electronics, thus the reduction of conductivity may assist in minimizing this effect.

Introducing hydrophobicity and synergistically increasing the fiber-to-matrix adhesion (59–216%) was demonstrated by Arnold et al. [17]. Covalently modifying the surface of carbon fibers through electrochemical reduction of nitro aryldiazonium salts to generate perfluorinated alkyl radicals (Fig. 11). The surface modification was confirmed through a combination of analysis techniques (XPS, ATR-IR, and TEM). This treatment resulted in achieving water contact angles (WCA) up to 135.5 ±

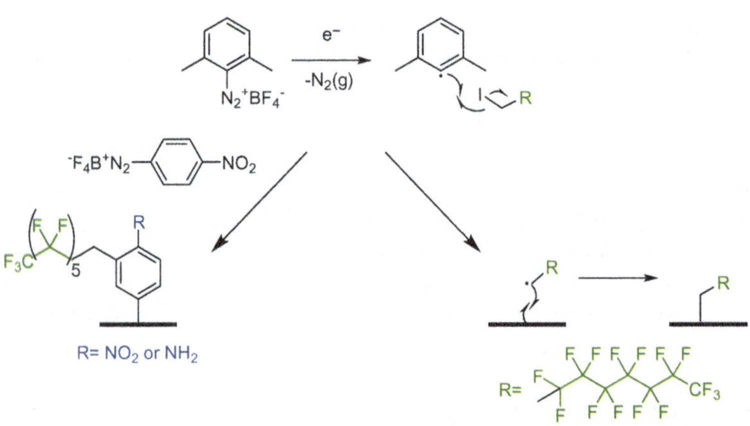

Fig. 11 Demonstrating the pathways of modification used in that research. Adapted with permission from reference [17] Arnold et al. Copyright (2019) Royal Society of Chemistry [17]

Fig. 12 Image showing the hydrophobic nature of the carbon fibers, labels refer (left to right) to: 9 M NaOH (with phenolphthalein indicator), 12 M HCl, neutral (tap) water. Adapted with permission from reference [17]. Copyright (2019) Royal Society of Chemistry [17]

0.2°, demonstrating the fibers were extremely hydrophobic (Fig. 12). The chemistry involved in that work employed diazonium salts through two mechanisms, the first involved using a 4-nitrobenzene diazonium salt to graft the aryl ring to the surface of the fiber, providing a layer of reactive groups for subsequent modification. The second role of the aryldiazonium salts involved using 2,6-dimethylbenzene diazonium tetrafluoroborate, as a sacrificial diazonium salt. The sterically hindered diazonium salt is able to be reduced but is too sterically encumbered to graft-to any surface. Nevertheless, it is able to abstract an iodine from corresponding alkyl halides and generate the corresponding alkyl radicals for electrode modification.

3.3 Combining Grafting Methods

There is a scarcity of literature in this area that involves employing a combination of grafting methods. For example, Szabo et al. [18] involved the depositing a range of molecular entities on the surface of CF, through the initial grafting of an aromatic structure with appropriate functions via diazonium species. In this work, the diazonium species is used as an anchor to the fiber and cellulose derivatives. Increase in IFSS was obtained for the cellulose propionate-grafted carbon-fiber composite.

In 2018 Eyckens et al. [19] examined the impregnation CFs with aryl diazonium salts, subjecting them to external stimuli (such as heating to 100 °C or UV light) to induce grafting resulting in surface modification which can deliver improvements of up to 150% in IFSS in epoxy resins. Aryl diazonium salts with the trifluoromethyl moiety were used to facilitate detection via XPS. Interrogating the fiber-to-matrix interface using molecular dynamics simulations suggests the surface grafted molecules impart a "dragging effect" through the polymer phase and the surface concentration of these compounds is critical to enhancing IFSS.

3.4 Examining the Use of Diazonium Salt Functionalisation

Predominantly the literature reports diazonium salts and modification of carbon fiber to investigate the diazonium salts used as a priming layer or an anchor for which other compounds can be attached. A critical first step that allows access to many multi-step functionalization processes, allowing for the attachment of further boutique molecules and polymers. Due to the diazonium salts grafting readily to surfaces, often using mild conditions aqueous conditions, their reliability enables the use as a first step in a multi-step functionalization process (Fig. 13). This functionalization has been adapted to multiple applications, from controlling the interface and improving adhesion to using as a metal ion detector. The published processes will be discussed below.

Diazonium salts incorporation to allow for access to a highly selective molecular architectural design. Eyckens et al. [23] employed diazonium chemistry as the first step to tether to introduce phenylacetylene to the surface. Copper azide-alkyne click chemistry was next used to tether a polyethyleneoxide-derived non-ionic surfactant unit to the fiber surface (Fig. 14). Molecular dynamics simulations revealed that lateral association of hydrophobic section leads to a "hooked" conformation through which polymer chains become entangled. The predictions from molecular dynamics simulations were supported by the synthesis and grafting of such a molecular architecture to a carbon fiber surface, giving almost identical improvement in interfacial adhesion (+276%).

Several reports in the literature have used diazonium salts as a preliminary step, allowing further access to click chemistry reactions. The copper alkyne-azide

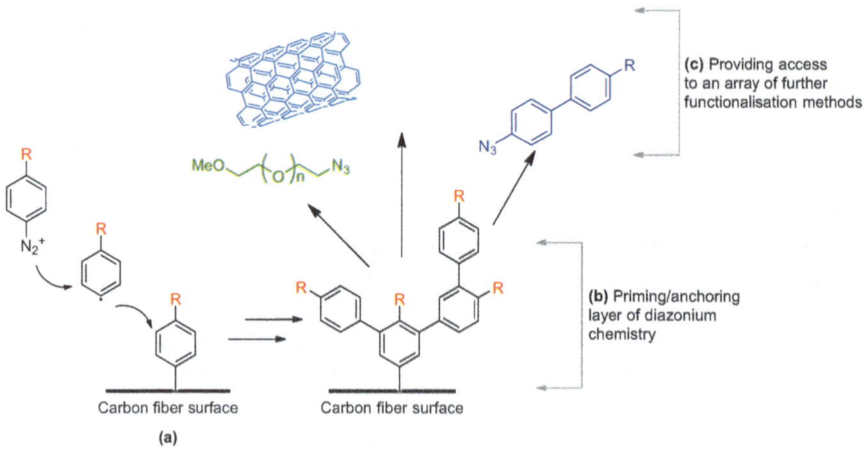

Fig. 13 Some examples of using diazonium salts as anchor/ priming layer on CF **a** First step: graft diazonium salt to fiber surface, **b** Generation of priming/anchoring layer **c** Last step: multi-purpose, use as an anchor for boutique chemistries and polymer layer. Adapted with permission from references [20–22]. Copyright (2015) Royal Society of Chemistry [20–22]

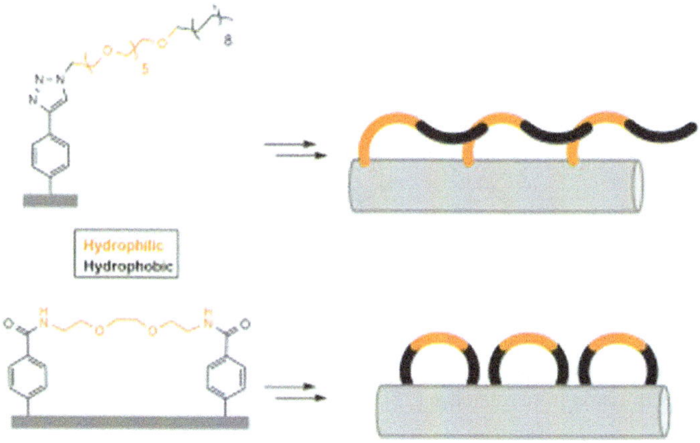

Fig. 14 Molecular architecture of a carbon-fiber surface. Reproduced with permission from reference [23]. Copyright (2020) Elsevier [23]

cycloaddition (CuAAC) and Sulfur-Fluoride Exchange (SuFEx) are two click chemistry reactions that have been reported to be used to assist in the functionalization of carbon fiber (Fig. 15). These reports will be further discussed below.

Yang et al. [21] provide a strategy to covalently modify carbon-fiber microelectrodes (CFMs), as a platform for creating selective adsorption sites for trace metal detection. A monolayer of acetylene-terminated scaffolds on CFMs through the electrochemical reduction of alkynyl aryl diazonium salt bearing sterically differentiated silyl groups. The deprotection of the silyl scaffolds was accomplished in only 5 min. Further functionalization is required to achieve azidomethyl ferrocene grafting. This

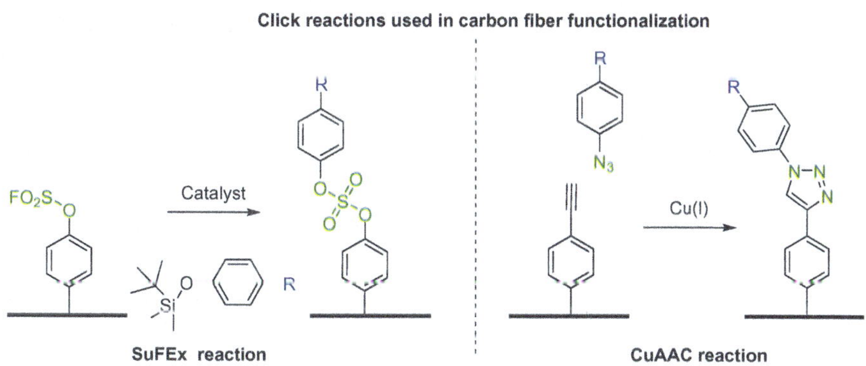

Fig. 15 Examples of click chemistries used in carbon-fiber functionalization. Left: SuFEx ligation adapted with permission from reference [X]. Copyright (2018) Wiley and Sons. Right: Using CuAAC as a ligation strategy on carbon fibers. Adapted with permission from reference [21]. Copyright (2015) Royal Society of Chemistry

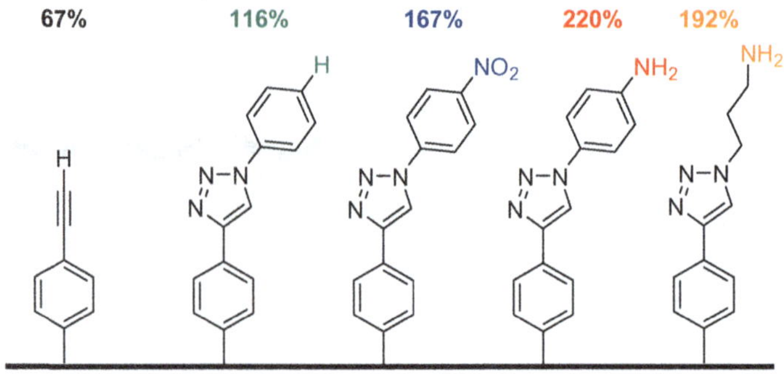

Fig. 16 Functionalisation introduced to the surface with corresponding IFSS improvements. Adapted with permission from reference [22]. Copyright (2016) Royal Society of Chemistry [22]

technology aims to allow for real-time-ultra-selective analysis of metals in complex ecological and biological systems [21].

In a similar strategy, but on a larger scale, Servinis et al. [22] developed an effective strategy to control the surface chemistry of carbon fibers, using a copper-mediated azide-alkyne cycloaddition approach (CuAAC or "Click Chemistry"). Electrochemical reduction of phenylacetylene-derived aryl diazonium salts onto the surface of carbon fiber was followed by click chemistry is used to tether desired surface characterization of choice. This approach demonstrates a small-molecular interface between carbon fiber and the matrix. Improving interfacial shear strength by 192 and 220%, for reactive amine (alkyl or aryl, respectively) (Fig. 16). The techniques do not affect the tensile strength or young's modulus of the fiber.

Research to investigate controlling the effect of molecular weight polymer, and thus the influence on interphase properties, was reported by Randall et al. [24] This paper explores how the carbon-polymer interface is affected using controlled molecular weight polymers (1, 2, 5 and 10 kDa) (Fig. 17). The focus of the work was to graft polyethylene oxide (PEO) polymers of sequentially increasing length and

Fig. 17 Investigating controlling the effect of molecular weight polymers on IFSS. Adapted with permission from reference [24]. Copyright (2019) Elsevier [24]

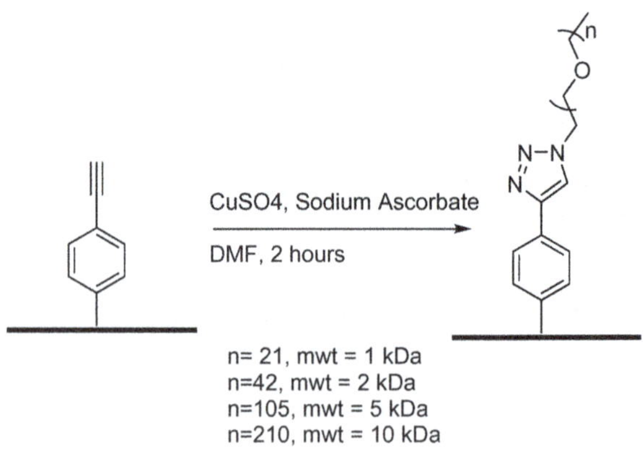

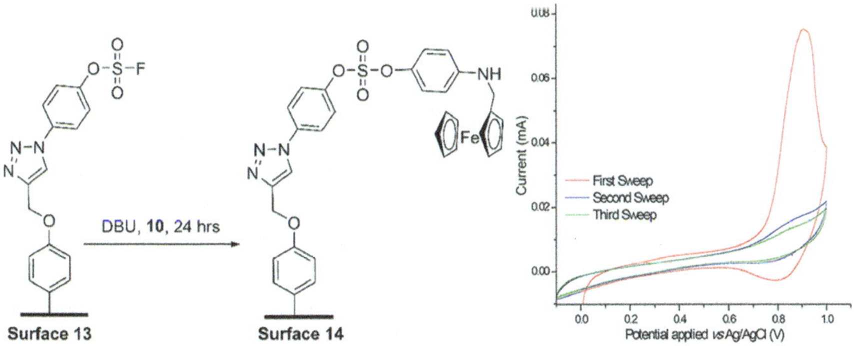

Fig. 18 SuFEx reaction used to install an electrochemically active molecule. Adapted with permission from reference [25]. Copyright (2018) John Wiley and Sons [25]

determine the effect on IFSS (large increases from 84 to 160%). Phenylacetylene groups were installed on the surface using diazonium electrochemistry, prior to the preparation of PEO polymers.

In this work, [25] surface modification was achieved using the sulfur-fluoride exchange (SuFEx) reaction [26, 27]. The sulfur (VI) fluoride moiety was directly installed to the surface via electrochemical deposition of fluorosulfate phenyl diazonium tetrafluoroborate salt. SuFEx-able surfaces can easily undergo exchange with aryl silyl ethers and that the subsequent sulfate linkages are themselves stable under electrochemical redox conditions (Fig. 18). Using this chemistry to install a pendant amino group to the fiber surface results in interfacial shear strength improvements of up to 130% in epoxy resin. The work demonstrates an ability to control and achieve chemoselective modification to the carbon fiber surface, while increasing interfacial adhesion.

Diazonium salts have been employed as a means to access the covalent grafting of carbon nanotubes onto carbon fiber. The process reported involves a two-step diazonium salt reduction, first the electrochemical grafting and second an electrochemical reduction to achieve the amine functionality [28]. The CNT layers were regularly anchored onto the carbon fiber surfaces ny using β-cyclodextrin molecular tubes for encapsulating the diazonium salts, a good approach to prevent immediate reduction of the nitro group. The final step involved immersing the amino-functionalized fiber in HCl solution to immobilize CNTs onto CFs. Carbon nanotube/carbon fiber hybrids exhibit a significant improvement in interfacial properties. 114% improvement and tensile strength were retained. Another approach reported to introduce CNTs to the carbon fiber surface involves using isoamyl nitrite and using the same *pseudo*-sandwiching strategy [20]. The aim of this research was to create a CNT/carbon fiber hybrid, achieved through a two-step aryl diazonium reaction (below). The addition of CNT to the interphase improved the roughness of the surface and improved adhesion. The fibers were characterized using FT-IR, XPS, SEM, TEM, and dynamic contact angle (DCA), the IFSS was increased by 104% compared to untreated fibers [20].

The authors attempt to introduce cellulose-based carbon fiber reinforced composites as a promising candidate to replace current resin systems [29]. Carbon fiber surfaces were modified by using diazonium chemistry to graft onto the surface, then click chemistry was employed to bind cellulose derivatives to the carbon fiber. Research aims to obtain a high-performing CF reinforced cellulose composite, tailoring the interphase. The functionalization involved grafting phenylacetylene to CF surface (using in situ diazotization). To which the cellulose derivative and aromatic azides can be covalently attached. SFFT demonstrated that lipophilicity and size of substituent play a role in determining molecular entanglement and mechanical interlocking effects, as penetration into the cellulose propionate matrix. Enhancement in IFSS was obtained for the functionalized fiber. Na_3PO_4.

3.5 One-Pot Functionalisation of Polymers to Carbon Fiber

The surface eletrco-initiated emulsion polymerization (SEEP) procedure is essentially a one-pot procedure for functionalizing polymers to a conductive surface [30]. This procedure has since been adapted to modify carbon fiber. Many methods that have been discussed above, require a complex mutli-step procedure to achieve the successful functionalization of polymers to the surface of carbon fiber. However, the SEEP procedure is an easily accessible method that ensures the radical grafting of polymers to the surface.

Eyckens et al. [31] present carbon fibers incorporating structural color, by modifying the fiber surface through in situ polymerization grafting. The striking blue color undergoes reversible color change through the visible spectrum upon exposure to a solvent and its subsequent evaporation (Fig. 19). The covalently bound polymer increases the tensile strength of the fiber while also improving its IFSS (311% improvement).

Stanfield et al. [32] employed diazonium chemistry to introduce multilayer interphases to carbon fiber composites. A first layer anthraquinone diazonium salt was grafted to the surface of the fiber and its redox response was observed electrochemically to confirm successful grafting. The conductivity of this diazonium layer was employed to graft a second layer (introducing a multilayer interface) (Fig. 20). The second layer was grafting using the SEEP process, intended to improve the adhesion between fiber and matrix, while also preserving the electrochemical redox of the anthraquinone priming layer.

4 Conclusion

In conclusion, diazonium chemistry has been transferred to carbon fiber, providing access to effective and irreversible functionalization of the surface. The introduction of diazonium salts has improved the properties of carbon fiber-reinforced composites;

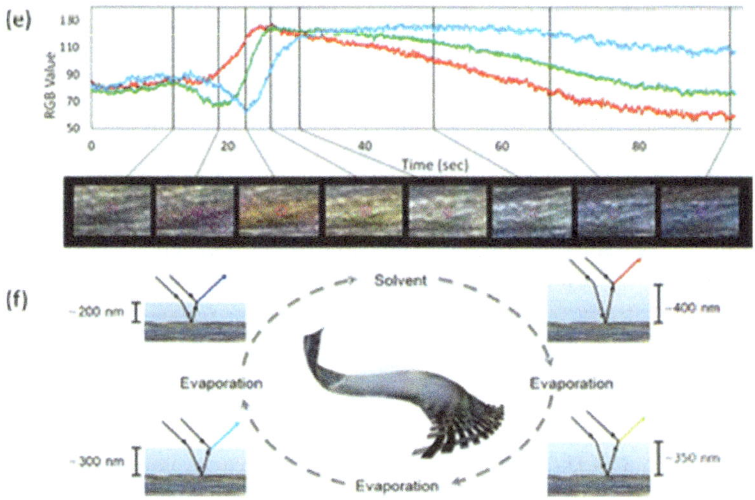

Fig. 19 (**e**) RGB plot with video frames at selected time points showing the wide range of fiber colors. (**f**) Schematic of the proposed process by which multiple colors are being generated. The dry fiber possesses a film of thickness between 220 and 240 nm, giving a blue color. Addition of a solvent causes the polymer to swell; subsequent evaporation, and thus progressive shrinkage of the polymer layer, results in refracted vs reflected path length changes, thereby smoothly transitioning through a range of colors until the original film thickness is once again present [31]. Reproduced with permission from reference [31]. Copyright (2019) American Chemical Society

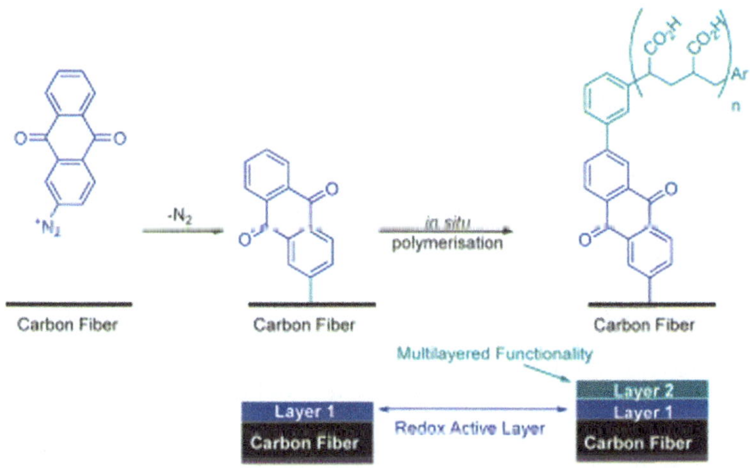

Fig. 20 A schematic representation of the investigation undertaken within this work. The use of a redox-active layer presents on carbon fibers as a mediator to develop multilayered functionality in a composite [32]. Reproduced with Permission from reference [32]. Copyright (2020) Elsevier

leading to increased interfacial shear strength. Moreover, this diverse modification approach has been used effectively to introduce multifunctionality to carbon fiber with the use of hydrophobic and redox-active molecules, increasing the diversity of sectors they could be applied, to in energy and marine infrastructure. The functionalization method is viable for scale-up to carbon fiber line, the solvent-free nature of the system is ideal for industrial scale, it is also rapid and provides significant cost–benefit to the final product. The functionalization via electrochemical reduction can be retrofitted to existing infrastructure on carbon fiber production lines, making the barrier of entry very low for adopting this approach for industrial manufacture.

References

1. Guitchounts G, Cox D (2020) 64-channel carbon fiber electrode arrays for chronic electrophysiology. Sci Rep 10(1):3830
2. Guitchounts G, Markowitz JE, Liberti WA, Gardner TJ (2013) A carbon-fiber electrode array for long-term neural recording. J Neural Eng 10(4):046016–046016
3. Hosford PS, Wells JA, Christie IN, Lythgoe MF, Millar J, Gourine AV (2019) Electrochemical carbon fiber-based technique for simultaneous recordings of brain tissue PO2, pH, and extracellular field potentials. Biosens Bioelectron: X 3:100034
4. Morgan P (2005) Carbon Fibers and Their Composites. Boca Raton
5. Fitzer E, Müller DJ (1975) The influence of oxygen on the chemical reactions during stabilization of pan as carbon fiber precursor. Carbon 13(1):63–69
6. Talreja R (2016) Physical modeling of failure in composites. Philosoph Trans Act Royal Soc A: Math Phys Eng Sci 374(2071):20150280
7. Feih S, Minzari KWD, Westermann P, Lilholt H (2004) Testing procedure for the single fibre fragmentation test. Forskningscenter Risø
8. Stojcevski F, Hilditch TB, Henderson LC (2019) A comparison of interfacial testing methods and sensitivities to carbon fiber surface treatment conditions. Compos A Appl Sci Manuf 118:293–301
9. Wang C, Ji X, Roy A, Silberschmidt VV, Chen Z (2015) Shear strength and fracture toughness of carbon fibre/epoxy interface: effect of surface treatment. Mater Des 85:800–807
10. Delamar M, Désarmot G, Fagebaume O, Hitmi R, Pinsonc J, Savéant JM (1997) Modification of carbon fiber surfaces by electrochemical reduction of aryl diazonium salts: application to carbon epoxy composites. Carbon 35(6):801–807
11. Servinis L, Henderson LC, Andrighetto LM, Huson MG, Gengenbach TR, Fox BL (2015) A novel approach to functionalise pristine unsized carbon fibre using in situ generated diazonium species to enhance interfacial shear strength. J Mater Chem A 3(7):3360–3371
12. Beggs KM, Servinis L, Gengenbach TR, Huson MG, Fox BL, Henderson LC (2015) A systematic study of carbon fibre surface grafting via in situ diazonium generation for improved interfacial shear strength in epoxy matrix composites. Compos Sci Technol 118:31–38
13. Li N, Wu Z, Huo L, Zong L, Guo Y, Wang J, Jian X (2016) One-step functionalization of carbon fiber using in situ generated aromatic diazonium salts to enhance adhesion with PPBES resins. RSC Adv 6(74):70704–70714
14. Wang X, Huang Z, Lai M, Jiang L, Zhang Y, Zhou H (2020) Highly enhancing the interfacial strength of CF/PEEK composites by introducing PAIK onto diazonium functionalized carbon fibers. Appl Surf Sci 510:145400
15. Servinis L, Beggs KM, Scheffler C, Wolfel E, Randall JD, Gengenbach TR, Demir B, Walsh TR, Doeven EH, Francis PS, Henderson LC (2017) Electrochemical surface modification of carbon fibres by grafting of amine, carboxylic and lipophilic amide groups. Carbon 118:393–403

16. Beggs KM, Randall JD, Servinis L, Krajewski A, Denning R, Henderson LC (2018) Increasing the resistivity and IFSS of unsized carbon fibre by covalent surface modification. React Funct Polym 129:123–128
17. Arnold CL, Eyckens DJ, Servinis L, Nave MD, Yin H, Marceau RKW, Pinson J, Demir B, Walsh TR, Henderson LC (2019) Simultaneously increasing the hydrophobicity and interfacial adhesion of carbon fibres: a simple pathway to install passive functionality into composites. J Mater Chem A 7(22):13483–13494
18. Szabó L, Imanishi S, Kawashima N, Hoshino R, Takada K, Hirose D, Tsukegi T, Ninomiya K, Takahashi K (2018) Carbon fibre reinforced cellulose-based polymers: intensifying interfacial adhesion between the fibre and the matrix. RSC Adv 8(40):22729–22736
19. Eyckens DJ, Stojcevski F, Hendlmeier A, Arnold CL, Randall JD, Perus MD, Servinis L, Gengenbach TR, Demir B, Walsh TR, Henderson LC (2018) An efficient high-throughput grafting procedure for enhancing carbon fiber-to-matrix interactions in composites. Chem Eng J 353:373–380
20. Wang Y, Meng L, Fan L, Wu G, Ma L, Huang Y (2015) Preparation and properties of carbon nanotube/carbon fiber hybrid reinforcement by a two-step aryl diazonium reaction. RSC Adv 5(55):44492–44498
21. Yang Y, Ibrahim AA, Stockdill JL, Hashemi P (2015) A density-controlled scaffolding strategy for covalent functionalization of carbon-fiber microelectrodes. Anal Methods 7(17):7352–7357
22. Servinis L, Beggs KM, Gengenbach TR, Doeven EH, Francis PS, Fox BL, Pringle JM, Pozo-Gonzalo C, Walsh TR, Henderson LC (2017) Tailoring the fibre-to-matrix interface using click chemistry on carbon fibre surfaces. J Mater Chem A 5(22):11204–11213
23. Eyckens DJ, Demir B, Randall JD, Gengenbach TR, Servinis L, Walsh TR, Henderson LC (2020) Using molecular entanglement as a strategy to enhance carbon fiber-epoxy composite interfaces. Compos Sci Technol 196:108225
24. Randall JD, Eyckens DJ, Servinis L, Stojcevski F, O'Dell LA, Gengenbach TR, Demir B, Walsh TR, Henderson LC (2019) Designing carbon fiber composite interfaces using a 'graft-to' approach: surface grafting density versus interphase penetration. Carbon 146:88–96
25. Randall JD, Eyckens DJ, Stojcevski F, Francis PS, Doeven EH, Barlow AJ, Barrow AS, Arnold CL, Moses JE, Henderson LC (2018) Modification of carbon fibre surfaces by sulfur-fluoride exchange click chemistry. ChemPhysChem 19(23):3176–3181
26. Dong J, Krasnova L, Finn MG, Sharpless KB (2014) Sulfur(VI) fluoride exchange (SuFEx): another good reaction for click chemistry. Angew Chem Int Ed 53(36):9430–9448
27. Barrow AS, Smedley CJ, Zheng Q, Li S, Dong J, Moses JE (2019) The growing applications of SuFEx click chemistry. Chem Soc Rev 48(17):4731–4758
28. Liu Y-T, Wu G-P, Lu C-X (2014) Grafting of carbon nanotubes onto carbon fiber surfaces by step-wise reduction of in-situ generated diazonium salts for enhancing carbon/epoxy interfaces. Mater Lett 134:75–79
29. Szabó L, Imanishi S, Kawashima N, Hoshino R, Hirose D, Tsukegi T, Ninomiya K, Takahashi K (2018) Interphase engineering of a cellulose-based carbon fiber reinforced composite by applying click chemistry. ChemistryOpen 7(9):720–729
30. Deniau G, Azoulay L, Bougerolles L, Palacin S (2006) Surface electroinitiated emulsion polymerization: grafted organic coatings from aqueous solutions. Chem Mater 18(23):5421–5428
31. Eyckens DJ, Arnold CL, Randall JD, Stojcevski F, Hendlmeier A, Stanfield MK, Pinson J, Gengenbach TR, Alexander R, Soulsby LC, Francis PS, Henderson LC (2019) Fiber with butterfly wings: creating colored carbon fibers with increased strength, adhesion, and reversible malleability. ACS Appl Mater Interfaces 11(44):41617–41625
32. Stanfield MK, Eyckens DJ, Médard J, Decorse P, Pinson J, Henderson LC (2021) Using redox active molecules to build multilayered architecture on carbon fibers and the effect on adhesion in epoxy composites. Compos Sci Technol 202:108564

Diazonium Salts and Related Compounds in Electrochemical Energy Storage and Conversion

Daniel Bélanger

Abstract Chemical grafting of organic molecules by the diazonium chemistry on various materials used in electrochemical technologies has been shown to lead to significant improvement of the electrochemical performance of lithium-ion batteries, electrochemical capacitors, biofuel cells, and microbial fuel cells. For all these applications, the search of new materials is going at a very fast pace with the aim to increase their performance and increase the likelihood of applications in practical devices. The chemical modification of materials may not be considered to lead to real new materials. Nonetheless, appropriate surface modification may give rise to materials with improved stability that could reach a level that would be sufficient to find a specific application.

1 Introduction

Chemical modification of surfaces has emerged as an effective approach to control the reactivity at surfaces. Initial investigations date back to half a century ago when, in the 1970s, the concept of chemical modification was introduced [1]. Since then, a very large toolbox of methods have been developed and found applications in a vast number of areas [2]. Among various modification approaches that have been developed, the so-called diazonium chemistry has been used to modify a large panel of materials [3, 4]. Basic concepts related to the mechanism of grafting (e.g., chemical, electrochemical, etc.) have been presented in earlier chapters of this book.

Population growth as well as population prosperity is strongly correlated with energy consumption. Recent demographic forecasts predict a significant increase in emerging countries. This means that global energy demand will grow strongly. Currently, oil is the world's main source of energy. A significant fraction of the energy consumed also comes from fossil fuel derivatives, mainly for the transportation market. Oil reserves are not inexhaustible, and the use of fossil fuels generates

D. Bélanger (✉)
Département de Chimie, Université du Québec À Montréal, Case Postale 8888, Succursale Centre-Ville, Montréal, QC 3C 3P8, Canada
e-mail: belanger.daniel@uqam.ca

M. M. Chehimi et al. (eds.), *Aryl Diazonium Salts and Related Compounds*, Physical Chemistry in Action, https://doi.org/10.1007/978-3-031-04398-7_22

significant amounts of greenhouse gases, which would be primarily responsible for global warming.

In order to limit greenhouse gas emissions, we are currently seeing a very clear increase in research and industrial development of clean and sustainable electricity generation such as wind, solar, and hydropower which could allow the production of electricity without greenhouse gas emissions. However, these means require the use of electrochemical energy storage devices such as rechargeable batteries. When energy is continuously produced over a period of several hours (solar for example), it can be stored regularly in rechargeable batteries via charging. During the discharge cycle, this energy is returned to users. Among the various types of batteries currently on the market, that are the subject of research, lithium batteries are particularly interesting. In addition to batteries, a great deal of interest has been focused toward the development of electrochemical capacitors as electrochemical energy storage systems since it has been shown that they would be suitable as high-power sources. Indeed, electrochemical supercapacitors bridge the critical performance gap between the high energy densities of batteries and very high-power densities offered by conventional electrolytic capacitors [5].

This chapter will focus on applications of materials modified by the diazonium chemistry in electrochemical technologies such as rechargeable batteries, electrochemical capacitors, and compared to earlier reviews [6–8], it will be expanded to include fuel cells and microbial fuel cells. In the first part of each sub-sections, basic concepts of these electrochemical systems will be presented. For each of these systems, in a second part, selected examples of modified materials will be presented as well as representative examples describing the improvement of the performance of lithium-ion batteries, electrochemical capacitors, and fuel cells.

2 Batteries

2.1 Basic Concepts

A typical example of a rechargeable lithium-ion battery consists of a negative graphite electrode as an anode material, a non-aqueous electrolyte containing lithium salt, and a positive electrode as a cathode that can be an olivine such as $LiFePO_4$ and other compounds such as $Li(Ni,Mn,Co)O_2$, $LiCoO_2$ and $LiMn_2O_4$ [9, 10]. The charge of this battery involves the insertion of lithium ions from the electrolyte in the anode, and the transfer of these ions in the electrolyte from the cathode occurs following the deinsertion of lithium ions from the cathode material (Scheme 1). During the discharge, the opposite reactions occur.

In a practical battery, both anode and cathode are composite electrodes that contain, in addition to the electroactive material, inactive materials such as binder and conductive carbon additive. The carbon additive is used to increase the electrical

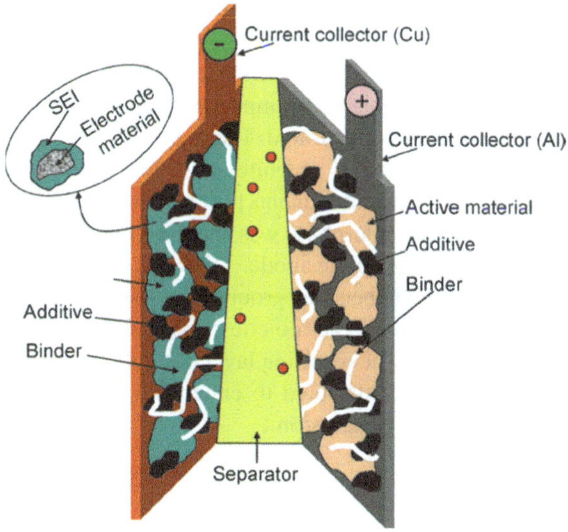

Scheme 1 Schematic representation of a lithium-ion battery and a common composite electrode. Reused with permission from Reference [9]. Copyright (2009) (The Royal Society)

percolation, thus the electronic conductivity of the electrode, whereas the binder insures mechanical stability to the whole electrode (Scheme 1).

2.2 *Anode*

2.2.1 Graphite

The electrochemical activity of graphite as anode in lithium-ion batteries is based on the intercalation and deintercalation of Li^+ ions into and from its layered structure. Despite that lithium-ion batteries with a graphite as anode has been industrially developed, their rate capability is limited because lithium ions diffusion within the solid-state electrode materials is slow. In addition, lithium ions insertion into graphite occurs at low potential in the range where reduction of the organic electrolyte also occurs and possibly lithium plating while overcharging the battery. The latter occurs during the first few charge/discharge cycles and gives rise to the formation of a passivating solid electrolyte interphase (SEI) layer on the surface of the graphite anode. The formation of SEI prevents further degradation of the electrolyte at the anode surface and is essential for the long-term operation of the graphite anode. However, the formation of SEI leads to irreversible capacity loss during initial cycles and loss of lithium ions from the electrolyte. Consequently, the formation of an artificial SEI by grafting could be attractive. In the following section, examples of surface modification by means of the diazonium chemistry will be presented.

The modification of graphite anode has been carried out by the covalent attachment of aryl multilayers of lithium benzoate [11], nitrophenyl [12], and 4-carboxyphenyl

[13, 14]. For the latter, the initial 4-carboxyphenyl layer was subsequently used as anchor layer for the formation of a polysiloxane layer [13]. These anodes were characterized by improved electrochemical performance in lithium-ion batteries. The formation of a stable and compact passive [11, 12] and flexible [13] solid electrolyte interphase was demonstrated. The presence of a grafted layer was also shown to mitigate the exfoliation of graphite and offered the possibility to control the reactivity toward electrolyte components [14]. An interesting approach to elaborate a functional solid electrolyte interphase was achieved by deposition of polar functional groups at the surface of a graphite anode. The layer was formed by electrochemical grafting of 4-(trimethylsilyl)-benzene groups from the corresponding diazonium species, deprotection to obtain alkyne moieties, and further modification by using thiol-yne click chemistry [15]. The resulting layer not only provided a stable solid electrolyte interphase but also contributed to enhancing lithium transport and preventing solvent migration by size exclusion.

2.2.2 Silicon

An attractive approach to increase the energy density of lithium-ion batteries is to replace the graphite anode by materials such as silicon that is characterized by a theoretical specific capacity about 10 times higher than that of graphite. The second advantage of silicon anode is its moderate operation potential versus lithium, which can prevent the safety issue of lithium deposition observed with a graphite anode. The large number of Li atoms that can be inserted into silicon to form various alloys, however, causes extremely large volume changes that reach 300% for full lithiation of Si. The irreversible volumetric and structural changes observed during Li insertion and extraction lead to mechanical fracture/decrepitation of silicon particles and a rapid fade of the specific capacity. Several approaches have been developed to mitigate these issues [16, 17]. Among these approaches, the morphology and size of silicon particles, the presence of electrolyte additive, using multifunctional binder, entrapment in conducting materials, and formation of a coating have been shown to yield improved electrochemical performance and stability. Similar to graphite anode discussed above, grafting can lead to the formation of an artificial SEI and prevent continuous electrolyte decomposition side-reactions as well as the electrical isolation of fractured active silicon particles.

The chemical modification of silicon anode can be classified into three distinct strategies. Firstly, silicon can be modified by the classical approach involving the grafting of aryl groups such as carboxyphenyl [18]. Secondly, silicon particles can be directly coupled to graphite [19], multiwalled carbon nanotubes [20], or graphene [21], by a two-step procedure involving the initial grafting of aminophenyl groups and subsequent bridging to Si with phenyl groups by diazotization of the grafted amine. This strategy was mainly developed to avoid losing electronic connection between Si and carbonaceous materials. Thirdly, in a representative example of another two-step procedure, silicon is initially modified by 1-(bromoethyl) benzene on the surface of hydrogenated silicon via diazonium chemistry and subsequently with polyacrylic

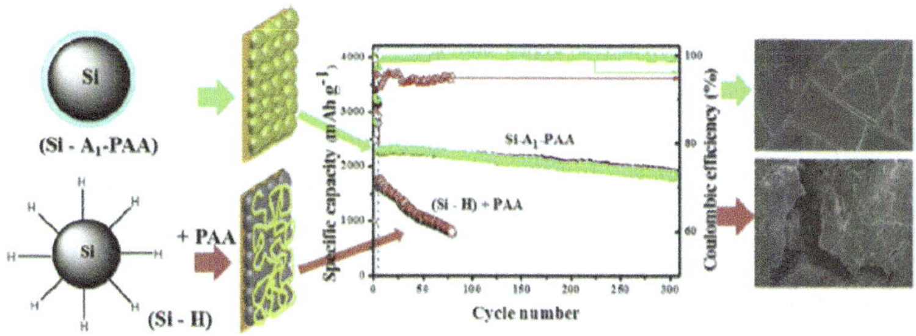

Scheme 2 Schematic representation of H-terminated Si (Si–H) and polyacrylic acid-modified Si (Si-A$_1$-PAA), their electrochemical performance and surface morphology evaluated by scanning electron microscopy. Reused with permission from Reference [22]. Copyright (2017) (Elsevier)

acid by surface-initiated atom transfer radical polymerization (ATRP) as shown in Scheme 2 [22–24]. The resulting materials (denoted Si-A$_1$-PAA) showed a significant improvement of gravimetric capacitance, capacity retention, and Coulombic efficiency with respect to the unmodified Si, as illustrated in Scheme 2 [22]. It is also noteworthy that no electrolyte additive was used in this case. A Si/graphene composite electrode was prepared by combining the polyacrylic acid-modified Si (Si-A$_1$-PAA) with graphene [24]. It is possible to control the loading of both 1-(bromoethyl) benzene and polyacrylic acid by changing the synthesis conditions. Interestingly, the layer of (1-(bromoethyl) benzene and polyacrylic acid) grafted on the surface of hydrogenated silicon nanopowder displayed binder-like properties that allowed the preparation of the Si/graphene composite without the need of using a binder. The mechanical properties of composite electrode using covalently grafted polyacrylic acid on Si enabled a superior elongation without breakage as well as stronger adhesion to the current collector than materials without the grafted layer. Also, graphene is permitted to obtain superior electrochemical performance, mechanical, and adhesion properties than other carbon additives such as carbon black [23].

2.3 Cathode

Several oxide and phosphate-based lithium intercalation compounds have been studied and used as an active cathode material in a rechargeable lithium-ion battery. These materials including $LiCoO_2$, $LiNiO_2$, $LiMn_{1.5}Ni_{0.5}O_4$, $LiMn_2O_4$, $LiFePO_4$, and others have been the subject of numerous studies. Although these electrode materials can deliver interesting energy density, several problems must be solved to improve performance, long-term stability, and safety. Thus, some materials are limited by a low electronic conductivity. To overcome this problem, one approach is to deposit a thin nanometric layer of carbon on the surface of the electrode material particles [25]. This approach has significantly increased the conductivity and the electrochemical performance of these electrodes in lithium-ion batteries. In terms of stability, the use of nanoparticles implies that a very large area of the electrode material is in contact with the electrolyte. This increases the risk of secondary reactions involving electrolyte decomposition, which causes high irreversibility, low Coulombic efficiency, and a reduced lifespan. Obviously, this phenomenon is greatly amplified at higher battery operating temperature. In addition, the dissolution of metal ions is a problem commonly faced with high-voltage (e.g., $LiMn_{1.5}Ni_{0.5}O_4$) cathode materials. The safety aspect stems from the very high reactivity of cathode materials operating at very high voltage. The general approach to minimize these problems is to form a protective coating on the surface of the particles of the cathode material [26]. An interesting approach to mitigate the problems mentioned above involves deposition of an organic layer with a specific functionality.

The initial study with cathode electrode materials dealt with the grafting of nitrophenyl groups on $Li_{1.1}V_3O_8$ [27]. The grafted layer prevented electrolyte decomposition without affecting charge transfer and charge storage. Subsequently, $LiFePO_4$ powders were functionalized with a wide variety of organic molecules by the diazonium chemistry [28–38]. The first study demonstrated that the grafted molecules can assist the insertion of lithium ions [28]. The spontaneous grafting of nitrophenyl groups at the surface of carbon-coated $LiFePO_4$ led to partial oxidation and delithiation of $LiFePO_4$ to form the $FePO_4$ phase [29, 30]. The chemical modification of $LiFePO_4$ with aminophenyl and bromobenzene groups resulted in a slight improvement of the electrochemical performance when an optimum loading of aryl molecules was used [31]. Indeed, a too high loading was found to be detrimental due to the resistive effect of the organic molecules. The chemical grafting of trifluoromethylsulfonimidebenzene ($-C_6H_4SO_2NHSO_2CF_3$) groups on $LiFePO_4$ enabled the formation of a more homogeneous film electrode and the electrical conductivity of the film was not affected despite that sp^3 defects are formed on the carbon coating as result of grafting [32]. In addition, wettability of the resulting composite electrode was improved due to the higher hydrophilicity of the modified powder. As substitute for a carbon coating, a polyphenylene conducting layer was formed on pristine $LiFePO_4$ by spontaneous reaction of the phosphate with $C_6H_5N_2^+$[33]. It was also shown that a carbon additive was not required for this polyphenylene-coated $LiFePO_4$ to obtain good electrochemical performance.

A detailed study of the mechanism of the spontaneous reaction of diazonium ions (from a diazonium salt or in situ generated from the corresponding amine) with $LiFePO_4$ was performed by gas chromatography coupled to mass spectrometry [34]. It was found that the grafting efficiency on carbon-coated $LiFePO_4$ was only slightly different for in situ generated diazonium ions (10%) and diazonium salt (14%), suggesting similar grafting mechanisms. These low grafting efficiencies are in line with the presence of by-products found in the synthesis solution following grafting. The influence of the substrate onto which grafting is attempted was investigated. The presence of the carbon coating on $LiFePO_4$ is essential for grafting [29, 34]. Moreover, the carbon coating prevents the reaction of the diazotization agent (tert-butyl nitrite) with $LiFePO_4$ that leads to the oxidation of the latter. This unwanted reactivity of $LiFePO_4$ was demonstrated by an almost quantitative yield for the formation of diazonium ions in the presence of only acetylene black [34]. The chemical grafting of trifluoromethylphenyl groups (about 0.5 wt%) on carbon-coated $LiFePO_4$ powder led to an increase of hydrophobicity as revealed by water contact angle measurements [35]. The presence of the trifluoromethylphenyl groups on the carbon-coated $LiFePO_4$ powder also led to an increase of its stability in water and reduced its iron dissolution into the electrolyte used for assembling the battery. This protection from moisture could be an advantage for the storage of modified-$LiFePO_4$ powder under less strict conditions and its use for the preparation of water slurry as well as for aqueous lithium-ion batteries [36]. Chemical modification of electrode materials by the diazonium chemistry is not limited to $LiFePO_4$. Similar to polyphenylene-modified $LiFePO_4$ [33], the same polymer was grafted on $LiNi_xCo_yMn_zO_2$ [37].

Organic batteries were also investigated with chemically modified materials by several methods including the diazonium chemistry. The main objective in several studies was to avoid the dissolution of the active organic materials encountered when a composite electrode was prepared by simple mixing with a carbon support. High surface area conductive Ketjenblack (KB) powder was modified by reaction with in situ generated diazonium cations from the corresponding diamino-aryl-1,4,5,8-naphtalenetetra-carboxylic diimide molecule [38]. The resulting organic electrode material was characterized by a redox potential of 2.45 V versus Li/Li^+ and an organic lithium-ion battery based on this material delivered a specific capacity of 100 mAh/g. Ketjenblack (KB) was also chemically functionalized with 9,10-phenanthrenequinone (PAQ), 9,10-anthraquinone and pyrene-4,5,9,10-tetraone [39]. The PAQ modified KB/Li cell when tested in (trifluoromethanesulfonyl) imide (Li(TFSI))/propylene carbonate was characterized by a sloping voltage between 2 and 3 V and a specific capacity of 75 mAh/g when a current density of 30 mA/g was used. However, a significant loss of capacity was observed during 500 charge/discharge cycles. Multiwalled carbon nanotubes were used to covalently immobilize anthraquinone and provide electronic conductivity for the composite electrode [40]. The resulting electrode when cycled in 1 M LiTFSI/tetraethylene glycol dimethyl ether (50/50) showed a specific capacity of about 110 $mAh/g_{electrode}$ that translates into a capacity of 260 mAh/g for anthraquinone alone, i.e., without carbon additives) and a loss of 20% of its specific capacity over 500 cycles.

In addition, the diazonium chemistry was proved useful in lithium-oxygen ($Li-O_2$) [41] and lithium-sulfur [42] batteries. A $Li-O_2$ battery consists of a Li anode and a porous O_2 cathode with an appropriate Li^+-conducting electrolyte. At the cathode, which could be a carbon-based material, on discharge O_2 is reduced to produce solid Li_2O_2 while on recharge oxidation of Li_2O_2 occurs. The passivation of the cathode due to the formation of a poorly conducting Li_2O_2 film leads to battery failure. The chemical grafting of hexylbenzene on super P carbon has been shown to slow down the adsorption of Li_2O_2 and significantly increase the lifetime of the $Li-O_2$ battery [41]. The chemical grafting of diazonium cations is containing electrochemically active disulfide groups onto multiwalled carbon nanotubes enabled the preparation of electrode of a lithium-sulfur battery for which the dissolution of the sulfur-based materials was mitigated in 0.1 M $LiClO_4$/acetonitrile electrolyte [42].

2.4 Carbon Additive

As mentioned above, a composite electrode contains, in addition to the active material, a polymeric binder and a conducting carbon additive. Unfortunately, when used with a high-voltage cathode requiring charging well above 4.0 V versus Li/Li^+, the carbon materials can degrade leading to a decrease of electronic conductivity. In addition, above 4.5 V versus Li/Li^+, intercalation of electrolyte anions in the graphitic layers of the carbon additive can occur resulting in structural modification and degradation of the conductive carbon. The functionalization of the carbon additive of $LiNi_{0.5}Mn_{1.5}O_4$ positive electrode by $C_6H_4NO_2$ [43], $C_6H_4CF_3$, $C_6H_4SO_3H$, C_6H_4COOH, and $C_6H_4N(C_2H_5)_2$ [44] groups has been shown to mitigate the reactivity of the electrode toward the electrolyte as well as the degradation of the carbon additive. A similar observation has been made for graphene modified with $C_6H_4CF_3$, C_6H_4COOH, and $C_6H_4N(C_2H_5)_2$ groups [45].

The acetylene black (AB) carbon additive of a $LiFePO_4$ composite cathode was modified by a redox active tris (1,10-phenanthroline) iron (II) complex [46]. A slight increase in the specific capacity of the $LiFePO_4$ electrode was observed at high cycling rate. The elaboration of a $LiFePO_4$-AB-$[Phen]_3Fe^{2+}$ electrode was motivated by the fact that the grafted electroactive species, with their redox potential slightly higher than that of $LiFePO_4$ could provide overcharge protection for the $LiFePO_4$ electrode, as is the case with electroactive species dissolved in the electrolyte. Unfortunately, overcharge protection was not observed presumably due to low amount of available grafted $[Phen]_3Fe^{2+}$ species [46]. The acetylene black carbon additive of a composite $LiMn_{1.5}Ni_{0.5}O_4$ cathode was modified with benzo-15-crown-5 ether, and the resulting cathode was investigated in $LiMn_{1.5}Ni_{0.5}O_4$/graphite full cell [47]. It was shown that the grafted crown ether enables the trapping of a fraction of metallic ions dissolved from the cathode as well as a decrease in the amount of transition metal deposited on the graphite anode (Scheme 3).

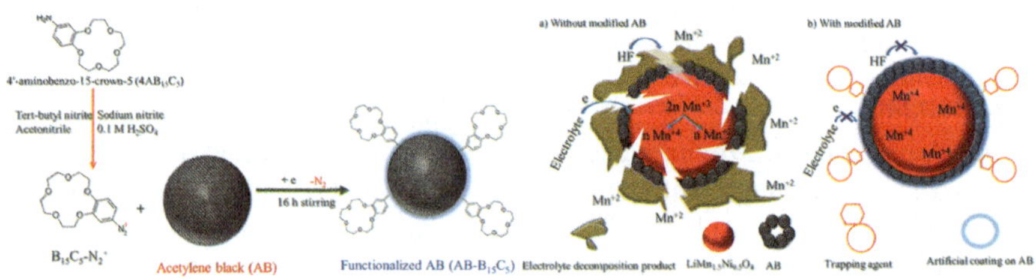

Scheme 3 Functionalization of acetylene black (AB) with 4′-aminobenzo-15-crown-5 in acetonitrile (Method 1) or in acidic media (Method 2) by in situ generation of aryl diazonium ions ($B_{15}C_5-N_2^+$) and (right) illustration of processes occurring at unmodified and modified carbons. Reused with permission from Reference [47]. Copyright (2020) (American Chemical Society)

3 Electrochemical Capacitors

3.1 Basic Concepts

The classical electrochemical double layer capacitor comprises of two porous carbon electrodes in the presence of an appropriate electrolyte and separated by a porous and inert separator (Scheme 4). At the electrode/electrolyte interface, charges of opposite sign accumulate at the electrode surface (within the porous structure) and in the electrolyte [48]. At equilibrium, the electronic charges in the electrode are counterbalanced by ionic charges from the electrolyte. This leads to the formation of an electrochemical double layer whose thickness is in the nanometer range. The energy density (Wh/kg) of an electrochemical double layer capacitor is determined by the potential (V) being developed and the capacitance (F/g) of the electrodes. The charge storage mechanism of electrochemical double layer capacitor only involves transport of ions in solution and since such process is fast, the power density (W/kg) of electrochemical double layer capacitor is high, in fact much higher than those of common rechargeable batteries (Scheme 5). On the other hand, redox reactions that occur at batteries electrodes are usually much slower processes than ion transport in the electrolyte. Consequently, the energy density of batteries is significantly higher than that of electrochemical double layer capacitors. Furthermore, charge storage in the electrical double layer at a surface/electrolyte interface is not limited to electrostatic charge storage. As illustrated in Scheme 5, pseudocapacitive charge storage, involving fast surface and near-surface redox reactions, provides enhanced charge storage properties [49]. Transition metal oxides such as RuO_2 and MnO_2 (in mild aqueous electrolytes) as well as conducting polymers are typical examples of pseudocapacitive materials that can provide enhanced charge storage capability, hence increasing the energy density of an electrochemical energy storage systems. An attractive strategy investigated to increase the energy density of carbon-based electrochemical capacitor involves the grafting of electroactive organic molecules on active carbon materials [6–8].

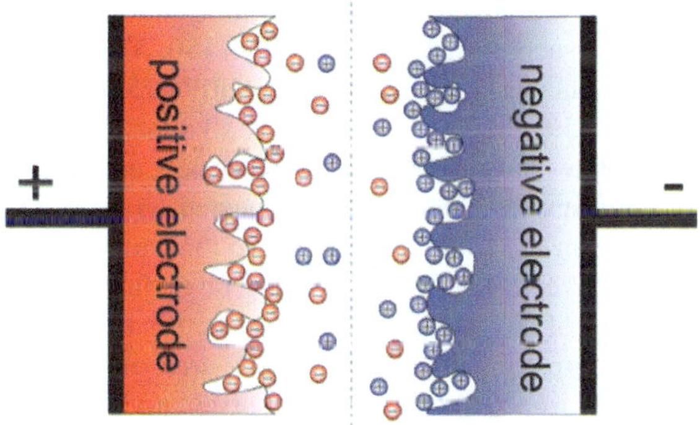

Scheme 4 Schematic representation of a classical electrochemical double layer capacitor. Reused with permission. From Reference [48]. Copyright (2014) (Wiley–VCH Verlag)

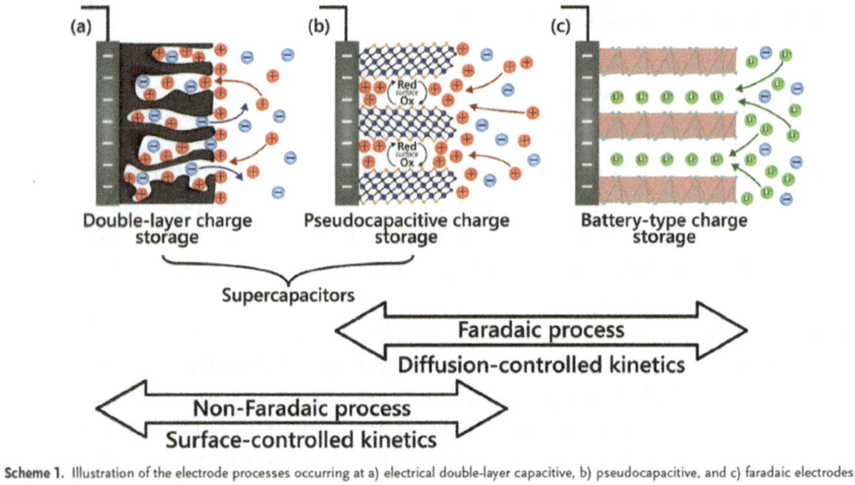

Scheme 5 Illustration of: **a** electrical double layer capacitive, **b** pseudocapacitive, and **c** Faradaic electrode processes. Reused with permission from Reference [49]. Copyright (2019) (Wiley–VCH Verlag)

3.2 *Grafting of Redox Active Molecules*

The simplest and more convenient strategies to use electroactive organic molecules is by mixing them with a carbon support, the latter usually being in the form of a powder or relying on slow adsorption [2]. Unfortunately, while these approaches have proved useful, they are commonly plagued by the slow desorption of the electroactive molecules with the consequence of loss/decrease of charge storage capabilities. To circumvent this issue, an interesting approach consists of the covalent immobilization of redox molecules by the diazonium chemistry [6–8]. Organic molecules that have been grafted include anthraquinone derivatives [50–59], catechol (or 1,2-dihydroxybenzene) [60–64], phenanthrenequinone [65, 66], rhodizonic acid [67], nitrobenzene [68], diaminonaphthalene derived film [69], and phenanthroline [70]. The aforementioned organic molecules have been grafted primarily on carbon supports such as carbon fabric/cloth [50, 57], carbon black [51], activated carbon, [52] graphene and carbide-derived carbon [55], Vulcan XCMax22 [69], and $Ti_3C_2T_x$ MXene [59]. In addition, aqueous alkaline [54], acid [51] as well as organic [71] electrolytes have been tested. Finally, less conventional protic ionic liquid has been recently investigated [56]. As expected, grafting of electroactive groups led to an increase in the stored charge due to their redox processes that occur in addition to double layer charging. The loading of grafted groups is controllable by appropriate selection of the grafting conditions such as the concentration of the reagents as well as the porosity and specific surface area of the carbon materials. The stability of these electroactive molecules was evaluated, and commonly, a significant decrease of the charge stored by the grafted molecules has been observed, but the double layer capacitance of the support remained usually barely affected [51]. The fade of the Faradaic charge was attributed to the desorption of weakly bonded or physisorbed molecules. Other factors that can contribute to the capacity fade include degradation of the organic moieties, loss of electrical connexion between the redox sites and the conducting substrate, and the nature of the electrolyte. In general, improved stability will be needed for possible applications of some organic-based materials in electrochemical energy storage systems.

3.3 Two-Electrode Systems

Grafting quinones with an appropriate redox potential (e.g., anthraquinone) allow to increase in the charge storage of the negative electrode. However, this leads to charge storage imbalance, and consequently, such organic molecule-modified negative electrode must be matched with a positive electrode displaying similar charge storage properties. In several studies, an anthraquinone-based negative electrode was coupled with organic [72] and inorganic [73–75] positive electrodes. Organic positive electrodes include catechol [72] and phenanthroline [70], whereas inorganic electrode materials such as RuO_2, [73] $Ni(OH)_2$ [74], and $LiMn_2O_4$ [75] were used. The improvement of the electrochemical performance of a classical electrochemical double layer capacitor by using a Faradaic positive electrode and a modified carbon as negative electrode is illustrated in Scheme 6 [74]. For these systems, aqueous electrolytes limited the cell voltage to the range of 1.5–2 V. Interestingly, the cell voltage of a hybrid system based on carbon electrodes modified by *N*-(2-aminoethyl)-1,8-naphthalimide and 2,2,6,6-tetramethylpiperidine-*N*-oxyl as negative and positive electrodes, respectively, was increased to 2.9 V by using an appropriate non-aqueous electrolyte [76]. Early on, the electrochemical performance of grafted electrodes was expressed in specific capacitance units (F/g) [6, 50, 51]. However, it is now starting to become accepted that it is more appropriate to report specific capacity (C/g or mAh/g) as recently explained [7, 74, 77]. Specific capacitance values are potentially prone to confusion because they strongly depend on the considered voltage window [78]. The terminology used to describe these systems was initially either hybrid electrochemical capacitors or asymmetric capacitor. However, it would be more appropriate to use a different terminology to describe, for example, a fully organic system based *N*-(2-aminoethyl)-1,8-naphthalimide and 2,2,6,6-tetramethylpiperidine-*N*-oxyl as negative and positive electrodes as well as those with a positive battery electrode such as $Ni(OH)_2$ [74] and $LiMn_2O_4$ [75]. These systems are similar to a rechargeable battery and consequently using the term electrochemical capacitor should be avoided.

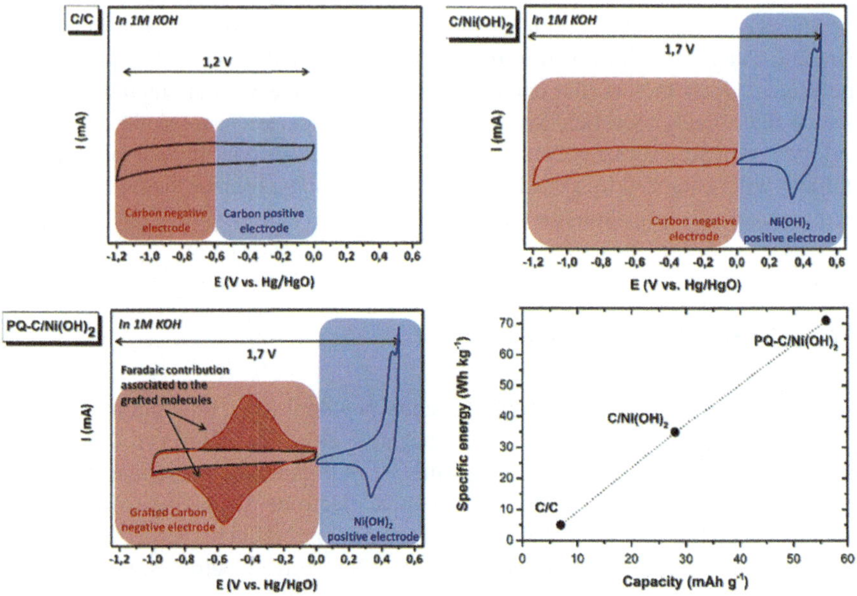

Scheme 6 Schematic representation of cyclic voltammograms of both positive and negative electrodes in a carbon/carbon symmetric (top left); carbon/$Ni(OH)_2$ hybrid (top right); phenanthrequinone-modified carbon/$Ni(OH)_2$ hybrid configuration (bottom left) and estimation of full device capacity and specific energy for the three configurations (bottom right) (dashed line is a guide for the eyes). Reused with permission from Reference [74]. Copyright (2017) (Elsevier)

3.4 Immobilization of Metal Oxide on Grafted Materials

An electrode assembly was prepared by chemical binding of amorphous manganese oxide and carbon particles by aminophenyl groups [79]. A two-step procedure consisted in the surface functionalization of carbon particles with aminophenyl groups that was followed by the linkage of manganese oxide particles through generated phenyl groups. The carbon/phenyl/MnO_2 nanocomposite electrode delivered a specific capacitance twice that of a simple mixture of MnO_2 and carbon. Graphene oxide modified with 4-carboxyphenyl groups was used as support for iron oxide nanoparticles [80]. The presence of homogeneously distributed carboxyphenyl groups allowed the preparation of a nanocomposite with well-distributed iron oxide particles. More favorable interactions between the electrode components led to improved electrochemical performance (in terms of capacitance and stability) of the nanocomposite electrode material.

3.5 *Grafting Electrochemically Inert Species*

Carbon materials were also chemically modified by electroinactive groups [81, 82]. Grafting sulfonate groups have permitted a change of the classical charge storage mechanism where the grafted groups influenced the nature of the ionic species present at the interface [81]. Functionalization of high surface area carbon black was achieved by a covalent attachment of polyacrylic acid by atom transfer radical polymerization using 1-(bromoethyl)benzene groups initially bonded to the carbon material [82] as described above for silicon [22]. The polyacrylic acid-modified carbon electrode films were more hydrophilic with better wettability in an aqueous electrolyte than that of the unmodified carbon. Surprisingly, the modified electrodes delivered a higher specific capacitance and a wider working potential window when tested in an aqueous electrolyte. Reduced graphene oxides flake interlayers were linked by different molecular species, such as phenyl, biphenyl), and *p*-terphenyl by using bis-diazonium salts to obtain graphene interlayer sub-nanopores of 0.49, 0.7, and 0.96 nm, respectively [83]. The modified materials with a gap of 0.7 nm yielded the best electrochemical performance in solution containing cations with size of 0.68 nm (1 M $TEABF_4$ in organic electrolyte) and 0.6 nm (aqueous 6 M KOH).

4 Fuel Cells

4.1 *Basic Concepts*

A fuel cell represented in Scheme 7 consists of an anode and a cathode that are separated by an electrolyte, commonly a polymer electrolyte [84]. A fuel (e.g., hydrogen, methanol, ethanol, formic acid) is oxidized at the anode while the reduction of an oxidant (oxygen) takes place at the cathode. In proton exchange membrane fuel cells, (PEMFCs), and specifically for a hydrogen–oxygen fuel cell, protons generated by oxidation of hydrogen move across the proton exchange (ionomer) membrane and electrons flow through the external circuit. At the cathode, oxygen is reduced by combining with protons to generate water. In fuel cells and unlike batteries, energy is generated as long as the fuel and the oxidant are supplied.

In the case of the PEMFC, the so-called membrane electrode assembly (MEA) is a very important part of a single cell since it is the site of the fuel cell reactions. The MEA is located between carbon flow field plates which feed humidified fuel (H_2 and O_2) to the MEA. The reactants permeate through the carbon-based electrode support structures into the platinum catalyst layers where they react to generate electricity, heat, and pure water.

An expanded view of the components of one side of a single PEM fuel cell presented in Scheme 8 shows the direction for the flow of reagents, the location for the oxygen reduction half-reaction and also highlights the hydrophobic and hydrophilic properties of the gas diffusion backing and catalyst layer, respectively.

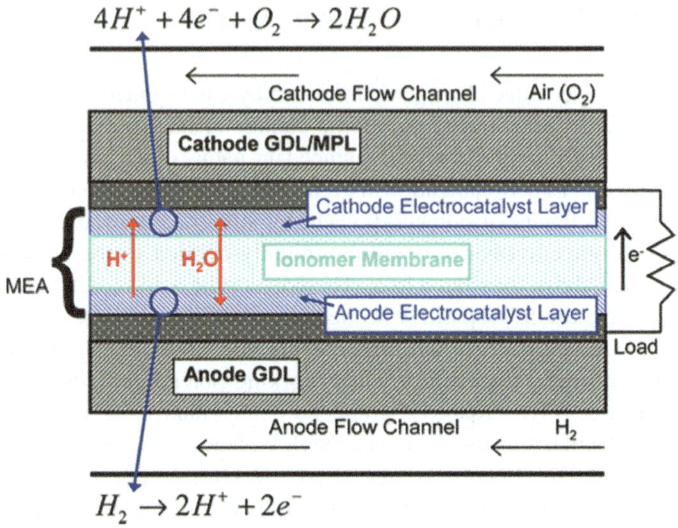

Scheme 7 Schematic representation of a fuel cell with its major components. Reused with permission from Reference [84]. Copyright (2007) (American Chemical Society)

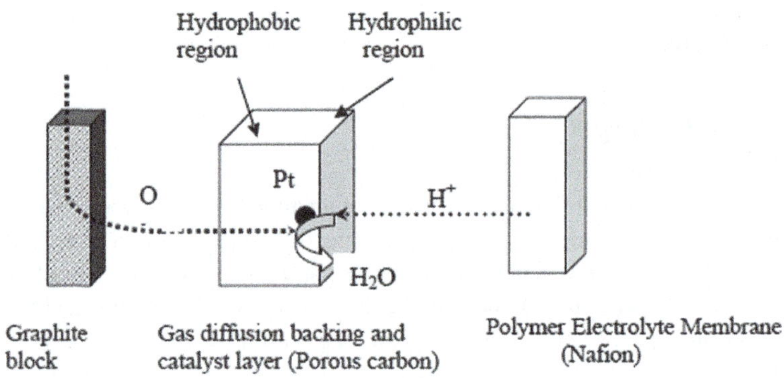

Scheme 8 Expanded view of one half-cell of a PEMFC

The diffusion backing layer should provide an effective oxygen gas supply to the catalyst layer and have adequate water handling capability where water is generated. It should also have a high electrical conductivity and show good corrosion resistance. The gas diffusion backing, most commonly a porous carbon paper or cloth, must combine a highly hydrophobic region (reactant gas inlet) with a hydrophilic region, the active layer where the catalyst is located. The hydrophobic part of the porous carbon (which is in contact with a graphite block which acts as current collector) allows for the easy diffusion of the reactant gas through the pores toward the catalytic sites and also prevents penetration of water (generated at the catalytic sites). The active layer where the catalyst is located should be hydrophilic to ensure sufficient wetting of the carbon and the electrocatalyst. This is crucial to obtain a low electrical resistivity and maximize catalyst utilization. The latter can be achieved by improving the dispersion of the metal particles and controlling their size and their distribution on the surface of the carbon substrate. One of the routes that can be followed to reach this goal is to functionalize the surface of the carbons to: (i) favor the nucleation and to limit the aggregation of the metal particles; (ii) to control the hydrophilicity of the catalyzed carbons; and iii) favor the deposition of metal particles at the surface of the carbon electrode. Modification of the MEA by the diazonium chemistry can potentially allow to provide suitable hydrophobic/hydrophilic character. Hence, carbon powders modified by hydrophilic functionalities can improve ionic conductivity, particle size distribution, and dispersion, whereas a hydrophobic functionality could be useful for the water management in a PEMFC.

4.2 *Functionalization of Carbon Supports*

Chemical grafting of sulfonated [85–88] fluorinated [89–93] and thiol [94] groups on various carbon supports has been performed with the goal to improve the performance of PEMFC. Chemical grafting of ethanesulfonic acid groups (-$CH_2CH_2SO_3H$) on 20% Pt/Vulcan XC-72 catalyst enabled an increase of performance relative to an unmodified catalyst [85]. In addition, a smaller amount of the expensive Nafion was needed. The grafting from an aliphatic diazonium generated by reaction of 2-aminoethanesulfonic acid with isobutyl nitrite is intriguing since unlike aryl diazonium salt, aliphatic diazonium salt are significantly less stable. Phenyl sulfonic acid-modified Vulcan XC-72 yielded Nafion composite with one order of magnitude higher ionic conductivity than a layer based on unmodified carbon [86]. A PEMFC with phenyl sulfonic acid-modified carbon-supported platinum catalyst electrodes delivered an enhanced power density relative to one elaborated with unmodified carbon [87]. The grafting of phenyl sulfonic groups on two carbon supports (XC-72 and Black Pearls 2000) loaded with Pt led to a significant increase of the corrosion resistance of the carbon and the stability of the Pt nanoparticles [88].

Carbon supports were rendered hydrophobic by chemical grafting of aryl groups such as 2,3,4,5,6-pentafluorophenyl [89], trifluoromethylphenyl [90], 2,3,4,5,6-pentafluorophenyl [91], 3,5-bis(trifluoromethyl)phenyl [92], and a set of fluorinated and nitrile compounds [93]. More specifically, grafting was performed on E-TEK 20% Pt/Vulcan XC-72 [89] graphitized carbon [90], Vulcan [91], mesoporous colloid imprinted carbon powder [91], gas diffusion layer [92], and carbon nanotubes [94]. The modified carbon materials led to a prevention of water flooding [89], improvement of the dispersion of Pt nanoparticles with a concomitant decrease of their sintering [90, 94], increase of the corrosion resistance of the carbon support due to surface passivation by loss of carbon active sites [91]. The resulting modified materials allowed to improve the performance and stability in conditions mimicking fuel cell operation [93] as well as in PEMFC [92].

4.3 Functionalization of Metal Particles

An alternative and interesting approach to prepare oxygen reduction electrocatalyst for application in PEMFC was to cap Pt nanoparticles by various para-substituted (R = $-CH_3$, $-F$, $-Cl$, $-OCF_3$, and $-CF_3$) phenyl groups [95]. The Pt nanoparticles modified with trifluoromethyl phenyl groups were the most active due to the effect of the electronegative ligands leading to weaker oxygen adsorption.

4.4 Functionalization of Carbon as Support for Nonprecious Metal Catalyst

Finally, due to the high cost and limited resources associated with Pt, a great deal of work has been done to develop nonprecious metal catalysts based on iron. To this end, carbon black support has been modified with 1,10-phenanthroline groups by the diazonium and benzimidazole chemistry [96]. The resulting catalysts show good activity for the oxygen reduction reaction and a very low production of hydrogen peroxide.

5 Biofuel and Microbial Fuel Cells

5.1 Basic Concepts

One important factor, which contributes to the motivation for the development of biofuel cells, is the relative scarcity and high cost of precious metal catalysts such as platinum. Similarly to fuel cells, a biofuel cell consists of two electrodes separated by a membrane and an appropriate electrolyte (Scheme 9). The bioanode,

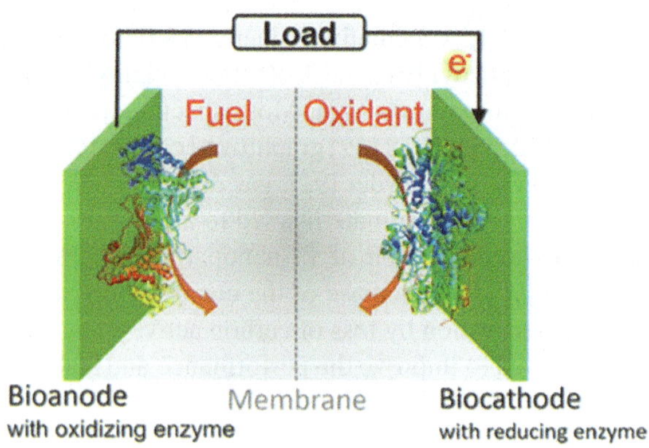

Scheme 9 Schematic representation of a biofuel cell with its major components that include a bioanode and a biocathode. Reused with permission from Reference [97]. Copyright (2019) (American Chemical Society)

on which appropriate redox enzyme is immobilized, allows the oxidation of a fuel (e.g., glucose). The resulting electrons flow in the external electrical circuit to the cathode, where, in a classical case, oxygen is reduced to water. In a full biofuel cell, a biocathode with an appropriate enzyme is immobilized at the electrode surface [97].

The development of biofuel cells is limited by their extremely low electrochemical performance relative to fuel cell. More specifically, biofuel cells are characterized by their low energy and power densities and poor stability [97]. The low energy density is mainly due to the incomplete reaction of the fuel while the low power density derives from the relatively long distance between the active site of the enzyme, located inside an insulating protein matrix, and the electrode surface. In addition, the enzyme loading is commonly much smaller than that of a metal catalyst of a fuel cell. The poor stability is associated with the denaturation of the biomolecules as well as its desorption from the electrode surface. An obvious approach to solve the issue of desorption is to form a covalent linkage between the biomolecules and the electrode surface.

5.2 *Immobilization of Enzyme on Aryl-Modified Electrode*

In an early study, glucose oxidase and bilirubin oxidase enzymes were immobilized on an aminophenyl modified gold electrode by the diazonium chemistry [98]. The aminophenyl groups serve as anchoring sites for the immobilization of an osmium-based redox polymer and the biocatalytic species. The initial grafting of aminophenyl groups led to enhanced stability of the bioelectrode assembly. Other systems that were characterized by improved stability following grafting of different aryl groups include cellobiose dehydrogenase immobilized on various substituted aryl-modified single-walled carbon nanotubes [99, 100], laccase on azido-modified carbon nanotubes [101], and laccase covalently grafted on carboxyphenyl-modified graphite via the formation of an amide bond [102].

5.3 *Biofuel Cell*

A glucose/oxygen biofuel cell consisting of glucose oxidase as the anode biocatalyst and either bilirubin oxidase or laccase as biocatalyst at the cathode showed improved stability when the enzymes and appropriate osmium redox polymers are immobilized at graphite electrode surface via aniline groups [103]. An H_2/air proton exchange membrane biofuel cell was combining a Pt/C anode with a cathode elaborated by grafting of anthraquinone and naphthoate on multiwalled carbon nanotubes onto which a laccase was immobilized. The biocathode under the form of a gas diffusion electrode showed very good activity for the oxygen reduction reaction [104]. In another biofuel cell, instead of the classical Pt/C anode, a [NiFeSe] hydrogenase was immobilized on anthraquinone modified carbon nanotubes to form a bioanode.

A gas diffusion electrode assembly exhibited excellent activity for the oxidation of hydrogen when used in combination with a bilirubin oxidase immobilized on naphthoate-modified carbon nanotubes biocathode [105].

5.4 Microbial Fuel Cell

In a microbial fuel cell, the chemical energy of organic material is converted to electrical energy via interactions between microbial cells and electrode. A microbial fuel cell could be used in environmental remediation by the degradation of waste. An important feature of microbial fuel cells is that in contrast to relatively short live biofuel cell, they can be operated over much longer period of time by using self-reproducing microorganisms.

A small loading of nitrogen introduced on carbon cloth anodes by modification with 4-(N,N-dimethylamino)benzene groups led to enhanced performance of a microbial fuel cell. The improvement was attributed to a better bacterial adhesion to the surface without bacterial degradation and negatively affects electron transfer behavior [106]. The graphite anode of a microbial fuel cell was modified with positively, neutral, and negatively charged groups by the diazonium chemistry. A microbial fuel cell used wastewater, acetate as substrate at the anode and the reduction of ferricyanide at the cathode. An optimum loading of grafted molecules is required to obtain the highest performance. Furthermore, the cell with negatively charged carboxylate groups delivered lower power density than that using positively charged grafted groups [107]. Phenylboronic acid groups grafted on the graphite anode of microbial fuel cell according to the procedure described in Scheme 10 presented both in higher power densities and anodic catalytic current densities compared to cell based on an unmodified electrode [108]. Anthraquinone-2-sulfonic acid covalently immobilized onto graphite felt electrode was shown to be efficient electron transfer mediator from bacteria to electrode and significantly enhanced the power production of a microbial fuel cell, which also displayed good stability [109]. Finally, it is interesting to note that an oxygen reducing mixed culture biocathode with complex bacterial biofilms was more activated when the graphite electrode support was not modified by the diazonium chemistry [110].

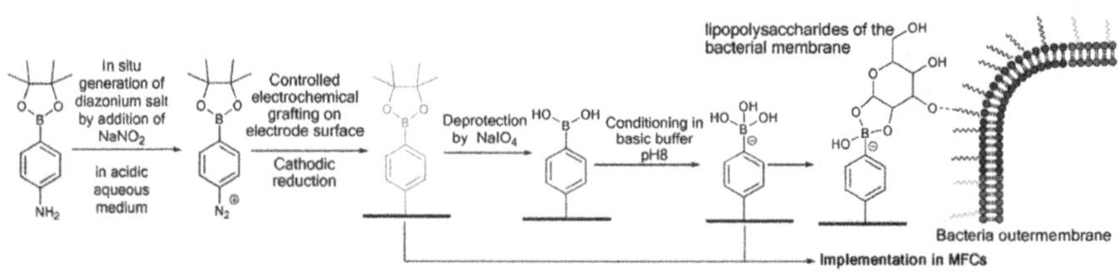

Scheme 10 Functionalization of electrode before implementation in microbial fuel cell [110]. Reused with permission from Reference [110]. Copyright (2013) (Wiley–VCH Verlag)

6 Conclusion

In the past two decades, the diazonium chemistry has become a popular method to modify various surfaces used in electrochemical technologies systems. Although this modification method has some disadvantages that have been described above, its relative simplicity to use makes it useful for scientists of various fields. In the area of electrochemical energy storage, its usefulness has been demonstrated in various systems as described above. For lithium-ion batteries, chemically grafted groups have yielded electrode materials with improved stability and less prone to degradation due to mitigation of the direct contact with the electrolyte. Electrochemical capacitors have benefited from the presence of grafted species that provided enhanced charge storage ability. The latter was associated with the covalent immobilization of redox species that prevented dissolution in the electrolyte. In the case of fuel cells, the diazonium chemistry was useful to tailor the hydrophobicity and hydrophilicity of the electrode assembly as well as the control on water management and catalyst utilization. Finally, in biofuel and microbial fuel cells, the adhesion of the biological components was promoted by appropriate selected grafted groups.

Acknowledgements This chapter is dedicated to Jean Pinson for his 80th birthday that should be considered as "the" pioneer of the field of chemical modification with the diazonium chemistry. Jean was a great inspiration for the work performed in our laboratory in the past 25 years. I also would like to acknowledge the contribution of graduate students, postdoctoral fellows, and collaborators that were productively involved in fundamental research (not mentioned in this chapter) and more applied projects that are presented in this chapter.

References

1. Watkins BF, Behling JR, Kariv E, Miller LL (1975) A chiral electrode. J Am Chem Soc 97:3549–3550
2. Murray RW (1980) Chemically modified electrodes. Acc Chem Res 13:135–141
3. Pinson J, Podvorica F (2005) Attachment of organic layers to conductive or semiconductive surfaces by reduction of diazonium salts. Chem Soc Rev 34:429–439
4. Bélanger D, Pinson J (2011) Electrografting: a powerful method for surface modification. Chem Soc Rev 40:3995–4048
5. Shao Y, El-Kady MF, Sun J, Li Y, Zhang Q, Zhu M, Wang H, Dunn B, Kaner RB (2018) Design and mechanisms of asymmetric supercapacitors. Chem Rev 118:9233–9280
6. Assresahegn BD, Brousse T, Bélanger D (2015) Advances on the use of diazonium chemistry for functionalization of materials used in energy storage systems. Carbon 92:362–381
7. Brousse T, Cougnon C, Bélanger D (2018) Grafting of quinones on carbons as active electrode materials in electrochemical capacitors. J Braz Chem Soc 29:989–997
8. Anothumakkool B, Guyomard D, Gaubicher J, Madec L (2017) Interest of molecular functionalization for electrochemical storage. Nano Res 10:4175–4200
9. Palacín MR (2009) Recent advances in rechargeable battery materials: a chemist's perspective. Chem Soc Rev 38:2565–2575
10. Winter M, Barnett B, Xu K (2018) Before Li Ion batteries. Chem Rev 118:11433–11456

11. Pan Q, Wang H, Jiang Y (2007) Covalent modification of natural graphite with lithium benzoate multilayers via diazonium chemistry and their application in lithium ion batteries. Electrochem Commun 9:754–760
12. Pan Q, Wang H, Jiang Y (2007) Natural graphite modified with nitrophenyl multilayers as anode materials for lithium ion batteries. J Mater Chem 17:329–334
13. Pan Q, Jiang Y (2008) Effect of covalently bonded polysiloxane multilayers on the electrochemical behavior of graphite electrode in lithium ion batteries. J Power Sources 178:379–386
14. Verma P, Novák P (2012) Formation of artificial solid electrolyte interphase by grafting for improving Li-ion intercalation and preventing exfoliation of graphite. Carbon 50:2599–2614
15. Moock DS, Steinmüller SO, Wessely ID, Llevot A, Bitterer B, Meier MAR, Bräse S, Ehrenberg H, Scheiba F (2018) Surface Functionalization of silicon, HOPG, and graphite electrodes: toward an artificial solid electrolyte interface. ACS Appl Mater Interfaces 10:24172–24180
16. Luo F, Liu B, Zheng J, Chu G, Zhong K, Li H, Huang X, Chen L (2015) Review-Nano-silicon/carbon composite anode materials towards practical application for next generation Li-ion batteries. J Electrochem Soc 162:A2509–A2528
17. Obrovac MN (2018) Si-alloy negative electrodes for Li-ion batteries. Curr Opin Electrochem 9:8–17
18. Yang S, Pan Q, Liu J (2010) Improving the cycleability of Si anodes by covalently grafting with 4-carboxyphenyl groups. Electrochem Commun 12:479–482
19. Martin C, Alias M, Christien F, Crosnier O, Bélanger D, Brousse T (2009) Graphite-grafted silicon nanocomposite as a negative electrode for lithium-ion batteries. Adv Mater 21:4735–4741
20. Martin C, Crosnier O, Retoux R, Bélanger D, Schleich DM, Brousse T (2011) Chemical coupling of carbon nanotubes and silicon nanoparticles for improved negative electrode performance in lithium-ion batteries. Adv Func Mater 21:3524–3530
21. Yang S, Li G, Zhu Q, Pan Q (2012) Covalent binding of Si nanoparticles to graphene sheets and its influence on lithium storage properties of Si negative electrode. J Mater Chem 22:3420–3425
22. Assresahegn BD, Bélanger D (2017) Synthesis of binder-like molecules covalently linked to silicon nanoparticles and application as anode material for lithium-ion batteries without the use of electrolyte additives. J Power Sources 345:190–201
23. Assresahegn BD, Bélanger D (2017) Effects of the formulations of silicon-based composite anodes on their mechanical, storage, and electrochemical properties. Chemsuschem 10:4080–4089
24. Assresahegn BD, Ossonon BD, Bélanger D (2018) Graphene nanosheets and polyacrylic acid grafted silicon composite anode for lithium ion batteries. J Power Sources 391:41–50
25. Wang J, Sun X (2012) Understanding and recent development of carbon coating on $LiFePO_4$ cathode materials for lithium-ion batteries. Energy Environ Sci 5:5163–5185
26. Li W, Song B, Manthiram A (2017) High-voltage positive electrode materials for lithium-ion batteries. Chem Soc Rev 46:3006–3059
27. Tanguy F, Gaubicher J, Gaillot A-C, Guyomard D, Pinson J (2009) Lowering interfacial chemical reactivity of oxide materials for lithium batteries. A molecular grafting approach. J Mater Chem 19: 4771–4777
28. Madec L, Humbert B, Lestriez B, Brousse T, Cougnon C, Guyomard D, Gaubicher J (2013) Covalent vs. non-covalent redox functionalization of C-$LiFePO_4$ based electrodes. J Power Sources 232:246–253
29. Madec L, Robert D, Moreau P, Bayle-Guillemaud P, Guyomard D, Gaubicher J (2013) Synergistic effect in carbon coated $LiFePO_4$ for high yield spontaneous grafting of diazonium salt. Structural examination at the grain agglomerate scale. J Am Chem Soc 135:11614–11622
30. Madec L, Seid KA, Badot J-C, Humbert B, Moreau P, Dubrunfaut O, Lestriez B, Guyomard D, Gaubicher J (2014) Redirected charge transport arising from diazonium grafting of carbon coated $LiFePO_4$. Phys Chem Chem Phys 16:22745–22753

31. Delaporte N, Perea A, Amin R, Zaghib K, Bélanger D (2015) Chemically grafted carbon-coated $LiFePO_4$ using diazonium chemistry. J Power Sources 280:246–255
32. Delaporte N, Perea A, Lebègue E, Ladouceur S, Zaghib K, Bélanger D (2015) Increasing the affinity between carbon-coated $LiFePO_4$/C electrodes and conventional organic electrolyte by spontaneous grafting of a benzene-trifluoromethylsulfonimide moiety. ACS Appl Mater Interfaces 7:18519–18529
33. Guo L, Zhang Y, Wang J, M, L, Ma S, Zhang Y, Wang E, Bi Y, Wang D, McKee WC, Xu Y, Chen J, Zhang Q, Nan C, Gu L, Bruce PG, Peng Z (2015) Unlocking the energy capabilities of micron-sized $LiFePO_4$. Nat Commun 6, art. no. 7898
34. Delaporte N, Zaghib K, Bélanger D (2016) In situ formation of bromobenzene diazonium ions and their spontaneous reaction with carbon-coated $LiFePO_4$ in organic media. New J Chem 40:6135–6140
35. Delaporte N., Trudeau ML, Bélanger D, Zaghib K (2020) Protection of $LiFePO_4$ against moisture. Materials 13(4), art. no. 942
36. Alloin F, Crepel L, Cointeaux L, Leprêtre J-C, Fusalba F, Martinet S (2013) The interest of diazonium chemistry for aqueous Lithium-ion battery. J Electrochem Soc 160:A3171–A3178
37. Sun Z, Li Z, Gao L, Zhao X, Han D, Gan S, Guo S, Niu L (2019) Grafting Benzenediazonium Tetrafluoroborate onto $LiNi_xCo_yMn_zO_2$ Materials Achieves Subzero-Temperature High-Capacity Lithium-Ion Storage via a Diazonium Soft-Chemistry Method. Adv Energy Mater 9(6), art. no. 1802946
38. Delaporte N, Belanger RL, Lajoie G, Trudeau M, Zaghib K (2019) Multi-carbonyl molecules immobilized on high surface area carbon by diazonium chemistry for energy storage applications. Electrochim Acta 308:99–114
39. Jaffe A, Saldivar Valdes A, Karunadasa HI (2015) Quinone-functionalized carbon black cathodes for lithium batteries with high power densities. Chem Mater 27:3568–3571
40. Charrier G, Desrues A, Barchasz C, Leroy J, Cornut R, Jousselme B, Campidelli S (2016) Covalently functionalized carbon nanotubes as stable cathode materials of lithium/organic batteries. J Mater Chem A 4:15036–15040
41. Guo L, Wang J, Gu F, Ma L, Zhao Z, Liu J, Peng Z (2019) Relieving the "sudden Death" of Li-O_2 batteries by grafting an antifouling film on cathode surfaces. ACS Appl Mater Interfaces 11:14753–14758
42. Charrier G, Kamaleddine H, Barchasz C, Cornut R, Jousselme B, Campidelli S (2018) Sulfur-containing molecules grafted on carbon nanotubes as highly cyclable cathodes for lithium/organic batteries. ChemElectroChem 5:1732–1737
43. Soon J, Chae S, Lee TJ, Jung J, Ryu JH, Oh SM (2018) Grafting nitrophenyl groups on carbon surfaces by diazonium chemistry to suppress irreversible reactions in high-voltage $LiNi_{0.5}Mn_{1.5}O_4$ positive electrodes. J Electrochem Soc 165:A1372-A1376
44. Saneifar H, Delaporte N, Zaghib K, Bélanger D (2019) Functionalization of the carbon additive of a high-voltage Li-ion cathode. J Mater Chem A 7:1585–1597
45. Saneifar H, Bélanger D (2021) Synthesis and characterization of aryl substituted functionalized graphene sheets and their electrochemical behavior. J Solid State Electrochem 25:149–158
46. Saneifar H, Delaporte N, Shul G, Bélanger D (2019) Attachment of redox active molecules on the carbon additive and its effect on the cycling performance of $LiFePO_4$ electrodes. Mater Chem Phys 235, art. no. 121739
47. Saneifar H, Zaghib K, Bélanger D (2020) Crown ether functionalized conductive carbon for high-voltage spinel $LiNi_{0.5}Mn_{1.5}O_4$/Graphite Cell. ACS Appl Energy Mater 3:647–657
48. Béguin F, Presser V, Balducci A, Frackowiak E (2014) Carbons and electrolytes for advanced supercapacitors. Adv Mater 26:2219–2251
49. Mathis TS, Kurra N, Wang X, Pinto D, Simon P, Gogotsi Y (2019) Energy storage data reporting in perspective—guidelines for interpreting the performance of electrochemical energy storage systems. Adv Energy Mater 9, art. no. 1902007
50. Kalinathan K, DesRoches DP, Liu X, Pickup PG (2008) Anthraquinone modified carbon fabric supercapacitors with improved energy and power densities. J Power Sources 181:182–185

51. Pognon G, Brousse T, Bélanger D (2011) Effect of molecular grafting on the pore size distribution and the double layer capacitance of activated carbon for electrochemical double layer capacitors. Carbon 49:1340–1348
52. Pognon G, Brousse T, Demarconnay L, Bélanger D (2011) Performance and stability of electrochemical capacitor based on anthraquinone modified activated carbon. J Power Sources 196:4117–4122
53. Comte AL, Pognon G, Brousse T, Bélanger D (2013) Determination of the quinone-loading of a modified carbon powder-based electrode for electrochemical capacitor. Electrochemistry 81:863–866
54. .Le Comte A, Brousse T, Bélanger D (2014) Simpler and greener grafting method for improving the stability of anthraquinone-modified carbon electrode in alkaline media. Electrochimica Acta 137:447–453
55. Brousse K, Martin C, Brisse AL, Lethien C, Simon P, Taberna PL, Brousse T (2017) Anthraquinone modification of microporous carbide derived carbon films for on-chip micro-supercapacitors applications. Electrochim Acta 246:391–398
56. Gorska B, Ratajczak P, Béguin F (2019) Faradaic processes on quinone-grafted carbons in protic ionic liquid electrolyte. Electrochimica Acta, 328, art. no. 135090
57. Malka D, Hanna O, Hauser T, Hayne S, Luski S, Elias Y, Attias R, Brousse T, Aurbach D (2018) Improving the capacity of electrochemical capacitor electrode by grafting 2-aminoanthraquinone over kynol carbon cloth using diazonium chemistry. J Electrochem Soc 165:A3342–A3349
58. Pameté Yambou E, Gorska B, Ratajczak P, Béguin F (2020) Hybrid capacitor with anthraquinone-grafted carbon as a battery-type electrode operating in a low pH aqueous salt solution. J Mater Chem A 8:13548–13557
59. Boota M, Urbankowski P, Porzio W, Barba L, Osti NC, Bleuel M, Keum JK, Mamontov E (2020) Understanding functionalization of titanium carbide (MXene) with quinones and their pseudocapacitance. ACS Appl Energy Mater 3:4127–4133
60. Pognon G, Cougnon C, Mayilukila D, Bélanger D (2012) Catechol-modified activated carbon prepared by the diazonium chemistry for application as active electrode material in electrochemical capacitor. ACS Appl Mater Interfaces 4:3788–3796
61. Lebègue E, Brousse T, Crosnier O, Gaubicher J, Cougnon C (2012) Direct introduction of redox centers at activated carbon substrate based on acid-substituent-assisted diazotization. Electrochem Commun 25:124–127
62. Lebègue E, Brousse T, Gaubicher J, Cougnon C (2013) Chemical functionalization of activated carbon through radical and diradical intermediates. Electrochem Commun 34:14–17
63. Cougnon C, Lebègue E, Pognon G (2015) Impedance spectroscopy study of a catechol-modified activated carbon electrode as active material in electrochemical capacitor. J Power Sources 274:551–559
64. Touzé E, Gohier F, Daffos B, Taberna P-L, Cougnon C (2018) Improvement of electrochemical performances of catechol-based supercapacitor electrodes by tuning the redox potential via different-sized O-protected catechol diazonium salts. Electrochim Acta 265:121–130
65. Le Comte A, Chhin D, Gagnon A, Retoux R, Brousse T, Bélanger D (2015) Spontaneous grafting of 9,10-phenanthrenequinone on porous carbon as an active electrode material in an electrochemical capacitor in an alkaline electrolyte. J Mater Chem A 3:6146–6156
66. Abbas Q, Ratajczak P, Babuchowska P, Le Comte A, Bélanger D, Brousse T, Béguin F (2015) Strategies to improve the performance of carbon/carbon capacitors in salt aqueous electrolytes. J Electrochem Soc 162:A5148–A5157
67. Legoupy S, Lebègue E, Cougnon C (2016) Preparation of a tetrahydroxyphenazine-modified carbon as cathode material for supercapacitor in aqueous acid electrolyte. Electrochem Commun 70:47–50
68. Menanteau T, Benoît C, Breton T, Cougnon C (2016) Enhancing the performance of a diazonium-modified carbon supercapacitor by controlling the grafting process. Electrochem Commun 63:70–73

69. Farquhar AK, Smith SR, Dyck CV, McCreery RL (2020) Large capacity enhancement of carbon electrodes by solution processing for high density energy storage. ACS Appl Mater Interfaces 12:10211–10223
70. Amiri M, Shul G, Donzel N, Bélanger D Aqueous Electrochemical energy storage system based on phenanthroline- and anthraquinone-modified carbon electrodes. In preparation
71. Lebègue E, Benoit C, Brousse T, Gaubicher J, Cougnon C (2015) Effect of the porous texture of activated carbons on the electrochemical properties of molecule-grafted carbon products in organic media. J Electrochem Soc 162:A2289–A2295
72. Algharaibeh Z, Pickup PG (2011) An asymmetric supercapacitor with anthraquinone and dihydroxybenzene modified carbon fabric electrodes. Electrochem Commun 13:147–149
73. Algharaibeh Z, Liu X, Pickup PG (2009) An asymmetric anthraquinone-modified carbon/ruthenium oxide supercapacitor. J Power Sources 187:640–643
74. Le Comte A, Brousse T, Bélanger D (2016) New generation of hybrid carbon/$Ni(OH)_2$ electrochemical capacitor using functionalized carbon electrode. J Power Sources 326:702–710
75. Hanna O, Luski S, Brousse T, Aurbach D (2017) Aqueous energy-storage cells based on activated carbon and $LiMn_2O_4$ electrodes. J Power Sources 354:148–156
76. Lebègue E, Brousse T, Gaubicher J, Retoux R, Cougnon C (2014) Toward fully organic rechargeable charge storage devices based on carbon electrodes grafted with redox molecules. J Mater Chem A 2:8599–8602
77. Balducci A, Belanger D, Brousse T, Long JW, Sugimoto W (2017) A guideline for reporting performance metrics with electrochemical capacitors: from electrode materials to full devices. J Electrochem Soc 164:A1487–A1488
78. Brousse T, Bélanger D, Long JW (2015) To be or not to be pseudocapacitive? J Electrochem Soc 162:A5185–A5189
79. Ramirez-Castro C, Crosnier O, Athouël L, Retoux R, Bélanger D, Brousse T (2015) Electrochemical performance of carbon/MnO_2 nanocomposites prepared via molecular bridging as supercapacitor electrode materials. J Electrochem Soc 162:A5179–A5184
80. Rebuttini V, Fazio E, Santangelo S, Neri F, Caputo G, Martin C, Brousse T, Favier F, Pinna N (2015) Chemical Modification of Graphene Oxide through Diazonium Chemistry and Its Influence on the Structure-Property Relationships of Graphene Oxide-Iron Oxide Nanocomposites. Chem Eur J 21:12465–12474
81. Pech D, Guay D, Brousse T, Bélanger D (2008) Concept for charge storage in electrochemical capacitors with functionalized carbon electrodes. Electrochem Solid-State Lett 11:A202–A205
82. Assresahegn BD, Bélanger D (2015) Multifunctional carbon for electrochemical double-layer capacitors. Adv Func Mater 25:6775–6785
83. Lee K, Yoon Y, Cho Y, Lee SM, Shin Y, Lei H, Lee H (2016) Tunable Sub-nanopores of graphene flake interlayers with conductive molecular linkers for Supercapacitors. ACS Nano 10:6799–6807
84. Borup R, Meyers J, Pivovar B, Kim YS, Mukundan R, Garland N, Myers D, Wilson M, Garzon F, Wood D, Zelenay P, More K, Stroh K, Zawodzinski T, Boncella J, McGrath JE, Inaba M, Miyatake K, Hori M, Ota K, Ogumi Z, Miyata S, Nishikata A, Siroma Z, Uchimoto Y, Yasuda K, Kimijima K-I, Iwashita N (2007) Scientific aspects of polymer electrolyte fuel cell durability and degradation. Chem Rev 107:3904–3951
85. Xu Z, Qi Z, Kaufman A (2003) Advanced fuel cell catalysts: Sulfonation of carbon-supported catalysts using 2-aminoethanesulfonic acid. Electrochem Solid-State Lett 6:A171–A173
86. Saab AP, Garzon FH, Zawodzinski TA (2003) The effects of processing conditions and chemical composition on electronic and ionic resistivities of fuel cell electrode composites. J Electrochem Soc 150:A214–A218
87. Selvarani G, Sahu AK, Choudhury NA, Sridhar P, Pitchumani S, Shukla AK (2007) A phenyl-sulfonic acid anchored carbon-supported platinum catalyst for polymer electrolyte fuel cell electrodes. Electrochim Acta 52:4871–4877
88. Xu F, Wang M-X, Sun L, Liu Q, Sun H-F, Stach EA, Xie J (2013) Enhanced Pt/C catalyst stability using p-benzensulfonic acid functionalized carbon blacks as catalyst supports. Electrochim Acta 94:172–181

89. Xu Z, Qi Z, Kaufman A (2005) Hydrophobization of carbon-supported catalysts with 2,3,4,5,6-pentafluorophenyl moieties for fuel cells. Electrochem Solid-State Lett 8:A492–A494
90. Kim JY, Lee S, Kim T-Y, Kim H-T (2014) A simple diazonium coupling reaction enhances durability of modified graphitic carbons used as catalyst supports for polymer electrolyte membrane fuel cell. Electrochim Acta 134:418–425
91. Forouzandeh F, Li X, Banham DW, Feng F, Joseph Kakanat A, Ye S, Birss V (2018) Improving the corrosion resistance of proton exchange membrane fuel cell carbon supports by pentafluorophenyl surface functionalization. J Power Sources 378:732–741
92. Thomas YRJ, Benayad A, Schroder M, Morin A, Pauchet J (2015) New method for super hydrophobic treatment of gas diffusion layers for proton exchange membrane fuel cells using electrochemical reduction of diazonium salts. ACS Appl Mater Interfaces 7:15068–15077
93. Park Y-B, You E, Pak C, Min M (2018) Preparation and characterization of durable catalyst via diazonium reaction in PEMFC. Electrochim Acta 284:242–252
94. Kim JY, Kim YS, Lee S, Pak C, Kim H-T (2015) Enhanced durability of linker-containing carbon nanotube functionalized via diazonium reaction. Electrochim Acta 154:63–69
95. Zhou Z-Y, Kang X, Song Y, Chen S (2012) Ligand-mediated electrocatalytic activity of Pt nanoparticles for oxygen reduction reactions. J Phys Chem C 116:10592–10598
96. Pauric AD, MacLean BJ, Easton EB (2011) Fe-N/C oxygen reduction catalysis prepared by covalent attachment of 1,10-phenanthroline to a carbon surface. J Electrochem Soc 158:B331–B336
97. Xiao X, Xia H-Q, Wu R, Bai L, Yan L, Magner E, Cosnier S, Lojou E, Zhu Z, Liu A (2019) Tackling the challenges of enzymatic (Bio)Fuel cells. Chem Rev 119:9509–9558
98. Boland S, Jenkins P, Kavanagh P, Leech D (2009) Biocatalytic fuel cells: a comparison of surface pre-treatments for anchoring biocatalytic redox films on electrode surfaces. J Electroanal Chem 626:111–115
99. Tasca F, Harreither W, Ludwig R, Gooding JJ, Gorton L (2011) Cellobiose dehydrogenase aryl diazonium modified single walled carbon nanotubes: Enhanced direct electron transfer through a positively charged surface. Anal Chem 83:3042–3049
100. Ortiz R, Ludwig R, Gorton L (2014) Highly efficient membraneless glucose bioanode based on corynascus thermophilus cellobiose dehydrogenase on aryl diazonium-activated single-walled carbon nanotubes. ChemElectroChem 1:1948–1956
101. Gentil S, Rousselot-Pailley P, Sancho F, Robert V, Mekmouche Y, Guallar V, Tron T, Le Goff A (2020) Efficiency of site-specific clicked laccase, carbon nanotubes biocathodes towards O_2 Reduction. Chem Eur J 26:4798–4804
102. Zheng M, Griveau S, Dupont-Gillain C, Genet MJ, Jolivalt C (2015) Oxidation of laccase for improved cathode biofuel cell performances. Bioelectrochemistry 106:77–87
103. Rengaraj S, Kavanagh P, Leech D (2011) A comparison of redox polymer and enzyme co-immobilization on carbon electrodes to provide membrane-less glucose/O_2 enzymatic fuel cells with improved power output and stability. Biosens Bioelectron 30:294–299
104. Sorrentino I, Gentil S, Nedellec Y, Cosnier S, Piscitelli A, Giardina P, Le Goff A (2019) POXC Laccase from Pleurotus ostreatus: a aigh-performance multicopper enzyme for direct oxygen reduction reaction operating in a proton-exchange membrane fuel cell. ChemElectroChem 6:1023–1027
105. Gentil S, Che Mansor SM, Jamet H, Cosnier S, Cavazza C, Le Goff A (2018) Oriented Immobilization of [NiFeSe] Hydrogenases on covalently and noncovalently functionalized carbon nanotubes for H_2/Air enzymatic fuel cells. ACS Catal 8:3957–3964
106. Saito T, Mehanna M, Wang X, Cusick RD, Feng Y, Hickner MA, Logan BE (2011) Effect of nitrogen addition on the performance of microbial fuel cell anodes. Biores Technol 102:395–398
107. Picot M, Lapinsonnière L, Rothballer M, Barrière F (2011) Graphite anode surface modification with controlled reduction of specific aryl diazonium salts for improved microbial fuel cells power output. Biosens Bioelectron 28:181–188

108. Lapinsonnière L, Picot M, Poriel C, Barrière F (2013) Phenylboronic acid modified anodes promote faster biofilm adhesion and increase microbial fuel cell performances. Electroanalysis 25:601–605
109. Tang X, Li H, Du Z, Ng HY (2014) Spontaneous modification of graphite anode by anthraquinone-2-sulfonic acid for microbial fuel cells. Biores Technol 164:184–188
110. Rothballer M, Picot M, Sieper T, Arends JBA, Schmid M, Hartmann A, Boon N, Buisman CJN, Barrière F, Strik DPBTB (2015) Monophyletic group of unclassified γ-Proteobacteria dominates in mixed culture biofilm of high-performing oxygen reducing biocathode. Bioelectrochemistry 106:167–176

Recent Patents and Industrial Applications

Guy Deniau, Cédric Zobrist, Denis Doizi, Aurélien Doublet, and Gaëlle Charrier

Abstract In this chapter, we will show how it is possible to modify the properties of material surfaces, thanks to the chemistry of aromatic diazonium salts applied to their functionalization. Overall, it is sufficient to play with the fragility of the C–N bond between the aromatic and the diazonium ($-N_2^+$) group. This bond breaks during a reduction reaction and leads to the formation of an extremely reactive aryl radical, which can react with a surface and form a covalently grafted coating by successive additions or initiate a radical polymerization reaction. This diazonium salt chemistry is simple, inexpensive, works in water and does not require energy input. It can be adapted to many substrates and is thus particularly suitable for industrial developments. We will see through the three following examples how to take advantage of these reactions to confer new properties to the surface of materials as varied as those encountered to biocompatibilize surgical implants (example 1), to make a robust optical sensor (example 2) or to treat light alloys for aeronautics (example 3).

1 Supporting Intraocular Lenses: The GraftFast® Process

1.1 Introduction

After working on intravascular implants (stents) and creating the start-up Alchimedics in the early 2000s (now SINOMED), we present here a work on intraocular implants. In this example, the problem was to enhance the biocompatibility of the surfaces of intraocular implants introduced into the eye of patients suffering from lens opacification, i.e. cataract. The aim was to avoid the proliferation of crystalline cells that have not been completely extracted, which, as they develop, cause an opacification of the implant or secondary cataract.

G. Deniau (✉) · C. Zobrist · D. Doizi · A. Doublet · G. Charrier
DRF-IRAMIS-UMR NIMBE, Université Paris-Saclay, CEA Saclay, 91191 Gif-sur-Yvette, France
e-mail: guy.deniau@cea.fr

M. M. Chehimi et al. (eds.), *Aryl Diazonium Salts and Related Compounds*, Physical Chemistry in Action, https://doi.org/10.1007/978-3-031-04398-7_23

In this case, the reduction of the diazonium cation cannot be induced by the insulating surface of the implant, usually an acrylate polymer. Only a reduction initiated in solution can lead to the desired reaction. This is the case with the GraftFast® process.

1.2 Summary of the Graftfast® Process

The Graftfast® process (http://www.Graftfast.com/) was launched by CEA in 2007 [1]. This process was extensively protected by patent deposits [2]. A basic research allowed understanding its mechanism [3–5] and a start-up, Pegastech (http://www.pegastech.com/) was created in 2009 to develop this process for specific markets including metallization of plastics.

The GraftFast® process is a simple, versatile and robust chemical method that allows the chemical modification of *any kind of material* by extremely thin polymer films (between 5 and 100 nm) which consequently do not modify the physical and mechanical properties of the pristine material. It operates at room temperature, with water-based reactants and without any external energy source. It relies on the chemical reduction of diazonium salts in the presence of vinylic monomers, either in solution or in suspension in water. The result is an organic polymer film chemically grafted (thus very robust) on the surface of the substrate previously put in the reacting solution. The control of coating thickness is obtained by adjusting the parameters of synthesis as the reactants concentration (diazonium salt and vinyl monomer) as well as the duration of the immersion. The diazonium salt may even be synthesized in situ from the corresponding aromatic amine (in a so-called "one-pot" experiment) that avoids using poorly stable commercial diazonium salts. A typical Graftfast® experiment is schematized in Fig. 1.

Firstly published in 2007 [6], the mechanism of the Graftfast® reaction relies on several steps, the first being the formation of aryl radicals (Fig. 2i) from the ascorbic acid-induced reduction of a diazonium salt in solution. These aryl radicals play a double role: they (i) graft on the surface immersed in the solution and form, by successive additions, a polyaryl film (Fig. 2ii) and (iii) initiate a radical polymerization reaction of a vinyl monomer introduced in solution (Fig. 2iii). Finally, some

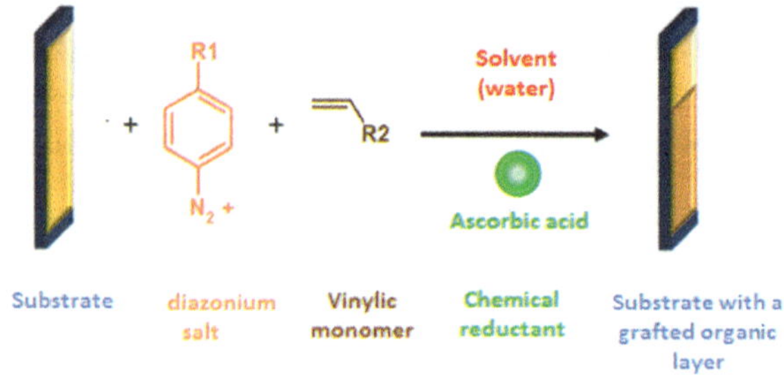

Fig. 1 Scheme of the Graftfast® reaction

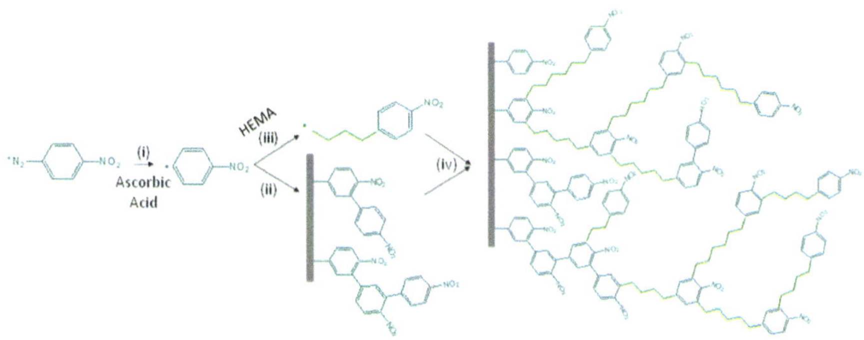

Fig. 2 Simplified schematic illustration of the mechanism of PHEMA (poly hydroxyethyl methacrylate) grafting to gold substrate via the Graftfast® process, using ascorbic acid as potent reducing compound. Reprinted from reference 5 with permission of the author

growing radical polymer chains graft by a "grafting to" pathway (or chain termination) on the aromatic rings present on the surface (Fig. 2iv) to form a thin grafted polymer coatings onto the surface of the substrate.

An advantage linked to the association between the diazonium cation and the vinyl monomer is that the two chemical functions carried by the diazonium cation on the one hand and by the vinyl compound on the other hand are combined on the coating obtained. This allows to widen the range of graftable chemical functions and thus the range of surface properties.

1.3 Heparin-Like Coatings

Glycosaminoglycans (GAGs) are among the most complex polysaccharide chains that are either covalently linked to protein cores (to form proteoglycans) or free as hyaluronan. They exist associated with virtually all cell surfaces and extracellular matrices of higher organisms, where their fine structure facilitates interactions with proteins, which underlie their myriad of biological functions [7]. Heparan sulfate is an important subset of these complex polysaccharides that represent the third major class of biopolymers and is present in the glycocalyx, a carbohydrate-rich layer surrounding virtually every cell in vertebrates and invertebrates. It is a copolymer of uronic acid (iduronic and glucuronic) and glucosamine; this copolymer bears sulfate groups directly attached to its amine and alcohol groups. The spatiotemporal regulation of this chemical structure is responsible of the interactions between cells and of the tissues development. A scheme for heparin is shown in Fig. 3.

When heparin (or heparan sulfate) is covalently attached to support materials, the conformation of binding proteins on such materials is modulated, which may lead to shielding the sites responsible for bonding to eukaryotic and prokaryotic cells, and result in the inhibition of the adhesion and the proliferation of those cells. Such effects

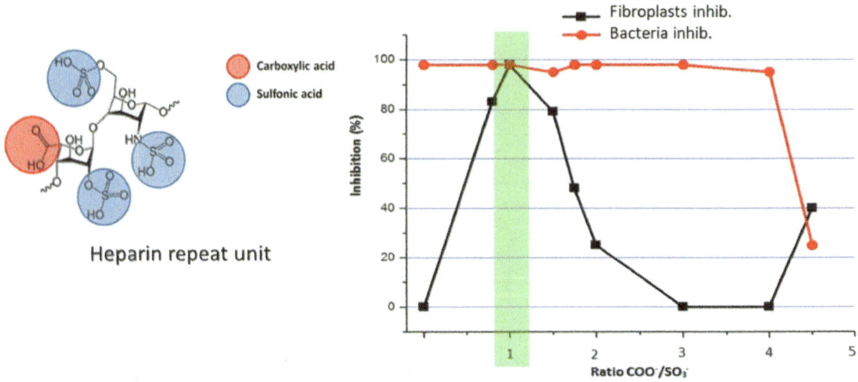

Fig. 3 Effect of the $COO^{-}/SO_3{}^{-}$ ratio (for bulk polymer) on cell and bacteria proliferation for an heparin-like model. Reused with permission from reference 9 Copyright 1997 Elsevier

have however only been demonstrated in vitro. Indeed, in vivo application of these mechanisms that control adhesion and cell proliferation was disappointing because, even attached to the surface of materials, heparin is still biodegradable: hence heparin is rapidly discarded from those surfaces in real conditions. Consequently, "heparin-like" biomimetic copolymers were proposed to overcome that difficulty.

1.4 Synthetic Heparin to Eradicate Secondary Cataracts

In ophthalmology, surgical implants treat many eye diseases, including intraocular lenses (cataract), trabecular drainage implants (glaucoma) and artificial retinas (pigmentary retinopathies). In this context, we worked in a collaborative research program that aims to develop new biomaterials and bioactive coatings that meet both the medical expectations of ophthalmologists and the constraints of the implant market. We chose to graft very thin biomimetic polymeric heparin-like coatings at the surface of intraocular lenses, in order to bring an antiproliferative coating against the epithelial lens cells responsible for the main post-operative complication of cataract surgery: the opacification of the posterior capsule.

Cataract, the first worldwide cause of blindness, is an ageing disorder of the crystalline cells (epithelial lens cells), which are becoming cloudy, causing a progressive decrease of vision by opacification of the natural lens inside the eye. Surgery, which extracts the natural crystalline lens from its capsule and replaces it with a polymer-based intraocular implant, is the only current treatment for cataracts. With more than 20 million implants per year, this surgery is by far the most practiced in the world. The proliferation of the few remaining crystalline cells inside the capsule, inducing secondary cataract, is the main post-operative complication, affecting up to ~20% of cataract surgeries; avoiding cell adhesion is thus a major area of research and a constant concern for manufacturers designing intraocular lenses. Studies have shown that this proliferation is related to the nature of the intraocular lens surface [8]. In fact, secondary cataract is now treated with a YAG laser shot in the eye to open up the posterior side of the capsule, a procedure that, due to the large number of patients who need it, is costly to society and time-consuming for ophthalmologists. In addition, this treatment cannot be given to a patient suffering from another eye disease (glaucoma or macular degeneration).

The in vitro biological properties of biomaterials are generally demonstrated by measurements of their hemocompatibility and the adhesion and the proliferation of bacteria and cells deposited on them. In order to show the antiproliferative characteristic of heparin-like polymers, Jozefowicz et al. [9] designed bulk random copolymers consisting of statistically distributed methylmethacrylate (MMA), methacrylic acid (MA) and styrene sulfonate (SS) monomeric units. MA and SS are present as ions, respectively carboxylate (COO^-) and aryl sulfonate (SO_3^-) groups; it was shown [9] that the biological properties of these polymers depend on the ratio between sulfonate and carboxylate groups (Fig. 3). Hence, it is clear from that preliminary study that, depending on the COO^-/SO_3^- ratio, very significant inhibition of the fibroblast proliferation can be observed on those copolymers. It is noteworthy that antibacterial activity was also observed for a given COO^-/SO_3^- range, which raises hope to be able to gather both properties on the same material. The main objective of the present work is to reproduce such behavior on the surface of the polymer materials commonly used to make ophthalmic implants. To avoid optical aberrations and to preserve the design and the transparency of the implant, ultrathin surface modification by COO^-/SO_3^- groups is the goal to reach.

The first results were acquired by using 4-aminobenzoic acid as precursor of the 4-diazobenzoic acid and 2-acrylamido-2-methylpropane sulfonic acid (AMPS, see insert Fig. 4) as vinylic monomers (in a "one-pot" experiment). Water was used as

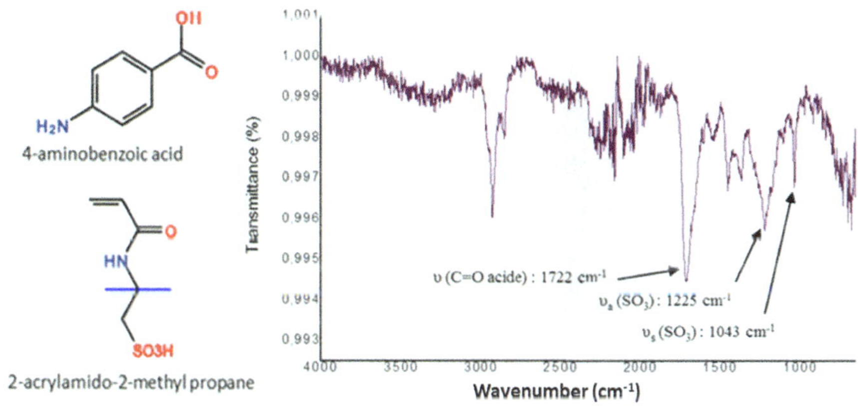

Fig. 4 Left: Chemical structures of the 4-aminobenzoic acid and the 2-acrylamido-2-methylpropane sulfonic acid. Right: IR ATR spectra of a thin film grafted on the surface of a gold substrate

solvent and ascorbic acid as reducer. The treated substrates were blocks of hydrophilic methacrylate (poly hydroxyethylmethacrylate, PHEMA, from which are obtained the artificial lenses). No surface pre-treatments were realized. A gold sample is added to the solution in order to perform IR analysis (it is impossible to conduct IR analysis of a nanometric organic film deposited on an organic substrate). The resulting coating was checked by IR analysis. The characteristic absorption bands $\upsilon C{=}O$ and υSO_3 are present on the spectrum presented Fig. 4.

XPS analysis performed on the same film also demonstrated the presence of carboxylate and sulfonate groups, showing that Graftfast® was able to graft a "heparin-like" film on PHEMA. Biological tests demonstrated that this coating is very efficient in limiting epithelial lens cell proliferation, with less than 4% of counted cells on the coated substrates *versus* the reference uncoated ones (Fig. 5).

The coatings obtained are of nanometric thicknesses, so they do not alter the optical properties of the implant. Covalently grafted, the coatings obtained are stable and robust. Physico-chemical characterizations validated our hypotheses and in vitro culture tests showed Fig. 5 demonstrated that cell proliferation was well inhibited.

While the results presented are convincing, problems of reproducibility have emerged. After analysis of the coatings (IR and XPS) we found that the problems were related to the non-stability of the COO^-/SO_3^- ratio, which is in fact difficult to control, solely by the Graftfast® process, the two molecules do not have the same reactivity. To overcome this problem, we have synthesized bifunctional molecules adaptable to the process.

The molecules that were synthesized are shown in Fig. 6. The precursor molecules of the diazonium cations (aromatic amine) are used alone, the vinylic monomer in

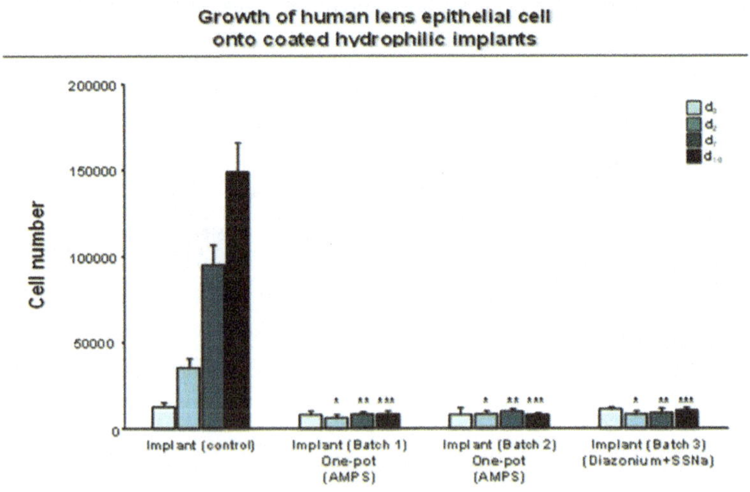

Fig. 5 Cell proliferation study on hydrophilic blocks (d1 = after one day) after grafting with AMPS (batch 1 and 2) and with sodium styrene sulfonate (SSNa, batch 3). Asterisks indicate statistically significant differences from the control

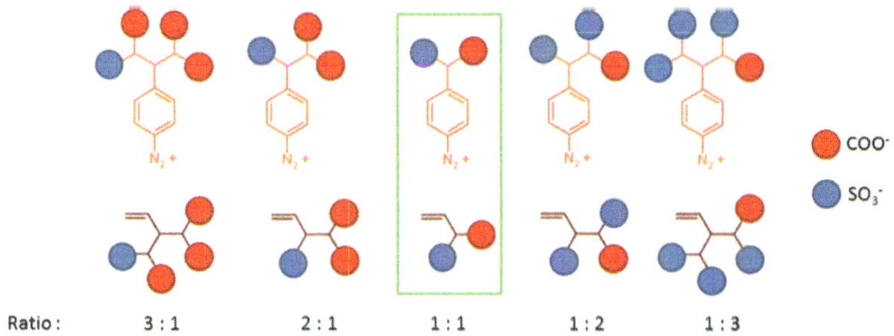

Fig. 6 Synthesized molecules with a fixed COO^-/SO_3^- ratio

association with a neutral salt such as nitrobenzene tetrafluoroborate for example. The COO^-/SO_3^- ratio thus depends only on the molecule chosen during the GraftFast® synthesis. For example, in the case of an aromatic amine carrying one SO_3^- and one COO^- the ratio obtained in the coating can only be one. Using this molecule, we obtained coatings on the surface of the implants, which, after characterization, were tested in vitro. However, the results shown in Fig. 7 (in the middle) are not as good as those obtained previously, as cell proliferation is observed upon days. In fact, the cellular antiproliferation property of heparin results from the concomitant presence of the charged groups COO^- and SO_3^- randomly distributed on the GAG chain. However, our synthesized heparin-like coating does not have this random distribution. We then added a neutral monomer, hydroxyethyl methacrylate (HEMA), to the synthesis medium in order to space these charged groups. The results of the in vitro tests are shown in Fig. 7 on the right.

This strategy has removed the problems of reproducibility of ratio values and made it possible to obtain stable, robust coatings that significantly limit the proliferation

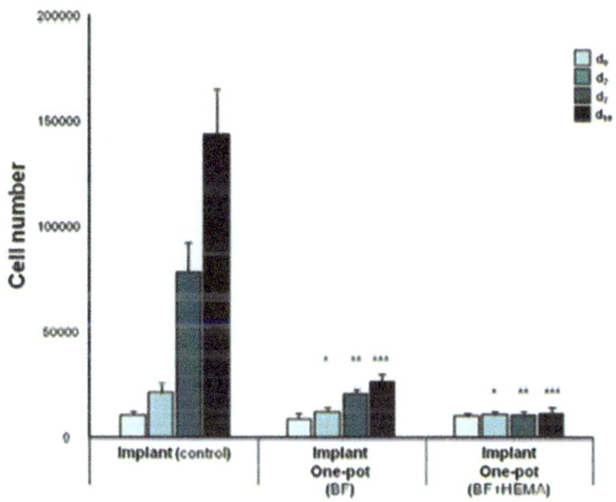

Fig. 7 Growth of human lens epithelial cells onto coated hydrophilic implant. (d1 = after one day)

of crystalline cells and thus the appearance of secondary cataract. This work was patented in 2013 [10].

2 GraftFast® to Improve the Robustness of pH Sensors

2.1 Introduction

This study deals with the field of chemical sensors for pH measurement, and more precisely of fiber optic chemical sensors, also called optodes. Although this kind of optical sensor already exists, the performances of the existing optodes can be improved, especially their stability over time and their lifetime. To achieve this goal, an interesting idea is to replace the conventional immobilization by physical entrapment or electrostatic binding of the pH indicator(s) on the optical fiber by a chemical covalent grafting. The purpose of this work is to study the covalent grafting of dyes for the design of a new type of optodes.

2.2 Optode

A conventional commercial pH measuring optode consists in a computer-controlled spectrometer, a light source, an optical fiber and a probe. This device is shown in Fig. 8.

The probe, where the optical fiber ends, consists in the active area, usually a sol–gel matrix containing a pH sensitive dye, and a light-reflecting mirror. The mirror

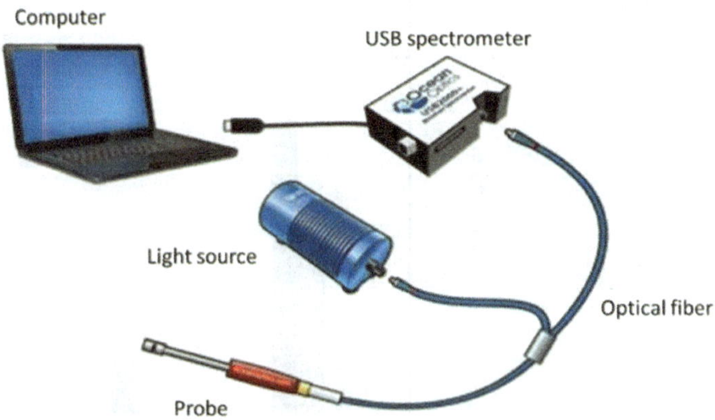

Fig. 8 Schematic diagram of a commercial optode

reflects the light coming from the fibered light source back to the spectrometer. Therefore, the light travels twice through the pH sensor (see Fig. 9).

These commercial optodes suffer from a major problem, the non-stability over time during measurements. Figure 10 shows the response of such an optode over time.

Initially, a drift toward acidic pH is observed during the first hours (about 60 h).Then, a great instability appears after 110 h. The fact that the measured pH value remains essentially the same when the medium is changed proves that the pH drift in the first phase is not due to a change in the measuring medium and therefore not to carbonation of the water by air. The most likely hypothesis to explain this instability is the progressive degradation of the dye patch. Noiré et al. [11] observed this phenomenon during the design of an optode for pH measurement made by immobilization of the dye in a sol–gel matrix through simple impregnation.

Fig. 9 Detail of the probe tip

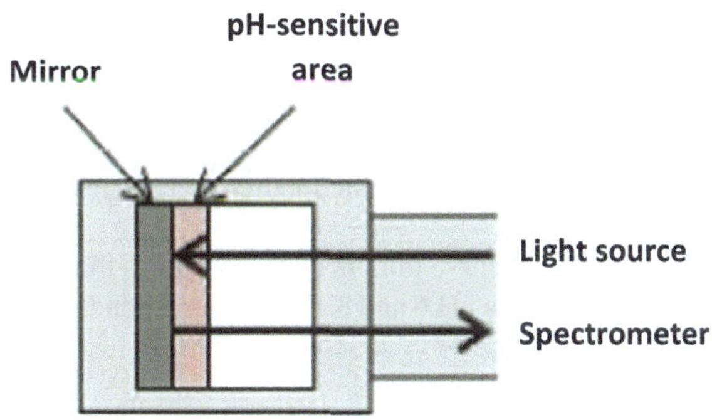

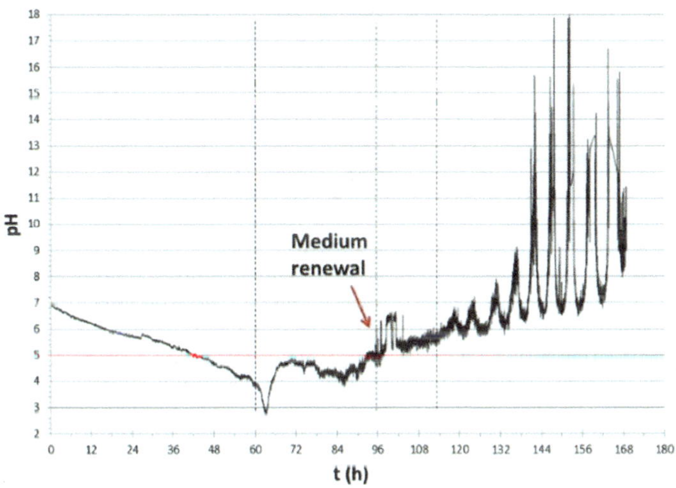

Fig. 10 pH monitoring of an aqueous solution as a function of time

As a result, the choice to study the covalent grafting of dyes for such an application is thus quite relevant, since this covalent grafting of the dye on the optical fiber should make it possible to improve their stability and durability.

To carry out this grafting, the GraftFast® process is used (see example 1 above). After some unsuccessful tests concerning this grafting on the optical fiber material (acrylate or silica), adding a semi-transparent metallic layer, i.e. gold (about 20 nm thick) on silica surfaces was chosen which allows the visible absorption study by transmission and makes infrared spectroscopy analysis easier. Neutral red was used as a colored indicator.

2.3 Neutral Red

Neutral Red was chosen because it is a pH indicator and it is diazotizable thanks to the presence of an aromatic primary amine (see Fig. 11). Red in acidic medium, it becomes yellow in basic medium, with a turning zone between pH 6.8 and pH 8 (pKa = 7.1).

The study of neutral red by UV–visible absorption shows a maximum absorption at 520 nm (in aqueous acidic medium). This absorbance is plotted as a function of pH in Fig. 12.

The sigmoid curve obtained clearly shows that the dye has a zone of quasi-linear variation between pH 6 and 8, which corresponds well to the turning zone of the dye.

Fig. 11 Acido-basic equilibrium of neutral red

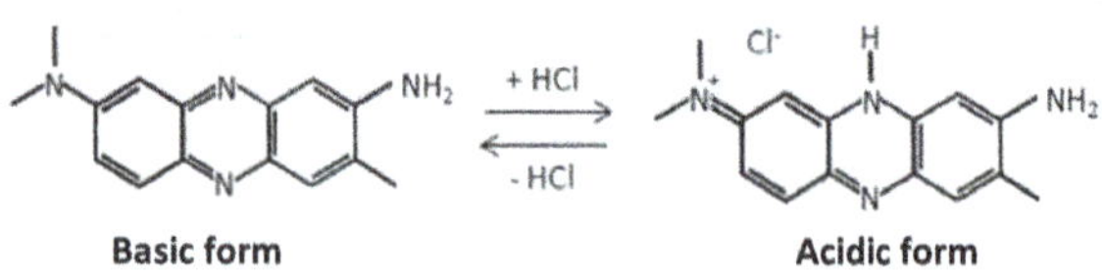

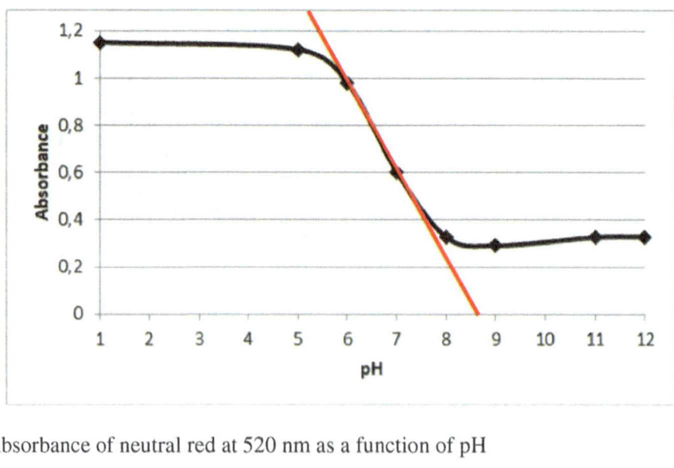

Fig. 12 Absorbance of neutral red at 520 nm as a function of pH

2.4 Neutral Red Grafting

The optimization of the coating synthesis by Graftfast® is carried out by adjusting the different parameters: the initial concentration of neutral red, the oxidant concentration of $NaNO_2$ to form the diazonium cation and the duration of the reaction. The monitoring is performed by infrared spectroscopy with the objective of following the increasing of the main absorption band intensity (around 1600 cm^{-1}) and thus the thickness of the coating and finally the sensitivity of the sensor.

An important parameter for the GraftFast® process is the grafting reaction time, i.e. the time during which the substrate is immersed in the reaction bath after the addition of the reducer (here iron powder). In order to evaluate the influence of the grafting time on the amount grafted, seven golden slides were grafted simultaneously with neutral red ([neutral red] = $6.9.10^{-3}$ M; 1 equivalent $NaNO_2$; HCl 0.1 M) and were removed from the reaction bath respectively 1′, 5′, 10′, 15′, 30′, 45′ and 60′ after the addition of the iron powder. These slides were then analyzed by infrared spectroscopy in ATR mode (Fig. 13), and the intensities of the band due to aromatic bonds at 1610 cm^{-1} were measured (Fig. 14).

In a grafting reaction with a reducing agent in solution (GraftFast® type), a large part of the radicals formed dimerize. The dimers, generally insoluble in the medium, can precipitate on the surface to form a physisorbed component on the film. Beyond 30 min of reaction, the growth of the film stops (see plateau Fig. 14). It is the physisorption that blocks the access of radicals to the chemisorbed part of the coating; they only thicken this physisorbed component. This part is eliminated during the ultrasonic rinsing procedure, which explains the plateau reached for the intensity of the infrared band.

A sample synthesized after an immersion time of 30 min is analyzed using UV–visible spectroscopy. The maximum absorbance for neutral red in solution is obtained

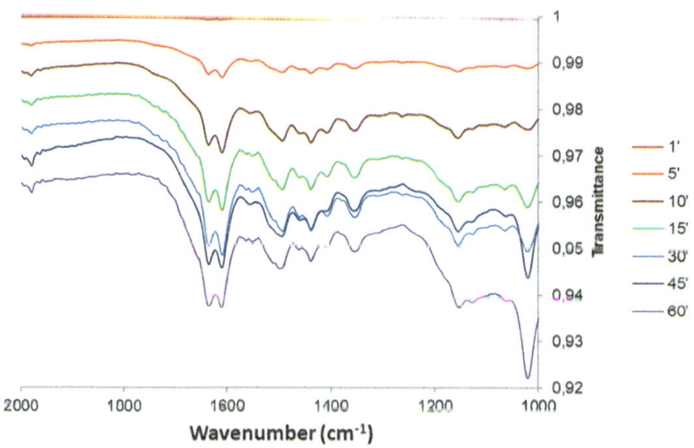

Fig. 13 IR spectra obtained on golden slides for different grafting times

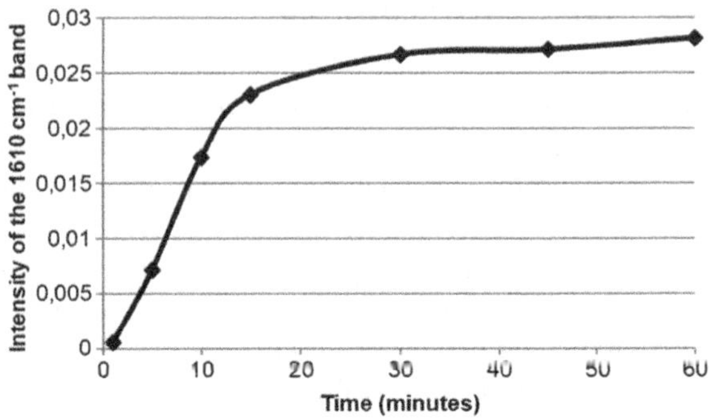

Fig. 14 Intensity of the infrared band due to the aromatic bonds at 1610 cm^{-1} *versus* the grafting reaction time

for a wavelength of 520 nm (see Fig. 12), whereas it is reached here for a wavelength of 550 nm. This difference can be explained by the disappearance of the aromatic primary amine function, which was involved in the conjugated system of the dye and was therefore partly responsible for the optical properties of the dye.

This sample was immersed in different solutions of known pH: 1, 5, 6, 7, 8, 9 and 12. These pH values have been chosen according to the range of variation of the ungrafted neutral red (between pH 5 and 8). The maximum absorption is centered at 550 nm and the corresponding intensities are plotted as a function of pH (see Fig. 15).

This result does not allow to conclude about the optical properties of the grafted neutral red. The shape of the curve is indeed very different from that obtained in solution (sigmoid in Fig. 12), the maximum absorption is shifted. Moreover, the absorption intensities are very (too) low to extract any information. The conclusion

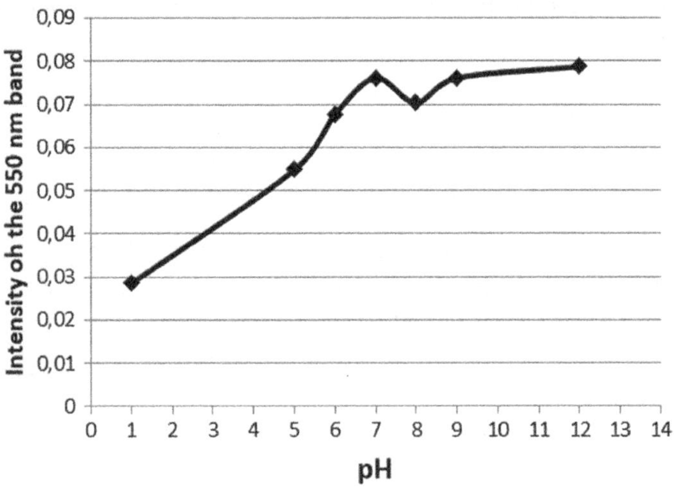

Fig. 15 Height of the band at 550 nm for a golden slide grafted with neutral red, immersed in solutions of different pH values

is that although the synthesis parameters were optimized, the grafted layer remains too thin.

2.5 *Thickening of the Grafted Layer*

In order to increase the thickness of the coating and thus most probably the intensity of the absorption signal, we tried to limit the presence of the physisorbed component at the origin of its stagnation in thickness (plateau of Fig. 14). To do this, our idea was to combine rather long immersion times with successive rinsings. We called this technique sequencing as schematized on Fig. 16.

Different sequencing schemes were used with a constant immersion time of 30 min (1 × 30′, 2 × 15′ …). The results obtained by IR spectrometry (band intensity at 1610 cm^{-1}) are shown on Fig. 17.

It can clearly be seen that the 6 × 5′ synthesis gives the maximum signal in IR spectrometry. The optical properties of the layer obtained from this configuration are then measured in transmittance by UV–visible spectroscopy as a function of pH. The results are shown in Fig. 18.

The curve is similar to that obtained for the neutral red in solution but the pH range is changed from pH 5–8 in solution to pH 1–5 (see Fig. 12) most probably for the same reason that is at the origin of the wavelength shift of the absorption maximum: the loss of the aromatic amine group during grafting. On the other hand, the sigmoid curve obtained clearly shows that this layer is functional for pH measurements.

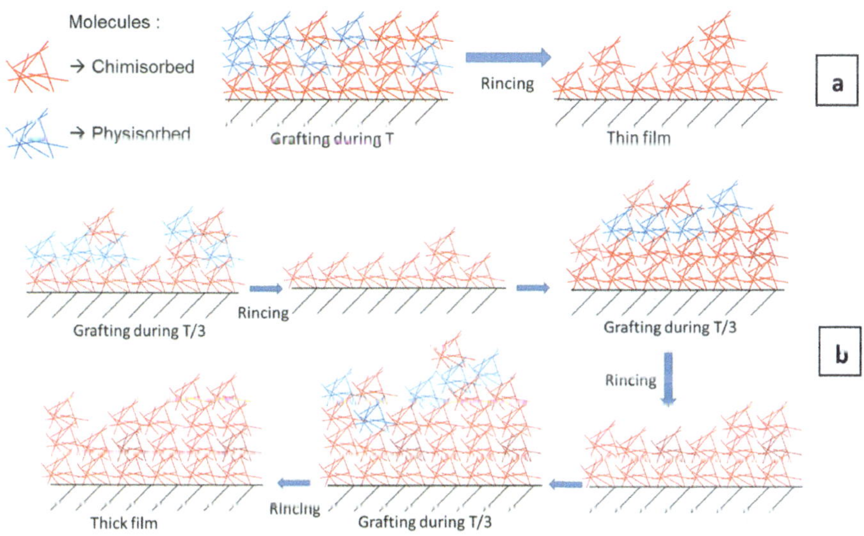

Fig. 16 Schematic diagram of the method: grafting without (**a**) and with (**b**) sequencing

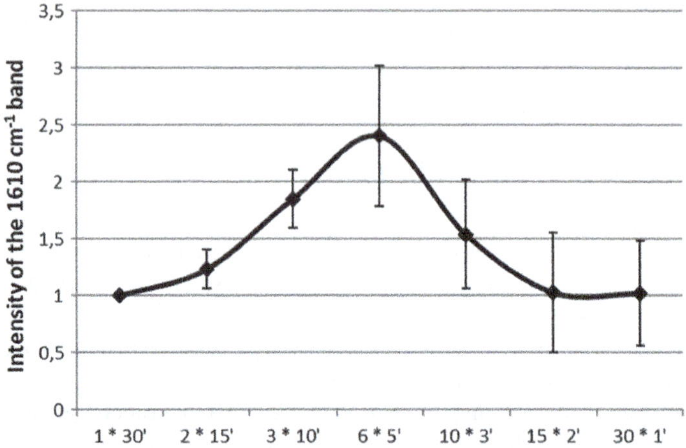

Fig. 17 Normalized intensities of the IR band at 1610 cm^{-1} for different sequences (point = mean over four measurements; bars = standard deviation)

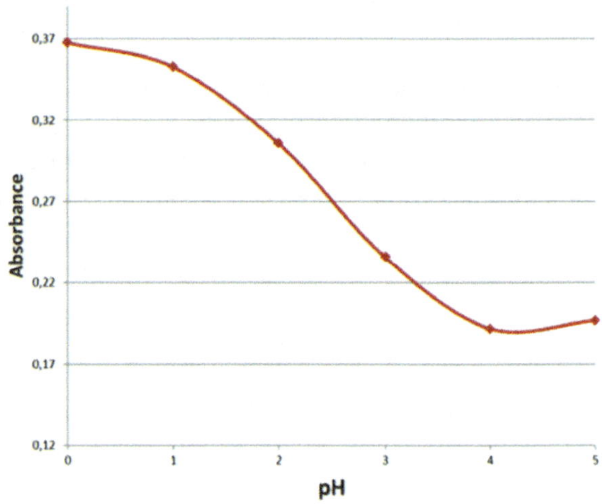

Fig. 18 Absorbance at 550 nm versus pH on golden substrates after neutral red grafting

Thus, it was shown that it is possible to obtain a robust pH-sensitive layer (resistant to solvent/ultrasound rinsing) and thus to improve the stability of the sensor. This work was patented in 2015 [12]. The development is now in progress to test the stability of the layer and to test it in a real optode configuration.

3 Treatment of the Aerospace Alloy Al2024

3.1 Introduction

A new challenge concerning the aeronautic alloys is to provide protection against corrosion together with paint adhesion without the use of chromium treatment. Thanks to its density, its affordable price and its ability to protect against lightning (Faraday cage), aluminum is an essential material in the aeronautical industry [13]. Several alloys can be used depending on the desired properties [14–16]. For the AA2024-T3 alloy, with copper as the primary alloying element (4–5%), the main characteristic is the very good fatigue resistance but a poor corrosion resistance [17, 18]. The most used protection treatment is the anodic oxidation, which consists in thickening the native oxide layer (alumina). This layer then acts as a barrier toward the corrosion [19, 20]. The alloy is polarized as anode and the electrolytic current oxidizes the metallic surface. The growth of the oxide layer is mainly (80%) toward the bulk of the metal. The thicker the oxide layer, the better the protection of aluminum against corrosion. For more than half a century, this anodization has been carried out in chromic acid (Cr^{VI}) electrolytes. The obtained layer is porous which confers an excellent adhesion but weakens the anti-corrosion. A pore sealing step reinforces the barrier effect but weakens the adhesion. Nevertheless, the Reach has classified the Cr^{VI} as a CMR (carcinogenic, mutagenic and reprotoxic) and it will be definitively banned in the next few months [21]. Today, there are no industrialized alternatives to the use of Cr^{VI} that would provide both anticorrosion and paint adhesion on aluminum alloys.

In this work, we have developed a chromium free treatment meeting these two requirements. In addition, this treatment will be the most likely to be industrially applied as it will be inexpensive and green.

3.2 Diazonium Coating as Adhesion Primer

3.2.1 Special Case of Conductive Surfaces

As we know, to obtain a grafted coating on a surface with diazonium cation, the latter must be transformed into an aryl radical. Such radical was easily obtained after reduction of the cation (see Eq. (1)):

$$R{-}ArN_2^+ + 1e^- \rightarrow R{-}Ar^{\cdot} + N_2 \quad (1)$$

The electron generally comes from a conductor (or semiconductor) polarized at the cathode electrode [22, 23], or from a reducing agent added in the solution (see previous paragraph). This electron transfer can be "spontaneous" when the standard potential of the substrate is lower than the reduction potential of the diazonium cation

($E^\circ < -0.4$ V/ENH), which is the case of aluminum ($E^\circ = -1.66$ V/ENH) and its alloys.

After degreasing the surface of the alloy with ethanol, the substrate was simply immersed in a solution containing the diazonium salt in sulfuric acid (1 M) for 2 min. In that case, sulfuric acid is necessary to dissolve the insulating native aluminum oxide layer (Al_2O_3), allowing the passage of electrons and thus enabling the "spontaneous" reduction of the diazonium cation. Without sulfuric acid, no spontaneous reaction occurs. After reduction by the metal released from its oxide, the radical formed grafts onto the metallic surface, following Eq. (2):

$$Al + R{-}Ar^{\cdot} \rightarrow Al - Ar - R \quad (2)$$

Then, a reaction of polymerization type starts by successive additions of the new radicals created [24, 25]. A grafted polymeric thin film is finally obtained on the alloy surface [26, 27].

After this step, a fast, simple and green (no added energy and the solution contains 95% tap water) low cost process is settled. The properties of the coatings can be modified by modulating the chemical groups carried by the diazonium salt.

3.2.2 Choice and Synthesis of Coatings

In the vast majority of cases, aeronautical paints are water-based two-component epoxies containing a cross-linking agent, triethylenetetramine (TETA) and an epoxy pre-polymer, bisphenol A diglycidyl ether (DGEBA). After application of the paint, cross-linking was initiated by heating the painted sample to 80 °C (amine groups of TETA react with epoxy groups of DGEBA). Once cross-linking is complete, a polyepoxide was obtained on the surface of the sample.

In our case, the process described in paragraph 2.1 may be used to increase the poor adhesion of a sealed anodized layer. By using diazonium salts carrying appropriate chemical functions, it should be possible to mimic the cross-linking reactions of the paint directly on the surface. Two solutions can be considered:

- Solution 1: Grafting a diazonium bearing a nucleophilic function (ester or amide) that can attack the epoxy functions carried by the prepolymer (DGEBA).
- Solution 2: Grafting of a diazonium bearing an electrophilic function (carbonyl) that can be attacked by the nucleophilic functions of the cross-linking agent (TETA).

Both solutions were tested, both gave good results especially with the ester and amide functions. Here are presented the results obtained with 4-carbamoyl benzene diazonium (amide) which can preferentially react as an electrophile (–C=O reacting with the amine of the TETA crosslinker) as shown in Fig. 19.

The films were obtained in a one-pot from 4-amino benzamide and sodium nitrite (which oxidizes the amine to diazonium) in a 1 M solution of sulfuric acid by immersing the alloy for a few minutes. They were characterized by IR. The main

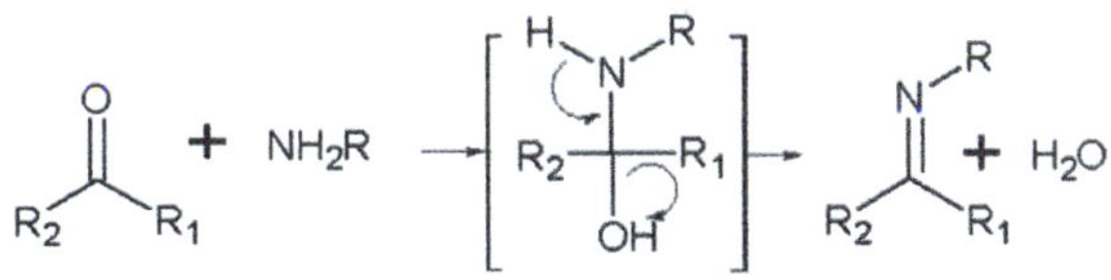

Fig. 19 Reaction between the carbonyl of the amide and the primary amine of TETA

bands of the amide function are well present: wide bands centered at 3200 and 1606 cm^{-1} for the NH bond and at 1664 cm^{-1} for the carbonyl group.

3.2.3 Coating Properties

After covering the surface with paint (cross-linked at 80 °C for 12 h), the samples were tested for adhesion by the standard grid test. The Paint Adhesion Test is a grid test carried out using an Elcometer 107 device, with 1 mm spacing claws, then applying a standard adhesive that will be quickly removed (according to NF EN ISO 2409). There are different levels of acceptance (see the illustration provided by the Elcometer manufacturer in Fig. 20). To show the positive impact of the applied organic coating, it is desirable to achieve a grade 0 or 1 for both wet and dry adhesion. This level of adhesion corresponds to a departure of less than 5% of the paint from the total grid area after removal of the adhesive (Fig. 21).

The grid adhesion test (GAT) shows that amide films can be used as an adhesion primer for paints used in the aerospace industry. Indeed, the results obtained (grade 0) are very encouraging and reveal that the film, due to its porosity and chemical composition, allows a strong anchoring with the paint (see Fig. 22).

Fig. 20 Adhesion criteria according to the NF EN ISO2409 standard

	0%	Grade 0
	5%	Grade 1
	5 à 15%	Grade 2
	15 à 35%	Grade 3
	35 à 65%	Grade 4
	>65%	Grade 5

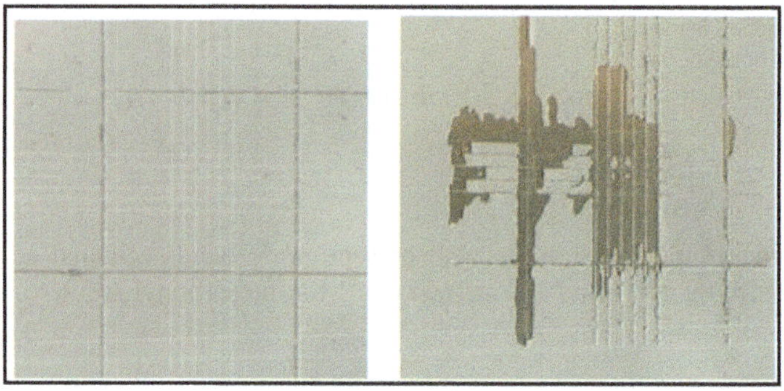

Fig. 21 Example of a dry adhesion test, grade 0 on the left, 5 on the right

Fig. 22 Grade 0 obtained in dry adhesion on the grafted Al2024-T3 sample (4-aminobenzamide—2 min immersion)

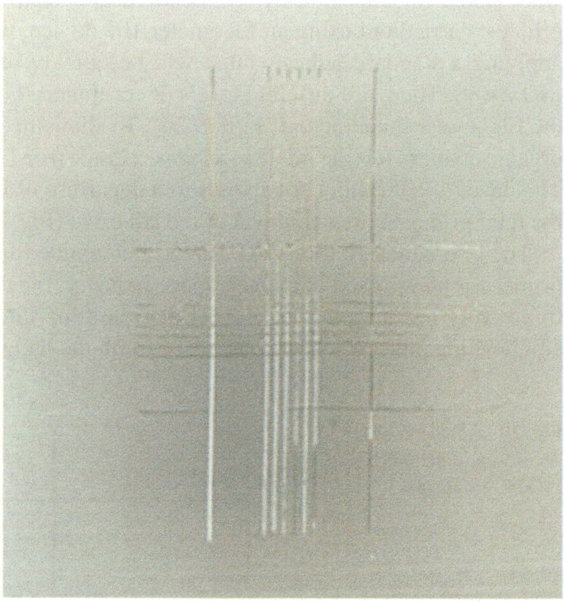

A green low-cost process able to promote paint adhesion on aluminum alloy 2024 is now operating; the last step is to characterize the anticorrosion performance of the coating.

The industry uses two tests for the qualification of anti-corrosion treatments: the salt spray test according to NF EN ISO 9227 and the filiform corrosion (FC) test, the most severe, according to EN NF 3665. The latter has been used in this work.

The FC test is a specialized corrosion test for painted pieces. It consists in making an incision on the surface of the sample (1 mm wide scratch) and applying hydrochloric acid on this incision for 1 min. The samples were then placed in a

Fig. 23 Samples having undergone the filiform corrosion test: **a** Al2024-T3 **b** Al2024-T3 after 4-amidoaryldiazonium grafting **c** Al2024-T3 after anodization treatment

controlled atmosphere (40 °C at 82% relative humidity) for 960 h and the propagation of corrosion is evaluated at the end. The average spreading of cracks on both sides of the incision was measured and must not exceed 2 mm to validate the test (in line with the requirements of many customers). Figure 23 shows the results obtained for filiform corrosion of the bare alloy (a), treated by diazonium grafting (b) and treated by sulfuric anodization (c).

It clearly appears that the organic film is much less effective than the reference treatments currently used in aeronautics to protect aluminum from corrosion.

In order to understand why these grafted films provide little or no protection against corrosion, we have studied the surfaces being treated by scanning electrochemical microscopy (SECM) in "feedback" acquisition mode [28]. This analysis reveals that just after grafting, the native oxidized surface layer (Al_2O_3) was indeed destroyed by sulfuric acid but that it reforms within a few hours. This reflects the porosity of the film [29]. These results explain why this type of coating is not relevant for the protection of the surface of this alloy against corrosion. On the other hand, it suggests that it should be possible to carry out a classical anticorrosion treatment such as anodizing through the organic film.

3.3 Anodization Under Coating

To test this hypothesis, a classical anodization treatment (15 V, 20 min in 1 M sulfuric acid) was performed on samples previously coated by spontaneous grafting with a commercial diazonium (nitrobenzene diazonium).

The formation of the thick layer of alumina through the film was verified by SECM [29]. To ensure that the organic layer was not altered by the anodic treatment, the film was then studied by IR. The results were compared to those obtained with non anodized grafted films. The results of the IR analysis were shown in Fig. 24. After the

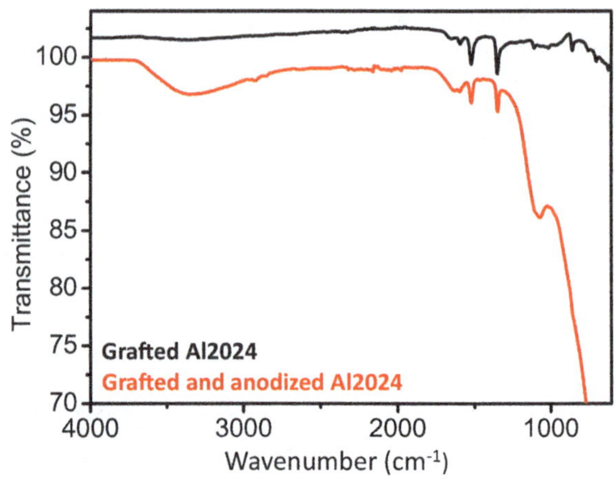

Fig. 24 Infrared spectrum obtained on a surface of Al2024-T3 grafted with 4-nitrobenzenediazonium and a surface of Al2024-T3 grafted and then anodized

anodization, the IR study revealed two intense bands at 1350 and 1520 cm^{-1}, characteristic of the NO_2 group, and another band at 1600 cm^{-1} indicating the presence of phenyl groups. A very wide band was also visible at 700 cm^{-1}. This band, not visible on the spectra of grafted and non-anodized films, was attributed to the vibrations of the Metal-O bonds and thus to the aluminum oxide formed during anodization [30].

This analysis confirms that the anodization has indeed taken place under the film and that the film was not altered and seems to emerge on the surface. This result can be explained, on the one hand, by the fact that the film itself is formed in 1 M sulfuric acid and, on the other hand, it is known that anodic oxidation provokes a growth of alumina mostly (three quarters) toward the interior of the metal. Cross-sectional EDX map analysis confirmed the presence of the aluminum substrate, then aluminum oxide and a polymeric film on top.

The complete process of treatment of the alloy can thus be schematized as shown in Fig. 25.

The last step is to check both paint adhesion and anticorrosion properties.

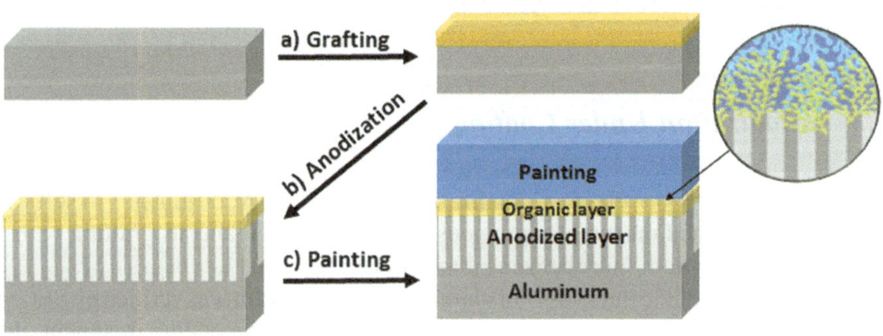

Fig. 25 Illustration of the complete process

For this study, we have synthesized a 4- substituted ester diazonium. The polymeric coating obtained was tested as primer for the water-based bi-component epoxy paint. After grafting on alloy surfaces, a sealed anodization was performed and the samples were finally painted (see Fig. 25). The two classical industrial tests previously described (FC and GAT) were carried out to qualify the new treatment.

The FC test shows that less than 0.5 mm average propagation was obtained for sealed anodized samples, either coated or not with the polymeric film (versus 5 mm for bare alloy). These results show that the anti-corrosive properties of the sealed anodization layer were not damaged by the presence of the organic coating (see Fig. 26a–c and g).

The GAT evaluates the adhesion of the paint on the substrate. For sealed anodized samples, a grade 2 was obtained. When the polymeric film was grafted before the sealed anodization treatment, a grade 0 was obtained, reflecting the quality of the adhesion primer of the grafted coating with respect to the epoxy paint (see Fig. 26d–g).

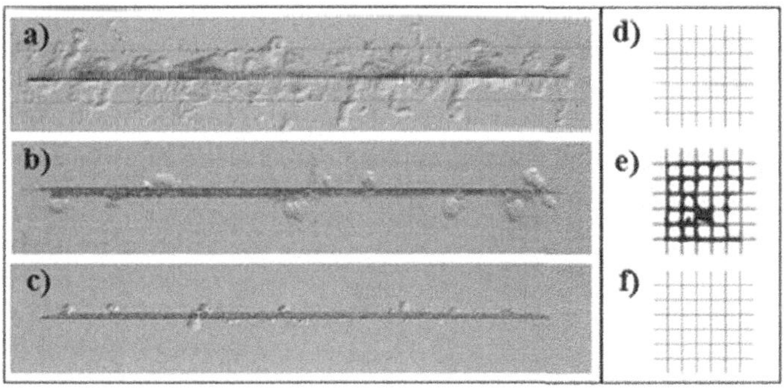

g)

Sample	Filiform corrosion	Grid adhesion
Expectation	<2 mm	Grade 0
Raw aluminum	5 mm	Grade 0
Sealed anodization	<0.5 mm	Grade 2
Organig coating (with $-COOCH_3$) + sealed anodization	<0.5 mm	Grade 0

Fig. 26 Filiform corrosion (FC) result for raw **a** aluminum; **b** sealed anodized aluminum; **c** grafted, anodized and sealed aluminum; **d** Adhesion grade 0 on raw aluminum; **e** Adhesion grade 2 on sealed anodized aluminum; **f** Adhesion grade 0 on grafted, anodized and sealed aluminum **g** Results of the industrial tests (FC and GAT) for the three previous samples

4 Conclusion

A new chromium-free surface treatment has been developed on AA2024-T3 by combining an organic coating acting as an adhesion primer (produced from diazonium salts) and a sealed anodization layer, which ensures anticorrosive properties. The process is simple, cost effective and meets the industrial requirements both in painting adhesion and protection against corrosion. The method can easily be adapted to any painting composition by changing the chemical groups carried by the diazonium salt. This work was patented in 2018 [31].

5 Chapter Conclusion

Through these three examples of applications illustrating very different fields, diazonium salt chemistry has shown that it is able to propose simple and robust solutions. This reflects a very good maturity of this technique of surface functionalization. As it can be easily implemented in water, does not use any toxic compound and does not require any energy input, this chemistry is particularly well suited to solve concrete problems and to its industrialization. The obtained coatings, generally thin, are very suitable as primer (ex. 3), but also find applications as such as illustrated by examples 1 (biomedical) and 2 (sensor).

References

1. Patent PCT 2007/052556
2. Patent PCT 2009/053977
3. Mesnage A et al (2010) Chem Mater 22:6229–6239 622.9. https://doi.org/10.1021/cm1014702
4. Deniau G et al (2012) Wiley, Weinheim, pp 159–180
5. Mesnage A (2011) Thesis of the doctoral school of the French Polytechnic School
6. Mévellec V, Roussel S, Tessier L, Chancolon J, Mayne-L'Hermite M, Deniau G, Viel P, Palacin S (2007) Chem Mater 19:6323–6330
7. Nugent MA, Zaia J, Spencer JL (2013) Biochemistry (Moscow) 78(7):726–735
8. Bozukova D, Pagnoulle C, Jerome R, Jerome C (2010) Mater Sci Eng R 69:63–83
9. Jozefowicz M, Jozefonvicz J (1997) Biomaterials 18:1633
10. Patent EP13172180.5 filed on June 14/2013, CEA/BWT co-ownership (50/50)
11. Noiré MH, Couston L, Douarre E, Pouyat D, Bouzon C, Marty P (2000) J Sol-Gel Sci Technol 17(2):131–136
12. Patent FR1553201 filed on Avril 13/2015 CEA Saclay
13. Altenpohl D (1998) Aluminum: technology, 6th edn. Appl Environ, Wiley-TMS
14. Schneider M, Yezerska O, Lohrengel MM (2008) Corros Eng Sci Technol 43:304
15. Vargel C (1999) Corrosion de l'aluminium. ed. Dunod p 87
16. Association T (2015) International alloy designations and chemical composition limit for wrought aluminum and wrought aluminum alloys
17. Campestrini P, Van Westing EPM, Van Rooijen HW, De Wit JHW (1853) Corros Sci 2000:42
18. Parvizi R, Hughes AE, Tan MY, Marceau RKW, Forsyth M, Cizek P, Glenn AM (2017) Corros Sci 116:98

19. Brace AW (1968) Technology of anodizing aluminum
20. Kulinich SA, Akhtar A (2012) Russ J Non-Ferrous Metals 53:176
21. Chaskiel P (2017) Environnement. Risques Santé 16:472
22. Delamar M, Désarmot G, Fagebaume O, Hitmi R, Pinson J, Savéant JM (1997) Carbon 35:801
23. Bahr JL, Yang J, Kosynkin DV, Bronikowski MJ, Smalley RE, Tour JM (2001) J Am Chem Soc 123:6536
24. Tessier L, Deniau G, Charleux B, Palacin S (2009) Chem Mater 21:4261
25. Mesnage A, Lefèvre X, Jègou P, Deniau G, Palacin S (2012) Langmuir 28:11767
26. Berisha A, Hazimeh H, Galtayries A, Decorse P, Kanoufi F, Combellas C, Pinson J, Podvorica FI (2016) RSC Adv 6:78369
27. Hurley BL, McCreery RL (2004) J Electrochem Soc 151:B252
28. Bard AJ, Mirkin MV, Unwin PR, Wipf DO (1861) J Phys Chem 1992:96
29. Doublet A, Kjelberg M, Jousselme B, Pinault M, Deniau G, Cornut R, Charrier G (2019) RSC Adv 9:24043
30. Magnacca G, Morterra C (1996) Catal Today 27:497
31. Patent FR1872018 filed on November 28/2018, CEA/PROTEC co-ownership (50/50)

Index

M. M. Chehimi et al. (eds.), *Aryl Diazonium Salts and Related Compounds*, Physical Chemistry in Action, https://doi.org/10.1007/978-3-031-04398-7

The manufacturer's authorised representative in the EU is Springer Nature Customer Service Centre GmbH, Europaplatz 3, 69115 Heidelberg, Germany. If you have any concerns regarding our products, please contact ProductSafety@springernature.com

Printed and bound by CPI Group (UK) Ltd, Croydon, CR0 4YY
22/09/2026
02220954-0008